A. Pignedoli (Ed.)

Some Aspects of Diffusion Theory

Lectures given at a Summer School of the
Centro Internazionale Matematico Estivo (C.I.M.E.),
held in Varenna (Como), Italy,
September 9-27, 1966

Springer

FONDAZIONE
CIME
ROBERTO CONTI

C.I.M.E. Foundation
c/o Dipartimento di Matematica "U. Dini"
Viale Morgagni n. 67/a
50134 Firenze
Italy
cime@math.unifi.it

ISBN 978-3-642-11050-4 e-ISBN: 978-3-642-11051-1
DOI:10.1007/978-3-642-11051-1
Springer Heidelberg Dordrecht London New York

Reprint of the 1st Ed. C.I.M.E., Ed. Cremonese, Roma 1967.
With kind permission of C.I.M.E.

Printed on acid-free paper

Springer.com

CENTRO INTERNAZIONALE MATEMATICO ESTIVO
(C.I.M.E.)

4° Ciclo - Varenna - dal 9 al 27 settembre 1966

"SOME ASPECTS OF DIFFUSION THEORY"

Coordinatore : Prof. A. PIGNEDOLI

CENTRO INTERNAZIONALE MATEMATICO ESTIVO

(C. I. M. E.)

V. C. A. FERRARO

DIFFUSION OF IONS IN A PLASMA WITH APPLICATIONS TO THE IONOSPHERE

Corso tenuto a Varenna dal 19 al 27 settembre 1966

DIFFUSION OF IONS IN A PLASMA
WITH APPLICATIONS TO THE IONOSPHERE

by
V. C. A. Ferraro
(Queen Mary College, University of London)

I. Derivation of the diffusion equations in plasmas

1. The term 'plasma' was first used by Langmuir for the state of a gas which is fully ionised (for example, the high solar atmosphere) or only partially ionised, (for example, the ionosphere). Our main interest in this course will be the diffusion of ions in such a plasma, arising from non-uniformity of composition, of pressure gradients or electric fields.

We begin by considering the simple case of a fully ionised gas and for simplicity restrict ourselves to the case when only one type of ion and electrons are present.

2. The velocity distribution function

We make the familiar assumption of molecular chaos, in which it is supposed that particles having velocity resolutes lying in a certain range are, at any instant, distributed at random. It is therefore most convenient to use six dimensional space in which the coordinates are the resolutes of the position vector $\underline{r}$ and velocity $\underline{v}$. The state of the plasma can then be specified by the distribution functions $f_\alpha(t, \underline{r}, \underline{v})$, where t is the time, that characterise each particle component α, for example, the ions or the electrons The quantity

$$f_\alpha(t, \underline{r}, \underline{v})\ d\underline{r}\ d\underline{v} \tag{1}$$

then represents the number of particles in the six dimensional volume element $d\underline{r}\ d\underline{v}$. In the simplest case, the plasma consists of single ions (α = i) and electrons (α = e) . In more complicated cases, the plasma may consist of several ion species in addition to neutral particles (α = n) such as atoms, molecules, exited atoms, etc. The total number of particles of constituent α in the element $d\underline{r}$ is obtained by integrating (1) throughout the velocity space. This number is, by hypothesis, $n_\alpha \, d\underline{r}$ and thus $n_\alpha = \int f_\alpha(t, \underline{r}, \underline{v}_\alpha) \, d\underline{v}_\alpha$ (2) . The behaviour of the ionised gas is described by a system of equations (Boltzmann equations) which can be derived as follows. Suppose that each particle of mass m_α is acted on by force $m_\alpha \underline{F}_\alpha$, then in a time dt in which the particles of constituent α suffer no collisions, the same particles that occupy the volume of phase space $d\underline{r} \, d\underline{v}_\alpha$ at time t would occupy the volume of phase $(\underline{r} + \underline{v}_\alpha dt)(\underline{v} + \underline{F}_\alpha dt)$ at time t + dt . The number in this set is

$$f_\alpha(t + dt, \; r + v_\alpha dt, \; v + \underline{F}_\alpha dt)$$

and the difference

$$f_\alpha(t + dt, \; \underline{r} + \underline{v}_\alpha dt, \; \underline{v}_\alpha + \underline{F}_\alpha dt) - f_\alpha(t, \underline{r}, \underline{v}_\alpha) \; d\underline{r} \; d\underline{v}_\alpha$$

therefore represent the difference in the gain of particles by collisions to this final set and the loss of the particle to the original set in time dt. This must be proportional to $d\underline{r} \; d\underline{v}_\alpha \, dt$; and we denote it by $C_\alpha \, d\underline{r} \, . \, d\underline{v}_\alpha \, dt$. Taking the limit as $dt \rightarrow 0$, we arrive at Boltzmann's equation for f_α , viz

$$\frac{f_\alpha}{t} + (\underline{v}_\alpha . \nabla) f_\alpha + (\underline{F}_\alpha . \nabla_{\underline{v}_\alpha}) f_\alpha = C_\alpha \tag{3}$$

where $\nabla_{\underline{v}_\alpha}$ stands for the gradient operator $\frac{\partial}{\partial u_\alpha}, \frac{\partial}{\partial v_\alpha}, \frac{\partial}{\partial w_\alpha}$ in velocity space.

3. Charge neutrality and the Debye distance

In general a plasma will rapidly attain a state of electrical neutrality; this is because the potential energy of the particle resulting from any space charge would otherwise greatly exceed its thermal energy. Small departures from strict neutrality will occur over small distances whose order of magnitude can be obtained as follows. The electrostatic potential V satisfies Poisson's equation

(4) $$\nabla^2 V = -4\pi(Zn_i - n_e)e$$

Here Ze is the charge on an ion and $-e$ that of the electrons. In thermodynamic equilibrium, the number densities of the ions and electrons respectively are given by

(5) $$n_i = n_i^{(o)} \exp(-ZeV/kT_i), \quad n_e = n_e^{(o)} \exp(eV/kT_e),$$

where k is the Boltzmann constant, T_i, T_e are the ion and electron temperatures and $n_i^{(o)}$ and $n_e^{(o)}$ are the values of n_i and n_e for strict neutrality so that $n_e^{(o)} = Zn_i^{(o)}$. In general, departures from neutrality are small so that we may expand the exponential to the first power of the arguments only. We have approximately

$$Zn_i - n_e \simeq Zn_i^{(o)} \left(1 - \frac{ZeV}{uT_i}\right) - n_e^{(o)} \left(1 + \frac{eV}{uT_e}\right)$$

and hence

(6) $$\nabla^2 V = \frac{V}{D^2},$$

where

(7) $$D = \left\{\frac{kT_eT_i}{4\pi Ze^2(n_i^{(o)}T_i + n_e^{(o)}Te)}\right\}^{1/2}$$

The quantity D has the dimensions of a length and is called the Debye distance. The solution of (6) for spherical symmetry is

$$V = \frac{e_\alpha}{r} \exp\left(-\frac{r}{D}\right) \tag{8}$$

where e_α is the charge on the particle. For small distances r from the origin ($r << D$), (8) reduces to the pure Coulomb potential of the charged particle. For large distances($r >> D$), $V \to 0$ exponentially . Thus in a neutral plasma in thermodynamical equilibrium the Coulomb field of the individual charge is cut off (shielded) at a distance of order D. Hence, we may assume that the particles do not interact in collisions for which the impact parameter is greater than D. The Debye shielding is not established instantaneously ; oscillations of the space charge will have a frequency $\omega_o = \left(4\pi n_e^2/m_e\right)^{\frac{1}{2}}$(since the displacemente of the electrons (or ions) bodily by a distance x gives rise to an electric field of intensity $4\pi n_e ex$ lending to restore neutrality) . Thus the time required to establish shielding is of the order

$$\tau \sim \frac{1}{\omega_o}$$

4. Diffusion of test particles in a plasma

A particular particle, which we call 'test particle', in a plasma will suffer collisions with the other particles in the plasma, which we call 'field particles' . Electrostatic forces between the particles have a greater range than the forces between neutral molecules in an ordinary gas. Consequently, the cumulative effect of distant encounters will be far more important than the effect of close collisions, which change completely the particle velocities. We shall therefore suppose that the deflections which the test particles undergo are mostly small. The motion of the test particle is most conveniently descibed in the velocity space, i.e., a space in which the velocity vector $\underline{v}$ is taken as the position vector and the apex of this vector is called the velocity point of the particle. Referred to Cartesian coordinates the coordinates of these points will be denoted by v_x, v_y, v_z.

As the test particle changes its position in ordinary space, its position in velocity space changes either continuously or discontinuously due to encounter with fixed particles. In general the displacement is complicated. (Fig. 1)

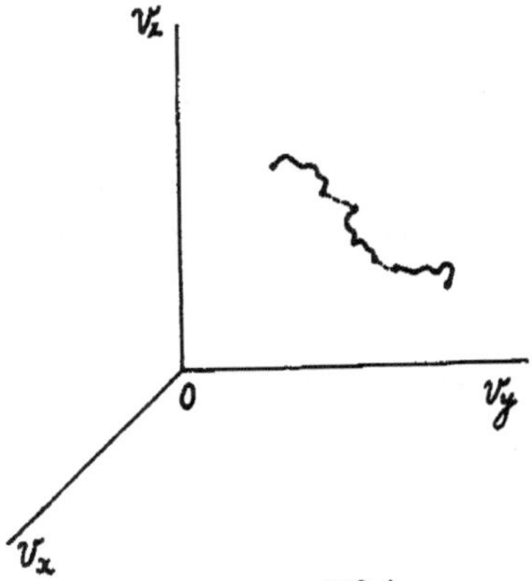

FIG. 1.

It is clearly impossible, and indeed futile, to trace the motion of a single particle and we are forced to consider a statistical description of the motion. In this, instead of a single particle, we consider an assembly containing a large number of test particles which have the same velocity $\underline{v}_0$ initially.

Suppose these are concentrated around the point $\underline{v}_0$ in the velocity space. At subsequent times the cloud will spread, changing both its size and shape, as a result of successive encounters.

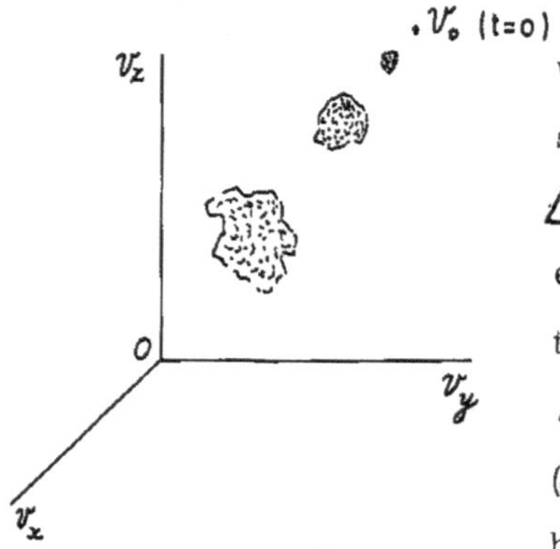

FIG. 2.

We now require to find quantities which will adequately describe the process. One such quantity is the change in velocity $\Delta \underline{v}$ of a test particle produced by the encounters. Suppose that $\underline{v}_0$ is parallel to the z-axis and consider the resolutes $\Delta v_x, \Delta v_y, \Delta v_z$ of $\Delta \underline{v}$. Suppose that $(\Delta v_x)_i$ is the change in Δv_x produced by the ith encounter. Then after N encounters,

$$\Delta v_x = \sum_{i=1}^{N} (\Delta v_x)_i$$

We assume that all the encounters are random, but as we have already seen, we cannot predict the change Δv_x for a single test particles. However, we can define an average value of Δv_x, say $\overline{\Delta v_x}$ for the large assembly of particles under consideration. If the distribution of velocities is isotropic, then $\overline{\Delta v_x} \equiv 0$, by symmetry , and likewise $\overline{\Delta v_y} \equiv 0$. But $\overline{\Delta v_z}$ need not vanish since the assembly (or cloud) has an initial velocity in the z-direction. However the mean square of $\overline{\Delta v_x^2}$ will not vanish . This mean value will contain terms of the form $\overline{(\Delta v_x)_i^2}$ and $\overline{(\Delta v_x)_i(\Delta v_x)_j}$: If the collisions are small we may expect that successive collisions will produce, on the average, the same average change as the first collisions. Thus the N terms $\overline{(\Delta v_x)_i^2}$ are all equal . But the mixed products $(\Delta v_x)_i(\Delta v_x)_j$ will vanish when averaged over all particles considered since successive collisions are uncorrelated . Hence

$$(9) \qquad \overline{\Delta v_x 2} = N\overline{(\Delta v_x)_i^2}$$

The dispersion of the points in Fig. 2 will therefore increase like $\sqrt{N}$, but not, in general, equally in all directions. But the centre of gravity may be displaced by an amount proportional to N. (Fig. 2)

The dispersion of the points in the velocity space produced by collisions of the test particles with the field particles is analogous to the diffusion of particles in an ordinary gas. To measure the rate of diffusion in the v_x direction, we consider the average value of (9) per unit time. The resultant value of $\overline{\Delta v_x 2}$, measuring the increase of velocity of dispersion of a group of particles per second, will be denoted by $\langle \Delta v_x^2 \rangle$ and called a 'diffusion coefficient', a term due to Spitzer . If the velocity distribution of the field particles is isotropic, the diffusion coefficients $\langle \Delta v_x \rangle$ and $\langle \Delta v_x \Delta v_y \rangle$ vanish identically,

The encounters between test and field particles which we are considering are assumed to be binary encounters only .(*) Let $\underline{v}$ be the velocity of a field particle relative to a test particle. Then there will be only three independent diffusion coefficients, namely, $\langle \Delta v_{\parallel} \rangle$, $\langle \Delta v_{\parallel}^2 \rangle$ and $\langle \Delta v_{\perp}^2 \rangle$, where $v_{\parallel}$ and $v_{\perp}$ are measured respectively parallel and perpendicular to $\underline{v}$. Their values will depend on the velocity distribution function of the field particles.

(*) The justification for this will be given in Section 7.

5. Binary encounter of two charged particles

(Hyperbolic orbit)

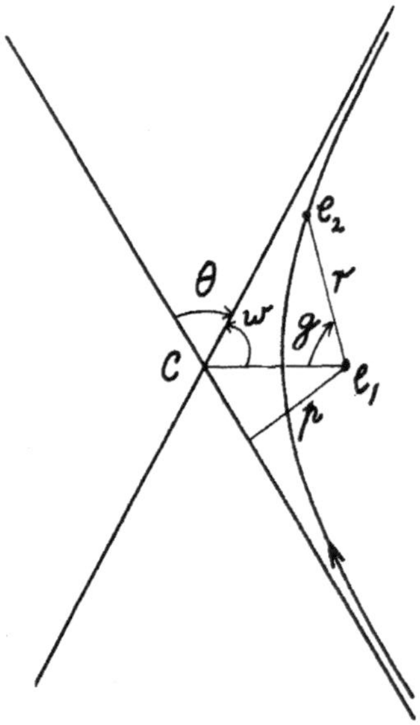

FIG. 3.

Consider the motion of charge e_2 relative to charge e_1 ; let $\underline{r}_1$ and $\underline{r}_2$ be the position vectors of e_1 and e_2 relative to a Newtonian origin. Then the equation of motion of the charges are respectively

$$m_1\ddot{\underline{r}}_1 = + \frac{e_1 e_2 \underline{r}}{r^3}, \quad m_2\ddot{\underline{r}}_2 = - \frac{e_1 e_2 \underline{r}}{r^3}$$

where $\underline{r} = \underline{r}_2 - \underline{r}_1$ and m_1 and m_2 are the masses of the charges. Hence

$$\ddot{\underline{r}} = \ddot{\underline{r}}_2 - \ddot{\underline{r}}_1 = -e_1 e_2 \left(\frac{1}{m_1} + \frac{1}{m_2}\right) \frac{\underline{r}}{r^3}$$

that is, the relative motion is the same as that of a particle under a central force at A varying inversely as the square of the distance whose strength is $\frac{e_1 e_2}{m_{12}}$, where $m_{12} = \frac{m_1 m_2}{m_1 + m_2}$ is the reduced mass. (Fig. 3).

Let v_∞ be the relative velocity of the charges at infinity and p the impact parameter. The energy integral is, with the usual notation,

$$v^2 = \frac{e_1 e_2}{m_{12}} \left(\frac{2}{r} + \frac{1}{a}\right)$$

whence

$$v_\infty^2 = \frac{e_1 e_2}{m_{12} a} \tag{10}$$

The polar equation of the orbit is

(11) $$r = \frac{\ell}{1 + e \cos\varphi}$$

where ℓ is the semi-latus rectum and e the eccentricity. As $r \to \infty$, $\varphi \to \pi - w$ so that (11) gives

$$\cos w = \frac{1}{e}$$

Also AC = ae ; hence

$$\sin w = \frac{p}{ae}$$

Thus $1 = \sin^2 w + \cos^2 w = \frac{1}{e^2} + \frac{p^2}{a^2 e^2}$ or $e^2 = 1 + \frac{p^2}{a^2}$ giving

$$\cos w = \frac{1}{\sqrt{1 + \frac{p^2}{a^2}}}, \quad \sin w = \frac{p}{a\sqrt{1 + \frac{p^2}{a^2}}}, \quad \tan w = \frac{p}{a}$$

or using (1)

(12) $$\tan w = \frac{p v_\infty^2 m_{12}}{e_1 e_2}$$

6. Calculation of diffusion coefficients

Consider the scattering of test particles (α) by a flux of field particles (β) . The spatial density of the latter is

$$dn_\beta = f_\beta(\underline{v}')\, d\underline{v}'$$

where $\underline{v}'$ is the velocity of the particles and f_β the distribution function of the field particles. Consider the collision of a test particle α with a field particle β of this flux. Then the velocity $\underline{v}_\alpha$ of the test particle is related to the velocity $\underline{v}_g$ of the centre of mass of the two particles and their relative velocity $\underline{u}$ by

$$\underline{v}_\alpha = \underline{v}_g + \frac{m_\beta}{m_\alpha + m_\beta}\,\underline{u}\ ;$$

hence , since $\underline{v}_g$ is unaltered by the encounter,

(13) $$\Delta \underline{v}_\alpha = \frac{m_{\alpha\beta}}{m_\alpha}\Delta \underline{u}$$

where $m_{\alpha\beta}$ is the reduced mass of m_α and m_β, and $\Delta\underline{u}$ the change in $\underline{u}$ produced by the encounter.

Also, in taking the average of the change of velocity over the test particles in the assembly, the summation reduces to a summation over all particles of the flux incident on a fixed scattering centre. The number of particles moving through an area $dA = pdpd\varphi$ of a plane π perpendicular to $\underline{u}$ in unit time is

(14) $$dn_\beta \,|\underline{u}|\, d\,A = f_\beta\,(\underline{v}')\,d\underline{v}'\,u\,dA$$

Multiply this by the components of the vector $\Delta\,\underline{v}_\alpha$ given by (13) and integrate over all the plane π and then over the velocities of the field particles we find

(15) $$<\Delta\,v_k> = \int f_\beta\,(\underline{v}')\,w_k\,d\underline{v}' \quad (k = x, y, z)$$

where

$$w_k = \frac{m_{\alpha\beta}}{m_\alpha}\int_{plane}\Delta u_k\ \ udA\ ,$$

(16) $$<\Delta v_k\,\Delta\,v_l> = \int f_\beta\,(\underline{v}')\,w_{kl}\,d\underline{v}'\ , \quad (k, l = x, y, z)$$

where

$$w_{kl} = \left(\frac{m_{\alpha\beta}}{m_\alpha}\right)^2 \int_{plane}\Delta u_k\,\Delta\,u_l\ \ ud\,A$$

(15) and (16) are the diffusion coefficients . It will be convenient to compute these integrals relative to a coordinate system in which the z-axis

is along $\underline{u}$.

For referring to Fig. a , and Fig. 4b .

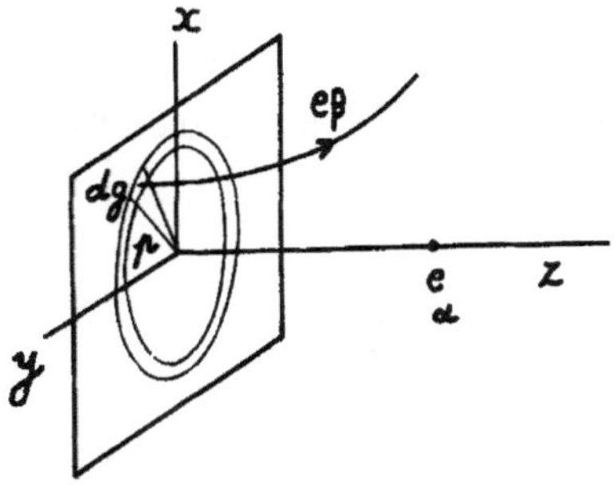

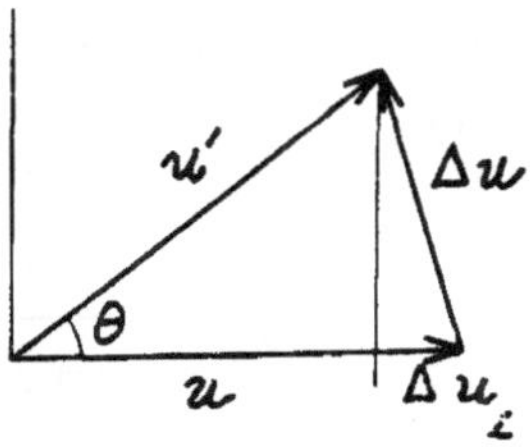

FIG. 4b

FIG. 4 a

we have

$$\Delta u_x = u \sin\theta \cos\varphi$$
$$\Delta u_y = u \sin\theta \sin\varphi$$
$$\Delta u_z = -u(1 - \cos\theta)$$

Also from (12) and the fact that $\theta = \pi - 2w$ we have

(17) $$\tan\frac{\theta}{2} = \frac{p_\perp}{p}$$

where $$p_\perp = \frac{e_\alpha e_\beta}{m_{\alpha\beta} u^2}$$

Then $$\Delta u_x = 2u \frac{p p_\perp}{p^2 + p_\perp^2} \cos\varphi , \quad \Delta u_y = 2u \frac{p p_\perp}{p^2 + p_\perp^2} \sin\varphi$$
$$\Delta u_z = -2u \frac{p_\perp^2}{p^2 + p_\perp^2}$$

Integration with respect to p and φ over the plane gives

$$w_x = 0 = w_y$$

whilst

$$w_z = \frac{m_{\alpha\beta}}{m_\alpha} \int_{\text{plane}} \Delta u_z \, u \, dA .$$

If the limits of integration for p are 0 and ∞, the integral diverges; however, we have already seen that the Coulomb field of individual charges is cut off at distance of order D, the Debye distance. Hence we can take D as the upper limit for p in the integral. We then find

$$\mathcal{W}_z = - \frac{1+\frac{m_\alpha}{m_\beta}}{4\pi u^2} \left(\frac{4\pi e_\alpha e_\beta}{m_\alpha}\right)^2 \int_0^D \frac{p\,dp}{p^2+p_\perp^2}$$

(18)

$$= -\lambda \frac{1+\frac{m_\alpha}{m_\beta}}{4\pi u^2} \left(\frac{4\pi e_\alpha e_\beta}{m_\alpha}\right)^2$$

where

(19) $$\lambda = \int_0^D \frac{p\,dp}{p^2+p_\perp^2} = \log_e \frac{(D^2+p_\perp^2)^{\frac{1}{2}}}{p_\perp}$$

It has been tacitly assumed that $D \gg p_\perp$; to illustrate that this is likely to be the case in general, let us take $T_i = T_e =$ $= 1$ keV ($10^{7\circ}$K) , $n_i = n_e = 10^5$ cm^{-3}, and Z = 1 , then

$$D = \left(\frac{kT}{8\pi ne^2}\right)^{\frac{1}{2}} \sim \frac{1}{2} \times 10^{-3} \text{ cm}$$

and

$$p_\perp = \frac{e^2}{3kT} \sim \frac{1}{2} \times 10^{-10} \text{ cm}$$

so that $D/p_\perp \sim 10^7$. Hence in (19) we may neglect $p_\perp$ compared with D in the numerator and write

(19a) $$\lambda \sim \log_e \frac{D}{p_\perp} = \log_e \left\{ \frac{3}{2e^3} \left(\frac{k^3T^3}{2\pi n}\right)^{\frac{1}{2}} \right\}$$

Likewise

$$w_{xx} = \left(\frac{m_{\alpha\beta}}{m_\alpha}\right)^2 \int (\Delta u_x)^2\, u\, dA = \left(\frac{m_{\alpha\beta}}{m_\alpha}\right)^2 \int_{\text{plane}} \left(2u \frac{p\, p_\perp}{p^2+p_\perp^2} \cos\varphi\right)^2 u p\, dp\, d\varphi$$

$$= \frac{1}{4\pi u} \left(\frac{4\pi e_\alpha e_\beta}{m_\alpha}\right)^2 \int_0^D \frac{p^3\, dp}{(p^2+p_\perp^2)^2}$$

$$\simeq \left(\lambda - \frac{1}{2}\right) \frac{1}{4\pi u} \left(\frac{4\pi e_\alpha e_\beta}{m_\alpha}\right)^2 \quad \text{neglecting terms of order} \left(\frac{p_\perp}{D}\right)^2$$

Again $w_{yy} = w_{xx}$ and $w_{\alpha\beta} = 0$, $\alpha \neq \beta$: Finally

$$w_{zz} = \left(\frac{m_{\alpha\beta}}{m_\alpha}\right)^2 \int (\Delta u_z)^2\, u\, dA = \left(\frac{m_{\alpha\beta}}{m_\alpha}\right)^2 \int_{\text{plane}} \left(-2u \frac{p_\perp^2}{p^2+p_\perp^2}\right)^2 u p\, dp\, d\varphi$$

and the integration with respect to p can, in fact, be carried out from 0 to ∞ since the integral is finite ;

we find

$$w_{zz} = 4\pi \left(\frac{e_\alpha e_\beta}{m_\alpha u}\right)^2$$

Since this is λ times smaller than w_{xx} or w_{yy} it may be set to zero . We can now express w_k and $w_{k\ell}$ as a vector or tensor respectively. In fact,

$$w_k = -\lambda \frac{1 + \frac{m_\alpha}{m_\beta}}{4\pi u^2} \left(\frac{4\pi e_\alpha e_\beta}{m_\alpha}\right)^2 \frac{u_k}{u} \tag{20}$$

$$w_{k\ell} = \begin{pmatrix} A & 0 & 0 \\ 0 & A & 0 \\ 0 & 0 & 0 \end{pmatrix} = \begin{pmatrix} A & 0 & 0 \\ 0 & A & 0 \\ 0 & 0 & A \end{pmatrix} - \begin{pmatrix} 0, & 0 & 0 \\ 0 & 0 & 0 \\ 0 & 0 & A \end{pmatrix}$$

$$= A\left(\delta_{k\ell} - \frac{u_k u_\ell}{u^2}\right) \tag{21}$$

where

(22) $$A = (\lambda - \frac{1}{2}) \frac{1}{4\pi u} \left(\frac{4\pi e_\alpha e_\beta}{m_\alpha} \right)^2$$

Finally

(23) $$< \Delta v_k > = - (1 + \frac{m_\alpha}{m_\beta}) Q_{\alpha\beta} \int \frac{u_k}{u^3} \quad f_\beta(\underline{v}') \, d\, \underline{v}'$$

(24) $$< \Delta v_k \, \Delta v_\ell > = \frac{Q_{\alpha\beta}}{4\pi} \int (\frac{\delta_{k\ell}}{u} - \frac{u_k u_\ell}{u^3}) \, f_\beta(\underline{v}') \, d\, \underline{v}'$$

where $\underline{u} = \underline{v} - \underline{v}'$

(25) $$Q_{\alpha\beta} = \lambda (\frac{4\pi e_\alpha e_\beta}{m_\alpha})^2 \quad .$$

It can be shown that the third and higher diffusion coefficients $<\Delta v_k \Delta v_\ell \Delta v_m ..>$ are smaller than the first two diffusion coeffients by a factor of λ . This means that the motion of Coulomb particles can be visualized as a diffusion in velocity space. The approximation in which only the first two diffusion coefficients are considered is called the Fokker- Planck approximation.

7. <u>Justification of the assumption of binary encounters in the theory.</u>

The assumption is certainly justified for short-range force. If the interaction range d(effective diameter of the molecule) is much smaller thatn the mean distance between the particles, $n^{-1/3}$, where n is the density of the gas, the sphere of action , of volume $\sim d^3$, will contain only a small number of particles N_d , that is

$$N_d = nd^3 << 1.$$

Under these conditions the probability of multiple collisions, involving three or more particles simultaneously, is very small. A description in terms of binary collisions is adequate.

Coulomb forces acting between particles of a plasma are not short-range forces. The potential energy between two such charges e_1 and e_2 is

$$\frac{e_1 e_2}{r} \exp\left(-\frac{r}{D}\right) \tag{26}$$

where r is the distance apart of the charges. Thus the interaction between them extends at least as far as the Debye distance D, and for conditions in which we are interested $D \gg n^{-1/3}$ and the sphere of action contains many particles, i.e.,

$$N_D = nD^3 \gg 1 \tag{27}$$

In this case a given particle will interact simultaneously with many particles and the results derived earlier on the basis of binary collisions is suspect. A rigorous analysis shows that the formulae derived yield logarithmic accuracy. However, a non-rigorous, but plausible discussion can be given along the following lines.

Let us consider a test particle moving through the plasma and suppose that it is so massive that its velocity can be treated as constant. Draw a cylinder of radius p with the trajectory as axis (Fig. 5)

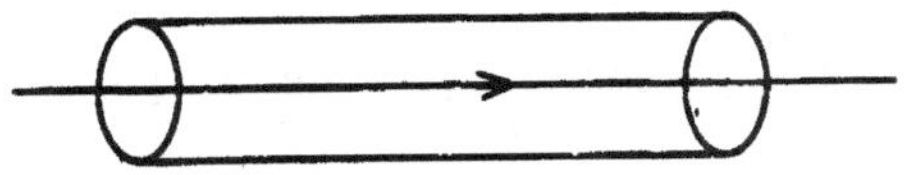

FIG. 5.

Collisions of the test particles with field particles for which $p >> n^{-1/3}$ will be many-body collisons. Those characterized by immact parameters $p << n^{-1/3}$ are binary collisions . We shall show that the method used to treat binary collisions need not be restricted to collisions with impact parameters $p << n^{-1/3}$, but can be extended to parameters $>> n^{-1/3}$.

Now, when $r << D$, the potential energy between the charges is simply $e_1 e_2 / r$ so that the presence of other particles has no effect on the interaction between two particles separated by a distance smaller than D. Thus results derived on the hypothesis of binary collisions apply for all impact parameters smaller than the Debye radius, i.e., $p << D$. Because $D >> n^{-1/3}$, in the present case the collisions can be regarded as binary interactions even when $p >> n^{-1/3}$ as long as $p << D$. Accordingly, even if $p \sim D$, the difference between the exact interaction formulae which takes account of other particles, and a pure Coulomb interaction, is small (by a factor of order 1) . Thus, cutting off the Coulomb interaction for the impact parameter $p = D$ provides an approximate method of taking into account the effect of multiple collisions for which $p >> n^{-1/3}$.

8. Diffusion in velocity space.

From a microscopic point of view , the change of spatial coordinates of a particle during a collision can always be neglected . Hence, as far as the spatial part of the phase space is concerned, the motion of a particle corresponds to a continuous point to point variation.

On the other hand, collisions have a marked effect on the continuity of motion in the velocity space. The velocity can be changed abruptly by a single near collision, essential in a vanishingly small time interval. Hence, a particular velocity point $\underline{v}$ in a cloud of particles in velocity space can be 'annihilated' by a collision and 'recreated' at some remote point without passing through intermediate points in the velocity space. Thus in general the effect of collisions cannot be expressed in the kinetic theory by introducing a term describing the divergence of flux in velocity space. But this will certainly only be the case for near collisions in which the velocity of the particle is changed abruptly. In the case of coulomb forces, the change in velocity, characterised by the quantities $< \Delta v_k >$ and $< \Delta v_k \Delta v_\ell >$ is due to the effect of remote interactions and the changes in velocity are small. For example, if $\lambda = 15$, then the relative change in particle velocity

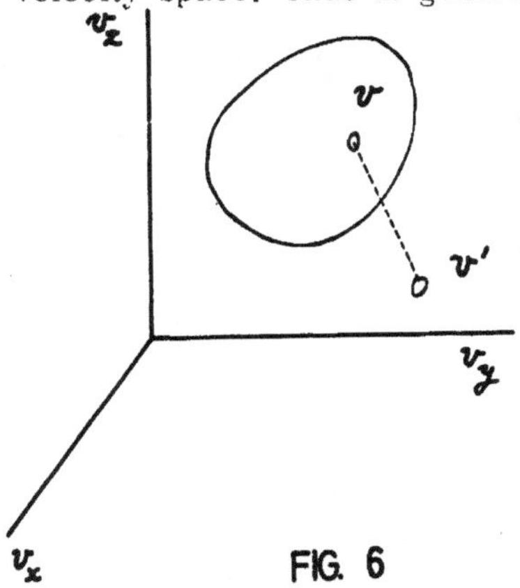

FIG. 6

$$\frac{|\Delta \underline{v}|}{v} = \frac{p_\perp}{p} = e^{-\lambda}\frac{D}{p} \sim 10^{-6}\left(\frac{D}{p}\right),$$

and so very small.

If these interactions are referred in velocity space, the whole process may be regarded as a form of diffusion. The motion can be regarded as nearly continuous.

9. Calculation of the diffusion coefficient for a Maxwellian distribution of velocities.

The expressions (23) and (24) may be expressed more conveniently by introducing the super-potentials. In fact, since

$$u = \sqrt{(\underline{v}_k - \underline{v}'_k)(\underline{v}_k - \underline{v}'_k)}$$

we have

$$\frac{\partial}{\partial v_k} \frac{1}{u} = - \frac{u_k}{u^3}$$

$$\frac{\partial^2}{\partial v_k \partial v_\ell} u = \frac{\partial}{\partial v_k} \frac{1}{2} \frac{2}{u} (v_\ell - v'_\ell) = \frac{\partial}{\partial v_k} \frac{v_\ell - v'_\ell}{u} = \frac{\partial}{\partial v_k} \frac{u_\ell}{u}$$

$$= - \frac{u_k u_\ell}{u^3} + \delta_{k\ell} \frac{1}{u}$$

Hence (23) and (24) can be written

(28) $$< \Delta v_k > = - (1 + \frac{m_\alpha}{m_\beta}) Q_{\alpha\beta} \frac{\partial \varphi_\beta}{\partial v_k}$$

(29) $$< \Delta v_k \Delta v_\ell > = - 2 Q_{\alpha\beta} \frac{\partial^2 \psi_\beta}{\partial v_k \partial v_\ell}$$

where

(30) $$\varphi_\beta = - \frac{1}{4\pi} \int \frac{f_\beta(\underline{v}') \, d\underline{v}'}{|\underline{v} - \underline{v}'|}$$

(31) $$\psi_\beta = - \frac{1}{8\pi} \int |\underline{v} - \underline{v}'| f_\beta(\underline{v}') \, d\underline{v}'$$

which have been termed 'super-potentials' by Rosenbluth et al. For a Maxwellian distribution function

$$f(\underline{v}') = n(\frac{m}{2\pi n T})^{3/2} e^{-\frac{m v'^2}{2kT}}$$

Chandrasekhar found that

(32) $$< \Delta v_{\parallel} > = - \frac{1}{2} n_\beta Q_{\alpha\beta} (1 + \frac{m_\alpha}{m_\beta}) G (\frac{m_\beta v}{2kT_\beta})$$

$$\langle \Delta v_{\perp}^2 \rangle = \langle \Delta v_{xx}^2 \rangle + \langle \Delta v_{yy}^2 \rangle$$

(33) $$= \frac{1}{2v} n_{\beta} Q_{\alpha\beta} \left\{ \Phi\left(\frac{m_{\beta} v}{2kT_{\beta}}\right) - G\left(\frac{m_{\beta} v}{2kT_{\beta}}\right) \right.$$

(34) where $$\Phi(x) = \frac{2}{\sqrt{\pi}} \int_0^x e^{-\xi^2} d\xi$$

is the usual error function and

(35) $$G(x) = \frac{\Phi(x) - x\Phi'(x)}{2x^2}$$

Values of G and $\Phi - G$ are given by Spitzer and others.

10. Relaxion times. (Collision interval)

The term "relaxation time" is used to denote the time in which collisions will alter the original velocity distribution; or again , the time that the ions and electrons in a gas will attain, through collisions, a Maxwellian distribution.

Various relaxation times can be defined ; the time between collisions (collision interval or the reciprocal of the collision frequency) may be defined as the time in which small deflections will deflect test particles through 90^{0} . More precisely, if τ_D is the 'deflection time', we have

(36) $$\langle \Delta v_{\perp}^2 \rangle \tau_D = v^2$$

Substituting from (33) we find

(37) $$\tau_D = \frac{2v^3}{n_{\beta} Q_{\alpha\beta} (\Phi_{\beta} - G_{\beta})}$$

An energy exchange time τ_E can likewise be defined by the relation

(38) $$<\Delta E^2> \tau_E = E^2 ;$$

the change of energy

(39) $$\Delta E = \frac{1}{2} m(2v\Delta v_{11} + \Delta v^2_{11} + \Delta v_{\perp}^2)$$

If only dominant terms are required

$$< \Delta E^2 > = m^2 v^2 < \Delta v^2_{11} >$$

and (38) gives

(40) $$\tau_E = \frac{v^3}{4 n_\beta Q_{\alpha\beta} G_\beta}$$

An important special case is that of a group of ions, or a group of electrons, interacting amongst themselves. If we consider such a group whose velocity has the root mean square value for the group , then $\left(\frac{mv^2}{2nT}\right)^{\frac{1}{2}} = 1.225.$

In this case we find that $\tau_D / \tau_E = 1{,}14$ so that $\tau_D \sim \tau_E$ and is a measure of both the time required to reduce substantially any anisotropy in the velocity distribution function and the time for the kinetic energies to approach a Maxwellian distribution. We shall call this particular value of τ_D the 'self-collision interval' for a group of particles and will be denoted by τ_c From (37) we have

(41) $$\tau_c = \frac{m^{\frac{1}{2}} (3 k T)^{3/2}}{5.71 \pi n e^4 Z^4 \log_e \lambda}$$

where T is in degrees K, m is the mass of a typical particle of the group. It may be written Am_H where m_H is the mass of a proton . For electrons, $A = \frac{1}{1825}$ so that the self-collision time for electrons is $\frac{1}{43}$ that for protons, provided the ions and electrons have the same temperatures.

We consider next the approach to equilibrium of a two component plasma; to fix our ideas we consider the case when the constituents are ions and electrons. There are three stages involved in the process. First, collisions between ions and electrons lead to an isotropic velocity distribution of electrons, and the same time collisions between electrons themselves establishes a Maxwellian distribution. Secondly, collisions between the ions themselves establishes an isotropic velocity distribution amongst the ions. Thirdly, the ions and electrons which have already attained Maxwellian distribution, but possibly at different temperatures T_i and T_e, will be brought to the same temperature by collisions between the ions and electrons.

To consider the last process we require the equation of energy ($\mathcal{E}_\alpha = \frac{1}{2} m_\alpha v^2$)

(42) $$\frac{d\mathcal{E}_\alpha}{dt} = \frac{1}{2} m_\alpha \frac{d}{dt} \overline{v_i v_i} = m_\alpha \left(\frac{1}{2} \langle \Delta v_i \Delta v_i \rangle + v_i \langle \Delta v_i \rangle\right)$$

using

(43) $$\overline{\Delta v_i \Delta v_i} = \overline{v_i v_i} - \bar{v}_i \bar{v}_i$$

Using (28) and (29) we find

(44) $$\frac{d\mathcal{E}_\alpha}{dt} = - m_\alpha Q_{\alpha\beta} \left[\varphi_\beta + \left(1 + \frac{m_\alpha}{m_\beta}\right) \underline{v} \cdot \nabla \varphi_\beta \right]$$

since

(45) $$\langle \Delta v_i \Delta v_i \rangle = - 2Q_{\alpha\beta} \nabla^2 \psi_\beta = - 2Q_{\alpha\beta} \varphi_\beta$$

Since the distribution of velocities are Maxwellian, this may be rewritten

(46) $$\frac{d\mathcal{E}_\alpha}{dt} = -\frac{2\mathcal{E}_\alpha}{\tau_{\alpha\beta}(\mathcal{E}_\alpha)}\left\{\frac{m_\alpha}{m_\beta}\mu(x_\beta) - \mu'(x_\beta)\right\}$$

where $x_\beta = \frac{1}{2}\frac{m_\beta v^2}{kT_\beta} = \frac{m_\beta}{m_\alpha}\frac{\mathcal{E}_\alpha}{kT_\beta}$, and $\mu(x) = \Phi(x) - x\,\Phi'(x)$

Also

(47) $$\tau_{\alpha\beta}(\mathcal{E}_\alpha) = \frac{4\pi v^3}{n_\beta Q_{\alpha\beta}} = \frac{(\frac{1}{2}m_\alpha)^{\frac{1}{2}}\,\mathcal{E}_\alpha^{3/2}}{\pi e_\alpha^2 e_\beta^2 n_\beta \log\lambda}$$

where $\mathcal{E}_\alpha = k(T_\alpha + \frac{m_\alpha}{m_\beta}T_\beta)$. After some algebra, (46) can be reduced to

(48) $$\frac{dT_\alpha}{dt} = \frac{T_\beta - T_\alpha}{\tau^*_{\alpha\beta}}$$

where

(49) $$\tau^*_{\alpha\beta} = \frac{3\,m_\alpha m_\beta k^{3/2}}{8(2\pi)^{\frac{1}{2}} n_\beta e_\alpha^2 e_\beta^2 \log\lambda}\left(\frac{T_\alpha}{m_\alpha} + \frac{T_\beta}{m_\beta}\right)^{3/2}$$

It is easily verified that

$$\tau^*_{ee} : \tau^*_{ii} : \tau^*_{ei} : \tau^*_{ie} = 1 : \sqrt{\frac{M}{m}} : \frac{M}{m} : \frac{M}{m}$$

where $T_\alpha \sim T_\beta$ and where M is the mass of the ion and m the electronic mass.

Equation (48) was first given by Spitzer ; it shows that if the mean square relative velocity, which is $\propto \left(\frac{T_\alpha}{m_\alpha} + \frac{T_\beta}{m_\beta}\right)$, does not change appreciably, , $\tau^*_{\alpha\beta}$ is nearly constant and departure from equipartitions decrease exponentially.

11. Relaxation towards the steady state

The solution of Boltzmann's equation for non-uniform gases is found by successive approximation. We write

$$f = f_o(1 + \mathcal{E}),$$

where f_o is the Maxwellian distribution function and $\mathcal{E}$ is small compared with unity. We have seen that in a plasma of two constituents each constituent will approach its Maxwellian distribution in a time equal to the relaxation time $\tau^*_{\alpha\alpha}$ and the two constituents will attain equal temperatures in a relaxation time $\tau^*_{\alpha\beta}$. As a first approximation, therefore, we can take the collision term C_α to be of the form

(50) $$-\frac{f - f_o}{\tau^*}$$

so that if f is the distribution function at time $t = 0$ and f_o the Maxwellian distribution function, then departure from a Maxwellian state $f - f_o \rightarrow 0$ with time as $e^{-t/\tau}$.

12. Equations of continuity and motion for a fully ionized gas

We consider the plasma to be a mixture of positive ions (i) and electrons (e) and denote their number densities by n_i and n_e, and their velocities by $\underline{v}_i$ and $\underline{v}_e$ respectively. Then

(51) $$n_i = \int f_i \, d\underline{v}_i , \qquad n_e = \int f_e \, d\underline{v}_e$$

where f_i and f_e denote the velocity distribution functions for the ions and electrons respectively and $d\underline{v}_i$ and $d\underline{v}_e$ denote an element of volume in the velocity space for ions and electrons, respectively. Denoting their masses by m_i and m_e and the densities of the ion and electron gas by ρ_i and ρ_e respectively, we have

(52) $$\rho_i = n_i m_i , \qquad \rho_e = n_e m_e$$

Denote by $\bar{\underline{v}}_i$ and $\bar{\underline{v}}_e$ the mean velocities of the ion and electron gas in a volume element of the plasma, then

$$n_i \overline{\underline{v}_i} = \int \underline{v}_i f_i \, d\underline{v}_i, \qquad n_e \overline{\underline{v}_e} = \int \underline{v}_e f_e \, d\underline{v}_e \tag{53}$$

It is convenient to introduce the total number density n_o and total mass density ρ_o, defined as $n_o = n_i + n_e$

$$\rho_o = \rho_i + \rho_e \tag{54}$$

and the mean velocity $\underline{v}_o$ of the plasma element defined by

$$\rho_o \underline{v}_o = \rho_i \overline{\underline{v}_i} + \rho_e \overline{\underline{v}_e} \tag{55}$$

Let $\underline{V}_i$ and $\underline{V}_e$ be the <u>peculiar</u> or <u>thermal</u> velocities of the ions and electrons, respectively, defined by

$$\underline{V}_i = \underline{v}_i - \underline{v}_o, \qquad \underline{V}_e = \underline{v}_e - \underline{v}_o \tag{56}$$

Then it follows from (55) that

$$\rho_i \overline{\underline{V}_i} + \rho_e \overline{\underline{V}_e} = 0 \tag{57}$$

The partial pressure for the ion and electron gases, and total pressures defined in a frame of reference moving with the mean velocity $\underline{v}_o$ are respectively given by

$$p_i = \rho_i \overline{\underline{V}_i \underline{V}_i}, \qquad p_e = \rho_e \overline{\underline{V}_e \underline{V}_e}, \qquad p_o = p_i + p_e \tag{58}$$

The hydrostatic partial pressures for ions and electrons are defined by

$$p_i = \frac{1}{3} \rho_i \overline{V_i^2}, \qquad p_e = \frac{1}{3} \rho_e \overline{V_e^2} \tag{59}$$

and the corresponding mean kinetic temperature by

(60) $$p_i = k n_i T_i, \quad p_e = k n_e T_e$$

Boltzmann's equation for the two distribution functions f_i and f_e are

(61) $$\frac{\partial f_\alpha}{\partial t} + (\underline{v}_\alpha . \nabla) f_\alpha + (\underline{F}_\alpha . \nabla_{\underline{v}_\alpha}) f_\alpha = C_\alpha \ , \alpha = i, e$$

where $m_i \underline{F}_i$ and $m_e \underline{F}_e$ are the forces acting on an ion and electron respectively. If these are produced by an electric field $\underline{E}$ and magnetic field $\underline{B}$, then

(62) $$\underline{F}_i = \frac{e_i}{m_i} (\underline{E} + \underline{v}_i \times \underline{B}), \quad \underline{F}_e = \frac{e_e}{m_e} (\underline{E} + \underline{v}_e \times \underline{B})$$

where e_i and e_e are the charges carried by an ion and electron respectively.

We next form the moment equations; if $\varphi_\alpha(\underline{v}_\alpha)$ be any function of molecular properties for the constituent α of the plasma, then by multiplying equation (61) by φ_α, integrating partially and remembering that

(63) $$n_\alpha \overline{\varphi}_\alpha = \int \varphi_\alpha f_\alpha d\underline{v}_\alpha$$

we find

(64) $$\frac{\partial (n_\alpha \overline{\varphi}_\alpha)}{\partial t} + \nabla \cdot (n_\alpha \overline{\varphi_\alpha \underline{v}_\alpha}) - n_\alpha \underline{F}_\alpha . \overline{\nabla_{v_\alpha} \varphi_\alpha} = \int \varphi_\alpha C_\alpha d\underline{v}_\alpha$$

The right-hand side represents the change of the mean value of φ_α due to collisions. This vanishes if $\varphi_\alpha = 1$ and (64) gives

(65) $$\frac{\partial n_\alpha}{\partial t} + \nabla \cdot (n_\alpha \underline{v}_o) + \nabla \cdot (n_\alpha \overline{\underline{V}}_\alpha) = \frac{\partial n_\alpha}{\partial t} + \nabla \cdot (n_\alpha \overline{\underline{v}}_\alpha) = 0$$

which is the equation of continuity for the component α. Multiplying the equations of continuity for the ions and electrons (65) by m_i and

m_e respectively and adding we have the equation of continuity for the plasma as whole,

$$\frac{\partial \rho_o}{\partial t} + \nabla \cdot (\rho_o \underline{v}_o) = 0 \tag{66}$$

If we set $\underline{\wp}_\alpha = m_\alpha \underline{v}_\alpha$, then , after some simplification and using (65), one obtains

$$\rho_\alpha \frac{d\underline{v}_o}{dt} + (\nabla \cdot p_\alpha - \rho_\alpha \underline{F}_\alpha) + \frac{d(\rho_\alpha \overline{\underline{V}}_\alpha)}{dt} + \rho_\alpha (\overline{\underline{V}}_\alpha \cdot \nabla)\underline{v}_o$$
$$+ \rho_\alpha \overline{V}_\alpha \nabla \cdot \underline{v}_o = \int m_\alpha \underline{v}_\alpha C_\alpha \, d\underline{v}_\alpha \tag{67}$$

Adding the equations for the ions and electrons and noting that the total momentum of the ions and electrons in the element is unaltered by collisions, we get

$$\rho_o \frac{d\underline{v}_o}{dt} = - \nabla \cdot p_o + \rho_i \underline{F}_i + \rho_e \underline{F}_e \tag{68}$$

which is the equation of mass motion. Equation (67) refers to an element of either constituent following the mass-motion of the plasma. An equation can also be obtained referred to the local mean velocity of the constituent, $\overline{\underline{v}}_\alpha$. Denoting by d_α / dt the time derivative in this case, so that

$$\frac{d_\alpha}{dt} = \frac{\partial}{\partial t} + \overline{\underline{v}}_\alpha \cdot \nabla \tag{69}$$

we find after some rearrangement of terms that

$$\rho_\alpha \frac{d_\alpha \overline{\underline{v}}_\alpha}{dt} + \nabla \cdot (p_\alpha - \rho_\alpha \overline{\underline{V}}_\alpha \overline{\underline{V}}_\alpha) - \rho_\alpha \underline{F}_\alpha = \int m_\alpha \underline{v}_\alpha C_\alpha \, d\underline{v}_\alpha \tag{70}$$

where

$$P_\alpha = p_\alpha - \rho_\alpha \overline{\underline{V}}_\alpha \overline{\underline{V}}_\alpha = \rho_\alpha \overline{\underline{V}_\alpha \underline{V}_\alpha} - \rho_\alpha \overline{\underline{V}}_\alpha \overline{\underline{V}}_\alpha \tag{71}$$

is the relative pressure tensor. It is easily shown that this is equal to

(72) $$P_{\alpha} = \rho_{\alpha} \overline{(\underline{v}_{\alpha} - \bar{\underline{v}}_{\alpha})(\underline{v}_{\alpha} - \bar{\underline{v}}_{\alpha})} = \rho_{\alpha} \overline{\underline{u}_{\alpha}\, \underline{u}_{\alpha}}$$

when $\underline{u}_{\alpha} = \underline{v}_{\alpha} - \bar{\underline{v}}_{\alpha}$ is the velocity of a particle relative to the mean velocity of the element. Thus (70) can now be written

(73) $$\rho_{\alpha} \frac{d_{\alpha} \bar{\underline{v}}_{\alpha}}{dt} + \nabla \cdot P_{\alpha} - \rho_{\alpha} \underline{F}_{\alpha} = \int m_{\alpha} \underline{v}_{\alpha} C_{\alpha} \, d\underline{v}_{\alpha}$$

which is the equation of motion of the constituent α referred to the mean velocity $\bar{\underline{v}}_{\alpha}$ of this constituent.

13. Approximate calculation of the collision term

Since particles of one constituent can collide with each other and with particles from another constituent, the collision term in Boltzmann's equation (3) may be written

(74) $$C_{\alpha} = \sum_{\beta} C_{\alpha\beta}(f_{\alpha}, f_{\beta})$$

where $C_{\alpha\beta}$ gives the change per unit time in the distribution function for particles of the constituent α due to collisions with particles of contituent β, $C_{\alpha\beta}$ depend on the respective distribution functions f_{α}, f_{β}. Certain properties of the collision terms are immediately obvious and do not depend on the explicit form of the $C_{\alpha\beta}$. Thus

(75) $$\int C_{\alpha\beta} \, d\underline{v}_{\alpha} = 0$$
$$\int m_{\alpha} v_{\alpha} C_{\alpha\alpha} \, d\underline{v}_{\alpha} = 0$$

neglecting processes which may convert particles of one constituent into that of another, e.g., ionization, dissociation , etc.

We have alredady noted in section 11 that as a rough approximation we may write

$$C_\alpha = \frac{f_\alpha^{(o)} - f_\alpha}{\tau} \tag{76}$$

where $f_\alpha^{(o)}$ is the Maxwellian distribution. On account of the second relation in (75), we may take τ to be $\tau^*_{\alpha\beta}$ and that, if departures from equilibrium are small, we may treat $\tau^*_{\alpha\beta}$ as constant. We can now evaluate approximately the collision term in (73). Write

$$\underline{v}_\alpha = \underline{v}_o + \underline{V}_\alpha , \tag{77}$$

where $\underline{v}_o$ is the mean mass velocity of the two constituent plasma. Then

$$\int m_\alpha \underline{v}_\alpha C_\alpha d\underline{v}_\alpha = \int m_\alpha \underline{v}_o C_\alpha d\underline{v}_\alpha + \int m_\alpha \underline{V}_\alpha C_\alpha d\underline{v}_\alpha \tag{78}$$

Since $\underline{v}_o$ is a constant in the first integral, this vanishes by virtue of the first equation in (75). Hence (78) reduces to

$$\int m_\alpha \underline{V}_\alpha C_\alpha d\underline{v}_\alpha . \tag{79}$$

Substituting (76) in (79) and noting that $d\underline{v}_\alpha = d\underline{V}_\alpha$, this reduces to

$$\int m_\alpha \underline{V}_\alpha \frac{f_\alpha^{(o)}}{\tau} d\underline{V}_\alpha - \int m_\alpha \underline{V}_\alpha \frac{f_\alpha}{\tau} d\underline{V}_\alpha . \tag{80}$$

But $\int m_\alpha \underline{v}_\alpha f_\alpha^{(o)} d\underline{V}_\alpha$ vanishes identically; thus (80) reduces to

$$-n_\alpha m_\alpha \overline{\underline{V}}_\alpha / \tau \tag{81}$$

Here $\tau \sim \tau^*_{\alpha\beta}$ denotes effectively the electron-ion scattering time and we may interpret this result as follows. The electrons lose their ordered velocity with respect to the ions in a time of the order τ and hence lose momentum $m_\alpha \underline{V}_\alpha$ per particle α which is communicated to the particle β . This implies that the particles are subjected to a frictional force $n_\alpha m_\alpha \bar{V}_\alpha / \tau$. This is equal and opposite to the force exerted on the particle β . In fact, since $\tau^*_{\alpha\beta} = \tau^*_{\beta\alpha}$ we have, adding to (81) the corresponding equation for the particle β ,

$$n_\alpha m_\alpha \bar{\underline{V}}_\alpha + n_\beta m_\beta \bar{\underline{V}}_\beta = 0$$

Using this relation, (81) may be expressed in terms of the mean relative velocity , namely,

$$-\frac{\rho_\alpha \rho_\beta}{\rho_0} (\bar{\underline{v}}_\alpha - \bar{\underline{v}}_\beta)/\tau \tag{82}$$

since $\bar{\underline{v}}_\alpha - \bar{\underline{v}}_\beta = \bar{\underline{V}}_\alpha - \bar{\underline{V}}_\beta$. Hence equation (73) can finally be written

$$\rho_\alpha \frac{d_\alpha \bar{\underline{v}}_\alpha}{dt} + \nabla . P_\alpha - \rho_\alpha \underline{F}_\alpha = -\frac{\rho_\alpha \rho_\beta}{\rho_0 \tau} (\bar{\underline{v}}_\alpha - \bar{\underline{v}}_\beta) \tag{83}$$

14. Rate of diffusion of the two constituents

Dividing this equation by ρ_0 and substracting from it the corresponding equation for the β-constituent we obtain an expression for the differential or diffusion velocity

$$\bar{\underline{v}}_\alpha - \bar{\underline{v}}_\beta = -\tau \left\{ \frac{d_\alpha \bar{\underline{v}}_\alpha}{dt} - \frac{d_\beta \bar{\underline{v}}_\beta}{dt} + \frac{1}{\rho} \nabla . P_\alpha - \frac{1}{\rho_\beta} \nabla . P_\beta - (\underline{F}_\alpha - \underline{F}_\beta) \right\} . \tag{84}$$

It is convenient at this stage to introduce the coefficient of diffusion of the other two constituents, namely,

$$D_{\alpha\beta} = kT \left(\frac{m_0}{m_\alpha \; m_\beta} \right) \tau \tag{85}$$

where $m_0 = m_\alpha + m_\beta$. We find, after some algebra, that (84) can be written

$$\underline{\bar{v}}_\alpha - \underline{\bar{v}}_\beta = D_{\alpha\beta}\left\{\frac{m_\alpha m_\beta}{k T m_o}(\underline{f}_\alpha - \underline{f}_\beta) - \frac{m_\alpha - m_\beta}{m_o}\nabla \log P_o \right.$$

(86)
$$\left. + \frac{n_o}{n_\alpha n_\beta}\nabla \cdot \frac{n_\alpha}{n_o} - \frac{m_\alpha m_\beta}{k\ T m_o}(\underline{F}_\alpha - \underline{F}_\beta)\right\}$$

where $P_o = P_\alpha + P_\beta$ is the total pressure, and we have written $\underline{f}_\alpha$ for $\frac{d_\alpha \underline{v}_\alpha}{dt}$, etc. The four terms inside the bracket (86) correspond to components of the relative velocity of diffusion due respectively to (1) the relative acceleration, (2) the pressure gradient, (3) a concentration gradient, and (4) external forces. Note that gravitational forces do not contribute to the velocity of diffusion. These component velocities of diffusion tend to have the following effects: (1) and (4) have indeed the same effect and tend to separate the constituents in the direction of the relative acceleration or forces. (2) tends to make the composition uniform and (3) tends to increase the proportion of the heavier constitution in the regions of higher pressure.

15. Three-constituent plasma. (Partially ionized gas) .

We shall consider a partially ionized gas consisting of electrons one kind of ions and one kind of neutral particles. The velocity of each constituent will be denoted by $\underline{v}_e$, $\underline{v}_i$, $\underline{v}_n$ respectively. Because of their much smaller mass, the momentum of the electrons may be neglected in defining the mean mass velocity $\underline{v}_o$, which is thus approximately

(87)
$$\underline{v}_o = \frac{1}{\rho_o}(n_i m_i \underline{v}_i + n_n m_n \underline{v}_n),$$

where $\rho_o = n_i m_i + n_n m_n$.

However, as in the case of a two-constituent plasma, it is more convenient to derive the moment equation of the Boltzmann equation of each constituent relative to axes moving with the mean velocity $\bar{v}$ of that component. The equation of continuity (65) will hold as before, and the equation of momentum will likewise be the same as before, except for the collision term C_α.
It is clear that the collision of particles of each constituent with those of the other two constituents will yield a collision term of the form (82); however, we can no longer drop the suffixes so that denoting by ρ_e, ρ_i, ρ_n the mass derivatives of the electrons, ions, and neutrals, we have

$$(88) \quad \int m_e \underline{v}_e C_e \, d\underline{v}_e = -\frac{\rho_e \rho_i}{\rho_i + \rho_e} (\bar{\underline{v}}_e - \bar{\underline{v}}_i)/\tau_{ei} - \frac{\rho_e \rho_n}{\rho_e + \rho_n} (\bar{\underline{v}}_e - \bar{\underline{v}}_n)/\tau_{en}$$

$$(89) \quad \int m_i \underline{v}_i C_i \, d\underline{v}_i = -\frac{\rho_i \rho_e}{\rho_i + \rho_e} (\bar{\underline{v}}_i - \bar{\underline{v}}_e)/\tau_{ie} - \frac{\rho_i \rho_n}{\rho_i + \rho_n} (\bar{\underline{v}}_i - \bar{\underline{v}}_n)/\tau_{in}$$

$$(90) \quad \int m_n \underline{v}_n C_n \, d\underline{v}_n = -\frac{\rho_n \rho_e}{\rho_n + \rho_e} (\bar{\underline{v}}_n - \bar{\underline{v}}_e)/\tau_{ne} - \frac{\rho_n \rho_i}{\rho_n + \rho_i} (\bar{\underline{v}}_n - \bar{\underline{v}}_i)/\tau_{ni}$$

The τ' s are called the 'collision intervals' by analogy with what has been said previously. We have $\tau_{ei} = \tau_{ie}$, $\tau_{in} = \tau_{ni}$, $\tau_{en} = \tau_{ne}$, so that there are effectively only three 'collision intervals' ; writing

$$(91) \qquad \theta_{\alpha\beta} = \frac{\rho_\alpha \rho_\beta}{\rho_\alpha + \rho_\beta} \frac{1}{\tau_{\alpha\beta}} \qquad (\alpha, \beta = e, i, n)$$

the equation of motions for the ions, electrons and neutral particles are respectively

$$(92) \quad \rho_i \frac{d_i \bar{\underline{v}}_i}{dt} = -\nabla \cdot P_i + Z n_i e (\underline{E} + \bar{\underline{v}}_i \times \underline{B}) + \rho_i \underline{F}_i - \theta_{ie} (\bar{\underline{v}}_i - \bar{\underline{v}}_e) - \theta_{in} (\bar{\underline{v}}_i - \bar{\underline{v}}_n)$$

(93) $$\rho_e \frac{d\underline{v}_e}{dt} = -\nabla\cdot P_e - n_e e(\underline{E} + \underline{v}_e \times \underline{B}) + \rho_e \underline{F}_e - \theta_{ie}(\underline{v}_e - \underline{v}_i) - \theta_{en}(\underline{v}_e - \underline{v}_n)$$

(94) $$\rho_n \frac{d\underline{v}_n}{dt} = -\nabla\cdot P_n + \rho_n \underline{F}_n - \theta_{in}(\underline{v}_n - \underline{v}_i) - \theta_{en}(\underline{v}_n - \underline{v}_e)$$

where n_i, n_e, n_n are the number densities of the ions, electrons and neutrals, P_n is the relative partial pressure of the neutral gas, and Ze, $-e$ are the charges on a positive ion and electron respectively.

16. Diffusive equilibrium in a fully ionized plasma in a magnetic field

The equations of motions for the ions (assumed to be singly ionized for simplicity) and electrons may be written respectively (dropping the bar over the velocities),

(95)
$$\frac{d\underline{v}_i}{dt} + \frac{1}{\rho_i}\nabla\cdot P_i - \underline{F}_i - \frac{e}{m_i}\underline{E} - \underline{v}_i \times \underline{\omega}_i = -\frac{\rho_e}{\rho_0 \tau}(\underline{v}_i - \underline{v}_e)$$
$$\frac{d\underline{v}_e}{dt} + \frac{1}{\rho_e}\nabla\cdot P_e - \underline{F}_e + \frac{e}{m_e}\underline{E} + \underline{v}_e \times \underline{\omega}_e = -\frac{\rho_i}{\rho_0 \tau}(\underline{v}_e - \underline{v}_i)$$

where $\underline{\omega}_i = e\underline{B}/m_i e$ and $\underline{\omega}_e = e\underline{B}/m_e c$ are the cyclotron frequencies for the ions and electrons. Now a plasma is electrically neutral to a high degree of approximation so that $n_i/n_e \simeq 1$. Also $m_e/m_i \ll 1$ so that $\rho_e/\rho_0 \simeq m_e/m_i$ and $\rho_i/\rho_0 \simeq 1$. With these approximations the equations may be rewritten

(96) $$\frac{d\underline{v}_i}{dt} + \frac{m_e}{m_i \tau}\underline{v}_i - \frac{m_e}{m_i \tau}\underline{v}_e - \underline{v}_i \times \omega_i \underline{u} = \frac{e}{m_i}\left(\underline{E} - \frac{1}{n_i e}\nabla\cdot P_i\right) + \underline{F}_i \equiv \underline{G}_i$$

(97) $$\frac{d\underline{v}_e}{dt} + \frac{1}{\tau}\underline{v}_e - \frac{\underline{v}_i}{\tau} + \underline{v}_e \times \omega_e \underline{u} = -\frac{e}{m_e}\left(\underline{E} + \frac{1}{n_e e}\nabla\cdot P_e\right) + \underline{F}_e \equiv \underline{G}_e$$

where $\underline{u}$ is a unit vector along $\underline{B}$, $\omega_i = |\underline{\omega}_i|$, $\omega_e = |\underline{\omega}_e|$, and $\underline{F}_i$ and $\underline{F}_e$ are extraneous forces other that the electric and magnetic forces or pressure gradients.

An inspection of these equations shows that there exists a transient part of the solution for $\underline{v}_i$ and $\underline{v}_e$ which decays exponentially with time approximately as $e^{-t/\tau}$, i.e., in a time of the order of the relaxation time. The plasma therefore attains what is termed a state of <u>diffusive equilibrium</u> in which the acceleration terms in (96) and (97) can be neglected, yielding the approximate diffusive equations

(98) $$\underline{v}_i - \underline{v}_e - \underline{v}_i \times \Omega \underline{u} = \underline{G}_i \frac{m_i}{m_e} \tau$$

(99) $$\underline{v}_e - \underline{v}_i + \underline{v}_e \times \Omega \underline{u} = \underline{G}_e \tau$$

where $\Omega = \omega_e \tau$. Solving these vector equations we obtain

(100) $$\underline{v}_i = \lambda_i \underline{u} + \frac{1}{\Omega^2}\left[\left(\frac{m_i}{m_e} \underline{G}_i + \underline{G}_e\right)\tau + \Omega\tau \frac{m_i}{m_e} \underline{G}_i \times \underline{u}\right]$$

(101) $$\underline{v}_e = \lambda_e \underline{u} + \frac{1}{\Omega^2}\left[\left(\frac{m_i}{m_e} \underline{G}_i + \underline{G}_e\right)\tau + \Omega\tau \underline{G}_e \times \underline{u}\right]$$

where λ_i and λ_e are arbitrary parameters. However, multiplying (98) and (99) scalarly by $\underline{n}$ and adding the resulting equations have

(102) $$\left(\frac{m_i}{m} \underline{G}_i + \underline{G}_e\right) \cdot \underline{u} = 0$$

Using the approximation $n_i/n_e \simeq 1$ this reduces to

(103) $$(\nabla \cdot P_o + \rho_i \underline{F}_i + \rho_e \underline{F}_e) \cdot \underline{u} = 0$$

Thus , along a magnetic line of force, the pressure gradient in that direction balances the total external force in the same direction. In the case of the protonosphere this implies that, along a line of force, the pressure decreases exponentially with height, the temperature being approximately constant in this region. The effect of the magnetic field depends roughly on the magnitude of Ω, that is, the product $\omega_e \tau$, the ratio of the cyclotron frequency to the collision frequency. If the electrons are able to spiral many times between collisions, then $\Omega \gg 1$ and we see from (100) and (101) that, whilst motion of the ions and electrons along the lines of force is unimpeded, the component at right angles is of order Ω^{-1} and thus becomes vanishingly small as $\Omega \to \infty$; we note also that the drift of the ions at right angles to the magnetic field due to the pressure gradient is greater than the corresponding drift for the electrons.

These results, of course, are to be expected from general consideration of the motion of charged particles in a magnetic field.

V. C. A. Ferraro

II. Application to the Ionosphere

17. The atmosphere

The scientific study of the upper atmosphere is nowadays called aeronomy - a term due to Chapman. The discussion of the atmosphere is based largely on chemical composition and temperature and the various layers into which the atmosphere can so be divided are referred to as 'spheres' . The upper boundaries are referred to as '-pause'. Thus the troposphere denotes the layer extending from ground level upwards in which the temperature decreases with height. Above this layer is the stratosphere in which , for many kilometres the temperature remains constant. The tropopause is the upper level of the troposhere and separates it from the stratosphere. The height of the the tropopause varies with latitude, being about 11 km in mid-latitudes. Above the stratosphere the temperature increases and the region of higher temperature is the mesosphere; the temperature then decreases again and reaches its lowest value (about 180° K) at a height of 80-85 km. The temperature rapidly rises above this level (mesopause) to about 300 km where it attains a temperature of over 1000° K. This region is called the thermosphere. Above this level the atmosphere is maintained in isothermal equilibrium. This is the exosphere and here collisions between the molecules are so rare that they move in free orbits under gravity. The various regions are illustrated in figure 7 .

18. The ionosphere, heliosphere and protonosphere

Sunlight of wave lengths less than about 2900 A (1 A = 10^{-8} cm) is capable of ionizing oxygen and nitrogen. This radiation cannot be observed at ground level owing to strong absorption by atmospheric ozone. Its emission from the sun has been detected by satellites and other space vehicles and, prior to this, its presence was indicated by the existence of several ionized layers in the atmosphere.

Radio methods of observations of the ionosphere have revealed the existence of two main layers, the E and F layers, the former at about 120 km height and the other at 180-300 km. The F layer is thicker than the E layer and separates into two parts during daytime, the lower called F_1 and the upper F_2. At night they partly merge and become indistinguishable. Below the E-layer there is another at about 70 km called the D-layer . The thickness of the F-layer diminishes rather slowly with height and has no well defined upper boundary. It merges into the heliosphere where neutral and ionized helium are present to a height of about 1000 km and above this we have many ionized hydrogen or protons. This highest part of the ionosphere is called the protonosphere but its limits are difficult to define. It seems likely that this region is in diffusive equilibrium. The various layers are illustrated in fig. 8 .

19. Processes in the ionosphere

We shall here deal only with the large-scale structure of the ionospheric layers, and particularly our attention will be directed towards the effect of diffusion on the distribution of ionization of the F_2 region.

The ionization in the ionosphere is due essentially to the production of ion-electron pairs by the absorption of solar U. V. and X-ray radiation - at least in middle and low latitudes. At higher latitudes , ionization can also be produced by collisions between high-energy charged particles , precipitated in the atmosphere, with neutral atmospheric molecules

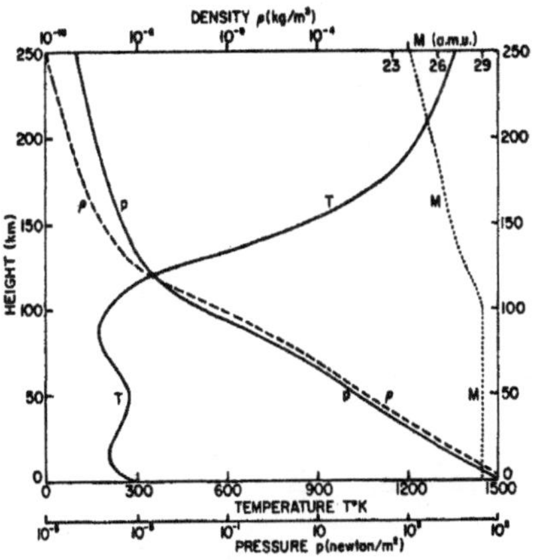

FIG. 7 . U.S. STANDARD ATMOSPHERE (Government Printing Office, Washington, D.C., 1962). Vertical distribution of pressure p, density ρ, temperature T and mean molecular mass M to 250 km. The composition is assumed constant up to 100 km.

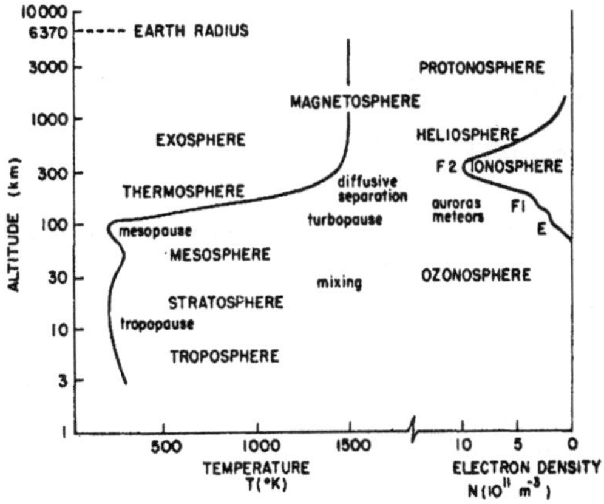

FIG. 8. REGIONS OF THE ATMOSPHERE, SHOWING CONVENTIONAL NAMES DESCRIPTIVE OF LEVELS, PHYSICAL REGIMES, AND CHARACTERISTIC CONSTITUENTS. The temperature profile is taken from the U.S. Standard Atmosphere and the electron density profile represents average daytime conditions for middle latitudes, high solar activity.

where

(2.3.61) $$J_n^*(\Gamma) = \int_{-1}^{1} E(\Gamma, \tau)(1-\tau^2)^n \, d\tau.$$

Bergman considers a partial differential equation of the form:

(2.3.62) $$\Delta_3 \psi + F(y, z)\psi = 0 .$$

We introduce the variables : $X = x$, $Z = (z+iy)/2$, $Z^* = -(z-iy)/2$, and express the function $F(y, z)$ appearing in Eq. (2.3.62) as a function of Z and Z^*; we also use the symbol F for this new function. The equation (2.3.62) then assumes the form :

(2.3.63) $$\psi_{XX} - \psi_{ZZ^*} + F(Z, Z^*)\psi = 0$$

We proceed to obtain particular solutions of (2.3.63) which are polynomials in X, as follows. Let $\tilde{\gamma}(Z, Z^*)$ be any solution of the equation:

(2.3.64) $$-\tilde{\gamma}_{ZZ^*} + F\tilde{\gamma} = 0$$

and let the polynomials $P^{(N, k, \alpha)}$ be defined as follows:

$$P^{(N, k, k-2\nu)} \equiv \binom{N}{k-\nu} \cdot \binom{k-\nu}{\nu} Z^{N-k+\nu} Z^{*\nu} ,$$

(2.3.65) $N=0, 1, 2, \cdots ;\quad k=0, 1, 2, \cdots, 2N;\ \nu=k, k-2, \cdots, k-2\left[\frac{k}{2}\right].$

Let the functions $\Pi^{(N, k, \alpha)}(Z, Z^*)$ satisfy the equations :

(2.3.66) $$-N\tilde{\gamma}_{Z^*} P^{(N-1, k, k)} - \Pi^{(N, k, k)}_{ZZ^*} + F\,\Pi^{(N, k, k)} = 0$$

$$- N\tilde{\gamma}_{Z^*} P^{(N-1, k, \nu)} - N\tilde{\gamma}_Z P^{(N-1, k-2, \nu)} +$$

(2.3.67) $$+ (\nu+2)(\nu+1)\Pi^{(N, k, \nu+2)} - \Pi^{(N, k, \nu)}_{ZZ^*} + F\,\Pi^{(N, k, \nu)} = 0, \quad \nu < k.$$

Negative ions may be formed in the lower ionosphere by attachment of elec trons.

The important losses of ionization arise from atomic ion-electron (radiative) recombination, molecular ion and electron (dissociative) recombination and, in the lower ionosphere, by the attachment of an electron to a neutral molecule.

Ionization may also be affected by transport processes; the ions and electrons (plasma) in the ionosphere may be thought as a minor constituent of the atmosphere. It is acted on by gravity and by pressure gradients in the plasma. Unlike the neutral constituent, the ions and elec trons are acted on by electric and magnetic forces. As we have seen in section 14, the plasma tends to diffuse through the neutral air if the forces acting on it are not in equilibrium. Ions and electrons diffuse together, since any tendency to separate the positive and negative charges w: give rise to a large electric field opposing this separation. This is called 'ambipolar' or 'plasma' diffusion and proceeds rapidly in the F region but not in the lower ionosphere. In this region the plasma tends to be set in motion by movements of the neutral air, which may be due to large scale wind-system or to temperature.

The various processes outlined above which modify the ionization in the layers of the ionosphere must balance and this balance can be expressed as an equation of continuity. If the transport processes (wind and diffusion) result in a net drift velocity $\underline{v}$, and we denote the electron density by n, the rate of production and loss of ionization by q and L respectively, the equation of continuity is

$$\frac{\partial n}{\partial t} + \text{div}\,(n\underline{v}) = q - L . \tag{104}$$

In the absence of production and loss of ionization (104) reduces to (65) already found in of section 12 .

Various wavelengths in the radiation from the sun are responsible for the production of ionization. It would be outside the scope of these lectures to go into more than a few details. A major part of the E-region ionization arises from the wavelength band 911-1027 A which ionizes 0_2 to $0_2^{+\cdot}$. In the F-region the wavebands 170-796A and 796 A to 911A are mainly responsible for the ionization, in this case the ions formed being 0^+ and N_2^+ .

Amongst the loss processes we may note the following :

(a) Ion-ion recombination (coefficient α_i)

$$X^+ + Y^- \rightarrow X + Y .$$

(b) Electron-ion recombination (coefficient α_e)

(i) Three-body : $X^+ + e + M \rightarrow X + M$

Here M denotes a neutral particle which exchanges energy and momentum but does not take part in the chemical reaction.

(ii) Radiative : $X^+ + e \rightarrow X^* \rightarrow X + h\nu$.

(iii) Dissociative : $XY^+ + e \rightarrow X^* + Y^*$

Here X^* denotes an atom left in the excited state.

Process (i) can occur in the lower D region but is rare at greater heights.

Process (ii) is likely to be the fastest loss process only in the uppermost levels of the F regions. Elsewhere in the E and F regions the dissociative recombination process are important.

(c) Ion-atom interchange

$$A^{+} + XY \rightarrow XY^{+} + A \ .$$

Ion-atom interchange (c) followed by dissociative recombination b (iii) is the principal loss process in the E and F regions. There is still considerable controversy as to precisely which reactions are important.

The rates for processes b (iii) and (c) are given , in terms of the reaction constants K_b and K_c , by the expressions

(105) $$\frac{dn(e)}{dt} = - K_b n(XY^{+})\, n(e)$$

(106) $$\frac{dn(A^{+})}{dt} = -K_c n(A^{+}) n(XY) \ .$$

If we suppose that the atmosphere is electrically neutral,

(107) $$n(A^{+}) + n(XY^{+}) = n(e)$$

and we suppose further that the ionization is in equilibrium . Then if the electrons and positive ions are produced by incident radiation at the rate q per unit volume

(108) $$q = K_b n(XY^{+}) n(e) = K_c n(A^{+}) n(XY) \ .$$

Eliminating $n(A^{+})$ and $n(XY^{+})$ by using (107) we find

(109) $$q = \frac{K_b K_c n(XY) n^2(e)}{K_c n(XY) + K_b n(e)} \ .$$

If $K_e n(XY) >> K_b n(e)$, this reduces to

(110) $$q = K_b n^2(e)$$

which corresponds to a quadratic law of recombination αn^2, the coefficient of recombination α being equal to K_b. This law holds very nearly in the E region where $\alpha \simeq 10^{-8}$ cm^3 sec^{-1}.

If $K_c n(XY) \ll K_b n(e)$, then (109) reduces to

(111) $$q = K_c n(XY) n(e)$$

which corresponds to an 'attachment' law of the form β n(e) with an 'attachment' coefficient

(112) $$\beta = K_c n(XY)$$

If, as is usually the case, n(XY) decreases upwards, so will β

20. Chapman's Theory

We consider the simple case of ionization by absorption of monochromatic radiation in an atmosphere of uniform composition and temperature. Such an atmosphere will be distributed exponentially; in fact if h denotes the height, n the number density, g the acceleration of gravity, m the mean molecular mass of the gas, the statical equation is

(113) $$\frac{dp}{dh} = -nmg \quad .$$

(114) Also $$p = knT$$

where k is Boltzmann's constant (1.38×10^{-16} cgs) and T is the temperature, so that eliminating p between these equations we find

(115) $$\frac{d \log n}{dh} = -\frac{mg}{kT} = -\frac{1}{H}$$

where H is a quantity having the dimensions of a length, called the scale height. Integration of (115) now gives

(116) $$n = n_o e^{-h/H}$$

that is, an exponential distribution of density. Let I denote the intensity of the radiation at the height h and I_∞ the intensity of the incident solar radiation. Let χ denote the zenith distance of the sun at the height h at any time, then the decrease in I by absorption over the path of length $ds = (\sec\chi)dh$ between the levels $h + dh$ and h is given by

(117) $$dI = -I\sigma n (\sec\chi) dh \,,$$

where σ is the absorption cross-section of the molecules. Using (116) we have

$$\frac{dI}{I} = -(\sigma n_o \sec\chi) e^{-h/H} dh$$

which can be integrated to give

(118) $$\log_e\left(\frac{I}{I_\infty}\right) = -(\sigma n_o H \sec\chi) e^{-h/H} dh$$

since $I \to I_\infty$ as $h \to \infty$. Hence

(119) $$I = I_\infty \exp(-\sigma n_o H e^{-h/H} \sec\chi) \,.$$

The absorption of radiant energy per unit volume of the atmosphere is $dI/ds = (dI/dh)\cos\chi$ and if β ions are produced by the absorption of unit quantity of energy, the rate of production of ions per unit volume is

$$q(h) = \beta I_\infty n_o \sigma \exp(-h/H - n_o \sigma H e^{-h/H} \sec\chi) \quad (120)$$

The total number of ions produced in a vertical column of air of unit area of cross-section by the complete absorption of the incident radiation is clearly $\beta I_\infty \cos\chi$.

The rate of ion-production q has a maximum q_m at a height h_m where

$$e^{h_m/H} = n_o \sigma H \sec\chi \quad (121)$$

giving

$$q_m = (\beta I_\infty \cos\chi)/He \quad . \quad (122)$$

Denote the values of h_m and q_m for the overhead sun ($\chi = 0$) by h_o and q_o ; then

$$e^{h_o/H} = n_o \sigma H, \qquad I_o = \beta I_\infty/eH \quad (123)$$

whence

$$h_m = h_o + H \log_e \sec\chi \quad (124)$$

$$q_m = q_o \cos\chi \quad . \quad (125)$$

In terms of q_o and h_o, we may now write (120) as

$$q(h) = q_o \exp\left(1 - \frac{h-h_o}{H} - e^{-\frac{h-h_o}{H}} \sec\chi\right) \quad (126)$$

or measuring heights in terms of H as unit from the level h_o, writing

$$z = \frac{h - h_o}{H} \quad (127)$$

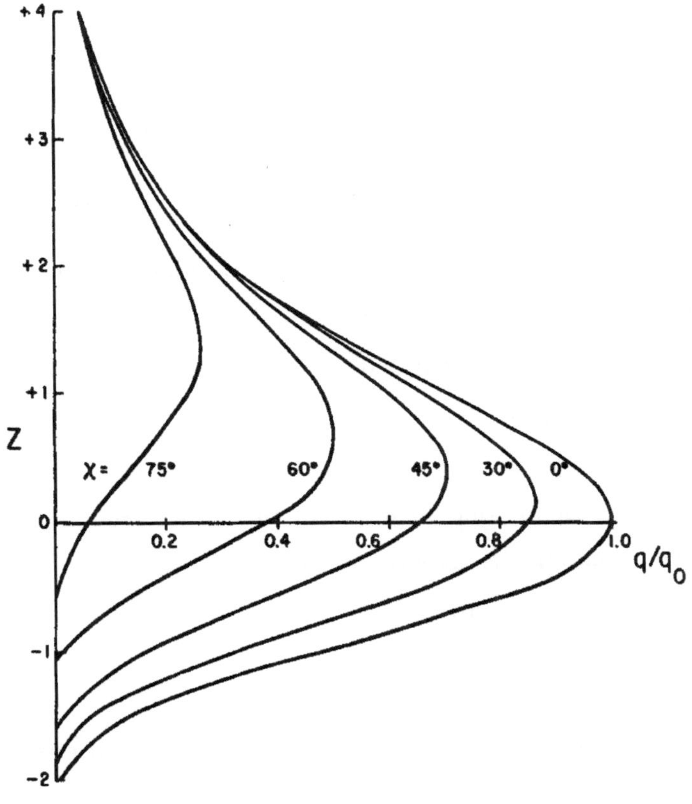

FIG. 9. NORMALIZED CHAPMAN PRODUCTION FUNCTION $q(z,\chi)/q_o = \exp\,(1 - z - e^{-z}\sec\chi)$. Values at several reduced heights are shown as a function of zenith angle. The broken line is the envelope, $q_m/q_o = \cos\chi$.

Once we know the relation between the density ρ and velocity q, we can integrate Eqs. (2.6.17) and (2.6.18) to yield the sonic line in the physical (x, y)-plane.

For stream functions ψ defined by corresponding $\chi_1(\theta)$, $\chi_2(\theta)$ given in (2.6.16), the sonic line in the physical (x, y)-plane is given in terms of the parameter θ by :

(2.6.19) $$x = (1/2)\,(6/5)^3 \left[a-(6b/5) \right] \cos^2 \theta ,$$

(2.6.20) $$y = (6/5)^3 \left\{ (1/2)\left[a+(6b/5)\right](\theta - \pi/2) + (1/4)\left[a-(6b/5)\right] \sin 2\theta \right\} ,$$

provided we take the origin in the physical plane as the image of the point $q=(5/6)^{1/2}$, $\theta = \pi/2$ on the sonic line in the hodograph plane. The value of q is calculated by $M^2 = q^2/\left[1-(\gamma-1)\,q^2/2\right]$ with $\gamma = 1.4$.

Example (I) : sonic line in physical plane is circular.

If we consider the stream function ψ defined by *(2.6.16) with* a, b related according to :

(2.6.21) $$a + (6b/5) = 0 ,$$

then (2.6.19) and (2.6.20) yield for parametric equations of the sonic line in the physical plane :

(2.6.22) $$x = a(6/5)^3 \cos^2 \theta , \qquad y = (1/2)a(6/5) \sin 2\theta .$$

Elimination of θ gives :

(2.6.23) $$x^2 + y^2 - a(6/5)^3\, x = 0 ,$$

(called the reduced height), we have

$$q(h) = q_o \exp\,(1 - z - e^{-z} \sec \chi\,) . \tag{128}$$

This is called the Chapman-function (Chapman, 1931). It has the interesting property that as χ varies, its shape is unchanged, its peak is shifted to the level $z_m = \log_e(\sec\chi)$ and its amplitute is scaled by the factor $\cos\chi$. This can be seen by writing the above equation in the form

$$q = (q_o \cos\chi) \exp\left[\,1 - (z-z_m) - e^{z_m - z}\right]. \tag{129}$$

The ratio q/q_o is shown in Fig. 9 .

In the actual ionosphere the production formula is considerably more complicated, partly because there are different atmospheric gases, differently distributed, and the ionizing radiation is not monochromatic but consists of a range of wave lengths and σ depends on the wavelength .

The above theory neglects the curvature of the earth ; Chapman has considered the modifications introduced by taking this into account. This correction is only important near sunrise and sunset.

The theory can also be extended to deal with gases which are not at the same temperature at all heights. If the temperature is proportional to the height, so that

$$H = K_o + \gamma h \tag{130}$$

then it can be shown that (125) takes the modified form

$$q_m = q_o (\cos\chi\,)^{1+\gamma} \tag{131}$$

21. Plasma diffusion

The F2 peak of ionization is observed at about 300 Km. No mechanism seemed capable of causing a peak of production at such a height and Bradbury (1938) suggested that the production peak occurred at a lower height (the F1 region, in fact) and he attributed the upward increase in electron density to a rapid upward decrease of a linear loss coefficient. The hypothesis is unsatisfactory, for even if β varies as

$$\exp\left[-(h-h_o)/H_\beta\right]$$

where H_β is the scale height, then the electron density well above the production peak is approximately $q/\beta \propto \exp\left[+(h-h_o)\left(\frac{1}{H_\beta}-\frac{1}{H_a}\right)\right]$, where H_a is the scale height of the ionizable gas. Since $H_a < H_\beta$, then q/β increases indefinitely upwards, and we must find some other explanation for the peak in the electron density. We require some transport process to limit the value of the electron density at great heights. One such process is plasma, or ambipolar diffusion Attention to the probable importance of diffusion in the ionosphere was first directed by Hulburt in 1928. I considered the problem in greater detail in 1945 and showed that diffusion was unimportant in the E and F_1 region of the ionosphere but that it might become important in the F_2 region and above it. Mariani (1956) drew similar conclusions but Yonezawa first discussed in detail the problem of the formation of the F_2 region and showed that diffusion could provide an explanation.

In deriving the equation of diffusion for the ionization, we shall neglect, for the present, the earth's magnetic field, so that the only forces acting are gravity $\underline{g}$, the electric field $\underline{E}$, and the frictional forces due to collisions. The full equations for a three constituent plasma,

(92-94), , have been derived in section 15. However, in the absence of any external electric fields, the slightest separation of the ions and electrons will give rise to large electrostatic fields opposing any further separation so that we many set $n_i = n_e$ and $\underline{v}_i = \underline{v}_e$ in these equations. Also, if we assume that the neutral air is at rest, $\underline{v}_n = 0$. Furthermore, the relaxation times in the F2 region are small , so that the steady state is quickly attained. That is, we may neglect the acceleration of the ions and electrons in (92-94) . Writing $\nu_{in} = 1/\tau_{in}$, etc. , for the collision frequencies, these become

$$-\nabla P_i + ne\,\underline{E} + nm_i\underline{g} - nm_i\nu_{in}\underline{v}_D = 0 \tag{132}$$

$$-\nabla P_e - ne\underline{E} + nm_e\underline{g} - nm_e\nu_{en}\underline{v}_D = 0 \tag{133}$$

$$-\nabla P_n + n_n m_n\underline{g} + nm_i\nu_{in}\underline{v}_0 + nm_e\nu_{en}\underline{v}_D = 0 \tag{134}$$

where we have written $\underline{v}_D$ for the common velocity of diffusion , of the the ions and electrons, and n for their number density.

Further simplifications can be made by noting that $m_e \ll m_i$, $m_i\nu_{in} \gg m_e\nu_{en}$ (i.e. collision with the neutral particles are important for ions but not for electrons). Also, if T_i and T_e are the temperature of the ionic and electronic constituent, we have

$$P_i = kn_iT_i\,, \qquad P_e = kn_eT_e\,. \tag{135}$$

Using these equations, on adding (132) and (133) and solving for the velocity of diffusion of the ion-electron component we find, if h denotes the height at any level,

(136) $$-v_D = \frac{1}{m_i \nu_{in}} \left\{ \frac{1}{n} \frac{\partial}{\partial n}\left[nk(T_i + T_e) + m_i g \right] \right\}$$

where v_D is measured positively upwards. In the F region, the ion, electron and neutral air temperatures may all be different. It seems likely that $T_i = T$, the temperature of the neutral air.

Then, if we introduce the neutral air scale height $H = kT/m_n g$, and write $\mu = m_i/2m_n$, $\tau = T_e/T_i = T_e/T$, and introduce the coefficient of diffusion of ions through the neutral gas

(137) $$D_{in} = kT/m_i \nu_{in}$$

equation (136) becomes

(138) $$-v_D = D_{in}(1+\tau)\left[\frac{1}{n}\frac{\partial n}{\partial h} + \frac{1}{T}\frac{\partial T}{\partial h} + \frac{2\mu}{(1+\tau)H} + \frac{\partial \tau/\partial h}{1+\tau}\right].$$

If $T_i = T_e = T$ at all heights, $\tau = 1$, and

(139) $$-v_D = D\left(\frac{1}{n}\frac{\partial n}{\partial h} + \frac{1}{T}\frac{\partial T}{\partial h} + \frac{\mu}{H}\right)$$

where we have written $D = 2D_{in}$.

The contribution of diffusion to the continuity equation (104) is to add the term $\partial(nv_D)/\partial h$ to the left hand side of equation. In general

(140) $$\frac{\partial(nv_D)}{\partial h} = D\mathcal{D}n$$

where $\mathcal{D}$ is a differential operator of the second order.

The value of D is of the form b/n_n, where the factor b depends on the temperature. It is now believed that in the F2 layer the neutral gas is mainly atomic oxygen and the ions mostly O^+. The value of b is affected by charge exchange and its value not exactly known. The value derived from kinetic theory, taking account of the effect of electrostatic induction by the charged ions on the neutral ions, gives n D of order 10^{19}. But this may well be too large by a factor of 10 because of charge exchange. Also $D \propto 1/n \propto e^{(h-h_o)/H}$; if we assume that the temperature of the gas is uniform, then

$$-v_D = D\left(\frac{1}{n}\frac{\partial n}{\partial h} + \frac{1}{2H}\right), \tag{141}$$

and the diffusion term in the continuity equation on the right hand side takes the form

$$D\mathscr{D} n = D\left(\frac{\partial^2 n}{\partial h^2} + \frac{3}{2H}\frac{\partial n}{\partial h} + \frac{n}{2H^2}\right), \tag{142}$$

(Ferraro 1945). The first term in this equation is characteristic of diffusion formulae. The second and third arise from the effect of gravity and the height dependence of D.

22. The equation of diffusion for the F2 region

We shall ignore for the present the geomagnetic field; we shall also assume that we are considering a locality on the equator so that $\chi = \varphi$, the time in radians measured from sunrise? It is related to the actual time by the equation

$$t = k\varphi, \tag{143}$$

where $k = 1.37 \times 10^4 \sec^{-1}$. The rate of production of electrons q will be taken to be represented by the Champan function

$$q = q_o \exp\left[1 - \frac{h-h_o}{H} - (\operatorname{cosec}\varphi)\, e^{-\frac{h-h_o}{H}}\right] \quad 0 \le \varphi \le \pi, \tag{144}$$

$$= 0 \qquad\qquad \pi \leq \varphi \leq 2\pi$$

where h_o refers to the level of maximum rate of production of ions q_o. Electrons are assumed lost in the F2 region according to an attachment-type law (119) with a coefficient which decreases at a rate proportional to the density of the neutral particles. Thus

(145) $$L = Kn$$

where

$$K = \beta_o \exp\left(-\frac{h-h_o}{H}\right)$$

Then the equation of diffusion to be solved is

(146) $$\frac{\partial n}{\partial t} = q - \beta_o n \exp\left(-\frac{h-h_o}{H}\right) + D\left(\frac{\partial^2 n}{\partial h^2} + \frac{3}{2H}\frac{\partial n}{\partial h} + \frac{n}{2H^2}\right)$$

(147) where $D = b/n_n$ and $n_n = N_o \exp\left(-\frac{h-h_o}{H}\right)$

Since $D = (b/n_n)$ increases exponentially upwards, whereas both q and L decrease exponentially with height, it follows that at some level the diffusion term will be the dominant term in equations (146) which then reduces to

(148) $$\frac{\partial^2 n}{\partial h^2} + \frac{3}{2H}\frac{\partial n}{\partial h} + \frac{n}{2H^2} = 0\ .$$

Solving this equation we find

(149) $$n = A_1 e^{-\frac{h}{2H}} + A_2 e^{-\frac{h}{H}}$$

and from equation (141) it follows that the first term corresponds to diffusive equilibrium, with $v_D = 0$, in which the ionization has a scale height twice that of the neutral gas. For the second term $v_D \neq 0$, and is positive if $A_2 > 0$. This represents a boundary condition of a <u>finite</u> flux of ionization at $h = \infty$, which is upwards if $A_2 > 0$. If ionization is gained or lost at the top of the ionosphere, then a term of this type must be included in the solution.

23. Solution of the diffusion equation

Writing

(150) $$x = \frac{h - h_o}{H}$$

and expressing t in terms of φ by (143), equation (146) can be written in the non-dimensional form

(151) $$\frac{\partial n}{\partial \varphi} = kq - \beta e^{-z} n + \frac{e^z}{\gamma}\left(\frac{\partial^2 n}{\partial z^2} + \frac{3}{2}\frac{\partial n}{\partial z} + \frac{n}{2}\right)$$

where

(152) $$\beta = k\beta_o \qquad \text{and} \qquad \gamma = \frac{N_o H^2}{Kb}$$

are non-dimensional parameters. The solution of (151) thus depends only on these two parameters and the boundary conditions. One of these, as we have already seen above, depends on the value of the flux of ionization at $z = +\infty$. The other requires that $n \to 0$ as $z \to -\infty$.

Assuming that q is given by the Chapman function (144), we require a solution of (151) which is periodic in φ, with period 2π. The method of solution of this, and related, equations has been given by Gliddon (1959) and consists in determining a suitable Green's function for this equation. The analysis is involved and reference must be made to the original papers . Typical solutions indicating the variation of the ionization at various heights from the level of maximum ion production are illustrated in Figs. 10 a and 10 b . Fig. 11 illustrates the variation of the level of maximum electron density for two distinct cases. From such solutions, the following general daytime characteristics of the behaviour of the ionization may be deduced.

(i) The F2 maximum electron density occurs at a level where diffusion and loss are comparable , i.e., where $K_m \simeq D_m/H^2$ and the subscript m refers to the maximum.

(ii) At the maximum and below it, the electron density is approximately given by $n \simeq q/K$, that is , balance between production and loss, as in the absence of diffusion.

(iii) Well above the maximum, diffusion becomes important and n varies as $e^{-(1/2)z}$ as already noted in section 22.

(iv) The level of maximum ionization falls rapidly at sunrise because of the rapid production of ionization in the lower F regions. It reaches a minimum height of one or one and a half scale heights <u>above</u> the level h_o of maximum ion-production which remains at about the same level until late afternoon. Therefore, the level rises again steadily to a height of about three scale heights above h_o after sunset.(See Fig. 11)

The night-time decay of the ionization has been studied by Martyn (1956) , Duncan (1956) and Dungey (1956) . This is also affected by vertical electromagnetic drifts .

24. Effect of a magnetic field on diffusion in the ionosphere

We now consider the effect of a magnetic field on diffusion of ions in the ionosphere. We are here dealing with a thernary mixture of neutral molecules, ions and electrons, of which the ionized particles form a minor

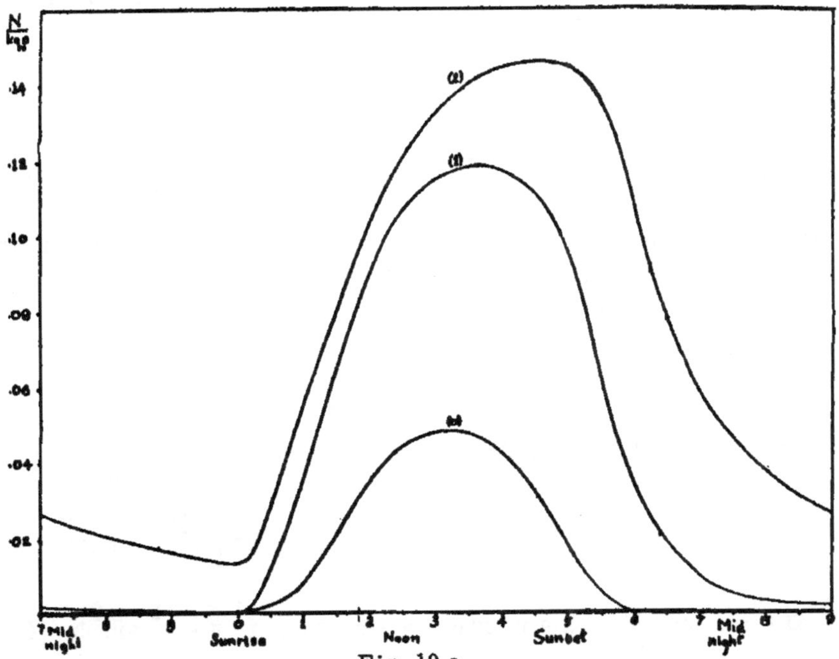

Fig. 10 a

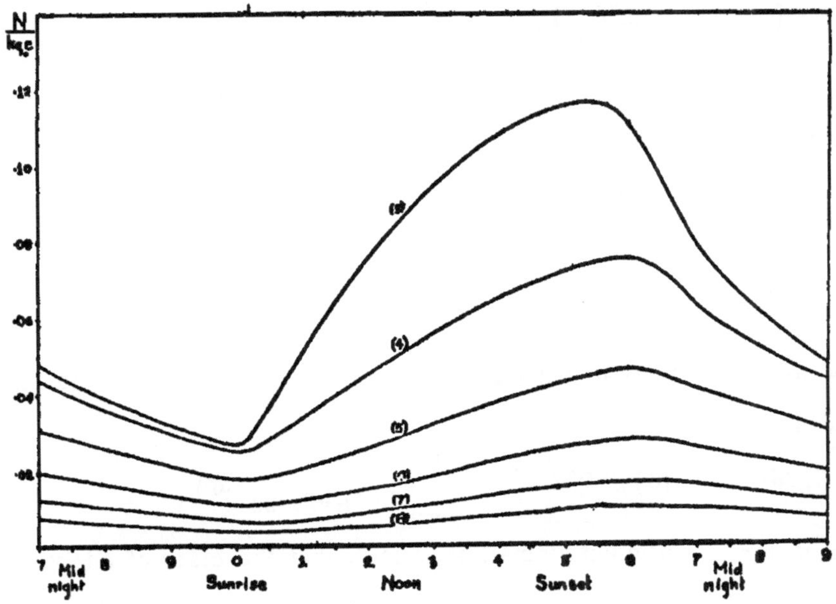

Fig.10 b Diurnal variation of electron density at intervals of one scale height. Case II ($\beta = 10$, $\gamma = 20$)

below (2.7.22) :

(2.7.24) $f_n(r, t) = P_n^{(\nu-1/2,\mu-1/2)}(t)\,(kr)^{-\mu-\nu} J_{\mu+\nu+2n}(kr), n = 0, 1, \cdots,$

where $t = \cos 2\theta$, $r^2 = x^2 + y^2$ and $P_n^{(\alpha,\beta)}$ stands for the Jacobi Polynomials (see [16]):

$$P_n^{(\nu-1/2,\mu-1/2)}(t)\,(kr)^{-\mu-\nu} J_{\mu+\nu+2n}(kr) \equiv f_n(r,t) =$$

$$= \Gamma(\nu+n+1/2)\left[\Gamma(\nu)\Gamma(1/2)\Gamma(n+1)\right]^{-1}\int_0^{\pi}\left\{(k\sigma)^{-\mu-\nu} J_{\mu+\nu+2n}(k\sigma)\right\}\cdot$$

$$\cdot\left\{(\sigma/x)^{\mu}\,\Phi_2(\mu, 1-\mu, \nu; \xi^1 - \eta^1)(\sin\phi)^{2\nu-1}\,d\phi\right\},$$

$$\sigma = x + iy\cos\phi, \quad x = r\cos\theta, \quad y = r\sin\theta, \quad \xi^1 = -y^2\sin^2\phi/(4x\sigma),$$

$$\eta^1 = -k^2y^2\sin^2\phi/4, \quad k, \mu, \nu, > 0, \eta = 0, 1, 2, \dots .$$

An arbitrary solution of the class S may be represented in a series form [39] :

$$w(r, \theta) = (kr)^{-\mu-\nu}\sum_{n=0}^{\infty} a_{2n}\, n!\left[\Gamma(n+\nu+1/2)\right]^{-1}.$$

(2.7.25) $\qquad \cdot P_n^{(\nu-1/2,\mu-1/2)}(\cos 2\theta)\, J_{\mu+\nu+2n}(kr),$

and an even analytic function regular about the origin may be expressed as

(2.7.26) $\quad f(\sigma) = \sigma^{-\mu-\nu}\sum_0^{\infty} a_{2n}\, J_{\mu+\nu+2n}(\sigma)$.

Hence for r sufficiently small it follows that the class of analytic functions (2.7.26) is mapped onto the class of solutions (2.7.25) by an operator of the form :

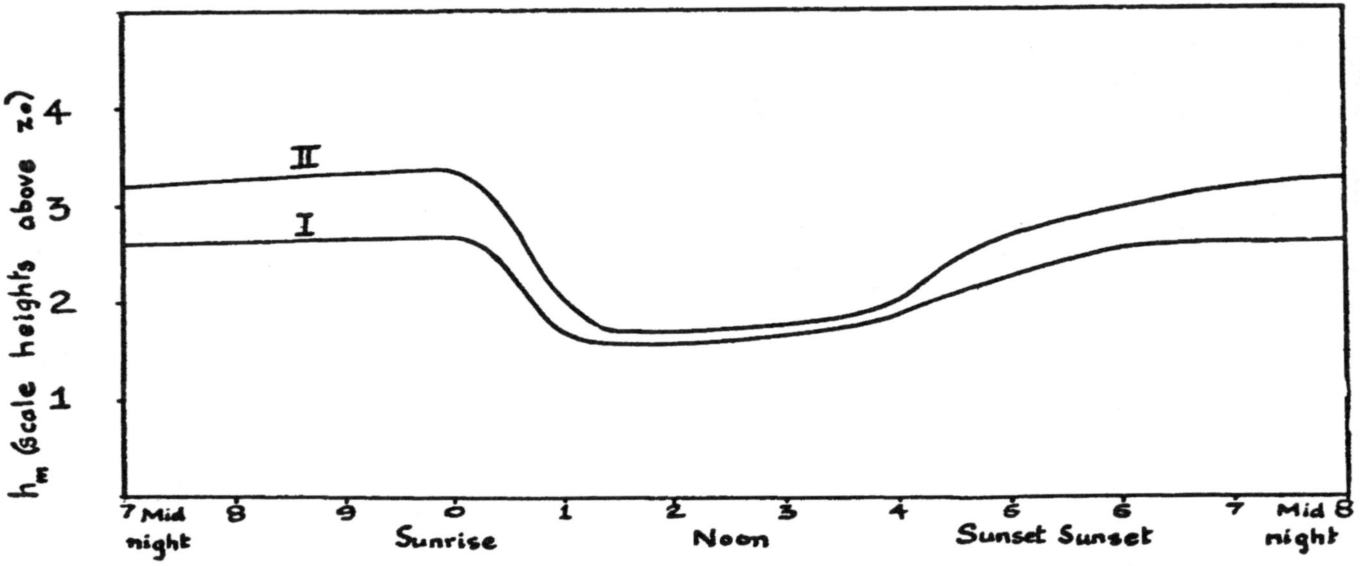

Fig. 11 . Comparison of height of maximum electron density for two distinct cases : case I for , case II for.

$$\int_{-1}^{+1} (1-\xi)^{\nu-1/2}(1+\xi)^{\mu+1/2} P_n^{(\nu-1/2,\mu-1/2)}(\xi)\, P_m^{(\nu-1/2,\mu-1/2)}(\xi)\, d\xi$$

$$= \delta_{nm}\, 2^{\mu+\nu}\Gamma(n+\nu+1/2)\Gamma(n+\mu+1/2)\,\cdot$$

(2.7.30) $$\cdot\ \left[(2n+\mu+\nu)\Gamma(n+1)\Gamma(n+\nu+\mu)\right]^{-1}.$$

Thus, if we define :

$$K_n(\sigma, r, \xi)\ \ 2^{-(\mu+\nu)} b_n (kr/k\sigma)^{\mu+\nu} J_{\mu+\nu+2n}(k\sigma)\,(J_{\mu+\nu+2n}(kr))^{-1}.$$

(2.7.31) $$\cdot\ P_n^{(\nu-1/2,\mu-1/2)}(\xi)\,(1-\xi)^{\nu-1/2}(1+\xi)^{\mu-1/2},$$

where
$$b_n = (2n+\mu+\nu)\,\Gamma(n+\mu+\nu)\ \ (\Gamma(n+\mu+1/2))^{-1},$$

we have
$$a_{2n}(k\sigma)^{-(\mu+\nu)} J_{\mu+\nu+2n}(k\sigma) = \int_{1}^{+1} K_n(\sigma, r, \xi)\,\cdot$$

(2.7.32) $$\cdot\ u\left(r((1+\xi)/2)^{1/2},\ r((1-\xi)/2)^{1/2}\right) d\xi.$$

Hence

(2.7.33) $$f(k\sigma) = \int_{-1}^{+1} K(\sigma, r, \xi)\, u\left[r((1+\xi)/2)^{1/2},\ r((1-\xi)/2)^{1/2}\right. d\xi,$$

where
$$K(\sigma, r, \xi) = 2^{-(\mu+\nu)}(r/\sigma)^{\mu+\nu}(1-\xi)^{\nu-1/2}(1+\xi)^{\mu-1/2}\,.$$

$$\cdot\ \sum_{0}^{\infty}(2n+\mu+\nu)\Gamma(n+\nu+\mu)\ \ (\Gamma(n+\mu+\nu))^{-1}$$

$$J_{\mu+\nu+2n}(k\sigma)\left[J_{\mu+\nu+2n}(kr)\right]^{-1} P_n^{(\nu-1/2,\mu-1/2)}(\xi).$$

To verify that these formal calculations are justified and that K

constituent. The relevant equations of the problem have been given in section 15, namely equations (92-94). However, for much the same reason as stated in section 21, we may neglect the acceleration of the ions and electrons, and assume that the neutral air is at rest. It should be stated, however, that Yonezawa (1958) and Dougherty (1960) have suggested the possibility that the neutral atoms and molecules are accelerated by the flow of the ions through the neutral air, in the comparatively short time of 20 minutes. Nevertheless we shall restrict ourselves to the case when $v_n \simeq 0$; Again quite apart from the fact that the collision frequencies for encounter between ions and electrons exceed those for collisions of the ions or electrons with the neutral atoms, the differential velocity $\underline{v}_i - \underline{v}_e$ between the ions and electrons must remain small otherwise large electric fields would develop because of the consequent large separation of charges of opposite sign. Likewise, we must have, very nearly, overall charge neutrality, so that $n_i \simeq n_e$. Assuming also that the ion and electron temperatures are equal, equations (92) and (93) then reduce to

$$(153) \qquad -\nabla P_i + ne(\underline{E} + \underline{v}_i \times \underline{B}/c) + \rho_i \underline{g} - \rho_i \nu_{in} \underline{v}_i = 0$$

$$(154) \qquad -\nabla P_e - ne(\underline{E} + \underline{v}_e \times \underline{B}/c) + \rho_e \underline{g} - \rho_e \nu_{en} \underline{v}_e = 0$$

where n is the number density of the ions or electrons, $\underline{B}$ the magnetic field intensity, ν_{in} and ν_{en} the collision frequencies for ion-neutral atom and electron-neutral atom encounters. Again, in the F2 region, $m_i \nu_{in} \gg m_e \nu_{en}$ so that collision with neutral particles are important for the ions but not for electrons. In fact, we may neglect this term and also the weight of the electrons in (146) because of the small mass of the electrons. This now becomes approximately,

$$(155) \qquad -\nabla P_e - ne(\underline{E} + \underline{v}_e \times \underline{B}/c) = 0\ .$$

Also $P_e = knT(\simeq P_i)$ and if, as is legitimate, T is assumed constant, this equation can be written

(156) $$kT\nabla \log n + e(\underline{E} + \underline{v}_e \times \underline{B}/c) = 0 \quad .$$

Hence ,

(157) $$\text{curl } \underline{E} + \text{curl } (\underline{v}_e \times \underline{B}/c) = 0$$

or, using Maxwell's equation, $\text{curl } E = -\frac{1}{c}\frac{\partial \underline{B}}{\partial t}$, this becomes

(158) $$\frac{\partial \underline{B}}{\partial t} = \text{curl } (\underline{v}_e \times \underline{B})$$

showing that the magnetic field lines are frozen in the electron gas. Since the magnetic pressure of the geomagnetic field greatly exceeds the electron gas pressure in the F2 region, the electrons can only move freely along the magnetic lines of force, but not at right angles to them. Thus $\underline{v}_e$ is parallel to $\underline{B}$ and (148) now gives the electric field as

(159) $$\underline{E} = -\frac{kT}{e}\nabla \log n$$

or

(160) $$n \propto e^{-eV/kT}$$

where V is the electrostatic potential. This implies that the electron density attains its thermodynamic equilibrium a each instant . Substituting (159) in (153) now gives an equation to determine the velocity of diffusion of the ions, namely,

(161) $$-2kT\nabla \log n + e\underline{v}_i \times \underline{B}/c + m_i\underline{g} - m_i \nu_{in}\underline{v}_i = 0.$$

The solution of this vector equation is

(162) $$\underline{v}_i(1+\Omega^2) = \underline{C} + (\underline{\Omega} \cdot \underline{C})\underline{\Omega} + \underline{C} \times \underline{\Omega}$$

where

(163) $$\underline{\Omega} = \frac{e\underline{B}}{m_i c \nu_{in}}$$

and

$$\underline{C} = \frac{1}{m_i \nu_{in}}(-2kT \nabla \log n + m_i \underline{g}) = D(\nabla \log n + \frac{1}{2H} \underline{k}) ,$$

$\underline{k}$ being a unit vector along the downward vertical, and D the coefficient of diffusion defined earlier, In general , $\nabla \log n$ will also be a vector directed along the vertical so that $\underline{C}$ is vertical.

Equation (162) shows that the velocity of diffusion of the ions has three components, one along $\underline{C}$ and the other two along and perpendicular to the magnetic field. Again, in the F2-region , $|\Omega|$ is large, being of the order of 200 at a level of 300 km. Hence, except over the magnetic equator where $\underline{C}$ and $\underline{\Omega}$ are in general perpendicular, the largest component of the diffusion velocity is along the magnetic lines of force and the smallest, that along $\underline{C}$. Hence, very nearly, we may write

(164) $$\underline{v}_i = (\underline{b} \cdot \underline{C})\underline{b} + \frac{1}{\Omega} \underline{C} \times \underline{b}$$

where $\underline{b}$ is a unit vector along $\underline{B}$ (or $\underline{\Omega}$) . This result is easily interpreted; a large value of Ω implies that the ions can spiral around a line of force many times between collisions and hence its velocity is along the line of force except when changed abruptly by a collision. The net results of such collisions is to give rise to the 'Hall' component of the velocity of diffusion given by the second term in (164) .

The equation of diffusion can now be obtained by using the equation of continuity for the ions

(165) $$\frac{\partial n}{\partial t} = q - L - \operatorname{div}(n\, \underline{v}_i)$$

and substituting for $\underline{v}_i$ the expression given in (164).
For a plane stratified atmosphere, I showed in 1945 that the diffusion term in (165) can be expressed in the approximate form $D\mathcal{D}n$, where

(166) $$\mathcal{D} n = D(\sin^2 I)\left(\frac{\partial^2 n}{\partial h^2} + \frac{3}{2H}\frac{\partial n}{\partial h} + \frac{n}{2H^2}\right) ,$$

I is the inclination of the lines of force above the horizontal (magnetic dip angle) and h the vertical height; (168) is the same as equation (142) except for the extra factor $\sin^2 I$. This approach is incorrect if the neutral air is in motion, as may well be the case. Near the magnetic equator the 'correction factor' $\sin^2 I$ fails to give the correct result, and a more careful derivation of the diffusion operator D is required. This was derived by Kendall (1962) and by Lyon (1963) . Assuming that the geomagnetic field is that of a centred dipole, Kendall found that one form of the operator is given by

$$\mathcal{D}' n = \frac{\sin^2 I}{2H^2} n + \frac{3\cot\theta}{aH(1 + 3\cos^2\theta)}\frac{\partial n}{\partial \theta} + \frac{1}{a^2\sin^2\theta(1 + 3\cos^2\theta)}\frac{\partial^2 n}{\partial \theta^2}$$

$$+ \left(\frac{15\cos^4\theta + 10\cos^2\theta - 1}{\sin^2\theta(1 + 3\cos^2\theta)^2}\right) n ,$$

where θ is the magnetic colatitude and a the distance of the highest point from the centre of the earth.

The geomagnetic field in effect reduces the coefficient of diffusion from D to $D\sin^2 I$ and this diminishes the velocity of diffusion in the ration $1 : \sin^2 I$. The correction is small, amounting to $\frac{1}{2}$ for $\theta = 45^o$; but near the equator, where I is small , vertical diffusion is clearly negligible.

One final remark needs to be made; although we have been able to determine the velocity of the ions, that of the electrons cannot be determined. However, as we have already mentioned, the velocities of the ions and electrons at any one point cannot differ greatly, and their components along a magnetic line of force must be very nearly equal.

25. Diffusive equilibrium in the upper ionosphere

The equations of diffusion of ions and electrons , (132) and (133), are also useful in discussing the equilibrium of charged particles in the topside F region. At these heights photochemical processes are negligible and the collision frequencies ν_{in}, ν_{en} so small that they can also be neglected. Assuming also that the ions are stratified horizontally, that they are singly ionized and all at the same temperature T_i, the equatior of equilibrium for the jth species of ions and electrons are respectively

(167) $$\frac{d(n_j kT_i)}{dh} = -n_j m_j g + n_j eE \qquad (j \neq 1, 2, \ldots)$$

(168) $$\frac{d(n_e kT_e)}{dh} = -n_e m_e g - n_e eE$$

where E is the elctric field and T the electron temperature. Since there must be very nearly overall charge neutrality we must have further

(169) $$\sum_j n_j = n_e$$

so that adding the equations (167) and (168) we can eliminate the electric field . Defining the mean ionic mass m_+ by the equation

(170) $$\sum_j n_j m_j = m_+ n_e$$

and neglecting the electronic mass, we find after elimination of E that

(171) $$-\frac{d \log n_e}{dh} = \frac{m_+ g}{kT_2(1+\tau)}$$

where $\tau = T_e / T_i$. Substituting this equation in (168) determines E which when inserted in (167) gives the equation for the distribution of ions as

(172) $$-\frac{d \log n_j}{dh} = (m_j - \frac{m_+ \tau}{1+\tau}) \frac{g}{kT_i} \quad .$$

If there is only one species of ion present, and $T_e = T_i$ (so that $\tau = 1$) , the mean ionic mass is equal to m_i so that the effective scale height is twice that of the neutral atomic mass, as before. However, a light atom for which $m_j < \frac{\tau}{1+\tau} m_+$ actually has a negative scale height, so that n_j increases upwards , as first pointed out by Mange (1957) . The solution of the equation (171) and (172) has been effected by Hanson (1962) taking account of the variation of g with height. A typical equilibrium distribution computed for a mixture of 0^+ , H_e^+ , H^+ ions is shown in Fig. 12 . The reduced scale height z refers to the level z = 0 at which the ionic concentrations of 0^+ and H_e^+ are equal.

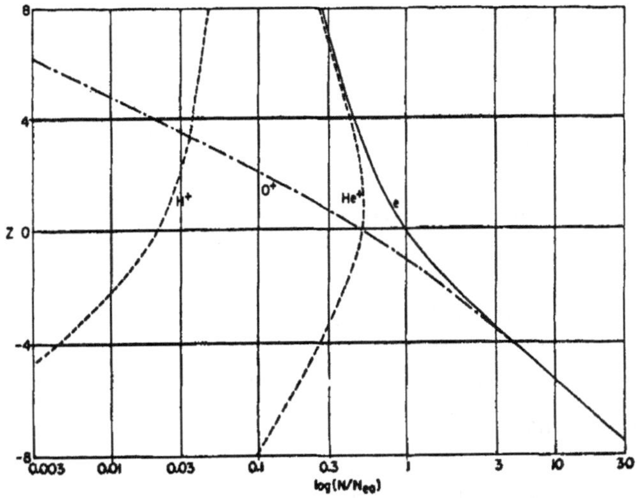

FIG. 12 IDEALIZED DISTRIBUTIONS OF ELECTRONS (e) AND OF O^+, He^+ and H^+ IONS, COMPUTED BY SOLVING EQS. (III-59), (III-60). Electron and ion concentrations are given in terms of the electron density N_{ea} at the level $z = 0$, at which height the ionic composition is taken to be $[O^+] = [He^+] = 49\%$, $[H^+] = 2\%$. The level at which the He^+ and H^+ concentrations are equal is near $z = 17$. The unit of reduced height z is the scale height of neutral atomic oxygen.

Fig. 12

$f_1 = a^m \cdot f_2 = a^n$; $g = x^p$; $n + mp = 0$; $p = - nm^{-1}$;

$f_1 = \exp(ma)$; $f_2 = \exp(n\,a)$; $g = x^p$; $n+mp=0$; $p = -nm^{-1}$;

(3.4.4) $f_1 = \exp(i\,ma)$; $f_2 = \exp(i\,na)$; $g=x^p$; $n+mp = 0$; $p = -nm^{-1}$;

where in the last proposition only the real parts $Re\{f_i\}$ should be taken into account. One may construct easily more complicated functions f_i (i = 1, 2) and g :

(3.4.5) $\bar{x} = ma + x$; $\bar{y} = y\exp(na)$; $\bar{y} = y\exp(p\bar{x}) = y\exp(px)$,

which results in $p = - n/m$, or :

(3.4.6) $\bar{x} = ma + x$; $\bar{y} = y\exp(a\exp n)$; $\bar{y} = y\exp(p\bar{x}) = y\exp(px)$,

which results in $p = - m^{-1}\exp n$. The absolute invariants of the group are :

(3.4.7) $\eta = y\,x^p$; $\eta = y\exp(px)$.

In a similar way we deal with the dependent variables; thus , as an example one may choose:

(3.4.8) $\bar{y}_\delta \equiv \bar{k}_\delta = f_3(a)\,y \equiv f_3 K_\delta$.

For illustrative purposes assume a function $g(x)$ equal to x^r, say . Then it should be :

(3.4.9) $k_\delta\, x^r = \bar{k}_\delta\, \bar{x}_\delta^r$ $\qquad f_3\, f_1^r = 1$, etc

The procedure is identical with the one, explained above. The absolute invariants of the group (3.4.8) are :

(3.4.10) $g_\delta = k_\delta\, x^r$.

Hence by virtue of equation :

V. C. A. Ferraro

ATMOSPHERIC DYNAMO

26. Introduction

Balfour Stewart in 1882 first put forward the hypothesis that the daily variations of the earth's magnetic field could be ascribed to electric currents induced in conducting regions of the upper atmosphere by its motion across the geomagnetic field. The motion of the atmosphere was attributed to tidal forces due to the sun and moon. The theory, generally known as the dynamo theory, was developed mathematically by Schuster in 1908 and later by Chapman. However, their theory encountered certain difficulties connected with the reduction of the electrical conductivity, by the geomagnetic field, the required value appearing to the too large to be reconciled with theoretical values of the estimates of the tidal motion. The difficulty was eventually resolved by Martyn and Hirono, almost simultaneously, who showed that because of the horizontal and vertical variations of the inducting polarization electric charges are set up which tend to restore the full electrical conductivity. We shall begin by calculating the electrical conductivity in a highly ionized gas .

27. The electrical conductivities

These were derived in formal manner by Baker and Martyn (1952-3) and reviewed by Chapman (1956) ; we shall denote, as before, all quantities referring to ions, electrons and neutral particles by suffixes i, e, n. Also the region of the ionosphere in which the dynamo currents flow are now known to lie at a height of about 110 km, that is, in the E-region . In this region, the number of ions and electrons is about 10^5 per cc. Thus, in the equations (92)-(94) we have $\rho_n \gg \rho_i \gg \rho_e$ so that

$\theta_{ie} \simeq \rho_e \nu_{ie}$, $\theta_{ie} \simeq \rho_i \nu_{in}$, $\theta_{en} \simeq \rho_e \nu_{en}$ and these equations become approximately

(173) $$\rho_i \frac{d\underline{v}_i}{dt} = -\nabla p_i + n_i e(\underline{E} + \underline{v}_i \times \underline{B}/c) + \rho_i \underline{F} - \rho_e \nu_{ie}(\underline{v}_i - \underline{v}_e) - \rho_i \nu_{in}(\underline{v}_i - \underline{v}_n)$$

(174) $$\rho_e \frac{d\underline{v}_e}{dt} = -\nabla p_e - n_e e(\underline{E} + \underline{v}_e \times \underline{B}/c) + \rho_e \underline{F} - \rho_e \nu_{ie}(\underline{v}_e - \underline{v}_i) - \rho_e \nu_{en}(\underline{v}_e - \underline{v}_n)$$

(175) $$\rho_n \frac{d\underline{v}_n}{dt} = -\nabla p_n + \rho_n \underline{F}_n - \rho_i \nu_{in}(\underline{v}_n - \underline{v}_i) - \rho_e \nu_{en}(\underline{v}_n - \underline{v}_e) .$$

In the dynamo region it is legitimate to neglect the acceleration of the particles; it is also convenient to introduce the mean mass velocity $\underline{v}_o$ and the electric current density $\underline{j}$ as new variables, where

(176) $$\rho_o \underline{v}_o = \rho_i \underline{v}_i + \rho_e \underline{v}_e + \rho_n \underline{v}_n ,$$
$$\rho_o = \rho_i + \rho_e + \rho_n$$
$$\underline{j} = \frac{e}{c}(n_i \underline{v}_i - n_e \underline{v}_e) ,$$

where ρ_o is the total mass density. The variables are thus $\underline{v}_o$, $\underline{v}_n$ and $\underline{j}$; assuming that $n_i \simeq n_e = n$ and neglecting the ratio m_e/m_i, we find approximately

(177) $$\underline{v}_i = \underline{v}_o + \frac{n_n}{n}(\underline{v}_o - \underline{v}_n) + \frac{m_e}{m_i}\frac{c}{ne}\underline{j}$$
$$\underline{v}_e = \underline{v}_o + \frac{n_n}{n}(\underline{v}_o - \underline{v}_n) - \frac{c}{ne}\underline{j} .$$

From (173) (175) we then find the approximate equation of mass equilibrium

(178) $$0 = -\nabla p + \underline{j}\times\underline{B} + \rho_i\underline{F} + m_e(\nu_{en} - \nu_{in})\frac{c}{e}\underline{j} - (\nu_{in} + \nu_{en}\frac{me}{mi})\rho_o(\underline{v}_o - \underline{v}_n)$$

whilst it can be shown that the approximate equation for the electric current density $\underline{j}$ is

(179) $$0 = -ne\underline{E}_o + \underline{j}\times\underline{B} + \frac{mec}{e}(\nu_{ie} + \nu_{en})\underline{j} - \frac{e}{c}\left(\frac{-\nabla p + \underline{j}\times\underline{B}}{m_i\nu_{in}}\right)\times\underline{B}$$

where we have taken $m_n \simeq m_i$ and $m_i\nu_{in} \gg m_e\nu_{en}$. Here

(180) $$\underline{E}_o = \underline{E}' + \underline{E}''$$

where $\underline{E}' = \underline{E} + \underline{v}_o\times\underline{B}/c$ is the electric field following the mean motion and $E'' = (\nabla p_e)/n_e$, the 'equivalent electric field produced by the electron pressure gradient. Equation (179) can be solved for $\underline{j}$ as a linear vector function of $\underline{E}_o$ and ∇p. Writing

(181) $$\underline{\omega}_e = \frac{e\underline{B}}{mc}, \qquad \underline{\omega}_i = \frac{e\underline{B}}{m_e c}, \qquad \underline{B} = B\,\underline{b}$$

so that $\underline{\omega}_i$ and $\underline{\omega}_e$ are the ion and electron cyclotron frequencies an $\underline{b}$ is a unit vector along $\underline{B}$, and defining the conductivity

(182) $$\sigma = \frac{ne^2}{m_e\nu_e}, \qquad \nu_e = \nu_{ie} + \nu_{en}$$

we find

(183) $$(A^2+D^2)\underline{j} = A\sigma\underline{E}_o + \sigma(AC+D^2)(\underline{E}_o\cdot\underline{b})\underline{b} + \sigma DE_o\times\underline{b} - AC\frac{\nabla p}{B}\times\underline{b} + CD\left(\frac{\nabla p}{B}\times\underline{b}\right)\times\underline{b}$$

where

(184) $$A = 1 + C, \qquad C = \frac{\omega_e \omega_i}{\nu_e \nu_{in}}, \qquad D = \frac{\omega_e}{\nu_e}.$$

In the case when $\nabla p = 0$, (173) can be simplified. In fact, if we write

(185) $$\underline{E}_o = \underline{E}_{11} + \underline{E}_{\perp}$$

where E_{11} and $E_{\perp}$ are the components of the total electric field $\underline{E}_o$ parallel and perpendicular to $\underline{B}$, we find

(186) $$\underline{E}_{11} = (\underline{E}_o \cdot \underline{b})\underline{b}, \qquad \underline{E} = \underline{b} \times (\underline{E}_o \times \underline{b})$$

so that (183) can now be written

(187) $$\underline{j} = \sigma_0(\underline{E}_o \cdot \underline{b})\underline{b} + \sigma_1 \underline{b} \times (\underline{E}_o \times \underline{b}) + \sigma_2 \underline{b} \times \underline{E}_o .$$

Here $\sigma_0, \sigma_1, \sigma_2$ are respectively the direct transverse and Hall conductivities and it is easily verified that

(188) $$\sigma_0 = \sigma$$

(189) $$\sigma_1 = \frac{A}{A^2 + D^2}\sigma$$

(190) $$\sigma_2 = \frac{D}{A^2 + D^2}\sigma$$

where σ is given by (182) . These formulae agree with those found by Chapman by a somewhat different approach . If B = 0, then C = D = 0 and $\sigma_1 = \sigma_0$ and $\sigma_2 = 0$ as should be the case. Equation (187) can be then written .

(191) $$\underline{j} = \sigma_0 \underline{E}_o$$

the usual form of Ohm's Law. If $B \neq 0$, then (187) shows that the electrical conductivity is anisotropic, and if we take cartesian axes $0(xyz)$ with $0x$ along $\underline{b} \times \underline{E}_0$, $0y$ along $\underline{E}$ and $0z$ along $\underline{E}_{11}$ we may wirte (187) in the form

(192) $$\underline{j} = \sigma . \underline{E}_0$$

where σ is the second order tensor

(193) $$\begin{bmatrix} \sigma_1 & -\sigma_2 & 0 \\ \sigma_2 & \sigma_1 & 0 \\ 0 & 0 & \sigma_0 \end{bmatrix}$$

(the right hand side of (192) denoting the contracted product of σ and $\underline{E}_0$). Formulae (189), (190) show that the transverse and Hall conductivities are reduced respectively in the ratio

(194) $$\frac{A}{A^2 + D^2}, \qquad \frac{D}{A^2 + D^2}$$

If the magnetic field is large, A and D will be large, the reduction of the electrical conductivities σ_1 and σ_2 will also be large, and the electric currents will flow mainly along the lines of force.

Let us suppose that an electric field is set up which prevents any further flow of the Hall current. Such a field must be in the direction of $\underline{b} \times \underline{E}_0$ so that the total electric field is now $\underline{E}_0 + \lambda \underline{b} \times \underline{E}_0$, where λ is a constant. Substituting in (187) we find that the Hall current (parallel to $\underline{b} \times \underline{E}_o$) vanishes provided $\lambda = \sigma_2/\sigma_1$, and that

(195) $$\underline{j} = \sigma_o(\underline{E}_o \cdot \underline{b})\underline{b} + \sigma_3 \underline{b} \times (\underline{E}_o \times \underline{b})$$

where

(196) $$\sigma_3 = \sigma_1 + \frac{\sigma_2^2}{\sigma_1}$$

is called the Cowling conductivity. In a fully ionized gas, $\sigma_3 = \sigma_o$, that is, the conductivity is the same as in the absence of a magnetic field. In a partially ionized gas; however, although σ_3 exceeds both σ_1 and σ_2, it is in general smaller than σ_o.

28. Numerical Illustrations

A review of the electrical conductivities in the ionosphere has been given by Chapman . Values were given for a hot (1500 °K) and a cool (850 °K) ionosphere and in the table below the values of ν_{ie}, ν_{in}, ν_{em} are taken from a corresponding table in Chapman's paper. In the table below we also give numerical values of the conductivities for the higher temperature. The conducting σ_o increases upwards over the range considered (100- 300 km) and the conductivities $\sigma_1, \sigma_2, \sigma_3$ each have one peak in this range.

29. Effective conductivities in the ionosphere

Baker and Martyn have shown that because of the limited vertical extent of the conducting layer of the E-region , the flow of current is nearly horizontal. In fact, if $\underline{j}$ contains a vertical component, charges will accummulate on the boundaries of the layer because this current cannot flow in the region of low conductivity. These charges are called polarisation

<u>Height in Km</u>

	90	100	125	150	175	200	250	300
ν_{ie}	1.43×10^{2}	9.35×10^{2}	7.63×10^{2}	6.66×10^{2}	4.84×10^{2}	3.74×10^{2}	3.08×10^{2}	2.68×10^{2}
ν_{en}	6.83×10^{5}	1.61×10^{5}	1.83×10^{4}	2.81×10^{3}	8.80×10^{2}	3.57×10^{2}	4.50×10	3.74×10
ν_{in}	4.12×10^{4}	8.59×10^{3}	6.16×10^{2}	1.12×10^{2}	3.20×10	1.21×10	2.90	1.08
w_e/ν_e	1.26×10	5.32×10	5.92×10^{2}	2.48×10^{3}	6.33×10^{3}	1.18×10^{4}	2.14×10^{4}	2.82×10^{4}
w_i/ν_i	3.92×10^{-3}	1.88×10^{-2}	.282	1.67	6.10	1.69×10	7.69×10	2.29×10^{2}
σ_o	4.12×10^{-15}	1.74×10^{-13}	2.91×10^{-12}	1.62×10^{-11}	4.14×10^{-11}	7.72×10^{-11}	1.76×10^{-10}	2.79×10^{-10}
σ_1	2.70×10^{-17}	1.23×10^{-16}	1.29×10^{-15}	2.88×10^{-15}	1.04×10^{-15}	3.87×10^{-16}	1.07×10^{-16}	4.31×10^{-17}
σ_2	3.25×10^{-16}	3.27×10^{-15}	4.54×10^{-15}	1.72×10^{-15}	1.71×10^{-16}	2.29×10^{-17}	1.38×10^{-18}	1.87×10^{-19}
σ_3	3.93×10^{-15}	8.70×10^{-14}	1.73×10^{-14}	3.91×10^{-15}	1.07×10^{-15}	3.88×10^{-16}	1.07×10^{-16}	4.31×10^{-17}

<u>Table I</u>

Collision frequencies, electric conductivities, and the ratio of cyclotron to collision frequencies for a model ionosphere at a temperature of 1480 ^{o}K .

charges. Because the flow is horizontal we can replace the 3×3 tensor σ by a 2×2 tensor σ^* representing the layer conductivity whose components depend on the magnetic dip angle I. It will be convenient to use coordinate x, y for the magnetic southward and eastward directions. Then we can write

$$\sigma^* = \begin{bmatrix} \sigma_{xx} & \sigma_{xy} \\ -\sigma_{xy} & \sigma_{yy} \end{bmatrix} \tag{197}$$

where

$$\sigma_{xx} = \frac{\sigma_0 \sigma_1}{\sigma_0 \sin^2 I + \sigma_1 \cos^2 I} \simeq \frac{\sigma_1}{\sin^2 I}$$

$$\sigma_{xy} = \frac{\sigma_0 \sigma_2 \sin I}{\sigma_0 \sin^2 I + \sigma_1 \cos^2 I} \simeq \frac{\sigma_2}{\sin I} \tag{198}$$

$$\sigma_{yy} = \frac{\sigma_2^2 \cos^2 I}{\sigma_0 \sin^2 I + \sigma_1 \cos^2 I} + \sigma_1 \simeq \sigma_1$$

The approximations given on the right arise generally since $\sigma_0 \gg \sigma_1$ or σ_2, but they are not valid near the magnetic equator where I=0. Here we have

$$\sigma_{xx} = \sigma_0, \quad \sigma_{xy} = 0, \quad \sigma_{yy} = \sigma_1 + \frac{\sigma_2^2}{\sigma_1} = \sigma_3 . \tag{199}$$

Two consequences follow immediately from (199) ; firstly , the high conductivity σ_0 along the magnetic lines of force ensures that these are very nearly electric equipotentials . Secondly, the east-west conductivity σ_{yy} at the equator is very large, being the Cowling conductivity σ_3

which is comparable with, though smaller than, σ_0. This highly conducting strip along the magnetic equator carries a large current, known as the equatorial electrojet and is confined to a belt a few degrees in widths about the magnetic equator, where $\sigma_0 \sin^2 I \ll \sigma_1 \cos^2 I$. Outside these belts the electric current falls rapidly. The simple model of the dynamo region is therefore a relatively horizontally stratified layer. The magnetic variations observed at the ground which are produced by these currents are best calculated by considering the layer as a current-sheet, with integrated conductivities,

$$\Sigma_1 = \int \sigma_1 \, dh, \qquad \Sigma_2 = \int \sigma_2 \, dh \quad ,$$

where the integrals are taken over the thickness of the horizontal layer. If Σ^* denotes the tensor $\int \sigma^* dh$, we can summarise the electrical equations as

(200) $$\underline{J} = \Sigma^* \cdot \underline{E}_t \, , \qquad \underline{J} = \int \underline{j} \, dh$$

where

(201) $$\underline{E}_t = \underline{w} \times \underline{B} - \nabla\varphi \, ,$$

$\underline{w}$ being the velocity of the neutral air and φ the electrostatic potential. The first term in (201) represents the induced field, whilst the electrostatic field $-\nabla\varphi$ forces the electric currents to flow horizontally.

cannot be zero . If in the last equation of (3.6.2), $f^4 = 0$, then $\frac{\partial \bar{u}}{\partial x} = \frac{\partial \bar{u}}{\partial y} = \frac{\partial \bar{u}}{\partial u} = 0$ in which case $\left|\frac{\partial(\bar{x}, \bar{y}, \bar{u})}{\partial(x, y, u)}\right| = 0$, contrary to the original assumption. Therefore it is necessary that $f^4 \neq 0$.

Similarly, the subgroup :

$$(3.6.10) \qquad S_{A_1} : \bar{x} = f^1(x, y ; a), \qquad \bar{y} = f^2(x, y ; a),$$

must have an inverse and therefore neither the Jacobian determinant associated with S_{A_1} :

$$(3.6.11) \qquad \left|\frac{\partial(\bar{x}, \bar{y})}{\partial(x, y)}\right| = \begin{vmatrix} \frac{\partial \bar{x}}{\partial x} & \frac{\partial \bar{x}}{\partial y} \\ \frac{\partial \bar{y}}{\partial x} & \frac{\partial \bar{y}}{\partial y} \end{vmatrix},$$

nor the Jacobian determinant :

$$(3.6.12) \qquad \left|\frac{\partial(x, y)}{\partial(\bar{x}, \bar{y})}\right| = \begin{vmatrix} \frac{\partial x}{\partial \bar{x}} & \frac{\partial x}{\partial \bar{y}} \\ \frac{\partial y}{\partial \bar{x}} & \frac{\partial y}{\partial \bar{y}} \end{vmatrix},$$

associated with the inverse transformation may be equal to zero . From (3.6.12), it follows that not both $\frac{\partial x}{\partial \bar{y}}$ and $\frac{\partial y}{\partial \bar{y}}$ are equal to zero.. If $\frac{\partial u}{\partial x} \neq 0$, then from (3.6.8) :

$$(3.6.13) \qquad \frac{\partial x}{\partial \bar{y}} = 0,$$

and therefore

$$(3.6.14) \qquad \frac{\partial y}{\partial \bar{y}} \neq 0.$$

Eqs. (3.6.6), (3.6.7), (3.6.8) and the above remarks imply that :

16. D.F. Martyn , Processes controlling ionization distribution in the F2 reginn, Aust J. Phys. , 9 , 161 -165 , (1956) .

17. H. Rishbeth, Further analogue studies of the ionospheric F layer, Proc. Phys. Soc. , 81, 65 - 77 (1963).

18. T. Yonezawa, A ŋew theory of the formation of the F2 layer, J. Radio Res. Lab. , 3, 1 - 16 , (1956).

19. T. Yonezawa, On the influence of electron-ion diffusion exerted upon the formation of the F2 layer, J. Radio Res. Lab., 5 , 165 - 187 (1958).

and

$$\text{(vi)}\qquad \rho C_v\left(\frac{\partial T}{\partial t} + u\frac{\partial T}{\partial x} + v\frac{\partial T}{\partial y} + w\frac{\partial T}{\partial z}\right) +$$

$$+ p\left(\frac{\partial u}{\partial x} + \frac{\partial v}{\partial y} + \frac{\partial w}{\partial z}\right) - K\left(\frac{\partial^2 T}{\partial x^2} + \frac{\partial^2 T}{\partial y^2} + \frac{\partial^2 T}{\partial z^2}\right) -$$

$$- 2\mu\left[\left(\frac{\partial u}{\partial x}\right)^2 + \left(\frac{\partial v}{\partial y}\right)^2 + \left(\frac{\partial w}{\partial z}\right)^2\right] - \mu\left[\left(\frac{\partial v}{\partial x} + \frac{\partial u}{\partial y}\right)^2 +\right.$$

$$+ \left(\frac{\partial w}{\partial y} + \frac{\partial v}{\partial z}\right)^2 + \left(\frac{\partial u}{\partial z} + \frac{\partial w}{\partial x}\right)^2 +$$

$$\text{(3.6.21)}\qquad \left. + \frac{2}{3}\left(\frac{\partial u}{\partial x} + \frac{\partial v}{\partial y} + \frac{\partial w}{\partial z}\right)^2\right] = 0 \quad \text{(energy)},$$

where μ , R , C_v and k are constants

The above system of six partial differential equations in six unknown functions of x, y, z and t will be referred to as the system " A " . In order to find the conditions on a group A_1 such that " A " is conformally invariant under A_1 , it is necessary to find the system corresponding to " A" and A_1. A_1 is a transformation group defined as :

$$A_1 : \bar{x} = f^1(x, y, z; a), \quad \bar{y} = f^2(x, y, z; a), \quad \bar{z} = f^3(x, y, z; a),$$
$$\bar{u} = f^4(u; a) \equiv f^5(u; a)\, u + f^6(a) .$$

Since no new techniques are involved in finding the system in question, this problem will be left to the reader. The present paper will confine its attention to one such group under which " A " is conformally invariant. It is the following group :

$$P^1 : \bar{x} = x + \gamma_1 a, \quad \bar{y} = y - \gamma_2 a, \quad z = \bar{z} + \gamma_3 a,$$

$$\tau = t - \gamma_4 a, \quad \bar{u} = u, \quad \bar{v} = v, \quad \bar{w} = w,$$

CENTRO INTERNAZIONALE MATEMATICO ESTIVO

(C.I.M.E)

. P.C. KENDALL

"ON THE DIFFUSION IN THE ATMOSPHERE AND IONOSPHERE "

Corso tenuto a Varenna dal 19-al 27 settembre 1966

$$+\gamma_1\gamma_2\frac{\partial^2 \nu_2}{\partial \eta_1^2} + \gamma_1 \gamma_3 \frac{\partial^2 \nu_2}{\partial\eta_1 \partial\eta_2} + \gamma_2^2 \frac{\partial^2 \nu_3}{\partial\eta_1\partial\eta_2} +$$

$$+\gamma_2\gamma_4 \frac{\partial^2 \nu_3}{\partial\eta_1\partial\eta_3} \Big) = 0 , \tag{3.6.24}$$

" B "

$$\text{(ii)}\ \nu_5 \Big(\gamma_3 \frac{\partial \nu_2}{\partial \eta_3} + \gamma_2 \nu_1 \frac{\partial \nu_2}{\partial \eta_1} + \cdots\Big) + \cdots = 0 , \tag{3.6.25}$$

$$\text{(iii)}\ \nu_5 \Big(\gamma_3 \frac{\partial \nu_3}{\partial \eta_3} + \gamma_2 \gamma_1 \frac{\partial \nu_3}{\partial \eta_3} + \cdots\Big) + \cdots = 0 ,$$

$$\text{(iv)}\ \nu_4 = R\nu_5 \nu_6 , \tag{3.6.27}$$

$$\text{(v)}\ \gamma_3 \frac{\partial \nu_5}{\partial \eta_3} + \gamma_2 \frac{\partial(\nu_5\nu_1)}{\partial \eta_1} + \gamma_1 \frac{\partial(\nu_5\nu_2)}{\partial \eta_1} + \gamma_3 \frac{\partial(\nu_5\nu_2)}{\partial \eta_2} +$$

$$+\gamma_2 \frac{\partial(\nu_5\nu_3)}{\partial \eta_2} + \gamma_4 \frac{\partial(\nu_5\nu_3)}{\partial \eta_3} = 0 , \tag{3.6.28}$$

$$\text{(vi)}\ C_v \nu_5 \Big(\gamma_3 \frac{\partial \nu_6}{\partial \eta_3} + \gamma_2 \gamma_1 \frac{\partial \nu_6}{\partial \eta_1} + \cdots\Big) + \cdots = 0 . \tag{3.6.29}$$

The system " B " is absolutely invariant under the group :

$$P^{(2)} : \bar\eta_1 = \eta_1 + \gamma_5 a , \quad \bar\eta_2 = \eta_2 - \gamma_6 a , \quad \bar\eta_3 = \eta_3 + \gamma_7 a ,$$

$$\bar\nu_j = \nu_j \qquad (j = 1, \cdots, 6) , \tag{3.6.30}$$

whose invariants , as in the previous case, may be chosen as follows:

C.I.M.E. lectures

by
P.C. KENDALL
(University of Sheffield)

Introduction

This set of lectures is not intended to be a comprehensive survey of the ionosphere. It is intended to lead the student with great speed to certain specialized research areas of possible interest to applied mathematicians . I apologise to the many ionospheric physicists (both experimental and theoretical) whose work is not mentioned here; the short list of references included at the end reflects current theory. Present day achievements have only been made possible through the patient experimental work and international collaboration of many scientists over decades.

P. C. Kendall

Lecture 1

In his lectures Professor Ferraro will show how diffusion of a plasma (a gas composed of equal numbers of ions and electrons) takes place through a neutral atmosphere. The electrons are often very mobile and would in absence of electrostatic forces disperse to infinity in a very short time. The ions are bigger and heavier and so cannot move as quickly through the neutral atmosphere. Thus, the tendency of the electrons to "boil off" is prevented by the electrostatic forces between ions and electrons, and an electric field is set up. This has the effect of increasing by a factor of two the pressure gradient which causes diffusion; thus causing faster diffusion, known as ambipolar diffusion The purpose of this lecture is to establish the equations of the problem including the effects of a uniform magnetic field H_o. In general these equations have not yet been fully solved. I will describe what can be done, and also mention the difficulties. Practical application of the equations may be made in the upper ionosphere, in diffusion of meteor trains and in the laboratory, wherever the plasma is a minot constituent.

Solution of the problem (ion-neutral collisions only)

The main problem is how to work out the electric field arising from attempted charge separation. In the case when the electrons are perfectly mobile (i.e. we can ignore collisions of electrons with both ions and neutral atoms) this can be done in full (provided that certain other minor simplifications are made). In particular, it will be seen that the external electric field gives rise to complicated drift motions of the joint ion-electron gas.

The equation of motion of the electrons is

(1) $$N\, m_e \frac{d\underline{V}_e}{dt} = - \underline{\nabla}\, p_e - e\, N\, (\underline{E} + \underline{V}_e \times \underline{B})$$

Here,

m_e = mass of electron

p_e = electron pressure

N = electron density = ion density

e = electronic charge

$\underline{V}_e$ = electron velocity

d/dt = mobile operator $\partial/\partial t + (\underline{V}_e . \underline{\nabla})$

$\underline{E}$ = total electric field in fixed axes

The magnetic field is assumed constant and uniform (i.e. the number of ions and electrons is so small that their associated electric currents may be neglected). Thus the electric field is a potential field, i.e.

(2) $$\text{curl}\ \underline{E} = 0 \qquad \therefore \quad \underline{E} = - \underline{\nabla}\, \Omega$$

Neglecting terms which involve m_e (which is small)

$$0 = - \underline{\nabla}\, p_e - e\, N\, (\underline{E} + \underline{V}_e \times \underline{B}),$$

(3) i.e. $$\underline{\nabla}\, p_e - e\, N \underline{\nabla}\, \Omega = - N\, e\, \underline{V}_e \times \underline{B}$$

So, in an atmosphere at <u>uniform temperature</u> T, if k is Boltzmann constant such that

$$p_e = N\, k\, T,$$

equation (3) becomes

(4) $$\underline{\nabla}\left(\frac{kT}{e}\log N - \Omega\right) = -\underline{V}_e \times \underline{B}$$

Write

(5) $$\Omega = \frac{kT}{e}\log N + \Omega_H$$

Then

(6) $$\underline{B}\cdot\underline{\nabla}\,\Omega_H = 0,$$

showing that the potential Ω_H is constant along a magnetic field line. Further, if

$$\underline{E}_H = -\underline{\nabla}\,\Omega_H$$

we obtain

(7) $$\underline{V}_e = \underline{E}_H \times \underline{B}\,/B^2$$

This is known as the Hall drift or Hall motion of electrons. The electric field $\underline{E}_H$ is at right angles to the magnetic field and if present must be regarded as an applied electric field, external in origin. The electric potential kT (log N) / e is internal in origin. One might well ask what happens when $N \to 0$; apparently the electric field becomes infinitely large. The answer lies in our assumption of electrical neutrality, which becomes invalid for small values of N.

The equation of motion of the ions is

$$N m_i \frac{d\underline{V}_i}{dt} + N m_i \nu_{ia}(\underline{V}_i - \underline{V}_a) = -\underline{\nabla}\,p_i + e N(\underline{E} + \underline{V}_i \times \underline{B}) .$$

Here,

m_i = mass of ion

p_i = ion pressure

N = ion density = electron density

$\underline{V}_i$ = ion velocity

$\underline{V}_a$ = neutral air velocity

ν_{ia} = collision frequency ions - neutral atoms

In the diffusion approximation it is usual to neglect the acceleration. (The student should find out why). Then

(8) $$N m_i \nu_{ia}(\underline{V}_i - \underline{V}_a) = - \underline{\nabla} p_i + e N (\underline{E} + \underline{V}_i \times \underline{B})$$

Adding (3) and (8) to eliminate $\underline{E}$.

(9) $$\dot{N} m_i \nu_{ia} (\underline{V}_i - \underline{V}_a) = - \underline{\nabla} (p_i + p_e) + e N \underline{V}_i \times \underline{B} - e N \underline{V}_e \times \underline{B}$$

Writing

(10) $$p_i = p_e = N k T .$$

and

(11) $$\underline{F} = \underline{V}_a - \frac{2 k T}{N m_i \nu_{ia}} \underline{\nabla} N - \frac{e}{m_i \nu_{ia}} \underline{V}_e \times \underline{B}$$

we see that

(12) $$\underline{V}_i + \frac{e}{m_i \nu_{ia}} \underline{B} \times \underline{V}_i = \underline{F}$$

This must be solved for $\underline{V}_i$. Then the diffusion equation is formed by using the continuity equation, which in absence of production and loss is

(13) $$\frac{\partial N}{\partial t} = - \operatorname{div} (N \underline{V}_i) = - \underline{\nabla} . (N \underline{V}_i)$$

Denote by $\parallel$ and $\perp$ components parallel to and perpendicular to $\underline{B}$. Then

$$\underline{V}_i^{\parallel} = \underline{F}^{\parallel} = \underline{V}_a^{\parallel} - \frac{2kT}{N m_i \nu_{ia}} \underline{\nabla}^{\parallel} N \tag{14}$$

and

$$\underline{V}_i^{\perp} + \frac{e}{m_i \nu_{ia}} \underline{B} \times \underline{V}_i^{\perp} = \underline{F}^{\perp} \tag{15}$$

There are many ways of solving this equation. A quick convenient method is to introduce coordinates x, y , z with the z-axis parallel to $\underline{B}$ (Fig. 1).

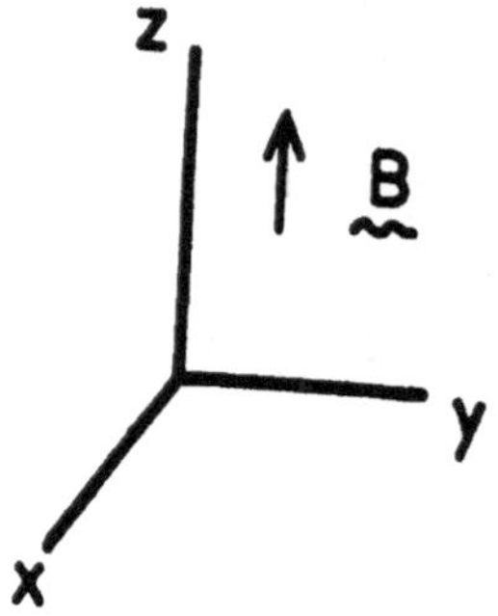

Fig. 1

Then if

$$\underline{V}_i^{\perp} = (V_1, V_2, 0)$$

we have

$$\underline{B} \times \underline{V}_i^{\perp} = (-BV_2, BV_1, 0) .$$

It follows that if we introduce

$$j = \sqrt{(-1)}$$

and write

$$\underline{V}_i^{\perp} = V_1 + jV_2$$

we obtain

$$\underline{B} \times \underline{V}_i^{\perp} = B\, j\, \underline{V}_i^{\perp} .$$

So equation (15) becomes

$$\underline{V}_i^{\perp} = \underline{F}^{\perp} \Big/ \left(1 + \frac{e\,B}{m_i\, \nu_{ia}}\, j\right), \tag{16}$$

where

$$\underline{F}^{\perp} = \underline{V}_a^{\perp} - \frac{2\,k\,T}{N\, m_i \nu_{ia}}\, \underline{\nabla}^{\perp} N + \frac{e\,B}{m_i\, \nu_{ia}}\, j\, \underline{V}_e^{\perp} \tag{17}$$

Equations (14) (16) and (17) describe the motion completely and with (13) give an equation governing N. First we discuss two special cases. These will help to clarify the general problem.

<u>Special case I B = 0 (no magnetic field)</u>

Equation (12) gives at once

$$\underline{V}_i = \underline{V}_a - \frac{2\,k\,T}{N\, m_i\, \nu_{ia}}\, \underline{\nabla} N$$

This is the case of ordinary ambipolar diffusion. We define the coefficient of ambipolar diffusion <u>in this case</u> to be

$$D_a = \frac{2\,k\,T}{m_i\, \nu_{ia}}\ \mathrm{cm}^2/\mathrm{sec}$$

so from (13) if T and ν_{ia} are constants,

(18) $$\frac{\partial N}{\partial t} = - \operatorname{div} (N \underline{V}_a) + D_a \nabla^2 N$$

This is the isotropic diffusion equation. The term $\operatorname{div} (N \underline{V}_a)$ is called a transport term. Its presence indicates that all diffusion takes place relative to the wind $\underline{V}_a$ in the neutral air. Note the factor 2 in D_a.

Special case II $eB/m_i \, \nu_{ia} \to \infty$

Physically

$$\nu_{ia} = \text{collision frequency}$$

$$\frac{e B}{\pi m_i} = \text{Larmor frequency} \; (*)$$

(*) In absence of electric fields the motion of a single ion is given by

$$m_i \frac{d\underline{V}}{dt} = e \, \underline{V} \times \underline{B}$$

i.e.

$$\frac{d \underline{V}}{dt} = \underline{\omega} \times \underline{V}$$

where

$$\underline{\omega} = - e\underline{B}/m_i \; .$$

Thus in absence of other forces the velocity vector rotates with angular velocity $\omega = eB/m_i$ in a certain sense. The motion is one of spiralling about the magnetic field lines.

Thus case II corresponds to many spirals between collisions and magnetic effects dominate.

Clearly $\underline{V}_i^{\parallel}$ is unchanged from (14) ; but from (16) and (17)

$$\underline{V}_i^{\perp} = \underline{V}_e^{\perp} = \underline{E}_H \times \underline{B}/B^2 \quad .$$

The diffusion equation then becomes

$$\frac{\partial N}{\partial t} = - \operatorname{div} \left(N \underline{V}_a^{\parallel}\right) - \operatorname{div} \left(N \underline{V}_e^{\perp}\right) + D_a \frac{\partial^2 N}{\partial z^2} \quad , \tag{19}$$

where z is the coordinate along the magnetic field. Note that (a) the only effective neutral wind component is along a field line (b) the Hall drift, caused by electric fields is at right angles to a field line.

In general

Rearranging equation (17) ,

$$\underline{F}^{\perp} = \underline{V}_a^{\perp} - \underline{V}_e^{\perp} - \frac{D_a}{N} \underline{\nabla}^{\perp} N + \left(1 + \frac{e B}{m_i \nu_{ia}} j\right) \underline{V}_e^{\perp}$$

Thus from (16)

$$\underline{V}_i^{\perp} = \underline{V}_e^{\perp} + \left(1 + \frac{e B}{m_i \nu_{ia}} j\right)^{-1} \left(\underline{V}_a^{\perp} - \underline{V}_e^{\perp} - \frac{D_a}{N} \underline{\nabla}^{\perp} N\right)$$

Also from (14)

$$\underline{V}_i^{\parallel} = \underline{V}_a^{\parallel} - \frac{D_a}{N} \underline{\nabla}^{\parallel} N$$

So, letting $s = e B/m_i \nu_{ia}$ we have

$$\frac{\partial N}{\partial t} = - \operatorname{div} (N \underline{V}_a^{\parallel}) - \frac{1}{1+s^2} \operatorname{div} (N \underline{V}_a^{\perp}) + \frac{s}{1+s^2} \operatorname{div} (N \underline{k} \times \underline{V}_a^{\perp})$$

$$- \operatorname{div} (N \underline{V}_e^{\perp}) + \frac{1}{1+s^2} \operatorname{div} (N \underline{V}_e^{\perp}) - \frac{s}{1+s^2} \operatorname{div} (N \underline{k} \times \underline{V}_e^{\perp})$$

$$+ D_a \frac{\partial^2 N}{\partial z^2}$$

$$+ \underline{\nabla}^{\perp} \cdot \left\{ \frac{D_a}{1+s^2} \underline{\nabla}^{\perp} N - \frac{s D_a}{1+s^2} \underline{k} \times \underline{\nabla}^{\perp} N \right\},$$

where $\underline{k} = (0, 0, 1)$.

Thus as $\quad \underline{\nabla}^{\perp} \cdot (\underline{k} \times \underline{\nabla}^{\perp} N) = \underline{\nabla} \cdot (\underline{k} \times \underline{\nabla} N) \equiv 0$, we have

$$\frac{\partial N}{\partial t} = - \operatorname{div} (N \underline{V}_a^{\parallel}) - \frac{1}{1+s^2} \operatorname{div} (N \underline{V}_a^{\perp}) + \frac{s}{1+s^2} \operatorname{div} (N \underline{k} \times \underline{V}_a^{\perp})$$

$$- \frac{s^2}{1+s^2} \operatorname{div} (N \underline{V}_e^{\perp}) - \frac{s}{1+s^2} \operatorname{div} (N \underline{k} \times \underline{V}_e^{\perp})$$

$$+ \frac{D_a}{1+s^2} \left(\frac{\partial^2 N}{\partial x^2} + \frac{\partial^2 N}{\partial y^2} \right) + D_a \frac{\partial^2 N}{\partial z^2} \tag{20}$$

Note how on putting $s = 0$ this equation reduces to (18) and how it reduces to (19) on putting $s = \infty$. The terms in $\underline{V}_e^{\perp}$ disappear when $s = 0$ and the first one remains when $s \to \infty$. They might be called "Hall terms" . Note also how complicated the effects of electric fields $\underline{E}_H$ (such that $\underline{V}_e = \underline{E}_H \times \underline{B}/B^2$) and winds $\underline{V}_a$ become when $s \sim 1$. The <u>effective</u> wind is not $\underline{V}_a$ but is

$$\underline{V}_{\text{effective}} = \underline{V}_a^{\parallel} + \frac{\underline{V}_a^{\perp} - s \underline{k} \times \underline{V}_a^{\perp}}{1+s^2}$$

P. C. Kendall

Asymptotic solutions of the ambipolar diffusion problem

Consider equation (20) in the case when $\underline{E}_H = 0$ and $\underline{V}_a = 0$ i.e. diffusion in a stationary atmosphere with no electric fields present other than internal ones. Then

$$\frac{\partial N}{\partial t} = \frac{D_a}{1+s^2}\left(\frac{\partial^2 N}{\partial x^2} + \frac{\partial^2 N}{\partial y^2}\right) + D_a \frac{\partial^2 N}{\partial z^2} \tag{21}$$

Write

$$x' = x\sqrt{[(1+s^2)/D_a]} \quad y' = y\sqrt{[(1+s^2)/D_a]} \quad z' = z/\sqrt{D_a}$$

Then (21) becomes

$$\frac{\partial N}{\partial t} = \frac{\partial^2 N}{\partial x'^2} + \frac{\partial^2 N}{\partial y'^2} + \frac{\partial^2 N}{\partial z'^2} \tag{22}$$

Suppose that diffusion takes place over a region of infinite extent in all directions, a given cloud of plasma being released at time $t = 0$. If the plasma is released as a point source such that if $r = \sqrt{(x'^2 + y'^2 + z'^2)}$,

$$N(r, 0) = \delta(r) \quad (*)$$

then it may be shown that

$$N(r, t) = \frac{1}{(4\pi t)^{3/2}} \exp - \frac{x'^2 + y'^2 + z'^2}{4t}.$$

It follows that for a general cloud of plasma such that

$$N = N(x', y', z', t)$$

(*) $\delta(r) \equiv 0$ if $r \neq 0$ and is so defined that $\iiint \delta(r)\, dx'\, dy'\, dz' = 1$

we have

$$N(x', y', z', t) \iiint \frac{N(x_o, y_o, z_o, 0)}{(4\pi t)^{3/2}} \exp\left[-\frac{(x'-x_o)^2+(y'-y_o)^2+(z'-z_o)^2}{4t}\right] dx_o\, dy_o\, dz_o\,.$$

So the asymptotic shape of any plasma cloud as $t \to \infty$ is

$$N = \frac{N_o'}{(4\pi t)^{3/2}} \exp - \frac{x'^2 + y'^2 + z'^2}{4t}\,,$$

where N'_o = total electron content $= \iiint N\, dx'\, dy'\, dz'$. Converting back we see that if

$$N_o = \iiint N\, dx\, dy\, dz\,,$$

as $t \to \infty$ the asymptotic form is obtained, namely,

$$N = \frac{(1+s^2)\, N_o}{(4\pi D_a t)^{3/2}} \exp\left[-\frac{(1+s^2)(x^2+y^2)+z^2}{4 D_a t}\right] \tag{23}$$

This shows that in a uniform magnetic field a discrete plasma cloud will take up the shape of an ellipsoid of revolution, elongated in the direction of the field.

Dispersion of "meteor trails"

A "meteor trail" is an infinitely long cloud of plasma left behind as the meteor passes through the ionosphere. (see Fig. 2 for notation) . At time $t = 0$ one might suppose the trail to be of small cross-section.

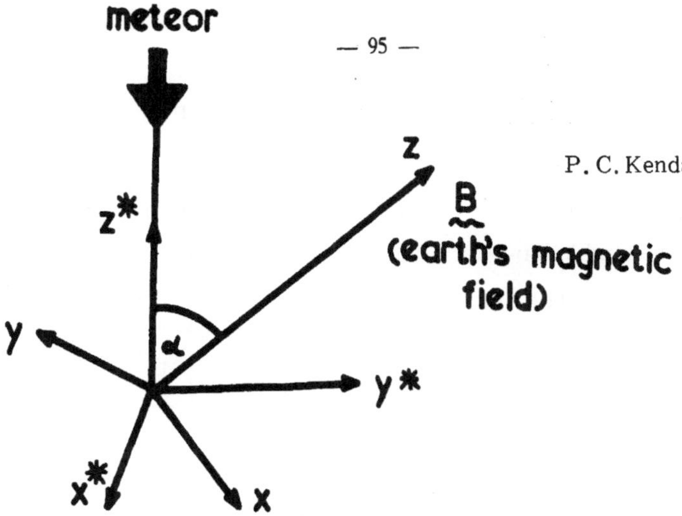

Fig. 2

Clearly the problem is in two dimensions, at right angles to the meteor train . Assuming that an equation of type (21) holds we may put

$$x^* = x \quad y^* = y \cos\alpha + z \sin\alpha \quad z^* = - y \sin\alpha + z \cos\alpha$$

and

$$N = N(x^*, y^*, t)$$

giving

$$\frac{\partial N}{\partial t} = \frac{D_a}{1+s^2}\frac{\partial^2 N}{\partial x^{*2}} + D_a \left(\sin^2\alpha + \frac{\cos^2\alpha}{1+s^2}\right)\frac{\partial^2 N}{\partial y^{*2}} \tag{24}$$

Unfortunately we have omitted electron collisions. Nevertheless, this serves as useful illustrative problem.

<u>Exercise</u> Show that a solution of the equation

$$\frac{\partial N}{\partial t} = \frac{\partial^2 N}{\partial x'^2} + \frac{\partial^2 N}{\partial y'^2}$$

is

$$N = \frac{n_o a^2}{(a^2 + 4t)} \exp - \frac{(x'^2 + y'^2)}{(a^2 + 4t)} \ .$$

Hence obtain a solution of equation (24) and discuss its nature.

Note . When electron-ion collisions and electron -neutral atom collisions are included the ambipolar diffusion equation can not be solved. A recent paper by Holway (1965) J. G. R. 70 3635 is known to be incorrect as curl $\underline{E} \neq 0$.

P. C. Kendall

Lecture 2

Diffusion in a plane stratified atmosphere

The diffusion equation

Figure 3

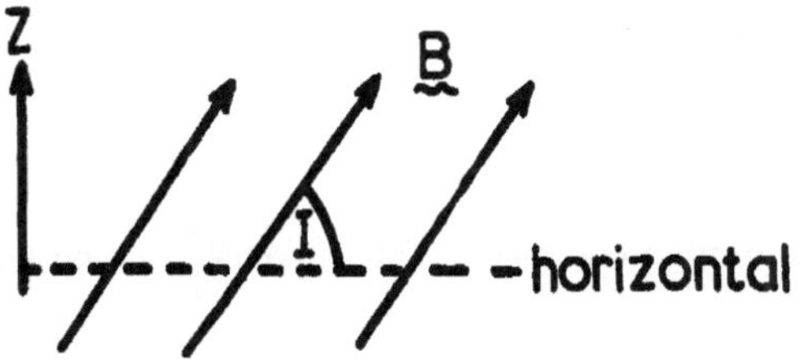

I = angle of dip

= angle between $\underset{\sim}{B}$ and horizontal

All variables are assumed to be functions of z only, where z = height above some fixed reference level (Fig. 3) . Roughly speaking we choose

$$z = 0 \text{ at base of ionospheric F layer}$$

The preceding results on ambipolar diffusion may be used to simplify the derivation, for in the F2 layer, 300 km above the ground

$$\nu_{ia} \sim 1 \text{ sec}^{-1}$$

Therefore

$$\frac{e\,B}{m_i \nu_{ia}} = \frac{1.6 \times 10^{-20} \times 0.3}{26 \times 10^{-24} \times 1} \approx 200$$

giving

$$s \gg 1$$

It follows that

(i) The only relevant equation of motion of the ions is that along a field line i.e. parallel to $\underline{B}$

(ii) The motion $\underline{V}_i^{\perp}$ at right angles to the field lines is determined by the Hall electric field. Thus

$$\underline{V}_i^{\perp} = \underline{E}_H \times \underline{B}/B^2$$

For a self consistent model we must take $\underline{V}_i^{\perp}$ to be uniform.

The physical meaning of (i) & (ii) is that the ions can spiral freely between collisions with neutral atoms, and do so many times. Their random motions across field lines are therefore considerably reduced (Fig. 4).

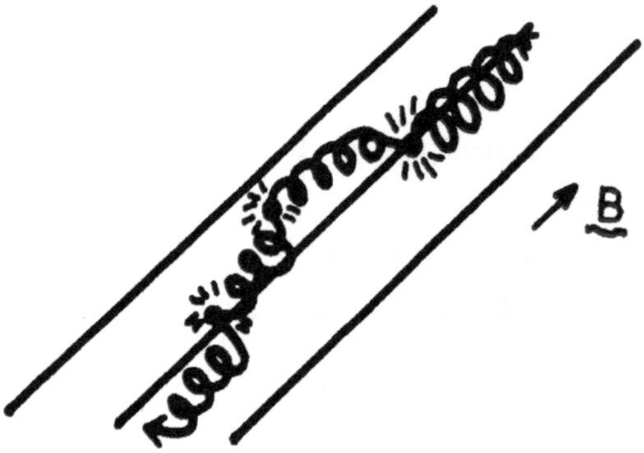

Figure 4

Schematic view of motion of ion

However, the Hall drift is as though the field lines were themselves moving with velocity $\underline{V}_i^{\perp}$, giving a transverse velocity $\underline{V}_i^{\perp}$ to the motion of the ions.

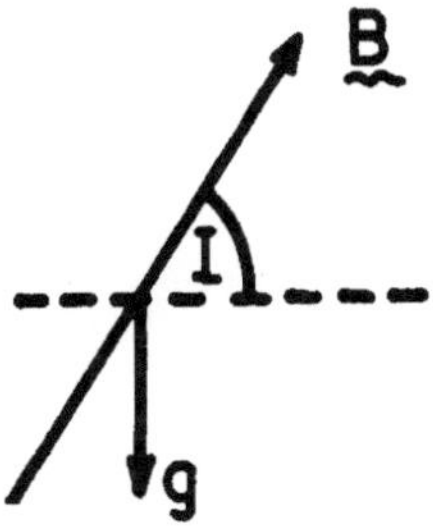

Figure 5

With $d/dt = 0$ the equation of motion of the plasma along a field line is (Fig. 5)

$$\frac{2NkT}{D}\left(\underline{V}_i^{\parallel} - \underline{V}_a^{\parallel}\right) = -\underline{\nabla}^{\parallel}(2NkT) + Nm_i\underline{g}^{\parallel}$$

(The partial pressure of the plasma is $p_i + p_e = NkT + NkT$) .
We assume that the temperature of all constituents is the same and is uniform and constant. Here

$$D = 2\,D_{ia} = 2 \times \text{coeff. diff. ions} \rightarrow \text{neutrals}$$

and from gas theory

$$D \propto 1/n_a \quad (n_a = \text{neutral air density})$$

The neutral atmosphere is in equilibrium. Thus the neutral air density is

$$n_a = n_o \exp(-z/H)\ ,$$

where

$$H = \text{scale height} = kT/m_a g$$

We shall assume that $m_a = m_i$.

Thus if $\hat{\underline{T}}$ denotes a unit tangent along a field line

$$\underline{\nabla}^{\parallel} = \hat{\underline{T}}(\hat{\underline{T}}.\underline{\nabla})$$

and

$$\underline{g}^{\parallel} = - g \sin I \hat{\underline{T}}$$

giving

$$\underline{V}_i^{\parallel} = \underline{V}_a^{\parallel} - \frac{D}{N}\left(\hat{\underline{T}}.\underline{\nabla} N + \frac{N}{2H} \sin I\right) \hat{\underline{T}}$$

The diffusion equation is formed from the continuity equation

$$\frac{\partial N}{\partial t} = Q - L - \underline{\nabla}.(N\underline{V}_i)$$

Here

Q = rate of electron production (by sun)

L = loss rate (by recombination)

The equation could be written

$$\frac{\partial N}{\partial t} = Q - L - W_H \cos I \frac{\partial N}{\partial z} - V_a \cos I \sin I \frac{\partial N}{\partial z}$$

(1) $$+ D \mathcal{D} N,$$

where

W_H = Hall drift $\perp^r$ to field lines (assumed uniform)

V_a = Horizontal velocity of neutral air (assumed uniform)

$\mathcal{D}$ = diffusion operator defined by :

$$D\mathcal{D}N = \underline{\nabla} . \left\{D(\hat{\underline{T}}.\underline{\nabla} N + \frac{N}{2H} \sin I) \hat{\underline{T}}\right\}$$

$$= (\hat{\underline{T}}.\underline{\nabla}) \left\{D(\hat{\underline{T}}.\underline{\nabla} N + \frac{N}{2H} \sin I)\right\}$$

But in this case

$$\hat{\underline{T}} \cdot \underline{\nabla} \equiv \sin \; I \; \frac{\partial}{\partial z}$$

Thus $$D \mathcal{D} N = \sin^2 I \; \frac{\partial}{\partial z} \, D\left(\frac{\partial N}{\partial z} + \frac{N}{2H}\right),$$

giving , as $\mathcal{D} \propto \exp(z / H)$

(2) $$\mathcal{D} N = \left(\frac{\partial^2 N}{\partial z^2} + \frac{3}{2H} \frac{\partial N}{\partial z} + \frac{N}{2H^2}\right) \sin^2 I$$

This equation is due to Professor V. C. A. Ferraro.

We have to solve equation (1) , where $\mathcal{D}$ is given by (2) . The rate of production may be taken to be the Chapman function

(3) $$Q = q_o \exp\left(1 - \frac{z}{H} - e^{-z/H} \sec \chi\right),$$

where χ is the zenith angle of the sun, during daylight hours. At night $Q = 0$. The constant q_o is the rate of maximum production.

Let δ = Northern declination of sun, ϕ' = local time in radians measured from noon ($\phi' = 0 \rightarrow$ noon) . Then spherical trigonometry gives (Fig. 6)

$$\cos \chi = \cos\left(\frac{\pi}{2} - \delta\right) \cos \theta + \sin\left(\frac{\pi}{2} - \delta\right) \sin \theta \cos \phi'$$

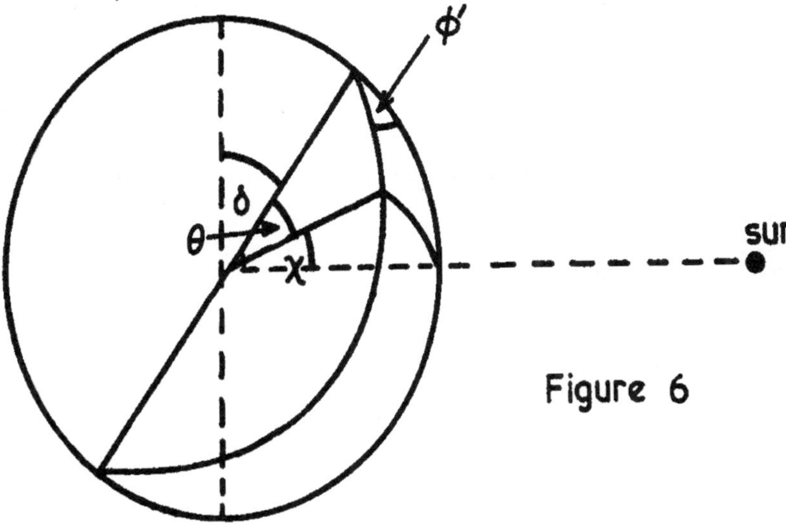

Figure 6

$$= \sin\delta \cos\theta + \cos\delta \sin\theta \cos\phi'$$

The length of day, ψ, is given by the equation

$$0 = \sin\delta \cos\theta + \cos\delta \sin\theta \cos\frac{1}{2}\psi$$

Thus time can be measured in radians from sunrise by writing

$$\phi = \frac{1}{2}\psi + \phi'$$

Then we have

sunrise at $\phi = 0$

sunset at $\phi = \psi$

and

$$\cos\chi = \sin\delta\cos\theta + \cos\delta \sin\theta \cos(\phi - \frac{1}{2}\psi) \tag{4}$$

$$= \text{function of time } \phi$$

The loss rate is proportional to N and is given by

$$L = \beta_o N \exp(-\lambda z/H) \tag{5}$$

The exact value of λ for the F2 layer is believed to be

$$\lambda = 1.75,$$

and numerical calculations are possible in this case. However, a wide range of analytic solutions is available in the case $\lambda = 1$. The constant β_o has dimensions $\sec^{-1}$.

Illustrative solutions

The problem may be formulated in dimensionless variables as follows. Note that for simplicity's sake we consider only the case

$$W_H = 0 \qquad V_a = 0$$

Write

(6) $$t = \tau \phi$$

where

$$\tau = \text{number of secs in a radian} = 1.37 \times 10^4$$

(7) $$\beta = \tau \beta_o$$

(8) $$\gamma = \frac{H^2}{\tau D_o \sin^2 I} \quad (D = D_o e^{z/H})$$

Then

$$\frac{\partial N}{\partial \phi} = \tau q_o \exp\left(1 - \frac{z}{H} - \frac{e^{-z/H}}{\cos\chi}\right) - \beta N e^{-\lambda z/H} + \frac{e^{z/H}}{\gamma} H^2 \mathfrak{D} N$$

Also put

$$\zeta = e^{-z/2H} \quad \text{and} \quad N = \zeta v$$

Then

(9) $$\frac{\partial v}{\partial \phi} = \tau q_o \, e \zeta e^{-\zeta^2 \sec\chi} - \beta \zeta^{2\lambda} + \frac{1}{4\gamma} \frac{\partial^2 v}{\partial \zeta^2}$$

The solution of equation (9) is required under the conditions

(i) Solution is periodic $\phi = 0, 2\pi$

(ii) There are no sources at $z = \infty$ ($\zeta = 0$)

(iii) $v \to 0$ as $z \to -\infty$ ($\zeta \to \infty$)

Gliddon (1958 a, b) has provided solutions of this equation in full for the case $\lambda = 1$ following earlier work by Ferraro and Özdogan for the case $\lambda = 0$. There are possible generalizations in the case $\lambda = 1$ which appear in the Journal of Atmospheric and Terrestrial Physic under the authorship of Gliddon or Gliddon and Kendall. The analysis and elegant mathematics are due to Gliddon.

Exercise Show by direct substitution that a periodic solution of equation (9) with $\lambda = 1$ and $\cos\chi = \sin\phi$ is

$$v(x,\phi) = \frac{1}{2}\tau q_o e (\beta\gamma)^{-\frac{1}{4}} \int_o^{\phi} x\{\Delta(\phi,\phi_o)\}^{-\frac{3}{2}} \exp\left\{-\frac{x^2\Gamma(\phi,\phi_o)}{4\Delta(\phi,\phi_o)}\right\} d\phi_o$$

$$+\frac{1}{2}\tau q_o e (\beta\gamma)^{-\frac{1}{4}} \sum_{n=1}^{\infty} \int_o^{\pi} x\{\Delta(\phi,\phi_o - 2n\pi)\}^{-\frac{3}{2}} \exp -\frac{x^2\Gamma(\phi,\phi_o - 2n\pi)}{4\Delta(\phi,\phi_o - 2n\pi)} d\phi_o$$

where

$$\Gamma(\phi_o,\phi) = (\beta\gamma)^{-\frac{1}{2}} \operatorname{cosec}\phi \cosh\left\{\left(\frac{\beta}{\gamma}\right)^{\frac{1}{2}}(\phi_o-\phi)\right\} + \sinh\left\{\left(\frac{\beta}{\gamma}\right)^{\frac{1}{2}}(\phi_o-\phi)\right\}$$

$$\Delta(\phi_o,\phi) = (\beta\gamma)^{-\frac{1}{2}} \operatorname{cosec}\phi \sinh\left\{\left(\frac{\beta}{\gamma}\right)^{\frac{1}{2}}(\phi_o-\phi)\right\} + \cosh\left\{\left(\frac{\beta}{\gamma}\right)^{\frac{1}{2}}(\phi_o-\phi)\right\}$$

and

$$x^2 = 4(\beta\gamma)^{\frac{1}{2}}\zeta^2$$

Note This solution does not satisfy condition (ii) ; however, Gliddon corrected this in his later papers.

There is scope for more analytical work on equation (9) for the case $\lambda = 1.75$ (or possibly $\lambda = 2$) .

The equilibrium problem

At noon and in the early afternoon the sun is nearly overhead for about 3 hours, so that near-equilibrium may be reached. Then

$$\partial N/\partial\phi \approx 0$$

and, choosing an overhead sun for convenience,

$$\cos\chi \approx 1 .$$

Thus

$$\frac{\partial^2 v}{\partial \zeta^2} - 4\beta\gamma\zeta^{2\lambda} v = -4\gamma\tau q_o e \zeta e^{-\zeta^2} \tag{10}$$

One boundary condition may be found from the expression for $\underline{V}_i^{\parallel}$.

If $\underline{V}_a = 0$

$$\underline{V}_i^{\parallel} \propto D\left(\frac{\partial N}{\partial z} + \frac{N}{2H}\right) \sin I$$

Thus

$$\underline{V}_i^{\parallel} \propto \partial v/\partial\zeta$$

and we require

(i) $\partial v/\partial\zeta = 0$ at $\zeta = 0$

Also

(ii) $v \to 0$ as $\zeta \to \infty$.

These are called <u>two point boundary conditions.</u> There are two methods of solution (a) the method of binary splitting and (b) the use of lattice techniques. In the next lecture I shall give <u>recommended formulae</u> and describe the numerical methods used in detail, and briefly, before describing how to integrate numerically equation (9) with $\partial/\partial\phi \neq 0$.

Note that $\tau q_o e$ is a scaling factor which may be omitted from the calculations and inserted afterwards.

Lecture 3

P.C. Kendall

Numerical and analytical methods

The binary splitting method

We present two methods of integrating the equation

$$\frac{\partial^2 v}{\partial \zeta^2} - 4\beta\gamma\zeta^{2\lambda} v = -4\gamma\zeta e^{-\zeta^2} \tag{1}$$

in preparation for the equation

$$\frac{\partial v}{\partial \phi} = \zeta e^{-\zeta^2 \sec\chi} - \beta\zeta^{2\lambda} v + \frac{1}{4\gamma}\frac{\partial^2 v}{\partial \zeta^2} \tag{2}$$

The first is the binary splitting method used by Rishbeth and Barron.

Briefly, we observe that equation (1) may be integrated under the initial conditions

$$v = v_0 , \quad \partial v/\partial\zeta = 0 \quad \text{at} \quad \zeta = 0 ,$$

where v_0 is a constant,. However, in general we see that the solution will not satisfy the condition

$$v \to 0 \quad \text{as} \quad \zeta \to \infty$$

except for a particular value of v_0, say , $v_0 = v_0^*$. In fact solutions fall into two types: either $v \to \infty$ or $v \to -\infty$. The different types of solution are shown in Fig. 7 .

Figure 7

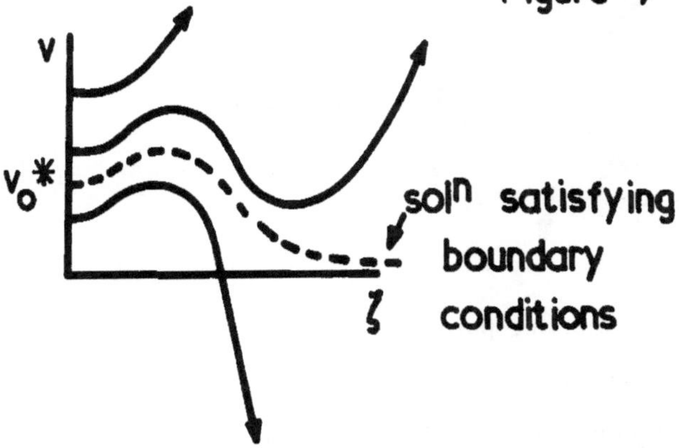

In practice as $N > 0$ always we know that a solution may be discarded as soon as v becomes -ve i.e. when $v < 0$. We also may assume that a solution is such that $v \to \infty$ when v reaches 10. We start with two solutions corresponding to $v_o = v_1$ and $v_o = v_2$ such that

$$\text{for } v_1 \qquad v \to -\infty$$

$$\text{for } v_2 \qquad v \to +\infty$$

Then test the solution for

$$v_3 = \frac{1}{2}(v_1 + v_2)$$

If it is such that $v \to -\infty$ we replace v_1 by v_3.
If it is such that $v \to +\infty$ we replace v_2 by v_3.
In this way we refine the value of v_o as far as the accuracy of the computer permits. The two final solutions give an approximation to the exact solution, and it is obvious where they fail to give a good approximation (Fig. 8);

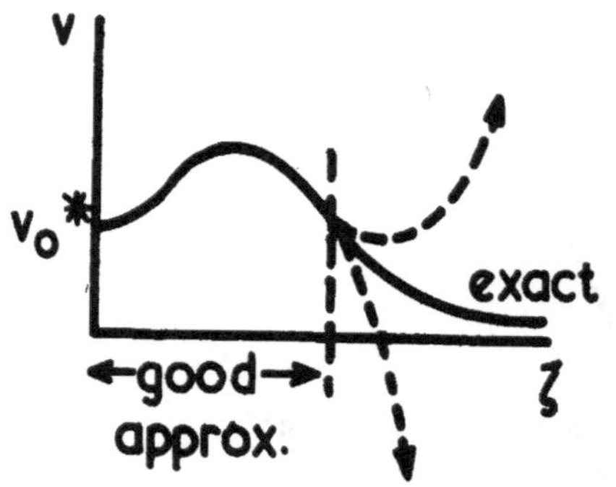

Figure 8

A suitable starting pair is $v_o = 0$ and $v_o = 9$ (say). The methods recommended are the Runge-Kutta method for starting the integration and Hamming's method for continuing it. These two integration formulae are equally accurate (of order h^5) but the Runge-Kutta method is unstable when v starts getting large. This can cause convergence to a false value of v_o^*. The reader should note that there are many other integration formulae, including Runge-Kutta formulae of different orders.

Runge-Kutta formula

To integrate the system of d.e.'s

$$\frac{dy_i}{dx} = f_i(x, y_1, y_2, \ldots, y_n) \qquad i = 1, 2, \ldots, n$$

we have for $n = 2$,

$$\frac{dy_1}{dx} = f_1(x, y_1, y_2)$$

$$\frac{dy_2}{dx} = f_2(x, y_1, y_2)$$

P. C. Kendall

The Runge-Kutta formula for the increments in y_1 and y_2 correspon-ding to an increment h in x is

$$\Delta y_i = \frac{1}{6}(K_{oi} + 2K_{1i} + 2K_{2i} + K_{3i}) \quad (i = 1, 2)$$

where

$$K_{oi} = h\, f_i(x^o, y_1^o, y_2^o)$$

$$K_{1i} = h\, f_i(x^o + \frac{1}{2}h,\ y_1^o + \frac{1}{2}K_{o1},\ y_2^o + \frac{1}{2}K_{o2})$$

$$K_{2i} = h\, f_i(x^o + \frac{1}{2}h,\ y_1^o + \frac{1}{2}K_{11},\ y_2^o + \frac{1}{2}K_{12})$$

$$K_{3i} = h\, f_i(x^o + h,\ y_1^o + K_{21},\ y_2^o + K_{22}) \qquad (i = 1, 2)$$

To use this in integrating (1) we have

$$n = 2 \qquad \zeta = x \qquad y_1 = v \qquad y_2 = \partial v / \partial \zeta$$

Then

$$\frac{dy_1}{dx} = y_2 \qquad = f_1(x, y_1, y_2)$$

$$\frac{dy_2}{dx} = 4\gamma x^{2\lambda}\, y_1 - 4\gamma x e^{-x^2} \qquad = f_2(x, y_1, y_2)$$

The procedure is quite straightforward, but we only make 5 steps, then start using Hamming's method (with same steplength).

Hamming's method (see Ralston and Wilf

To integrate numerically the d.e.

$$\frac{dy}{dx} = f(x, y)$$

let x_i $(i = 1, 2, \dots)$ be a sequence of values of x; spaced at equal intervals, h, and such that the values $y(x_i)$ $(o \leq i \leq n)$ are known. We require to find $y(x_{n+1})$.

To simplify the notation, write

$$y_i = y(x_i) \qquad , \left(\frac{dy}{dx}\right)_{x=x_i} = f_i$$

then y_{n+1} is evaluated by the following method :

(1) Predictor : $p_{n+1} = y_{n-3} + \frac{4h}{3}(2f_n - f_{n-1} + 2f_{n-2})$

(2) Modifier : $m_{n+1} = p_{n+1} - \frac{112}{121}(p_n - C_n)$; $m'_{n+1} = f(x_{n+1}, m_{n+1})$

(3) Corrector: $C_{n+1} = \frac{1}{8}\left[9y_n - y_{n-2} + 3h(m'_{n+1} + 2f_n - f_{n-1})\right]$

(4) Final Value : $y_{n+1} = C_{n+1} + \frac{9}{121}(p_{n+1} - C_{n+1})$

Lattice techniques

Assume that the solution at $\zeta = 0$ is v_o, at $\zeta = h$ is v_1, at $\zeta = 2h$ is v_2 and so on. Then

$$\frac{\partial^2 v}{\partial \zeta^2} \approx \frac{v_{n-1} - 2v_n + v_{n+1}}{h^2}$$

Suppose $\zeta = \zeta_n$ at $v = v_n$. The finite difference approximation to equation (1) is then

(3)
$$v_{n-1} - (2 + 4\beta\gamma h^2 \zeta_n^{2\lambda})v_n + v_{n+1} = -4\gamma h^2 \zeta_n e^{-\zeta_n^2}$$

At $\zeta = 0$ we have $\partial v/\partial \zeta = 0$. The correct procedure here is to introduce a spurious value v_{-1} outside the boundary. Then on the boundary $\zeta = 0$,

$$\frac{\partial v}{\partial \zeta} \approx \frac{v_1 - v_{-1}}{2h} = 0$$

(4)
$$\therefore \quad v_1 = v_{-1}$$

The d.e. is satisfied on the boundary, so we obtain from (4) with $n = 0$ in (3) ($\zeta_o = 0$)

(5)
$$-v_o + v_1 = 0$$

There is no easy way of exactly ensuring that $v \to 0$ as $\zeta \to \infty$. From (1) we observe that for large values of ζ ($\zeta \gg 1$)

$$v \approx \zeta^{1-2\lambda} e^{-\zeta^2} / \beta$$

Thus we choose for n large enough (= N, say)

(6)
$$v_N = \zeta_N^{1-2\lambda} e^{-\zeta_N^2} / \beta$$

and impose this as a boundary condition. In practice $\beta\gamma \approx 30$ and we could choose $\zeta_N = 1$ as a remarkably good approximation to ∞.

Equations (3) (5) and (6) give us the set of N-1 equations

$$-v_o + v_1 = 0$$

$$v_0 - A_1 v_1 + v_2 = F_1$$

$$v_1 - A_2 v_2 + v_3 = F_2$$

(7) $$\cdots$$

$$v_{N-3} - A_{N-2} v_{N-2} + v_{N-1} = F_{N-2}$$

$$v_{N-2} - A_{N-1} v_{N-1} = F_{N-1} - \gamma \zeta_N^{1-2\lambda} e^{-\zeta_N^2} / \beta$$

where

$$A_n = 2 + 4\beta\gamma h^2 \zeta_n^{2\lambda} \qquad F_n = -4\gamma h^2 \zeta_n e^{-\zeta_n^2}$$

This system of equations is called a <u>tridiagonal</u> system. It is straightforward enough to solve for $v_o \ldots v_{N-1}$ by using an algorithm given by Richtmyer.

<u>Algorithm for solving tridiagonal system</u> (see Richtmyer)

Suppose the general form is

(8) $$C_i v_{i-1} + B_i v_i + A_i v_{i+1} = K_i \qquad i = 0, 1, \ldots, N-1$$

Set (9) $$v_i = r_i v_{i+1} + H_i$$

(10) $$v_{i-1} = r_{i-1} v_i + H_{i-1}$$

Then

$$C_i r_{i-1} v_i + C_i H_{i-1} + B_i v_i + A_i v_{i+1} = K_i$$

(11) or $$v_i \; C_i r_{i-1} + B_i = K_i - C_i H_{i-1} - A_i v_{i+1}$$

Then comparing (9) and (11)

(12) $$r_i = \frac{-A_i}{C_i r_{i-1} + B_i} \quad ; \quad (13)\; H_i = \frac{K_i - C_i H_{i-1}}{C_i r_{i-1} + B_i}$$

Now from 1st equation ; i.e. 1st boundary condition

$$B_0 v_0 + A_0 v_1 = K_0$$

or $$v_0 = \frac{K_0}{B_0} - \frac{A_0}{B_0} v_1$$

so (14) $$r_0 = -\frac{A_0}{B_0} \quad ; \quad (15)\; H_0 = \frac{K_0}{B_0}$$

<u>Method of Solution</u> (1) Calculate H_i, r_i for $i = 1, \ldots, N-1$ using (12) - (1

(2) Using (10) repeatedly calculate v_i

<u>Numerical solution of $\partial N/\partial t = \partial^2 N/\partial x^2$</u> (Note that the methodes given here are only two of many possibilities)

<u>Well formulated problem</u> (Fig. 9)

h_t = distance between t lattice

h_x = distance between x lattice

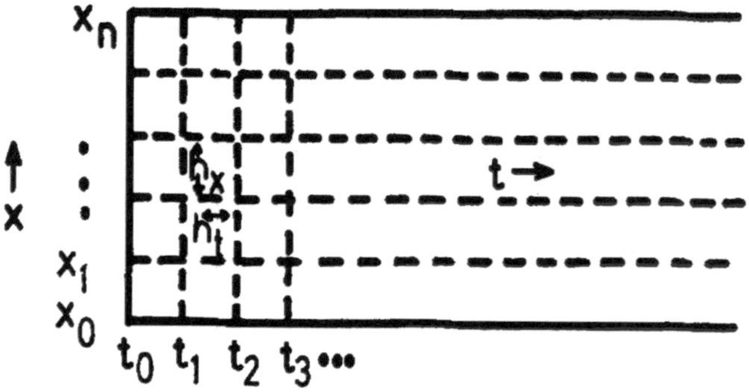

P. C. Kendall

Fig. 9

In a well formulated problem N will be given at time $t = t_o$ as a function of x and it is required to find $N = N(x, t)$ at all later times under given boundary conditions: one on $x = x_o$ and one on $x = x_n$.

Wrong numerical method (i.e. wrong for slow computers)

Denote by brackets $\{-\}_t$ the value of any function at time t. Then

$$\frac{\partial N}{\partial t} \approx \frac{\{N_o\}_{t+h_t} - \{N_o\}_t}{h_t}$$

and

$$\frac{\partial^2 N}{\partial x^2} \approx \frac{\{N_{-1} + N_1 - 2N_o\}_t}{h_x^2}$$

giving on substituting into $\partial N/\partial t = \partial^2 N/\partial x^2$

$$\{N\}_{t+h_t} = \{N\}_t + \frac{h_t}{h_x^2}\{N_{-1} + N_1 - 2N_o\}_t$$

Note It is common practice to refer suffixes to the central point as origin, thus at x_n we have the situation shown in Fig. 10

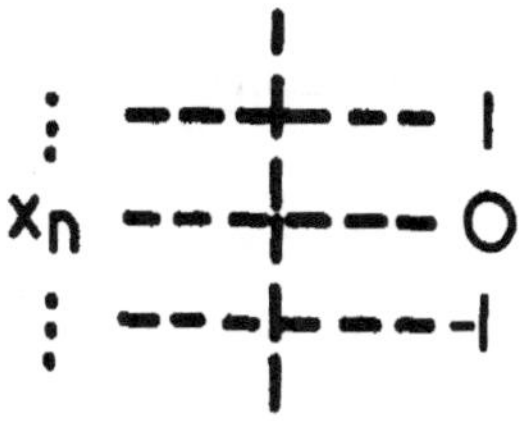

i.e. N_{-1} is N_{n-1}, N_0 is N_n and N_1 is N_{n+1}

Fig. 10

This gives quick and unstable integration from time t to time $t + h_t$. The method is <u>not recommended</u> (except for very fast computers where the steplength can be made small enough to achieve stability.

<u>Crank-Nicolson method</u> (See Smith)

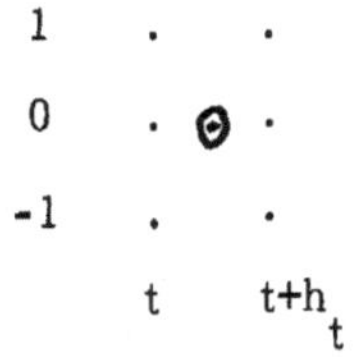

Evaluate all derivatives at a point ⊙ midway between the two t lattice points at $x = x_n$.

$$\frac{\partial N}{\partial t} \approx \frac{\{N_o\}_{t+h_t} - \{N_o\}_t}{h_t}$$

$$\frac{\partial^2 N}{\partial x^2} \approx \frac{\{N_{-1} + N_1 - 2N_o\}_{t+\frac{1}{2}h_t}}{h_x^2}$$

But we have no known values of N at $t + \frac{1}{2} h_t$, so we use the approximation

$$f_{t+\frac{1}{2}h_t} \approx \frac{1}{2}\left[\{f\}_t + \{f\}_{t+h_t}\right]$$

Then, substituting into $\partial N/\partial t = \partial^2 N/\partial x^2$ using the abbreviation

$$s = h_t/2h_x^2$$

gives

$$\{N_o\}_{t+h_t} - \{N_o\}_t =$$

$$s\left[\{N_{-1} + N_1 - 2N_o\}_{t+h_t} + \{N_{-1} + N_1 - 2N_o\}_t\right]$$

that is

$$-s\{N_{-1}\}_{t+h_t} + (1+2s)\{N_o\}_{t+h_t} - s\{N_1\}_{t+h_t}$$

$$= s\{N_{-1}\}_t + (1-2s)\{N_o\}_t + s\{N_1\}_t$$

Thus the integration proceeds from time t to $t+h_t$ by solving a system of tridiagonal equations like (7) . The first and last equations of the set are formed by introducing spurious points outside the boundary, if necessary, and by assuming that the differential equation is satisfied there.

Consider the operator $$\sum_{i=1}^{n} f_i(x_1, \ldots, x_n)\, \partial/\partial x_i \qquad (*)$$

This may be reduced to the form $\partial f_1/\partial x_1'$ by suitable transformations given below, thus making soluble the generalised diffusion equation

$$\sum_{i=1}^{n} f_i\, \partial N/\partial x_i = q - L + \mathfrak{D} N \qquad (**)$$

which can then be subjected to automatic numerical procedures.

In full (*) becomes

$$f_1 \frac{\partial}{\partial x_1} + f_2 \frac{\partial}{\partial x_2} + \ldots\ldots + f_n \frac{\partial}{\partial x_n}$$

Consider the system of ordinary differential equations

$$\frac{dx_2}{dx_1} = \frac{f_2}{f_1}$$

$$\frac{dx_3}{dx_1} = \frac{f_3}{f_1}$$

$$\vdots \qquad \vdots$$

$$\frac{dx_n}{dx_1} = \frac{f_n}{f_1}$$

with $x_2 = x_{2o}$, $x_3 = x_{3o}$, ..., $x_n = x_{no}$ at $x_1 = 0$.

This defines a series of functions

$$x_2 = x_2(x_1, x_{2o}, x_{3o}, \ldots, x_{no})$$

$$x_3 = x_3(x_1, x_{2o}, x_{3o}, \ldots, x_{no})$$

$$\vdots$$

$$x_n = x_n(x_1, x_{2o}, x_{3o}, \ldots, x_{no})$$

Consider the transformation from the variables $x_1, x_2, \dots x_n$ to variables $x_1', x_2' \dots x_n'$ defined by

$$x_1' = x_1$$

$$x_2' = x_{2o}(x_1, x_2, \dots x_n)$$

$$x_3' = x_{3o}(x_1, x_2, \dots x_n)$$

$$\dots\dots$$

$$x_n' = x_{no}(x_1, x_2, \dots x_n)$$

Then we know that

$$\frac{\partial}{\partial x_1'} = \frac{\partial}{\partial x_1} + \frac{\partial x_2}{\partial x_1}\frac{\partial}{\partial x_2} + \frac{\partial x_3}{\partial x_1}\frac{\partial}{\partial x_3} + \dots + \frac{\partial x_n}{\partial x_1}\frac{\partial}{\partial x_n}$$

$$= \frac{\partial}{\partial x_1} + \frac{f_2}{f_1}\frac{\partial}{\partial x_2} + \frac{f_3}{f_1}\frac{\partial}{\partial x_3} + \dots + \frac{f_n}{f_1}\frac{\partial}{\partial x_n}$$

and so (✱) becomes

$$f_1 \frac{\partial}{\partial x_1'} = \sum_{i=1}^{n} f_i \frac{\partial}{\partial x_i}$$

Note that all the transformations of variables can be carried out numerically, even if exact integrals cannet be found. The variables $x_{2o}, x_{3o}, \dots x_{no}$ retain the values of $x_2, x_3 \dots x_n$ at the start of the integration. At any subsequent point we obtain from (✱✱)

$$N = N(x_1', x_{2o}, x_{3o}, \dots x_{no} + \text{other coordinates})$$

The values of $x_1, \dots x_n$ at which this is the solution are obtained by integrating the system of equations. Note that in any case these would be needed in the computer at the correct stage of the calculation.

Exercise . Use the above analysis to derive the transformation which reduces the operator

$$\frac{\partial}{\partial t} + W_o \sin \overline{\omega t + \delta} \frac{\partial}{\partial a}$$

to the form $\frac{\partial}{\partial t'}$, where ω , δ and W_o are constants.

Answer $t' = t$, $a' = a + \frac{W_o}{\omega} \cos \overline{\omega t + \delta}$

P. C. Kendall

Lecture 4

The diffusion operator with curved magnetic field lines (neutral air at rest)

We choose to write the diffusion equation in the form

(1) $$\frac{\partial N}{\partial t} = Q - L + W\mathfrak{D}_1 N + D\mathfrak{D}_2 N$$

where W is a constant with dimensions of velocity, $\mathfrak{D}_1$ is the Hall drift operator defined by

(2) $$W\mathfrak{D}_1 N = -\underset{\sim}{\nabla}.(N\underline{V}_i^{\perp})$$

and $\mathfrak{D}_2$ is the diffusion operator defined by

(3) $$D\mathfrak{D}_2 N = -\underset{\sim}{\nabla}.(N\underline{V}_i^{\parallel})$$

Geometry of geomagnetic dipole field lines (Fig. 11)

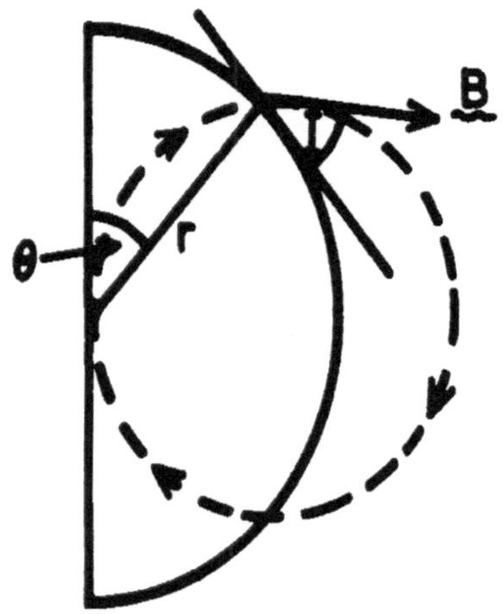

Field lines are family of curves

(4) $$r = a \sin^2 \theta$$

Thus

(5) $$\sin I = 2 \cos \theta / \Delta \qquad \cos I = \sin \theta / \Delta$$

where

(6) $$\Delta = \sqrt{(1 + 3 \cos^2 \theta)}$$

Also

(7) $$\tan I = 2 \cot \theta$$

Thus the tangent $\hat{\underline{T}}$ to a field line is

(8) $$\hat{\underline{T}} = (\sin I, \cos I, 0)$$

$$\underline{\nabla}^{\shortparallel} = \hat{\underline{T}} (\hat{\underline{T}} . \underline{\nabla})$$
$$= \hat{\underline{T}} \left(\sin I \frac{\partial}{\partial r} + \frac{\cos I}{r} \frac{\partial}{\partial \theta} \right)$$

<u>Form of</u> $\mathfrak{D}_2$

Thus as

(10) $$\frac{2NkT}{D} \underline{V}_i^{\shortparallel} = - \underline{\nabla}^{\shortparallel}(2NkT) + Nm_i \underline{g}^{\shortparallel}$$

we have

(11) $$N\underline{V}_i^{\shortparallel} = - D\hat{\underline{T}} \left(\sin I \frac{\partial N}{\partial r} + \frac{\cos I}{r} \frac{\partial N}{\partial \theta} + \frac{N \sin I}{2H} \right)$$

Hence $$\underline{\nabla} . N\underline{V}_i^{\shortparallel} = -(\hat{\underline{T}} . \underline{\nabla}) D \left(\sin I \frac{\partial N}{\partial r} + \frac{\cos I}{r^{+}} \frac{\partial N}{\partial \theta} + \frac{N \sin I}{2 H} \right)$$

(12) $$- D(\underline{\nabla} . \hat{\underline{T}}) \left(\sin I \frac{\partial N}{\partial r} + \frac{\cos I}{r} \frac{\partial N}{\partial \theta} + \frac{N \sin I}{2H} \right)$$

Using the fact that $D \propto \exp(z/H)$, where $z = r - r_o$ and $r = r_o$ is the base of the F2 layer, this gives

$$\mathfrak{D}_2 N = \left(\sin I \frac{\partial}{\partial r} + \frac{\cos I}{r} \frac{\partial}{\partial \theta} \right) \left(\sin I \frac{\partial N}{\partial r} + \frac{\cos I}{r} \frac{\partial N}{\partial \theta} + \frac{N \sin I}{2H} \right)$$

(13)

$$+ \left\{ \frac{\sin I}{H} + (\underline{\nabla} . \hat{\underline{T}}) \right\} \left(\sin I \frac{\partial N}{\partial r} + \frac{\cos I}{r} \frac{\partial N}{\partial \theta} + \frac{N \sin I}{2H} \right)$$

It may be verified that

(14) $$\underline{\nabla} . \hat{\underline{T}} = (9 \cos \theta + 15 \cos^3 \theta) / r \Delta^3$$

<u>Form of $\mathfrak{D}_1$</u>

We have

(15) $$\underline{V}_i^{\perp} = \underline{E}_H \times \underline{B} / B^2$$

Also, the electrostatic potential is constant along a field line so that

(16) $$\Omega_H = - F(\phi , a) ,$$

where ϕ is the longtitude ($t = \tau \phi$, where τ = number of seconds in a radian $= 1.37 \times 10^4$). Then

(17) $$\underline{E}_H = - \text{grad} \; \Omega_H ,$$

and so in spherical polar coordinates (r, θ, ϕ)

$$\underline{E}_H = \left(\frac{1}{\sin^2 \theta} , - \frac{2 \cos \theta}{\sin^3 \theta} , 0 \right) \frac{\partial F}{\partial a}$$

(18)

$$+ \left(0 , 0 , \frac{1}{r \sin \theta} \frac{\partial F}{\partial \phi} \right)$$

That is

$$\underline{E}_H = \frac{\Delta}{\sin^3\theta}(\cos I, \ -\sin I, \ 0)\frac{\partial F}{\partial a}$$

(19)

$$+ (0, \ 0, \ \frac{1}{r\sin\theta}\frac{\partial F}{\partial\phi})$$

Also if M = dipole moment, $B = M\Delta/r^3$ and

(20) $$\underline{B} = \frac{M\Delta}{r^3}(\sin I, \ \cos I, \ 0)$$

Thus
$$\underline{E}_H \times \underline{B} = \frac{M\Delta^2}{r^3\sin^3\theta}\frac{\partial F}{\partial a}(0, 0, 1)$$

(21)
$$- \frac{M\Delta}{r^4\sin\theta}\frac{\partial F}{\partial\phi}(\cos I, \ -\sin I, \ 0)$$

But
$$\underline{\nabla}.(\underline{E}_H \times \underline{B}) = \underline{B}.(\underline{\nabla}\times\underline{E}_H) - \underline{E}_H.(\underline{\nabla}\times\underline{B})$$

(22)
$$\equiv 0$$

So

$$W\mathfrak{D}_1 N \doteq -(\underline{E}_H \times \underline{B}).\underline{\nabla}\frac{N}{B^2}$$

(23)
$$= \frac{-M\Delta^2}{r^3\sin^3\theta}\frac{\partial F}{\partial a}\frac{1}{r\sin\theta}\frac{\partial}{\partial\phi}\left(\frac{r^6}{M^2\Delta^2}N\right)$$

$$+ \frac{M\Delta}{r^4\sin\theta}\frac{\partial F}{\partial\phi}\left(\cos I\frac{\partial}{\partial r} - \frac{\sin I}{r}\frac{\partial}{\partial\theta}\right)\left(\frac{r^6}{M^2\Delta^2}N\right)$$

And as $r = a\sin^2\theta$

(24)
$$W\mathfrak{D}_1 N = -\frac{a^2}{M}\frac{\partial F}{\partial a}\frac{\partial N}{\partial\phi} + \frac{r^2\,\partial F/\partial\phi}{M\Delta\sin\theta}\left(\cos I\frac{\partial N}{\partial r} - \frac{\sin I}{r}\frac{\partial N}{\partial\theta}\right)$$

$$+ \frac{6r(1+\cos^2\theta)}{M(1+3\cos^2\theta)^2}\frac{\partial F}{\partial\phi}N$$

Finally to make this dimensionless we may substitute

$$F = \frac{W\,M}{r_o^2}\; f\,(a, \Phi), \tag{25}$$

where f is dimensionless giving

$$\begin{aligned} \mathcal{D}_1 N = &- \frac{a^2}{r_o^2}\frac{\partial f}{\partial a}\frac{\partial N}{\partial \phi} + \frac{r^2\,\partial f/\partial\phi}{r_o^2\,\Delta\,\sin\theta}\left(\cos I\frac{\partial N}{\partial r} - \frac{\sin I}{r}\frac{\partial N}{\partial \theta}\right) \\ &+ \frac{6r(1+\cos^2\theta)}{r_o^2\,\Delta^4}\frac{\partial f}{\partial \phi}\,N \end{aligned} \tag{26}$$

Finally, we make the substitution

$$N = e^{-z/2H}\;V \tag{27}$$

to obtain the dimensionless form of (1), namely,

$$\begin{aligned} \frac{\partial V}{\partial \phi} = &\;\gamma q_o\; e\;\exp\left(-\frac{z}{2H} - \frac{e^{-z/H}}{\cos\chi}\right) - \beta\, e^{-\lambda z/H} V \\ &+ W'\,\mathcal{D}_1'\,V + \frac{e^{z/H}}{\gamma}\,\mathcal{D}_2'\;V \end{aligned} \tag{28}$$

$$\begin{aligned} H^{-1}\mathcal{D}_1'\;V = \Bigg\{ &-\frac{a^2}{r_o^2}\frac{\partial f}{\partial a}\frac{\partial V}{\partial \phi} + \frac{r^2\,\partial f/\partial\phi}{r_o^2\,\Delta\sin\theta}\left(\cos I\frac{\partial V}{\partial r} - \frac{\sin I}{r}\frac{\partial V}{\partial \theta}\right) \\ &+ \left[-\frac{r^2\cos I}{2Hr_o^2\,\Delta\sin\theta} + \frac{6r\,(1+\cos^2\theta)}{r_o^2\,\Delta^4 H}\right]\frac{\partial f}{\partial \phi}\;V \Bigg\} \end{aligned} \tag{29}$$

$$H^{-2}\mathfrak{D}_2' V = (\sin I \frac{\partial}{\partial r} + \frac{\cos I}{r} \frac{\partial}{\partial \theta})^2 V$$

(30) $$+ (\frac{\sin I}{2H} + \underline{\nabla}.\hat{\underline{T}}) (\sin I \frac{\partial}{\partial r} + \frac{\cos I}{r} \frac{\partial}{\partial \theta}) V$$

where

(31) $$W' = \gamma W/H ,$$

and β and γ are as before[1], except that a factor $\sin^2 I$ is omitted fro γ. We wish to integrate equation (28), which is a partial differential equation in 3 independent variables. This is now possible, and I first outline the difficulties.

Difficulties

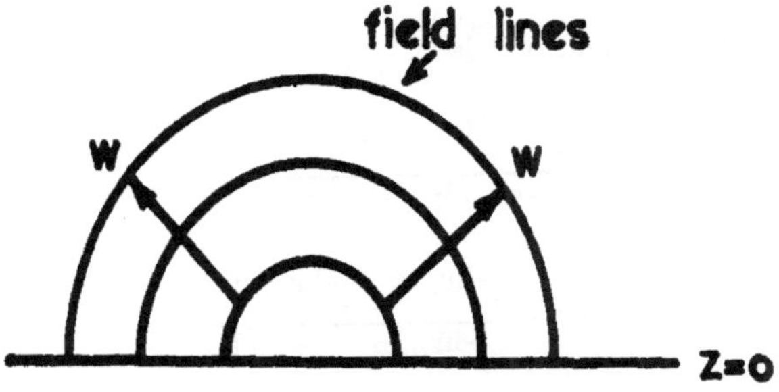

Fig. 12

Roughly speaking we are integrating outwards along lines $\perp^r$ to $\underline{B}$ (Fig. 12). The line z = 0 is not a natural boundary in our coordinate system (Fig. 13).

1. See Lecture 2, equations 7 and 8.

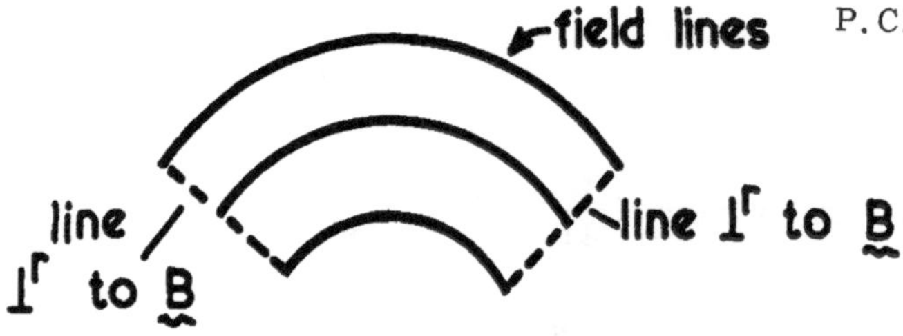

Fig. 13

Also as $r \to \infty$ the total length involved along a field line becomes infinite and the region of integration becomes infinite. We therefore need a coordinate transformation which

(a) transforms $z = 0$ into a natural boundary

(b) transforms $z = \infty$ into a finite point.

A suitable transformation is

$$x^2 = \frac{e^{-hr/H} - e^{-ha/H}}{e^{-hr_o/H} - e^{-ha/H}} \tag{32}$$

$$= 1 + \frac{e^{-hr/H} - e^{-hr_o/H}}{e^{-hr_o/H} - e^{-ha/H}} \tag{33}$$

where

h = dimensionless constant > 0

and

$x > 0$ in Northern hemisphere

$x < 0$ in Southern hemisphere

If $a \gg r$

$$x \approx e^{-hz/2H}$$

On equator $r = a$,

$$x = 0 .$$

Also at $r = r_0$,

$$x = 1 \text{ (Northern) or } -1 \text{ (Southern)}$$

We thus obtain the mapping shown in Fig. 14

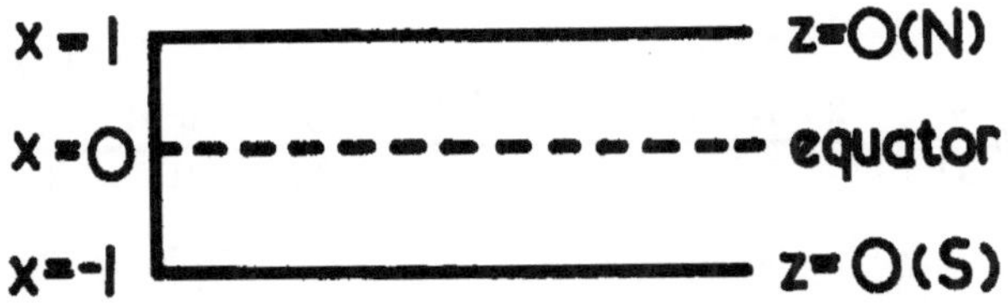

Fig. 14

Choose as independent variables

$$x^2 = 1 + \frac{e^{-hr/H} - e^{-hr_0/H}}{e^{-hr_0/H} - e^{-ha/H}} = 1 + \frac{\Lambda_1}{\Lambda_2} \tag{34}$$

and

$$a = \frac{r}{\sin^2\theta}$$

Then

$$\frac{\partial}{\partial r} = \frac{1}{\sin^2\theta}\frac{\partial}{\partial a} + \left(-\frac{h}{H}\frac{e^{-hr/H}}{\Lambda_2} - \frac{h}{H}\frac{\Lambda_1 e^{-ha/H}}{\Lambda_2^2}\frac{\partial a}{\partial r}\right)\frac{\partial}{\partial (x^2)} \tag{35}$$

$$\frac{1}{r}\frac{\partial}{\partial \theta} = -\frac{2\cos\theta}{\sin^3\theta}\frac{\partial}{\partial a} - \left(\frac{h}{H}\frac{\Lambda_1 e^{-ha/H}}{\Lambda_2^2}\frac{\partial a}{r\partial\theta}\right)\frac{\partial}{\partial (x^2)} \tag{36}$$

Thus

$$2\cos\theta\frac{\partial}{\partial r}+\frac{\sin\theta}{r}\frac{\partial}{\partial\theta}\equiv-\frac{2h}{H}\frac{e^{-hr/H}}{\Lambda_2}\cos\theta\frac{\partial}{\partial(x^2)} \tag{37}$$

giving

$$\sin I\frac{\partial}{\partial r}+\frac{\cos I}{r}\frac{\partial}{\partial\theta}\equiv\frac{-h\sin I\, e^{-hr/H}}{H(e^{-hr_o/H}-e^{-ha/H})}\frac{\partial}{\partial(x^2)} \tag{38}$$

Similarly

$$\sin\theta\frac{\partial}{\partial r}-\frac{2\cos\theta}{r}\frac{\partial}{\partial\theta}\equiv\frac{\Delta^2}{\sin^3\theta}\left(\frac{\partial}{\partial a}-\frac{h}{H}\frac{\Lambda_1 e^{-ha/H}}{\Lambda_2^2}\frac{\partial}{\partial(x^2)}\right)$$

$$-\frac{h\sin\theta\, e^{-hr/H}}{H\Lambda_2}\frac{\partial}{\partial(x^2)}$$

giving

$$\cos I\frac{\partial}{\partial r}-\frac{\sin I}{r}\frac{\partial}{\partial\theta}\approx\frac{\Delta}{\sin^3\theta}\frac{\partial}{\partial a}$$

$$-\frac{h}{H}\left\{\frac{(x^2-1)\Delta e^{-ha/H}}{\Lambda_2\sin^3\theta}+\frac{\sin\theta\, e^{-hr/H}}{\Delta\ \Lambda_2}\frac{\partial}{\partial(x^2)}\right\} \tag{39}$$

Therefore

$$\mathcal{D}_1{}'V=\frac{a^2H}{r_o^2}\frac{\partial f}{\partial\phi}\frac{\partial V}{\partial a}-\frac{a^2H}{r_o^2}\frac{\partial f}{\partial a}\frac{\partial V}{\partial\phi}$$

$$-\frac{h\,r^2\,\partial f/\partial\phi}{r_o^2\Delta\sin\theta}\left\{\frac{(x^2-1)\Delta e^{-ha/H}}{\Lambda_2\sin^3\theta}+\frac{\sin\theta\, e^{-hr/H}}{\Delta(e^{-hr_o/H}-e^{-ha/H})}\frac{\partial V}{\partial(x^2)}\right. \tag{40}$$

(40) $$+\left[\frac{6rH(1+\cos^2\theta)}{r_o^2\Delta^4} - \frac{r^2\cos I}{2r_o^2\Delta\sin\theta}\right]\frac{\partial f}{\partial\phi}V$$

and

$$\mathfrak{D}_2{}'V = \left(\frac{\frac{1}{2}h\sin I\, e^{-hr/H}}{e^{-hr_o/H}-e^{-ha/H}}\right)^2\left(\frac{1}{x^2}\frac{\partial^2V}{\partial x^2}-\frac{1}{x^3}\frac{\partial V}{\partial x}\right)$$

(41) $$+\frac{[h^2\sin^2 I + 2(hH/a\Delta^4)]}{2x(e^{-hr_o/H}-e^{-ha/H})}\left(e^{-hr/H}\frac{\partial V}{\partial x}\right)$$

$$-\left(\frac{\sin I}{2}+H\,\underline{\nabla}.\hat{\underline{I}}\right)\left(\frac{\frac{1}{2}h\sin I\, e^{-hr/H}}{x(e^{-hr_o/H}-e^{-ha/H})}\right)\frac{\partial V}{\partial x}$$

which simplifies to (see (14))

(42) $$\mathfrak{D}_2{}'V = \left\{\frac{h\sin I\, e^{-hr/H}}{2x\Lambda_2}\right\}^2\frac{\partial^2V}{\partial x^2}$$

$$-\frac{\frac{1}{4}h\sin^2 I\, e^{-hr/H}}{x\Lambda_2}\left\{\frac{(1+h)e^{-hr/H}-e^{-ha/H}}{e^{-hr/H}-e^{-ha/H}}+\frac{15\mu^4+10\mu^2-1}{r\,\mu^2\Delta^2/H}-2h\right\}\frac{\partial V}{\partial x}$$

<u>Check</u> : If $a \gg r_o$ and $h = 1$, $x \approx e^{-z/2H}$, $I = 90°$,

then $\mathfrak{D}_2{}'\ V \approx \frac{1}{4}e^{-z/H}\partial^2V/\partial x^2$ in agreement with Lecture 2, equation 9 ($\zeta = x$) .

We shall take

(43) $$f = (r_o^2/a^2)\quad\cos\phi$$

giving a Hall drift which is independent of <u>a</u> at the equator and $\propto \sin\phi$.

The full diffusion equation is thus

$$\left(1 - \frac{2\,W'\,H}{a}\cos\phi\right)\frac{\partial V}{\partial \phi} + W'H\sin\phi\,\frac{\partial V}{\partial a}$$

$$= \frac{e^{z/H}}{\gamma}\left\{\frac{h\sin I\ e^{-hr/H}}{2x\,\Lambda_2}\right\}^2\frac{\partial^2 V}{\partial x^2}$$

$$+\left[-\frac{\frac{1}{4}h\sin^2 I\ e^{(z-hr)/H}}{x\Lambda_2\gamma}\left\{\frac{(1+h)e^{-hr/H} - e^{-ha/H}}{e^{-hr/H} - e^{-ha/H}} + \frac{15\mu^4 + 10\mu^2 - 1}{r\mu^2\Delta^2/H} - 2h\right\}\right.$$

$$\left. + \frac{W'h\,r^2\sin\phi}{2xa^2\Lambda_2\Delta^2\sin^4\theta}\left\{(x^2-1)\ \Delta^2 e^{-ha/H} + \sin^4\theta\ e^{-hr/H}\right\}\right]\frac{\partial V}{\partial x}$$

$$- W'\sin\phi\left[\frac{6rH(1+\cos^2\theta)}{a^2(1+3\cos^2\theta)^2} - \frac{r^2}{2a^2(1+3\cos^2\theta)}\right]V$$

$$- \beta\, e^{-\lambda z/H}\ V$$

$$+ \tau\, q_0\ e\ \exp\left(-\frac{z}{2H} - \frac{e^{-z/H}}{\cos\chi}\right) \qquad (44)$$

We see from lecture (3) that this can be now integrate numerically.

P. C. Kendall

Lecture 5

Equilibrium solutions

Diffusive equilibrium (and comparison of theoretical and satellite results)

When the velocity of the ions is zero, i.e.

$$(1) \qquad \underline{V}_i = 0$$

we have a situation known as diffusive equilibrium.
You may be **surprised** that such a simple situation could occur at all; however all this means is that the ion gas is in hydrostatic equilibrium. Production Q and loss L of electrons could only be occurring if Q = L, but this is unlikely to occur simultaneously with (1). The ion-electron gas has half the mean molecular weight of the ion gas and so if the temperature is uniform

$$(2) \qquad N \propto \exp(-z/2H)$$

where z is the height above some fixed reference level and H is the scale height of the ions. It follows that (2) should satisfy equation (44) of the previous lecture. The student should verify this as an exercise. Note that this is also a useful check of complicated diffusion equations. Thus, what we are in fact obtaining when we use (1) is a solution of equation (44) of the previous lecture, for the case of no production (night-time) and no loss ($\beta = 0$). It follows that in investigating solutions of th full diffusion equation we first study diffusive equilibrium. There is another reason for doing this. We can include the effects of a height varying temperature field quite easily. Consider a constituent characterised by ionic mass m_i, partial pressures p_i, p_e for ions and electrons, and electric field $\underline{E}$. For equilibrium of the ions along a field line where I is th dip angle

$$\frac{dp_i}{ds} = - N m_i \; g \; \sin I \; + Ne \; E_s \; ,$$

where s denotes arc length and E_s is the component of $\underline{E}$ along a field line. For equilibrium of electrons, neglecting the electron mass,

$$\frac{dp_e}{ds} = - Ne \; E_s \; .$$

Thus, as $p_i = NkT = p_e$, where $T = T(z)$ is the temperature, z being the height, we have <u>along a field line</u>

$$\frac{dp_i}{ds} = - \frac{p_i \sin I}{2H(z)}$$

where

$$H(z) = kT(z)/m_i g$$

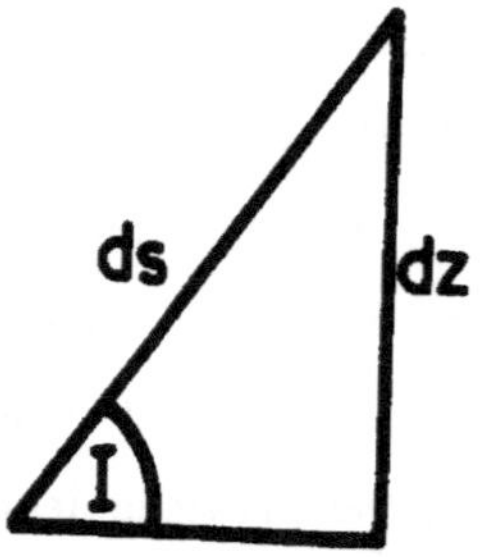

Fig. 15

But along a field line (Fig. 15)

$$dz = ds \; \sin \; I$$

Thus

$$\frac{dp_i}{dz} = - \frac{p_i}{2H(z)}$$

giving

$$p_i \propto \exp - \int \frac{dz}{2H} .$$

Thus <u>along a field line</u>

(3) $$N \propto \frac{1}{T} \exp - \int \frac{dz}{2H} .$$

If α is the latitude, the equation of a field line in spherical polar coordinates, r, θ, ϕ (θ = colatitude, ϕ = longtitude) is

(4) $$r = a \sin^2\theta = a \cos^2\alpha$$

we see that

(5) $$N = f(a) \quad g(z)$$

where f is some function of a only and

(6) $$g(z) = \frac{1}{T} \exp - \int \frac{dz}{2H} .$$

On these assumptions, (i.e. ignoring variations of H with respect to latitude) we can compare theory and experiment.

Satellite Aloutte I results* appear in the form of Fig. 16 , which shows electron density versus latitude, each curve being for a fixed height above the ground (shown in Km) .

* Taken from a private communication from Dr J.W. King

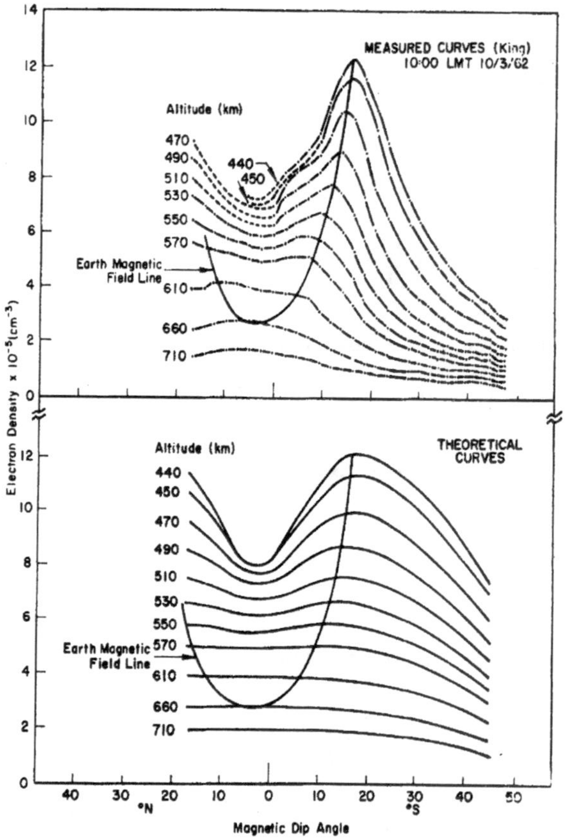

Figure 16

Figure 17

A clue as to the procedure is seen when we take K = constant and f(a) = any function with a single maximum. Then we obtain Fig. 17 (see Goldberg, Kendall and Schmerling, 1964, Journal of Geophysical Research <u>69</u> , 417-427) .

This is so similar to the previous diagram that further investigation is worthwhile. Baxter and Kendall, 1965 , Journal of Atmospheric and

Terrestrial Physics 27, 129-132 have made quantitative comparisons between theory and experiment. Thus, if the lowest of the constant height experimental curves is assumed to be given together with the equatorial profile, we can deduce all the other curves using this theory. As the experimental curves are also available this is an interesting comparison between theory and experiment. Thus, at fixed height $z = z_1$ (say) we have

$$N = N(z_1, \alpha) = \text{given function},$$

where α = latitude. Thus, putting $r_1 = r_e + z_1$, where r_e is the earth's radius and using (4), (5) gives

$$f(r_1 \sec^2\alpha) = N[z_1, \alpha]/g(z_1)$$

giving

$$f(a) = N[z_1, \cos^{-1}\sqrt{(r_1/a)}]/g(z_1).$$

We also know that at the equator

$$N = N[z, 0] = \text{given function}.$$

Thus from (5)

$$g(z) = N[z, 0]/f(r)$$

$$= N[z, 0]/f(r_e + z)$$

If follows that at a general point

$$N = f(a) \quad g(z)$$

$$= \frac{N[z_1, \cos^{-1}\sqrt{(r_1\cos^2\alpha/r)}]N[z, 0]}{N[z_1, \cos^{-1}\sqrt{(r_1/r)}]}$$

The results of the investigation are shown in Fig. 18 . The solid lines are observations, the broken lines are theory.

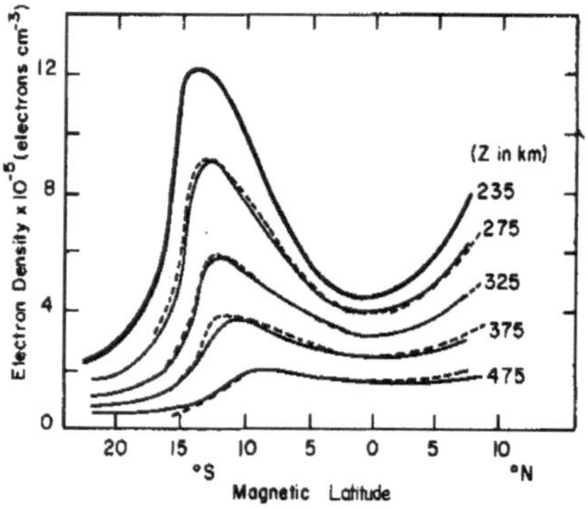

Figure 18

All we can conclude without further information on the temperature and composition is that mathematically the distribution N is almost separable in the variables z and a.

Effects of electrodynamic drift on the topside ionosphere

By expanding in a power series Baxter, Kendall and Windle, Journal of Atmospheric and Terrestrial Physics 1965 27 , 1263-1273 have studied the disturbance of this type of diffusive equilibrium by an upwards and outwards Hall drift. They find that the ionization adjusts itself so as to compensate for the Hall motion at right angles to the field lines. The diagrams will not be reproduced in these notes. The *ad hoc* methods they used to solve the diffusion equation show that there is room for mathematical work in this field.

Equilibrium solutions and the effects of electrodynamic drift

Bramley and Peart (1966) and Hanson and Moffett (1966) have both integrated the equilibrium equations with realistic production and loss terms. The word "equilibrium" is taken to mean $\partial/\partial t \equiv 0$. Thus we take

ϕ = constant and put $\partial/\partial\phi = 0$ in equation (44). Then, using obvious abbreviations

$$\frac{\partial V}{\partial a_0} = f_1(x, a_0) + f_2(x, a_0)V + f_3(x, a_0)\frac{\partial V}{\partial x} + f_3(x, a_0)\frac{\partial^2 V}{\partial x^2} ,$$

where

$$a_0 = a/H$$

It should be noted that these authors did not use the mathematical transformations of Lecture 4. The problem is clearly readily tractable numerically. The calculation is started from the field line $a_0 = (6550 \text{ km})/H$ and continued as far as may be required. The boundaries of the system are shown in Fig. 19

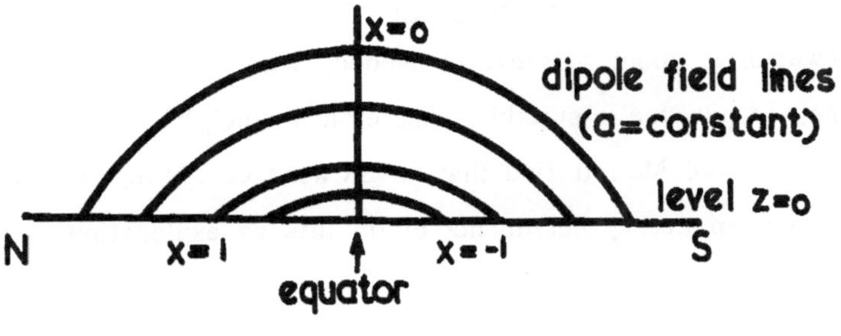

Fig. 19

The problem in the a_0,x plane is straightforward (Fig. 20)

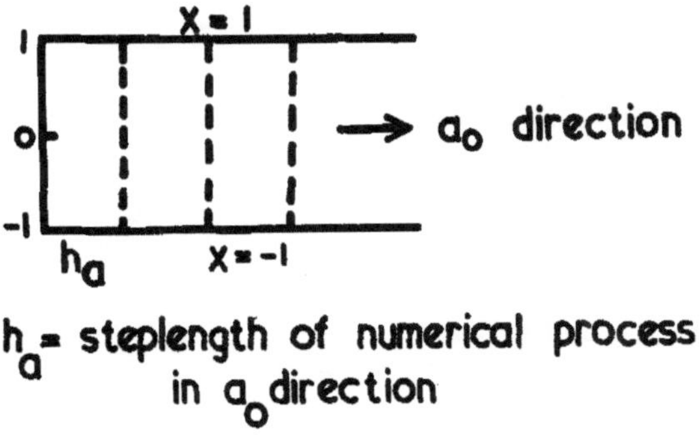

Figure 20

On the boundaries (which happen also to be $z = 0$, we put $V = - f_1(x, a_o) / f_2(x, a_o)$. The results obtained are sket ched in Fig. 21.

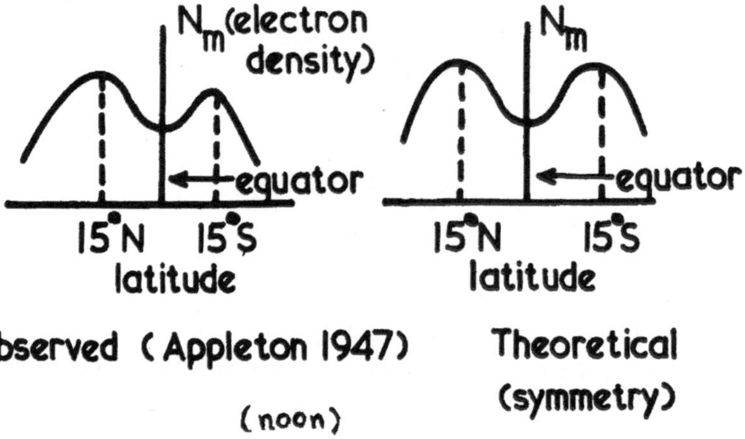

Figure 21

A drift upwards of about 10m/s is needed, with typical F2 layer parameters, to produce a suitable "Appleton anomoly" in the case of symmetry. Hanson and Moffett find that a SN wind of 60 m/s could cause the observed asymmetry, but do not claim this as explanation.

P. C. Kendall

Lecture 6

Time dependent solutions of the F2 region diffusion equation (with Dr. R.G. Baxter) Using obvious abbreviations, the full diffusion equation of Lecture 4 (equ.44) becomes approximately, if $W'H/a \ll 1$,

$$\frac{\partial V}{\partial \phi} + f(\phi)\frac{\partial V}{\partial a}$$

(1)
$$= f_1(x, a, \phi) + f_2(x, a, \phi)V + f_3(x, a, \phi)\frac{\partial V}{\partial x} + f_4(x, a, \phi)\frac{\partial^2 V}{\partial x^2}$$

The operator on the right hand side may be reduced by integrating the equation

$$\frac{da}{d\phi} = f(\phi)$$

Thus , as the integral is

$$a = a_o + \int_0^{\phi} f \, d\phi ,$$

where a_o is a constant of integration, the required transformation is

(2)
$$\phi' = \phi$$
$$a' = a - \int_0^{\phi} f \, d\phi$$

That this algorithm works can be seen at once, for

$$\frac{\partial}{\partial \phi} = \frac{\partial \phi'}{\partial \phi}\frac{\partial}{\partial \phi'} + \frac{\partial a'}{\partial \phi}\frac{\partial}{\partial a'}$$

$$= \frac{\partial}{\partial \phi'} - f\frac{\partial}{\partial a'}$$

Also

$$\frac{\partial}{\partial a} = \frac{\partial}{\partial a'}$$

Thus

$$\frac{\partial}{\partial \phi'} = \frac{\partial}{\partial \phi} + f\frac{\partial}{\partial a}$$

Using variables

$$x,\ a',\ \phi\ ,$$

so that

$$f_i(x, a, \phi) = f_i(x,\ a' + \int_0^{\phi'} f d\phi',\ \phi') \quad i = 1, 2, \dots 4,$$

equation (1) becomes

(3) $$\frac{\partial V}{\partial \phi'} = f_1 + f_2 V + f_3 \frac{\partial V}{\partial x} + f_4 \frac{\partial^2 V}{\partial x^2}$$

This is the form of equation dealt with successfully in lecture 3 by numerical methods. Mathematically we simply choose a value of a', then integrate in ϕ ' as usual, using the Crank-Nicolson method to advance

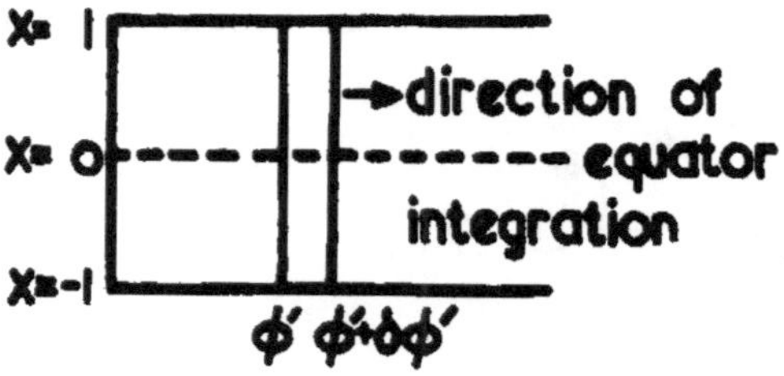

Fig. 22

the integration from ϕ' to $\phi' + \delta\phi'$ (Fig. 22) . On the boundaries we put

$$V = -f_1/f_2 \quad \text{(boundary condition)}$$

The integration is started at $\phi' = h_{\phi'}$, where $h_{\phi'}$, is the step length in the ϕ' direction. This avoids difficulty with the singularity at $\phi'=0$.

We can thus develop the values

$$V = V(x,\ a',\ \phi')$$

Hence

$$V = V(x,\ a, \phi)$$

Then the electron density is

$$N = e^{-z/2H}\ V,$$

where z denotes the height , and H is the constant scale height.

Physical explanation

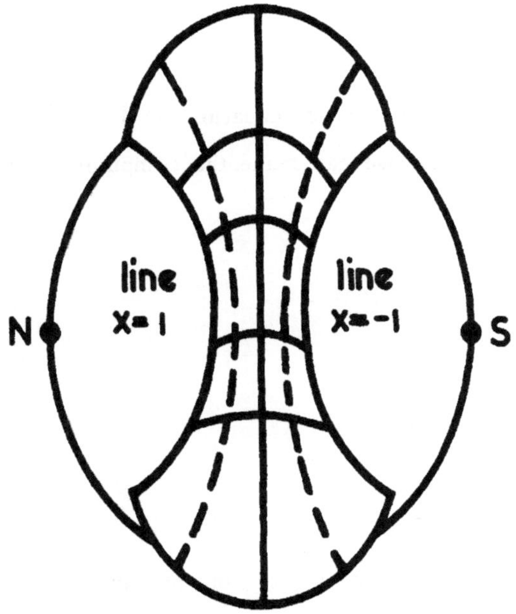

Figure 23

Fig. 23 shows how the Hall drift $\underline{E} \times \underline{B}/B^2$ combines with the earth's rotation so that a magnetic field line moves up and down and also rotates about the axis of symmetry. We have simply chosen coordinates moving with the field lines. The diagram shows the surface described by a moving field line as it passes round the earth.

Processing of results

The results appear as tables of N, the electron density, against x for a given ϕ' and a'. This immediately gives us

$$a = a' + \int_0^{\phi} f \; d\phi ,$$

so we then know which field line

$$r = a \sin^2 \theta$$

we are on. Knowing a, to each value of x equation (33) of lecture 4 gives one value of r. Thus we can make the computer print out

	x	N	r	θ
	0	-	-	-
	0.1	-	-	-
(say)	0.2	-	-	-
	...			
	1.0	-	-	-

The problem can thus be readily converted back into the original spherical polar coordinates r, θ, ϕ. (of course ϕ measures the time, because $t = \tau\phi$). As a final grace we can make the computer interpolate to $\theta = 90^o, 89^o, 88^o$ ... and also $\frac{z}{H} = 0, 0.5, 1.0$ ... This enables us to draw curves of N at fixed latitude or fixed height.

Physical data

The measurements of electron density in the ionosphere are made by ionosonde from either the ground, thus dealing with the 'bottomside' of the ionosphere, or from a satellite, thus dealing with the 'topside' ionosphere. A radio wave travelling into the ionosphere is reflected at a point where the electron density has a given value which depends on the wave frequency. Thus, in a layer with a single maximum of electron density there is a critical frequency f_oF_2 (in the ordinary mode of propagation) beyond which radio waves pass through without reflection. There is another mode of wave propagation, the extraordinary mode, with the critical frequency f_EF_2 . [See Ratcliffe's book]

Thus an analysis using the critical frequency f_oF2, from many ground based stations gives results such as those of Appleton (1947) (Fig. 24) and Martyn (1959) (Fig. 25)

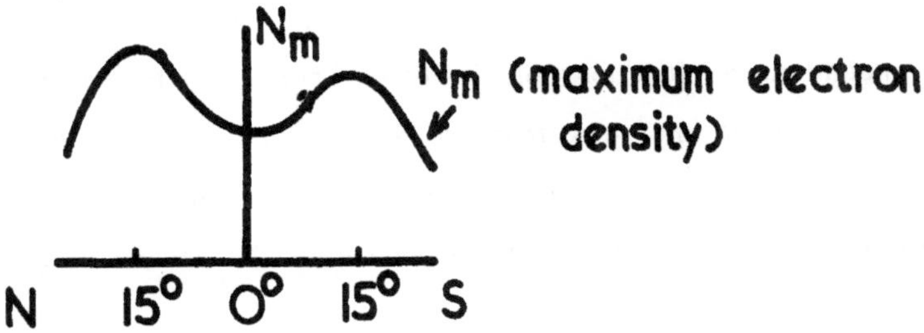

Appleton anomaly

(noon)

Figure 24

Figure 25

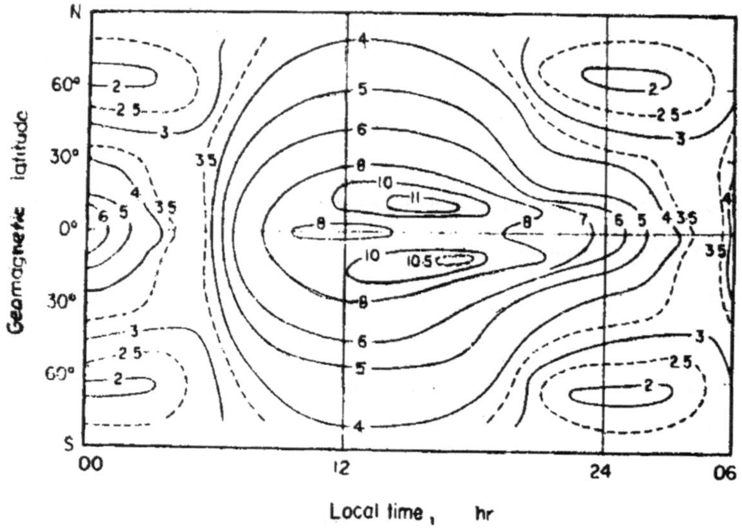

World curves of f_0F2 for the equinox, sunspot minimum year 1943–1944 (after Martyn).

In lecture 5 we saw some curves (resulting from a satellite) of N at fixed height as a function of latitude. These may be found in King et al (1964). There are now many of these results available as printed tables e.g. from the Canadian Defence Telecommunications Laboratory, Ottawa. Ground based results have been given by Croom et al for N at fixed height.

Results. Figures 26 and 27 show results for typical F2 region parameters and drift amplitudes of 7.3 m/s and 73 m/s respectively. Note the development of the Appleton anomaly. These results are only preliminary.

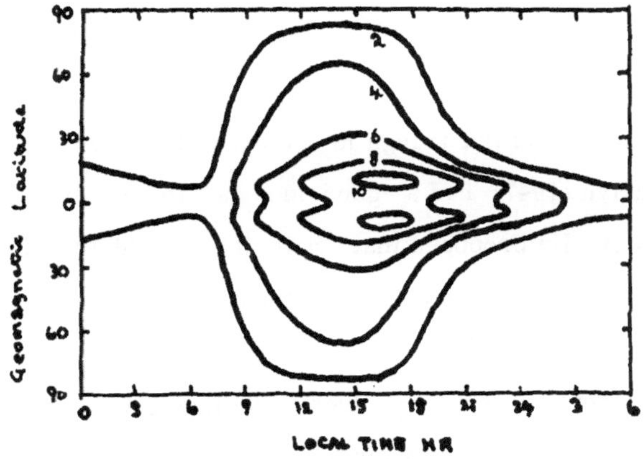

FIG 26 Nmax world curves - Isolines of Nmax $(100/\tau q_0 e)$
drift amplitude = 7.3 m/s

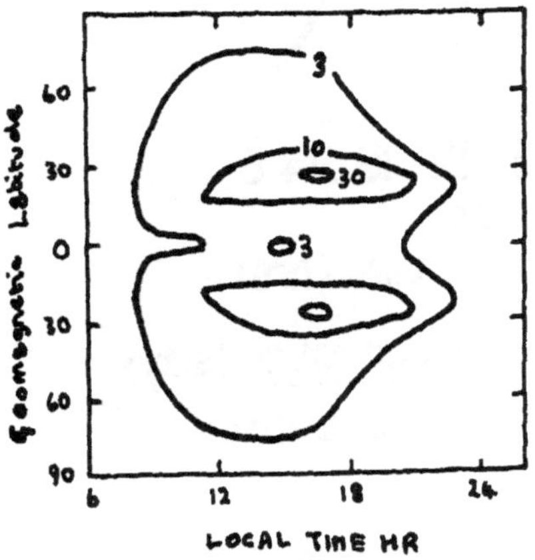

FIG 27 Nmax world curves
- Isolines of Nmax $(100/\tau q_0 e)$
drift amplitude = 73 m/s

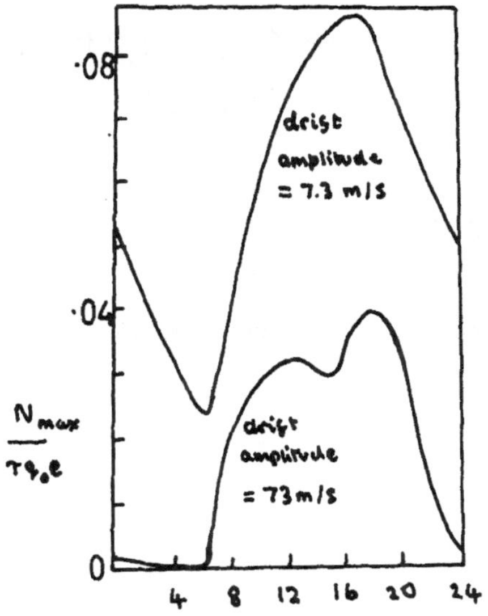

FIG 29
Diurnal variation of Nmax $/\tau q_0 e$
at equater

Clearly 73 m/s is too large, as the trough is too deep. However, the results do show that Martyn (1947) had a good idea as to the real cause of the Appleton anomaly. He proposed that Hall drift would transport ionization away from the equator, thus producing the trough.

Near the dip equator, at stations such as Huancayo , the diurnal variation of N_m has a double maximum, as sketched in Fig. 28

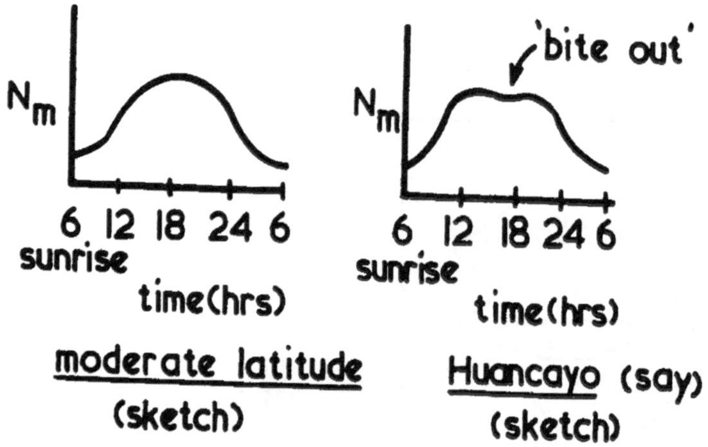

Fig. 28

This is reproduced by our theoretical results for a drift amplitude of 73 m/s as shown in Figure 29. It should be stressed that 73 m/s is too large and that these results are only preliminary.

P. C. Kendall

Lecture 7

Motions of the neutral air induced by ion-drag (Nith Mr. W. M. Pickering) The ion-drag problem Cowling (1945), Hirono (1953), Baker and Martyn (1953), Hirono and Kitamura (1956) and Dougherty (1961) have suggested that the motion of the plasma (in the F2 region) could itself set the neutral air in motion. Consider the case of moderate latitude, with a horizontally stratified atmosphere, as in lecture 2 (see Fig. 30)

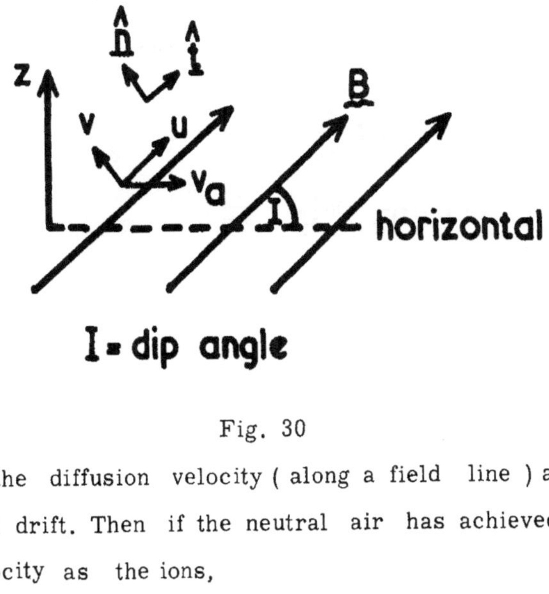

Fig. 30

Let u be the diffusion velocity (along a field line) and v be the specified Hall drift. Then if the neutral air has achieved the same horizontal velocity as the ions,

$$v_a = u \cos I - v \sin I$$

But along a field line the plasma equation of motion is

$$\frac{2NkT}{D}(u - v_a \cos I) = - \sin I \frac{\partial(p_i + p_e)}{\partial z} - Nm_i g \sin I$$

But $p_i = p_e = NkT$ (we assume) and $H = kT/m_i g$. Thus

$$u = - v \cot I - \frac{D}{N \sin I} \left(\frac{\partial N}{\partial z} + \frac{N}{2H} \right)$$

and if $\hat{\underline{t}}$ is the unit tangent along $\underline{B}$ and $\hat{\underline{n}}$ is the unit normal shown, the continuity equation becomes

$$\frac{\partial N}{\partial t} = Q - L - \underline{\nabla}.N(u\hat{\underline{t}} + v\hat{\underline{n}})$$

$$= Q - L - \hat{\underline{t}}.\underline{\nabla}(Nu) - v\,\hat{\underline{n}}.\underline{\nabla}\, N$$

But $\hat{\underline{t}} = (\cos\ I,\ \sin I)$ and $\hat{\underline{n}} = (-\sin I, \cos I)$. Thus

$$\frac{\partial N}{\partial t} = Q - L - \sin I \frac{\partial}{\partial z}(Nu) - v \cos I \frac{\partial N}{\partial z}$$

$$= Q - L - \frac{\partial}{\partial z} D\left(\frac{\partial N}{\partial z} + \frac{N}{2H}\right)$$

Thus as

$$D \propto \exp(z/H)$$

we have

$$\frac{\partial N}{\partial t} = Q - L + D\mathfrak{D}^* N,$$

where

$$\mathfrak{D}^* \equiv \left(\frac{\partial^2}{\partial z^2} + \frac{3}{2H}\frac{\partial}{\partial z} + \frac{1}{2H^2}\right)$$

Looking back to lecture 2 we there obtained

$$\mathfrak{D} \equiv \sin^2 I \ \ \mathfrak{D}^*$$

We also see that the electrodynamic (Hall) drift term v has been completely cancelled . The consequences of including neutral air coupling may thus be serious.

P. C. Kendall

The temperature gradient problem

Apart from the problem of ion-drag there is another problem. King and Kohl (1965) have suggested that as there is a temperature gradient in the upper atmosphere (see their references) it might drive a neutral wind through the pressure gradient . They showed that in the F-region the major force on the neutrals, apart from this pressure gradient, is due to their collisions with ions . It is clearly of interest to investigate the coupling between the plasma and neutral air motions. A given temperature gradient would certainly set the neutral air in motion; however, the velocity would be coupled with the ion velocity; giving rise to a coupled system of equations. If the ions were at rest, the problem of working out the neutral air velocity $\underline{v}_a$ would be relatively simple.The ions are, however, free to move along the field lines . In the F1 layer, for example, diffusion is regarded as unimportant and Geisler (1966) has shown how the problem of neutral air motion can be treated . We here consider the F2 layer where slipping between the moving plasma and the neutral atmosphere is believed to occur. The configuration in Fig. 3 will be used. The magnetic field $\underline{B}$ is taken to be horizontal. The x-axis is also horizontal. The y-axis is vertical. All variables are functions of x and y only, being, in particular, independent of time. The plane x = 0 is taken to be the equatorial plane, and there is supposed to be symmetry about this plane.

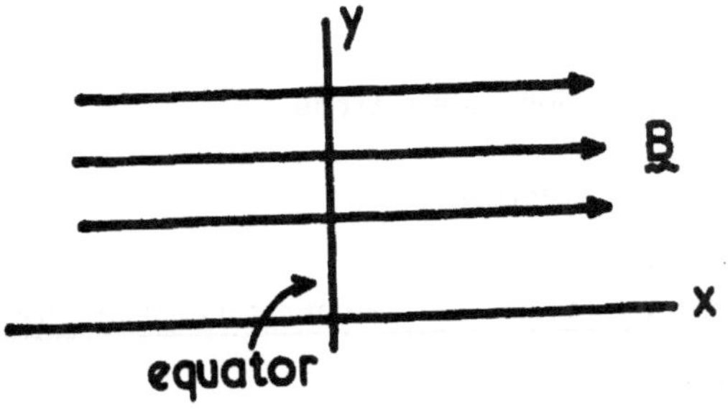

Fig. 31

Notation

n_a = equilibrium neutral air density

$= n_o \exp(-y/H)$

T = equilibrium temperature , assumed uniform

H = scale height $= kT/m_i g$

m_i = mass of ion (= mass of neutral atom)

k = Boltzmann constant

g = acceleration due to gravity

$\underline{v}_a$ = velocity of neutral air

v_{a1} = horizontal velocity of neutral air

v_{a2} = vertical velocity of neutral air

$\underline{v}_i$ = ion velocity

u = ion velocity along a field line

v = Hall drift $\perp^r$ to a field line = constant

T' = artificially maintained small temperature perturbation

D = twice coefficient of diffusion ions through neutrals

n'_a = perturbation in neutral air density

Q = rate of production of electrons

L = rate of loss of electrons

In the model we assume that all temperature gradients, velocities and their effects are small.

We also ignore the self viscosity of the neutral air.

Equations of the problem

The horizontal and vertical linearized equations of motion for the neutral air are

(1) $$\frac{2NkT}{D}(v_{a1} - u) = -\frac{\partial}{\partial x}(n'_a kT) - \frac{\partial}{\partial x}(n_a kT')$$

and

(2) $$\frac{2NkT}{D}(v_{a2} - v) = -\frac{\partial}{\partial y}(n_a' kT) - \frac{\partial}{\partial y}(n_a kT') - m_i n_a' g$$

The linearized horizontal equation of motion of the plasma is

(3) $$\frac{2NkT}{D}(u - v_{a1}) = -\frac{\partial}{\partial x}(2NkT)$$

We note that there is no vertical equation of motion of the plasma, as it moves vertically with the given Hall drift v, assumed constant. Adding (1) and (3) gives

$$\frac{\partial}{\partial x}(n'_a kT + 2NkT + n_a kT') = 0$$

Thus

$$n'_a + 2N + n_a T'/T = F(y) \quad ,$$

where F is an arbitrary function of y. We shall assume that

$$n'_a \to 0 \quad N \to 0 \quad \text{and} \quad T' \to 0 \quad \text{as} \quad x \to \infty \; ,$$

giving

(4) $$n'_a + 2N + n_a T'/T = 0$$

Substituting back we obtain

(5) $$v_{a1} = u + \frac{D}{N}\frac{\partial N}{\partial x}$$

(6) $$v_{a2} = v + \frac{D}{N}\frac{\partial N}{\partial y} + \frac{D}{H} + \frac{Dn_a T'}{2HNT}$$

Assuming that $N \ll n_a$, so that production and loss of neutral air molecules are negligible, the linearized continuity equation for the neutral air is

$$\operatorname{div} n_a \underline{v}_a = 0$$

Thus, substituting for $\underline{v}_a$ from (5) and (6),

$$\frac{\partial}{\partial x}(n_a u) + \frac{\partial}{\partial x}\left(\frac{n_a D}{N}\frac{\partial N}{\partial x}\right) + \frac{\partial}{\partial y}\left(\frac{n_a D}{N}\frac{\partial N}{\partial y}\right) + \frac{\partial}{\partial y}\left(\frac{n_a D}{H}\right)$$

$$+ \frac{\partial}{\partial y}\left(\frac{D n_a^* T'}{2HNT}\right) + \frac{\partial}{\partial y}(n_a v) = 0$$

Whence, under conditions of symmetry ($u = 0$ at $x = 0$), and assuming that

$$D \propto n_a^{-1}$$

We obtain an expression for the horizontal velocity of the ions, namely,

(7) $$u = -D\int_0^x \nabla^2 \log N \, dx + \frac{vx}{H} - D\int_0^x \frac{\partial}{\partial y}\left(\frac{n_a T'}{2HNT}\right) dx$$

This enables us to form the diffusion equation from the continuity equation for electrons, $Q - L = \operatorname{div} (N\underline{v}_i)$. Thus

$$0 = Q - L + \frac{\partial}{\partial x} ND \int_0^x \nabla^2 \log N \, dx - v\left(\frac{\partial N}{\partial y} + \frac{1}{H}\frac{\partial}{\partial x}(xN)\right)$$

(8) $$+ \frac{\partial}{\partial x} ND \int_0^x \frac{\partial}{\partial y}\left(\frac{n_a T'}{2HNT}\right) dx$$

Thus, even from a very simple model we have produced a <u>non-linear</u> diffusion equation.

3. Conclusion

It is of interest to compare equation (8) with the the simpler diffusion equation which might be obtained by making earlier approximations. For example, if we had argued that for the purpose of calculating $\underline{v}_a$ the ions might as well be assumed stationary, and the pressure gradient might as well be assumed to be $(\partial/\partial x)(n_a kT')$, we would have obtained

(9) $$v_{\alpha 1} = -\frac{D}{2N}\frac{\partial}{\partial x}\left(\frac{n_a T'}{T}\right) .$$

Then the diffusion equation would have become

(10) $$0 = Q - L + D\frac{\partial^2 N}{\partial x^2} - v\frac{\partial N}{\partial y} + \frac{\partial^2}{\partial x^2}\left(\frac{D n_a T'}{2T}\right)$$

This equation is linear, and has terms corresponding in turn to each term in (8). It is, nevertheless, incorrect. We note, however, that a temperature which is higher at the equator than alsewhere would give rise to terms of the same sign in (7) and (9), corresponding to transport of plasma away from the equator.

We conclude that although current F2 layer theory looks promising there are still questions to be answered concerning the coupling between the plasma and neutral air motions.

REFERENCES

Appleton, E.V. 1946 Nature , 157 , 691

Baker, W.G. and Martyn , D.F. 1953, Phil . Trans. Roy. Soc. A 246, 281

Bramley, E.N. and Peart, Margaret. 1965a, Nature 206 , 1245 and 1965b J. Atmos. Terr. Phys. 27 , 1201

Cowling, T.G. 1945, Proc. Roy. Soc. A 183, 453

Croom, Sheila, Robbins, Audrey and Thomas, J.O. 1959 , Nature , Lond. 184, 2003 also 185, 902 (1960)

Dougherty, J.P. 1961 , J. Atmos. Terr. Phys. 20 , 167

Ferraro, V.C.A. and Özdogan L. 1958, J. Atmos. Terr. Phys. 12, 140

Geisler, J.E. 1966, J. Atmos. Terr. Phys. 28 , 703

Gliddon, J.E.C. 1959a, b.Quart.Jour.Mech.& Appl.Math. 12, 340 and 12 , 347

Hanson, W.B. and Moffett, R.J. 1966 , J.Geophys. Res. 71 , 5559

Hirono, M. 1953 , J. Geomag. Geoelect. 5 , 22

Hirono, M. and Kitamura, T. 1956 , J. Geomag. Geoelect. 8 , 9

King, J.W. and Kohl, H. 1965, Nature, 206, 699

King, J.W.-et al. Proc. Roy . Soc. Lond. (A) 281, 464, 1964

Kunz, K.S. 1957, Numerical Analysis, McGraw Hill

Martyn, D.F. Proc. Roy .Soc. Lond. (A) 189 , 241, 1947

Martyn, D.F. Proc. Inst. Radio Eng.N.Y. , 47, 147, 1959

Moffett, R.J. and Hanson, W. B. 1965, Nature 206 , 705

Nawrocki, P. J. and Papa, R. Atmospheric Processes 1961 Geophysics Corporation of America

Ratcliffe, J.A. Magnetoionic Theory, Cambridge

Ralston, A. and Wilf, H.S. 1964 , Mathematical Methods for Digital Computers, Wiley

Richtmyer, R.D. 1957 , Difference Methods for Initial Value Problems, Interscience

Rishbeth, H. and Barron, D.W. 1960 , J. Atmos. Terr. Phys. 18 , 234

Smith, G.D. 1965, Numerical Solution of Partial Differential Equations, Oxford University Press

CENTRO INTERNAZIONALE MATEMATICO ESTIVO

(C.I.M.E.)

F. HENIN

"KINETIC EQUATIONS AND BROWNIAN MOTION"

Corso tenuto a Varenna (Como) dal 19 al 27 settembre 1966

KINETIC EQUATIONS AND BROWNIAN MOTION

by

F. Henin

(Université libre de Bruxelles)

Introduction.

In this series of lectures, we shall deal mainly with the microscopic theory of brownian motion.

Brownian motion owes its name to an English botanist, Robert Brown, who noticed in 1827 the fact that small particles suspended in fluids perform peculiarly erratic movements. The origin of this phenomenon is of course quite simple: we are dealing with a manifestation of the molecular motion.

The first satisfactory theory of brownian motion was produced by Einstein in 1905 who derived the diffusion equation. This result has been particularly important because the expression he obtained for the diffusion coefficient D allowed a determination of Avogadro's number N by Perrin.

Thereafter, the phenomenological theory has been widely developed. A very good presentation of the ideas used can be found in a review paper by Chandrasekhar [1]. We shall give a brief summary of these ideas in chapter I. The starting point is the Langevin equation which introduces as basic assumption the fact that the interactions of the particle with the medium have a twofold effect; first, an overall dynamical friction, then a fluctuating force. Intuitive assumptions about the statistical properties of this fluctuating force lead to the Fokker-Planck equation for the time evolution of the probability distribution of finding the particle at a given point in space with a given velocity. One of the most interesting features of the Fokker-Planck equation is that it is an irreversible equation; it predicts an irreversible evolution towards an equilibrium distribution.

The stochastic theory has been widely used and proved successful in the study of a great variety of phenomena. Neverthele , it requires a good deal of intuition to reach a phenomenological description of the effect of the medium on the particle. However, intuition can sometimes be misleading ; therefore, there has been much effort to understand this phenomenon on a microscopic level. Such an effort can be rewarding in several ways ; it will clarify the conditions under which the phenomenological theory will be valid, it will give us the phenomenological constants in terms of molecular parameters and finally, it may be hoped that it will show us the path to follow when the conditions for the validity of the phenomenological theory are not fulfilled.

An understanding of the brownian motion on a microscopic level necessarily requires the consideration of an N-body system. It is quite obvious that the detailed description provided by the laws of mechanics cannot be used directly and that one must resort to the methods of statistical mechanics. A most useful concept is the idea of an ensemble introduced by Gibbs. In classical statistical mechanics , such an ensemble is characterized by the N-particle distribution function which obeys the Liouville equation.

The Liouville equation has been the starting point for the study of non equilibrium many-body systems by Prigogine and his coworkers. An extensive presentation of the basic ideas can be found in the monographs by Prigogine [2], Balescu [3] and Résibois [4]. This method emphasizes strongly the role played by the correlations in the evolution of the distribution function. We really deal with a "dynamics of correlations". This formalism is particularly well suited to take account of the characteristic features of macroscopic systems : large number of degrees of freedom N, large volume Ω , finite concentration ; these features allow the consideration of the asymptotic case :

(1) $$N \longrightarrow \infty; \ \Omega \longrightarrow \infty, \ N/\Omega = C \ \text{finite}$$

which brings in several important simplifications.

Moreover, in general, we are interested in the asymptotic behavior in time of the system. Then , it can be shown that in many cases, the behavior of the system can be correctly described by the so called kinetic equations. (A simple example of kinetic equation is the Boltzmann equation for dilute gases) . The derivation of the kinetic equation for the velocity distribution function will be discussed in chaper II.

The kinetic equation is an irreversible equation : systems for which such an equation holds tend asymptotically to an equilibrium distribution , which is a function of the hamiltonian only.

Once we have equations for the description of the asymptotic behavior of the N-body system, we can introduce the special features of the brownian motion problem. There, we are interested in the motion of a single particle in a surrounding fluid. The simplest case will of course be that of a particle moving in a fluid at equilibrium . This is in fact the problem which , in microscopic theories, is often referred to as the brownian motion or test particle problem. The assumption that the fluid is at equilibrium introduces an enormous simplification in the kinetic equation: all the particles no longer play the same role. All of them, but one, are in the equilibrium state (strictly speaking, the fact that one particle is out of equilibrium prevents the others to stay in the equilibrium state; however, this departure from the equilibrium state is of order N^{-1} and can be neglected) .

There are two cases where the kinetic equation, particularized to the brownian motion problem, can be shown to lead to a Fokker-Planck equation. The simplest case is that of weakly interacting systems which will be discussed in chapter III. There, a Fokker-Planck equation is obtained whatever the mass of the brownian particle. The other case is that of brownian motion in systems interacting through short range forces where the brownian particle is much heavier that the particles of the fluid. This problem, in the absence of any external force, will be studied in chapter IV . In chapter V , we shall generalize it to the case where the brownian particle is charged and acted upon by a constant external electrical field . The interest of all these problems not only lie in the fact that they allow us to state the conditions of validity of the Fokker-Planck equation. They also enable us to obtain expressions of the diffusion coefficient which enters into the Fokker-Planck equation in therms of microscopic quantities . Moreover, they show us the way to obtain corrections to the Fokker-Planck equation when required . This will be briefly discussed in Chapter V.

So far, we have only considered classical systems. The same ideas can be extended to quantum mechanics, as we shall show in chapter VI . Here contact can be made with the results of recent experiments on the mobility of heavy ions in liquid He^4 and He^3 .

All the work which will be described in chapters III to VI concerns one special class of brownian motion: that of a particle moving in a fluid at thermal equilibrium (in chapters IV to VI , the brownian particle is supposed to be much heavier than the particles of the fluid) . Less specialized situations could of course be considered. We could for instance consider the motion of a test particle in a medium which is not at equilibrium . In this case, the problem is much less simple; we can no longer obtain a single closed equation for the distribution function of the test particle . However, in all systems where the kinetic equation is asymptotically valid, the basic features are preserved. The distribution function of the particle will obey an irreversible equation. Whenever, in the brownian motion in a fluid at equilibrium, we can derive a Fokker-Planck equation, the same kind of equation can be obtained if the fluid is out of equilibrium but the coefficients appearing in the equation will be functionals of the state of the system. A simple example of this is given in chapter III for the case of weakly coupled systems. More details can be found in a paper by Balescu and Soulet [5).

However, there are systems where the kinetic equation is not valid, even asymptotically. An important case is that of systems interacting through gravitational forces. For such systems, an entirely new approach seems necessary. We first have to derive an equation which will, in this case, play the same role as the kinetic equation for systems with short range interactions. In the last chapter (VII) we shall briefly describe a recent attempt by Prigogine and Severne [6) to obtain such an equation. This equation predicts a behavior which differs in many important aspects from the behavior predicted by the kinetic

REFERENCES

1. S. Chandrasekhar, Revs. Mod. Phys. 15, 1(1943) reprinted in "Noise and Stochastic Processes", ed N. Wax, Dover Publications, New York (1954)

2. I. Prigogine, Non equilibrium Statistical Mechanics, Interscience, New-York (1962)

3. R. Balescu, Statistical Mechanics of Charged Particles, Interscience, New-York (1963)

4. P. Résibois, in "Many Particle Physics", ed. E. Meeron, Gordon and Breach (to appear)

5. R. Balescu and Y. Soulet, J. Phys. 26 , 49 (1965)

6. I . Prigogine and G. Severne, Physica (to appear)

F. Henin

I

I. STOCHASTIC THEORY

I.I. Introduction

The stochastic theory does not make any attempt to describe in detail the interactions between the brownian particle and the particles of the fluid. Rather, it describes the effect of the medium on the heavy particle in a phenomenological way. From the beginning, one assumes that the influence of the medium on the particle can be split into two parts.

First, we have a systematic friction effect. Secondly, we have to account for the random motion. This is done by assuming that the medium exerts a fluctuating force on the particle. It is obvious that this force is not known exactly and that the best thing we can do is to make guesses about its statistical properties. The main question will then be: given the statistical properties of the fluctuating force, what is the probability that, if the brownian particle at $t = 0$ is at the point $\underset{\sim}{r}_0$ with velocity $\underset{\sim}{u}_0$, it will be at time t at the point $\underset{\sim}{r}$ with velocity $\underset{\sim}{u}$? The assumptions of the stochastic theory lead to the Fokker-Planck equation for this probability distribution.

We shall first discuss the assumptions which lead to the Langevin's equation of motion for the heavy particle (§ 2). Then we shall make some further assumptions about the statistical properties of the fluctuating force (§ 3) which will enable us to write down the probability distribution in velocity space (§ 4).
We shall then show how the problem of finding this distribution function can be reduced to the solution of a differential equation (§ 5), the Fokker-Planck equation in velocity space. The Fokker-Planck equation for the complete distribution function in phase space, with or without an

external field acting on the particle , will then be obtained by means of an easy generalization of the previous problem (§ 6) . Finally, we shall consider the case of an inhomogeneous system where the density gradient is small over distances of the order of the mean free path. Then, we shall see that for times much longer than the relaxation time, the spa-- tial distribution obeys a diffusion equation.

All this discussion will follow quite closely the excellent review paper by Chandrasekhar [1)]. An extensive bibliography can be found there.

I. 2. Langevin equation

The first step in the stochastic theory is to write down an e- quation of motion for the heavy particle. From the beginning , one assu- mes that the influence of the medium leads :

1. to a systematic slowing down effect ; the friction coefficient β is assumed to be independent of the velocity of the heavy particle . Usual- ly, one also assumes that it is given by Stokes' law. For a sperical particle of mass M and radius a, we then have :

$$\beta = \frac{6 \pi a \eta}{M} \tag{I.2.1}$$

where η is the viscosity of the fluid.

2. to the random motion of the particle; to account for this, we assume that, besides the dynamical friction, the medium exerts a a fluctuating force $\underset{\sim}{A}(t)$ on the particle. This fluctuating force is as- sumed to depend only on the time t. It is of course not known but

plausible assumtions can be made about its statistical properties Comparison with experiment will have to decide a posteriori of the validity of these assumptions.

If we now suppose that these two effects are additive, the motion of a brownian particle in the absence of an external field of force is given by the Langevin equation :

$$\frac{d\underset{\sim}{u}}{dt} = -\beta \underset{\sim}{u} + \underset{\sim}{A}(t) \tag{I.2.2}$$

where $\underset{\sim}{u}$ is the velocity of the particle :

$$\underset{\sim}{u} = \frac{d\underset{\sim}{r}}{dt} \tag{I.2.3}$$

$\underset{\sim}{r}$ being its position.

If an external field of force $\underset{\sim}{K}(\underset{\sim}{r}, t)$ acts on the particle, its effect has to be included in the equation of motion. This means that (I.2.1) has to be replaced by :

$$\frac{d\underset{\sim}{u}}{dt} = -\beta \underset{\sim}{u} + \underset{\sim}{K}(\underset{\sim}{r}, t) + \underset{\sim}{A}(t) \tag{I.2.4}$$

We may notice an important feature of the Langevin equation . The motion of the particle at time t is entirely independent of its motion at previous times. Whatever happened to the particle in the past does not matter to determine its future behavior at t+dt. This clearly corresponds to the assumption that the collisions between the brownian particle and the particles of the fluid are instantaneous.

I. 3. Statistical properties of the fluctuating force

Our next problem is now to specify the statistical properties of the fluctuating force $\underset{\sim}{A}(t)$. Of course, in the framework of a phenomenological theory, this amounts to the introduction of a certain number of a priori assumptions. These assumptions will be based on a very intuitive feeling of the phenomenon of brownian motion. Their justification and limitations certainly require a description on a microscopic level of the whole system.

First of all, we know empirically that the characteristic time for the variation of the macroscopic quantities (i.e. the quantities which we measure, as for instance the mean velocity) is much longer than the time interval between two successive collisions of the Brownian particle with particles of the fluid (which is of the order of 10^{-21} sec. in a normal liquid). Therefore, we shall assume that we can always find time intervals Δt such that during Δt macroscopic quantities change by a negligible amount :

$$\frac{\langle \underset{\sim}{u}(t+\Delta t)\rangle - \langle \underset{\sim}{u}(t)\rangle}{\langle \underset{\sim}{u}(t)\rangle} \lll 1 \tag{I.3.1}$$

while $\underset{\sim}{A}(t)$ undergoes a large number of fluctuations, such that $\underset{\sim}{A}(t+\Delta t)$ and $\underset{\sim}{A}(t)$ are completely uncorrelated. This assumption is quite reasonable if we take into account the fact that the brownian particle is much heavier than the particles of the surrounding fluid. Then, during the collisions with the fluid particles, the velocity of the brownian particle changes by a very small amount. During Δt, the net acceleration suffered by the brownian particle because of the action of the fluctuating force will be :

$$\text{(I.3.2)} \qquad \underset{\sim}{B}(\Delta t) = \int_t^{t+\Delta t} d\xi \, \underset{\sim}{A}(\xi)$$

We assume that this net acceleration depends only on the time interval Δt and not on the time t at which we start to compute it, i.e. we again neglect memory effects.

We shall now make an assumption about the probability of occurence of different values for $\underset{\sim}{B}(\Delta t)$. The net acceleration in Δt is due to the superposition of a large number of random accelerations. This is very much analogous to the situation one encounters when discussing random flight problems. There, one looks for the distribution function of the increment $\Delta \underset{\sim}{R}$ during Δt in the position of a particle which has performed a large number of random steps. If each displacement is governed by a probability distribution $\tau(|\underset{\sim}{r}|^2)$ which is spherically symmetric, one shows that : (see appendix I)

$$\text{(I.3.3)} \qquad W(\Delta \underset{\sim}{R}; \Delta t) = (4\pi D \Delta t)^{-3/2} \exp\left[-|\Delta \underset{\sim}{R}|^2 / 4 D \Delta t\right]$$

where D is the diffusion coefficient which depends on the average length of the step and on the time interval between steps (see A.I.1.17)Using the analogy between these problems, we shall assume that the probability distribution for $\underset{\sim}{B}(\Delta t)$ is given by :

$$\text{(I.3.4)} \qquad W\left[\underset{\sim}{B}(\Delta t)\right] = (4\pi q \Delta t)^{-3/2} \exp\left[-|\underset{\sim}{B}(\Delta t)|^2 / 4q\Delta t\right]$$

where q is a constant . The specification of this constant requires some additional assumptions about the equilibrium properties of the velocity distribution function (see § 4)

I.4. - Velocity distribution function.

Before dealing with the general problem of finding the complete probability distribution in phase space $W(\underset{\sim}{r}, \underset{\sim}{u}, t \mid \underset{\sim}{r}_o, \underset{\sim}{u}_o, o)$ to find the particle at $\underset{\sim}{r}$ with velocity $\underset{\sim}{u}$ at time t given the initial condition $\underset{\sim}{r}_o$, $\underset{\sim}{u}_o$, we shall consider a simpler problem. We shall try to find the probability $W(\underset{\sim}{u}, t \mid \underset{\sim}{u}_o)$ that the particle has a velocity $\underset{\sim}{u}$ at time t if its initial velocity is $\underset{\sim}{u}_o$.

The formal solution of the Langevin's equation is :

$$\underset{\sim}{u} - \underset{\sim}{u}_o \, e^{-\beta t} = e^{-\beta t} \int_0^t e^{\beta \xi} \underset{\sim}{A}(\xi)\, d\xi \tag{I.4.1}$$

Both sides of (I.4.1) have the same probability distribution. Now, if

$$\underset{\sim}{\alpha} = e^{-\beta t} \int_0^t e^{\beta \xi} \underset{\sim}{A}(\xi)\, d\xi \tag{I.4.5}$$

we may also write :

$$\underset{\sim}{\alpha} = \sum_{j=1}^{N} \exp\left[-\beta(t - j\,\Delta t)\right] \underset{\sim}{B}(\Delta t) = \sum_{j=1}^{N} \underset{\sim}{\alpha}_j \tag{I.4.3}$$

if we divide the interval (0, t) into N intervals Δt where Δt is of the kind defined above (i.e. such that $\underset{\sim}{A}$ suffers a large number of fluctuations while all other quantities, such as $e^{-\beta t}$, remain practically constant).

With our assumption (I.3.4) about the probability distribution of $\underset{\sim}{B}(\Delta t)$, i.e. of $\underset{\sim}{\alpha}_j$, we can, using the theory of random flights, obtain the distribution function of $\underset{\sim}{\alpha}$ (see appendix A. I. 2) :

$$W(\underset{\sim}{\alpha}) = \left[4\pi q \int_0^t d\xi \; e^{2(\xi - t)} \right]^{-3/2} \times$$

(I.4.4)

$$\times \quad \exp\left\{ - |\underset{\sim}{\alpha}|^2 \Big/ 4q \int_0^t d\xi \; e^{-2\beta(t-\xi)} \right\}$$

Therefore, the velocity distribution function is :

$$W(\underset{\sim}{u}, t \,;\, \underset{\sim}{u}_o) = \left[2\pi \frac{q}{\beta} (1 - e^{-2\beta t}) \right]^{-3/2} \times$$

(I.4.5)

$$\times \quad \exp\left\{ - \frac{\beta}{2q} \frac{|\underset{\sim}{u} - \underset{\sim}{u}_o e^{-\beta t}|^2}{1 - e^{-2\beta t}} \right\}$$

For long times ($\beta t \gg 1$), we obtain asymptotically :

(I.4.6) $$W(\underset{\sim}{u}, t \to \infty; \underset{\sim}{u}_o) = (2\pi q/\beta)^{-3/2} \quad \exp(-\beta u^2/2q)$$

Therefore, we have an irreversible evolution towards a gaussian distribution, independent of $\underset{\sim}{u}_o$. The system has forgotten its initial condition. A priori, nothing implies that this asymptotic distribution is the Maxwell-Boltzmann equilibrium distribution . If we add this condition as a further requirement, we must choose the diffusion coefficient in velocity space to be :

(I.4.7) $$q = kT\beta / M$$

Therefore, with the following set of assumptions :

1. Langevin equation
2. characteristic time for the variation of $\underset{\sim}{A}(t)$ much smaller that the characteristic time for the variation of macroscopic quantities
3. net acceleration between t and $t + \Delta t$ depends on Δt only

4. distribution function for the net acceleration during Δt is gaussian
5. asymptotic distribution for the velocity is Maxwell-Boltzmann distribution

the distribution function in velocity space for the brownian particle is is completely determined by (I. 4. 5) and (I. 4. 7)

I. 5. - Fokker-Planck equation in velocity space.

So far, we have obtained the probability distribution function corresponding to a well defined initial condition: at t=0, the velocity is $\underset{\sim}{u}_0$. To do this, we have introduced quite specific assumptions about the statistical properties of the fluctuating force. We shall now show that the problem of finding the distribution function can be reduced to the solution of a partial differential equation, the Fokker-Planck equation. In fact, this method will require less restrictive assumptions about the properties of the fluctuating force. When the same assumptions as above are made, the general Fokker-Planck equation takes a simple form and its solution reduces to (I. 4. 5) . Another interesting feature of this method is that when further restrictions on the problem are imposed, they can be expressed as boundary conditions for the solution of the Fokker-Planck equation. Also, this equation will appear as the most adequate tool for the comparison with the results of the microscopic theory.

Again , we assume the existence of time intervals Δt such that macroscopic quantities do not vary very much during these time intervals whereas the fluctuating force has changed several times.

If we consider brownian motion as a Markoff process, i.e. if we assume that the course which the brownian particle will take is entirely independent of its past history, we expect that the probability distribution function $W(\underset{\sim}{u}, t + \Delta t)$ will satisfy the following integral equation :

$$W(\underset{\sim}{u}, t+\Delta t) = \int d(\Delta \underset{\sim}{u}) W(\underset{\sim}{u} - \Delta \underset{\sim}{u}, t) \Psi(\underset{\sim}{u} - \Delta \underset{\sim}{u}; \Delta \underset{\sim}{u}) \tag{I.5.1}$$

where $\Psi(\underset{\sim}{u}; \Delta \underset{\sim}{u})$ is the transition probability for a velocity increase $\Delta \underset{\sim}{u}$ in Δt.

Let us now expand the lhs in a power series of Δt and the integrand in the rhs in a power series of $\Delta \underset{\sim}{u}$:

$$W(\underset{\sim}{u}, t) + \frac{\partial W}{\partial t} \Delta t + 0(\Delta t)^2 =$$

$$\int d(\Delta \underset{\sim}{u}) \left\{ W(\underset{\sim}{u}, t) - \frac{\partial W}{\partial u_i} \Delta u_i + \frac{1}{2} \frac{\partial^2 W}{\partial u_i \partial u_j} \Delta u_i \Delta u_j + \dots \right\} X \tag{I.5.2}$$

$$X \left\{ \Psi(\underset{\sim}{u}; \Delta \underset{\sim}{u}) - \frac{\partial \Psi}{\partial u_i} \Delta u_i + \frac{1}{2} \frac{\partial^2 \Psi}{\partial u_i \partial u_j} \Delta u_i \Delta u_j + \dots \right\}$$

With the notation :

$$\langle \alpha \rangle = \int d(\Delta \underset{\sim}{u}) \alpha \Psi(\underset{\sim}{u}; \Delta \underset{\sim}{u}) \tag{I.5.3}$$

this equation can be rewritten :

$$\frac{\partial W}{\partial t} \Delta t + 0(\Delta t)^2 = - \frac{\partial}{\partial u_i} \left[W \langle \Delta u_i \rangle \right] +$$

$$\frac{1}{2} \frac{\partial^2}{\partial u_i \partial u_j} \left[W \langle \Delta u_i \Delta u_j \rangle \right] + 0(\langle \Delta u_i \Delta u_j \Delta u_k \rangle). \tag{I.5.4}$$

Taking into account the fact that in the Langevin equation, all systematic effects are accounted for in the friction term and that the flucuating force is random, we have :

$$\left\langle \int_0^{\Delta t} d\tau \, \underset{\sim}{A}(\tau) \right\rangle = \underset{\sim}{0} \qquad \text{(I.5.5)}$$

$$\left\langle \int_0^{\Delta t} d\tau \int_0^{\Delta t} d\tau' \, A_i(\tau) \, A_j(\tau') \right\rangle$$

$$\sim \int_0^{\Delta t} d\tau \int_0^{\Delta t} d\tau' \, \delta(\tau - \tau') \sim \Delta t \qquad \text{(I.5.6)}$$

Therefore, if we take the limit $\Delta t \to 0$, we obtain the general Fokker-Planck equation for the velocity distribution function :

$$\frac{\partial W}{\partial t} = -\frac{\partial}{\partial u_i}\left[\frac{\langle \Delta u_i \rangle}{\Delta t} W\right] + \frac{1}{2} \frac{\partial^2}{\partial u_i \partial u_j}\left[\frac{\langle \Delta u_i \, \Delta u_j \rangle}{\Delta t} W\right] \qquad \text{(I.5.7)}$$

From the Langevin equation, we have :

$$\Delta \underset{\sim}{u} = -\beta \underset{\sim}{u} \, \Delta t + \underset{\sim}{B}(\Delta t) \qquad \text{(I.5.8)}$$

If we further assume that the probability distribution for the net acceleration $\underset{\sim}{B}(\Delta t)$ due to the fluctuating force is given by (I.3.4) and that the asymptotic distribution must be the Maxwell-Boltzmann distribution the transition probability becomes :

$$\psi(\underset{\sim}{u} ; \Delta \underset{\sim}{u}) = (4\pi \beta kT \, \Delta t / M)^{-3/2} \times$$
$$\times \exp\left[-M \left| \Delta \underset{\sim}{u} + \beta \underset{\sim}{u} \, \Delta t \right|^2 / 4\beta kT \, \Delta t\right] \qquad \text{(I.5.9)}$$

Then , we have :

$$\langle \Delta u_i \rangle = -\beta u_i \, \Delta t$$
$$\langle \Delta u_i \, \Delta u_j \rangle = (2\beta kT/M)\, \delta_{i,j} + 0(\Delta t)^2 \qquad \text{(I.5.10)}$$

and we obtain the special form of the Fokker-Planck equation :

(I.5.11) $$\frac{\partial W}{\partial t} = \beta \left[\frac{\partial W u_i}{\partial u_i} + \frac{kT}{M} \frac{\partial^2 W}{\partial u_i^2} \right]$$

One verifies easily that its fundamental solution, i.e. the solution which reduces at t=0 to a delta function :

(I.5.12) $$W(\underset{\sim}{u}, 0 ; \underset{\sim}{u}_o) = \delta(\underset{\sim}{u} - \underset{\sim}{u}_o)$$

is given by (I.4.5) . From it , it is of course trivial to derive the solution corresponding to an arbitrary initial distribution.

I.6.-Fokker-Planck equation in phase space.

The above procedure can be generalized to find an equation for the complete distribution function $W(\underset{\sim}{r}, \underset{\sim}{u}, t)$ in phase space . Instead of (I.5.1) , we have now :

(I.6.1) $$W(\underset{\sim}{r}, \underset{\sim}{u}, t + \Delta t) = \iint W(\underset{\sim}{r} - \Delta \underset{\sim}{r}, \underset{\sim}{u} - \Delta \underset{\sim}{u}, t) \psi(\underset{\sim}{r} - \Delta \underset{\sim}{r}, \underset{\sim}{u} - \Delta \underset{\sim}{u} ; \Delta \underset{\sim}{r}, \Delta \underset{\sim}{u}) \, d(\Delta \underset{\sim}{r}) \, d(\Delta \underset{\sim}{u})$$

From the Langevin equation (we directly consider the case where an external force is present), we obtain :

(I.6.2) $$\Delta \underset{\sim}{r} = \underset{\sim}{u} \Delta t$$
$$\Delta \underset{\sim}{u} = -(\beta \underset{\sim}{u} - \underset{\sim}{K}) \Delta t + \underset{\sim}{B}(\Delta t)$$

Therefore, we have :

(I.6.3) $$\psi(\underset{\sim}{r}, \underset{\sim}{u} ; \Delta \underset{\sim}{r}, \Delta \underset{\sim}{u}) = \psi(\underset{\sim}{u} ; \Delta \underset{\sim}{u}) \, \delta(\Delta \underset{\sim}{r} - \underset{\sim}{u} \Delta t)$$

We shall take for the transition probability $\psi(\underset{\sim}{u} ; \Delta \underset{\sim}{u})$ the assumption (I.5.9), in which we add a term $-\underset{\sim}{K} \Delta t$ in the exponential to take into account the effect of the external field. This assumption will lead us to a generalization for the complete phase space of equation (I. 5.11). If we do not make a special cho'ice for $\psi(\underset{\sim}{u};\Delta \underset{\sim}{u})$, we can follow the arguments of the previous paragraph and obtain a generalization of (I.5.7).

Integrating over $\Delta \underset{\sim}{r}$ and expanding again in a power series of Δt, $\Delta \underset{\sim}{u}$, one obtains :

(I.6.4)
$$\left(\frac{\partial W}{\partial t} + u_i \frac{\partial W}{\partial r_i}\right)\Delta t + 0\,(\Delta t)^2 = -\frac{\partial W\langle \Delta u_i\rangle}{\partial u_i} + \frac{1}{2}\frac{\partial^2 W\langle \Delta u_i\, \Delta u_j\rangle}{\partial u_i\, \partial u_j} + 0\,(\langle \Delta u_i \Delta u_j \Delta u_k\rangle)$$

If one computes the various averages and then takes the limit $\Delta t \rightarrow 0$, one obtains the Fokker-Planck equation :

(I;6.5)
$$\frac{\partial W}{\partial t} + u_i \frac{\partial W}{\partial r_i} + K_i \frac{\partial W}{\partial u_i} = \beta\left[\frac{\partial W u_i}{\partial u_i} + (kT/M)\frac{\partial^2 W}{\partial u_i^2}\right]$$

I.7. - Diffusion equation.

Let us now consider a spatially inhomogeneous system in which we have a certain number n of brownian particles. We assume that the dilution is such that we may neglect all interactions between these particles. Therefore, the probability distribution function $W(\{\underset{\sim}{r}\},\{\underset{\sim}{u}\},t)$ factorizes into a product of n factors :

(I.7.1)
$$W(\{\underset{\sim}{r}\}, \{\underset{\sim}{u}\}, t) = \prod_{i=1}^{n} W_i(\underset{\sim}{r}_i, \underset{\sim}{u}_i, t)$$

where W_i is the one particle distribution function which satisfies the Fokker-Planck equation .

We shall also assume that the density gradient is small.

In such a system, we have two kinds of processes : first , we have collisions with the particles of the surrounding fluid which insure that the velocity distribution function approaches the Maxwellian distribution ; next, we have a diffusion of the particles which will lead to spatial uniformization of the system. The time scale for the first process is given by the relaxation time β^{-1} ; this is much smaller than the time scale for the diffusion process . As a consequence, if we are interested in times long with respect to the relaxation time, we may expect that the distribution function for one particle will be of the form:

$$W_i(\underset{\sim}{r}_i, \underset{\sim}{u}_i, t) = n_i(\underset{\sim}{r}_i, t)\,(M_i/2\pi kT)^{3/2}\, e^{-M_i u_i^2/2kT} + \delta W_i(\underset{\sim}{r}_i, \underset{\sim}{u}_i, t) \tag{I.7.2}$$

The first term describes the local equilibrium distribution which is reached for times much longer than the relaxation time.
The second term is a small correction which takes into account the existence of the diffusion process; it is of the order of the density gradient .

We shall now show that, under these circumstances, the function $n_i(\underset{\sim}{r}_i, t)$ obeys a diffusion equation .

If we integrate the Fokker-Planck equation (I.6.5) over the velocity, we obtain :

$$\frac{\partial \int d\underset{\sim}{u}_i\, W_i}{\partial t} + \frac{\partial}{\partial \underset{\sim}{r}_i} \cdot \int d\underset{\sim}{u}_i\, \underset{\sim}{u}_i W_i = 0 \tag{I.7.3}$$

If we first multiply both sides of the Fokker-Planck equation by $u_{i\alpha}$ ($\alpha = x, y, z$) and then integrate over the velocity, we get :

$$\text{(I.7.4)} \qquad \frac{\partial \int d\underset{\sim}{u}_i W_i u_{i\alpha}}{\partial t} + \frac{\partial}{\partial \underset{\sim}{r}_i} \cdot \int d\underset{\sim}{u}_i \underset{\sim}{u}_i W_i u_{i\alpha} = -\beta \int d\underset{\sim}{u}_i W_i u_{i\alpha}$$

If we combine these two equations, we obtain :

$$\text{(I 7.5)} \qquad \frac{\partial}{\partial t}\left[\int d\underset{\sim}{u}_i W_i - \frac{1}{\beta} \frac{\partial}{\partial \underset{\sim}{r}_i} \cdot \int d\underset{\sim}{u}_i \underset{\sim}{u}_i W_i\right] = -\frac{1}{\beta} \frac{\partial^2}{\partial \underset{\sim}{r}_i \partial \underset{\sim}{r}_i} : \int d\underset{\sim}{u}_i \underset{\sim}{u}_i \underset{\sim}{u}_i W_i$$

Now , using (I.7.2) and keeping only lowest order terms, we easily, obtain the diffusion equation :

$$\text{(I.7.6)} \qquad \frac{\partial n_i(\underset{\sim}{r}_i, t)}{\partial t} = D \nabla^2_{\underset{\sim}{r}_i} n_i(\underset{\sim}{r}_i, t)$$

with

$$\text{(I.7.7)} \qquad D = kT/\beta M$$

The density of the particles at a given point $\underset{\sim}{x}$ of space will be :

$$\text{(I.7.8)} \qquad C(\underset{\sim}{x}, t) = \sum_{i=1}^{n} \int \delta(\underset{\sim}{x} - \underset{\sim}{r}_i)\, W(\{\underset{\sim}{r}\}, \{\underset{\sim}{u}\}, t)\, d\underset{\sim}{r}_i d\underset{\sim}{u}_i$$

Again, if we keep only lowest order terms, i.e. if we take :

$$\text{(I.7.9)} \qquad C(\underset{\sim}{x}, t) = \sum_{i=1}^{n} n_i(\underset{\sim}{x}, t)$$

we verify easily that this also obeys the diffusion equation

$$\text{(I.7.10)} \qquad \frac{\partial C(\underset{\sim}{x}, t)}{\partial t} = D \nabla^2 C(\underset{\sim}{x}, t)$$

Appendix I.1 - Proof of (I.3.3)

In the problem of random flights, one considers a particle which performs a sequence of steps $\underset{\sim}{r}_i \ldots \underset{\sim}{r}_i \ldots$ The magnitude and direction of all the different steps are independent of the preceding ones. One chooses a priori a distribution function $\tau_i(\underset{\sim}{r}_i)$ which gives the probability distribution that a given step $\underset{\sim}{r}_i$ lies between $\underset{\sim}{r}_i$ and $\underset{\sim}{r}_i + d\underset{\sim}{r}_i$. The problem is then to find the probability $W(\Delta \underset{\sim}{R}; \Delta t)$ that the particle has travelled a distance $\Delta \underset{\sim}{R}$ in the time interval Δt.

We shall give a proof of (I.3.3) for the simple case of one dimensional random walk with all steps of the same length and with equal a priori probability for a step to the left or to the right. Therefore, if the particle is at the origin at $t=0$, the probability that it will be at the point m after N steps $(-N \leq m \leq N)$, is given by :

(AI.1.1) $$W(m,N) = \frac{1}{2} W(m-1, N-1) + \frac{1}{2} W(m+1, N-1) \quad (N > 1)$$

(AI.1.2) $$W(1,1) = W(-1,1) = 1/2$$

Using Fourier transforms :

(AI.1.3) $$P_N(\ell) = \sum_{m=-\infty}^{+\infty} W(m,N) e^{-ilm}$$

we obtain from (AI.1.1) and (AI.1.2) :

(AI.1.4) $$P_N(\ell) = \cos l \quad P_{N-1}(\ell) \qquad N > 1$$
$$P_1(\ell) = \cos 1$$

Therefore,

(AI.1.5) $$P_N(\ell) = (\cos 1)^N$$

and hence inverting (AI. 1. 3) :

(AI. 1;6) $$W(m, N) = (2\pi)^{-1} \int_{-\pi}^{+\pi} (\cos l)^N e^{ilm} dl$$

Now, with

(AI. 1. 7) $$(\cos l)^N = 2^{-N} \sum_{p=0}^{N} \frac{N!}{p!(N-p)!} e^{-l(N-2p)i}$$

we obtain easily :

(AI. 1. 9) $$W(m, N) = 2^{-N} \sum_{p=0}^{N} \frac{N!}{p!(N-p)!} \frac{\sin\pi(N-m-2p))}{\pi(N-m-2p)}$$

The last factor vanishes unless $N-m-2p = 0$. Therefore :

$W(m, N) = 0$ if N even and m odd or vice versa (AI. 1. 9)

$W(m, N) = 2^{-N} N!\{[(N-m)/2]!\cdot[(N+m)/2]!\}^{-1}$ if both N and m even or odd (AI. 1. 10)

The first result is of course obvious. The second could have been obtained using combinatorial analysis. However, the method involving Fourier transforms can be generalized to more complicated problems and although exact results for arbitrary values of N cannot always be obtained, expressions such as (AI. 1. 6) are often useful to obtain an asymptotic result.

For our present problem, for $N \to \infty$ and m finite, using Stirling's formula :

(AI. 1. 11) $$\log n! = (n+\frac{1}{2}) \log n - n + \frac{1}{2} \log 2\pi \quad (n \to \infty)$$

we obtain :

$$\log W(m, N) = (N + \frac{1}{2}) \log N - \frac{1}{2}(N + m + 1) \log (\frac{N}{2} + m)$$

$$\text{(AI.I.12)} \quad - \frac{1}{2}(N - m + 1) \log (\frac{N}{2} - m) - \frac{1}{2} \log 2\pi - N\log 2$$

$$\simeq \log (2/N\pi)^{1/2} - m^2/2N$$

and hence the asymptotic expression :

$$\text{(AI.1.13)} \qquad W(m, N) \approx (2/\pi N)^{1/2} \exp(-m^2/2N)$$

If each step has a length 1 and if τ is the time lapse between two steps, introducing the variables :

$$\text{(AI.1.14)} \qquad x = ml \qquad \Delta t = N\tau$$

the probability $\overline{W}(x, \Delta t)\Delta x$ that the particle lies between x and $x + \Delta x$ after Δt is :($\Delta m = \Delta x/l$) :

$$\overline{W}(x, \Delta t)\Delta x = \sum_{m \in \Delta m} W(m, N)$$

$$\text{(AI.1.15)} \qquad = (1/2)\, W(x/l, \Delta t/\tau) \sum_{m \in \Delta m} 1$$

$$= (1/2l)\, W(x/l, \Delta t/\tau)$$

where the factor (1/2) takes into account the fact that for N given (odd or even), only one half of the values of m contribute (those which are odd or even).

Therefore, we obtain :

$$\text{(AI.1.16)} \quad \overline{W}(x, \Delta t) = (4\pi D \Delta t)^{-1/2} \exp(-x^2/4D\Delta t)$$

with

$$\text{(AI.1.17)} \qquad D = l^2/2\tau$$

(AI. 1;16) is nothing else than (I. 3. 3) for this simple one dimensional problem. This asymptotic formula can be generalized for several three dimensional random flight problems. For instance, for a gaussian probability distribution for the jth step :

(AI. 1. 18) $$\tau_j(\underset{\sim}{r}_j) = (2\pi l_j^2/3)^{-3/2} \exp(-3|\underset{\sim}{r}_j|^2/2l_j^2)$$

one obtains:

(A. I. 1. 19) $$W_N(\underset{\sim}{R}) = (2\pi N\langle l^2\rangle/3)^{-3/2} \exp(-3|\underset{\sim}{R}|^2/2N\langle l^2\rangle)$$

with

(AI. 1. 20) $$\langle l^2\rangle = N^{-1} \sum_{j=1}^{N} l_j^2$$

The same expression is obtained if the probability distribution is identical for each step and spherically symmetric. Then $\langle l^2\rangle$ is the average displacement in each step (the l_j 's are independent of the index j) .

Appendix I. 2. - Proof of (I. 4. 4)

We want the probability distribution of the quantity $\underset{\sim}{\alpha}$:

(AI. 2. 1) $$\underset{\sim}{\alpha} = \sum_{j=1}^{N} \underset{\sim}{\alpha}_j = \sum_j \psi_j \underset{\sim}{B}(\Delta t)$$

with

(AI. 2. 2) $$\psi_j = \exp\left[-\beta(t-j\Delta t)\right]$$

when the distribution function for $\underset{\sim}{B}(\Delta t)$ is given by (I. 3. 4) . This is again a random flight problem , the steps being the $\underset{\sim}{\alpha}_j$'s. The probability distribution for each $\underset{\sim}{\alpha}_j$ is a gaussian and corresponds to the function τ_j given in (AI.I.18) with :

(AI.2.3) $$\ell_j^2 = 6q\,\psi_j^2\,\Delta t$$

Therefore, for a large number of steps , using (AI.I.19) , we have

(AI.2.4) $$W(\underset{\sim}{R}) = \left[2\pi \sum_{j=1}^{N} l_j^2/3\right]^{-3/2} \exp\left[-3\left|\underset{\sim}{R}\right|^2/2 \sum_{j=1}^{N} l_j^2\right]$$

With

(AI.2.5) $$\sum_{j=1}^{N} l_j^2 = 6q\,\Delta t e^{-2\beta t} \sum_{j=1}^{N} e^{2\beta j \Delta t}$$

If we use the same approximation that led us from (I.4.2) to (I.4.3) , we may write :

(AI.2.6) $$\frac{2}{3} \sum_{j=1}^{N} l_j^2 = 4q\, e^{-2\beta t} \int_o^t d\xi\, e^{2\beta\xi}$$

Inserting (AI.2.6) into (AI.2.4) , we readily obtain (I.4.4)

References

1. S. Chandrasekhar : Revs. Mod. Phys. 15 , 1 (1943) reprinted in : Noise and Stochastic Processes, ed. Nelson Wax , Dover Publications Inc. N.Y. 19, New-York (1954)

F. Henin

II. KINETIC EQUATIONS

II.1 - Introduction.

In this chapter, we shall consider the microscopic description of an N-body system from the point of view of statistical mechanics. We shall restrict ourselves to classical systems which are homogeneous in space, although the formalism can be extended to include quantum systems and inhomogeneous situations, As this will be the most useful for us, we shall consider the case of a gas interacting through binary central forces. On the microscopic level, a complete description of the system is given by its hamiltonian, i.e. here :

$$H = \sum_j \frac{p_j^2}{2m_j} + \lambda \sum_{i<j} V_{ij} (|q_i - q_j|) = H_0 + \lambda V \tag{II.1.1}$$

where m_j is the mass of the jth particle, q_j and p_j its position and momentum. λ is a dimensionless coupling constant.

Once we have the hamiltonian and the initial conditions, the evolution of the system is of course completely determined by Hamilton's equations of motion.

However, for a large system, a set of 6N differential equations is not very practical. Moreover, we can only measure a few macroscopic quantities and we never have, even at $t = 0$, a detailed information about the positions and momenta of all the particles. Therefore, we shall use the idea of a representative ensemble in phase space. We imagine a large number of similar systems, with the same hamiltonian but differing by their initial states. If we take a sufficiently large set of equivalent systems, the ensemble will be characterized by a continuous density in phase space $\rho(\{p\},\{q\}, t)$. As all points in the ensemble move in time according to Hamilton's equations, the function ρ satisfies the Liouville equation :

$$\frac{\partial \rho}{\partial t} = [H, \rho] = -iL\rho \tag{II.1.2}$$

where $[H, \rho]$ is the Poisson bracket of the Hamiltonian and ρ, and, hence, the Liouville operator L is :

$$L = -i \sum_j \left\{ \frac{\partial H}{\partial p_j} \frac{\partial}{\partial q_j} - \frac{\partial H}{\partial q_j} \frac{\partial}{\partial p_j} \right\} \tag{II.1.3}$$

From this equation, one verifies easily that :

$$\int \rho(\{\underset{\sim}{p}\}, \{\underset{\sim}{q}\}, t) \{d\underset{\sim}{p}\, d\underset{\sim}{q}\}^N = \text{constant} \tag{II.1.4}$$

If we choose this normalization constant to be equal to unity :

$$\int \rho(\{\underset{\sim}{p}\}, \{\underset{\sim}{q}\}, t) \{d\underset{\sim}{p}\, d\underset{\sim}{q}\}^N = 1 \tag{II.1.5}$$

then, $\rho(\{\underset{\sim}{p}\}, \{\underset{\sim}{q}\}, t) \{d\underset{\sim}{p}\, d\underset{\sim}{q}\}^N$ is the probability of finding at time t a representative point in the volume element $\{d\underset{\sim}{p}\, d\underset{\sim}{q}\}^N$ of phase space.

A basic postulate in statistical mechanics is that all macroscopic quantities may be computed by taking the average value of the corresponding microscopic dynamical quantity over the distribution function of a suitable ensemble :

$$\langle A(t) \rangle = \int A(\{\underset{\sim}{p}\}, \{\underset{\sim}{q}\}) \, \rho(\{\underset{\sim}{p}\}, \{\underset{\sim}{q}\}, t) \{d\underset{\sim}{p}\, d\underset{\sim}{q}\}^N \tag{II.1.6}$$

This description has the advantage that the whole mechanical behavior is given by a single linear equation, the Liouville equation (II.1.2).

To the decomposition (II.1.1) of the hamiltonian into un unperturbed part H_o (kinetic term) and a perturbed part λV (interaction) corresponds a similar decomposition of the Liouville operator :

(II.1.7) $$L = L_o + \lambda \delta L$$

This feature, as well as the strong analogy between the Liouville equation (II.1.2) and the Schrödinger equation in quantum mechanics will enable us to develop easily a perturbation technique to study the time evolution of the distribution function. In this chapter, we shall concentrate ourselves on the evolution of the velocity distribution function :

(II.1.8) $$\rho_o(\{\underset{\sim}{p}\}, t) = \int \{dq\}^N \rho(\{\underset{\sim}{p}\}, \{\underset{\sim}{q}\}, t)$$

Essentially, we shall solve formally the Liouville equation and write the formal solution as a power series of the perturbation. The introduction of a diagram technique to represent the various contributions will enable us to rearrange the terms and to write the equation of evolution in a form suitable for further discussions :

(II.1.9) $$\frac{\partial \rho_o}{\partial t} = \int_o^t d\tau\, G(t-\tau)\, \rho_o(\tau) + \mathcal{D}(\{\rho_{\{\underset{\sim}{k}\}}(o)\}, t)$$

First, we have a non-markovian term which relates $\rho_o(t)$ to its value at an earlier time τ. G(t) is an operator which describes the effect of the collisions which occur in the system on the evolution of the velocity distribution function. The non-markovian character of the first contribution is due to the fact that the collisions last over a finite time interval τ_{coll}. The second term gives the contribution to the evolution of $\rho_o(t)$ due to the existence of initial correlations in the system , these being described by the functions $\rho_{\{\underset{\sim}{k}\}}(0)$.

We shall then show that, for systems interacting through short range forces and such that the initial correlations present at t=0 are over molecular distances, in the limit of a large system;

(II.1.10) $$N \to \infty , \quad \Omega \to \infty , \quad N/\Omega = C \text{ finite}$$

(Ω : volume of the system)

and for long times :

(II.1.11) $$t \gg \tau_{coll}$$

the second term in the rhs of (II.1.9) may be neglected and that $\rho_o(t)$ satisfies a closed equation , which may be written in a pseudomarkovian form :

(II.1.12) $$i \frac{\partial \rho_o}{\partial t} = \Omega \, \psi(0) \, \rho_o$$

where $\psi(z)$ is the Laplace transform of the collision operator G(t) and $\psi(0)$ its limit when $z \to 0$. Ω is a functional of ψ and its derivatives for $z \to 0$ and takes into account the finite duration of the collision.

In this chapter, we shall show in detail how the kinetic equation (II.1.12) can be derived. We shall then sketch briefly how the same formalism can be extended to discuss the evolution of space correlations in the system. We shall also indicate the necessary modifications when an external force is present . The equations so obtained will be our basic tools for the next chapters.

We shall be able to give here only a very short outline of the theory. More details , as well as references,to the original papers can be found in the monographs by Prigogine [1], Balescu [2] and Résibois [3].

II.2. Fourier analysis of the distribution function.

Let us expand the distribution function in a Fourier series with respect to the position variables :

$$\rho(\{\underset{\sim}{p}\},\{\underset{\sim}{q}\},t) = \Omega^{-N} \sum_{\underset{\sim}{k}_1 \ldots \underset{\sim}{k}_N} \rho_{\underset{\sim}{k}_1 \ldots \underset{\sim}{k}_N}(\{\underset{\sim}{p}\},t) \times$$

(II.2.1)

$$\times \exp\left[i \sum_{j=1}^{N} \underset{\sim}{k}_j \cdot \underset{\sim}{q}_j \right]$$

The factor Ω^{-N} is introduced to allow the normalization of ρ to unity:

(II.2.2) $$\int \{d\underset{\sim}{p}\, d\underset{\sim}{q}\}^N \rho(t) = \int \{d\underset{\sim}{p}\}^N \rho_o(t) = 1)$$

The formal expansion (II.2.1) is very interesting. Indeed, it is easily verified that the Fourier coefficients $\rho_{\underset{\sim}{k}_1 \ldots \underset{\sim}{k}_N}$ have a very simple physical meaning . First of all, we notice that, in a system which is homogeneous in space, i.e. such that the distribution function is invariant with respect to space translations :

(II.2.3) $$\rho(\{\underset{\sim}{q}_j + \underset{\sim}{a}\}\{\underset{\sim}{p}_j\},t) = \rho(\{\underset{\sim}{q}_j\},\{\underset{\sim}{p}_j\},t)$$

only those coefficients such that :

(II.2.4) $$\sum_{i=1}^{N} \underset{\sim}{k}_i = \underset{\sim}{0}$$

are different from zero. Therefore, Fourier coefficients such that (II.2.4) is not fulfilled are closely connected with the existence of spatial inhomogeneities in the system.

Also , we shall, most of the time, be interested in the average value of microscopic quantities which depend only on a small , finite number s of degrees of freedom. To compute these, all we need are the reduced distribution functions :

$$f_s(\underset{\sim}{p}_1 \dots \underset{\sim}{p}_s \underset{\sim}{q}_1 \dots \underset{\sim}{q}_s, t) = \int d\underset{\sim}{q}_{s+1} \dots d\underset{\sim}{q}_N d\underset{\sim}{p}_{s+1} \dots d\underset{\sim}{p}_N \rho(\{\underset{\sim}{p}\}, \{\underset{\sim}{q}\}, t) \tag{II.2.5}$$

In such reduced distribution functions, the only Fourier coefficients which play a role are obviously those which have at most s wave vectors different from zero.

One of the most important coefficients is that with all wave vectors equal to zero. It is the velocity distribution function :

$$\rho_o(\{\underset{\sim}{p}\}, t) = \int \{d\underset{\sim}{q}\}^N \rho(\{\underset{\sim}{p}\}, \{\underset{\sim}{q}\}, t) \tag{II.2.6}$$

From (II.2.2), we notice that this function is normalized to unity.

To find out the meaning of the other Fourier coefficients, let us consider for instance the average density.

$$\begin{aligned} \langle n(\underset{\sim}{x}, t) \rangle &= \int \{d\underset{\sim}{p}\, d\underset{\sim}{q}\}^N \sum_j \delta(\underset{\sim}{x} - \underset{\sim}{q}_j)\, \rho(\{\underset{\sim}{p}\}, \{\underset{\sim}{q}\}, t) \\ &= N/\Omega \left[1 + \sum_{\underset{\sim}{k}} e^{i\underset{\sim}{k} \cdot \underset{\sim}{x}} \int \{d\underset{\sim}{p}\}^N \rho_{\underset{\sim}{k}, \{\underset{\sim}{0}\}}(\{\underset{\sim}{p}\}, t) \right] \end{aligned} \tag{II.2.7}$$

In this way , we see that the Fourier coefficients with one wave vector different from zero are connected with the local deviations from the mean density N/Ω .

As another example, let us consider the binary correlation function :

$$g(\underset{\sim}{x},\underset{\sim}{x}',t)=\sum_{ij}\int d\underset{\sim}{q}_i d\underset{\sim}{q}_j\,\delta(\underset{\sim}{x}-\underset{\sim}{q}_i)\,\delta(\underset{\sim}{x}'-\underset{\sim}{q}_j)\Big\{\int d\underset{\sim}{p}_i d\underset{\sim}{p}_j f_2(\underset{\sim}{q}_i\underset{\sim}{q}_j\underset{\sim}{p}_i\underset{\sim}{p}_j t)$$

(II. 2. 8)

$$-\int d\underset{\sim}{p}_i f_1(\underset{\sim}{q}_i\underset{\sim}{p}_i t)\int d\underset{\sim}{p}_j\, f_1(\underset{\sim}{q}_j\underset{\sim}{p}_j\, t)\Big\}$$

$$=\sum_{ij}\sum_{\underset{\sim}{k}_i}\sum_{\underset{\sim}{k}_j}\Big\{\int\{d\underset{\sim}{p}\}^N\rho_{\underset{\sim}{k}_i,\underset{\sim}{k}_j}(\{\underset{\sim}{p}\},t)-\int\{d\underset{\sim}{p}\}^N\rho_{\underset{\sim}{k}_i}(\{\underset{\sim}{p}\},t)\times$$

$$\times\int\{d\underset{\sim}{p}\}^N\rho_{\underset{\sim}{k}_j}(\{\underset{\sim}{p}\},t)\Big\}\exp\left[i(\underset{\sim}{k}_i\cdot\underset{\sim}{x}+\underset{\sim}{k}_j\cdot\underset{\sim}{x}')\right]$$

In an homogeneous system, this reduces to :

(II. 2. 9)
$$g(\underset{\sim}{x},\underset{\sim}{x}',t)=\sum_{ij}\sum_{\underset{\sim}{k}}\int\{d\underset{\sim}{p}\}^N\rho_{\underset{\sim}{k},-\underset{\sim}{k}}(\{\underset{\sim}{p}\},t)\,e^{i\underset{\sim}{k}\cdot(\underset{\sim}{x}-\underset{\sim}{x}')}$$

Correlations among s particles therefore depend on the Fourier coefficients with at most s indices different from zero. For a large system, the spectrum of $\underset{\sim}{k}$ becomes continuous and the rhs of (II. 2. 9) vanishes for $|\underset{\sim}{x}-\underset{\sim}{x}'|\to\infty$ if $\rho_{\underset{\sim}{k},-\underset{\sim}{k}}$ is sufficiently regular.

Another interesting feature of (II. 2. 1) is that this is in fact an expansion in terms of the eigenfunctions of the unperturbed Liouville operator. Indeed, from (II. 1. 7), (II. 1. 3) and (II. 1. 1), we have :

(II. 2. 10)
$$L_o=-i\sum_j\frac{\underset{\sim}{p}_j}{m_j}\cdot\frac{\partial}{\partial\underset{\sim}{q}_j}$$

If we use the same notation for eigenfunctions as in quantum mechanics:

(II. 2. 11)
$$|\{\underset{\sim}{k}\}\rangle=\Omega^{-N/2}\exp\Big[i\sum_j\underset{\sim}{k}_j\cdot\underset{\sim}{q}_j\Big]$$

we have :

(II. 2. 12)
$$L_o|\{\underset{\sim}{k}\}\rangle=\Big[\sum_j(\underset{\sim}{k}_j\cdot\underset{\sim}{p}_j/m_j)\Big]|\{\underset{\sim}{k}\}\rangle$$

These eigenfunctions are orthogonal and normalized to unity :

$$\Omega^{-N}\int \{dq\}^{N} \exp\left[-i \sum_j (k_j - k_j').q_j\right] = \langle \{k\} | \{k'\} \rangle = \prod_j \delta_{k_j, k_j'} \qquad \text{(II. 2. 13)}$$

From these properties, the time dependence of the Fourier coefficients $\rho_{\{k\}}$ in a system of non interacting particles ($\lambda = 0$) is easily found :

$$\rho_{\{k\}}(\{p\}, t) = \bar{\rho}_{\{k\}}(\{p\}) \exp\left[i \sum_j (k_j . p_j / m_j) t\right] (\lambda = 0) \qquad \text{(II. 2. 14)}$$

When the particles are interacting , the time dependence of the $\rho_{\{k\}}(t)$ is of course much more complex. Besides the oscillating exponential factor corresponding to the free propagation of the particles, we have a further time dependence in the coefficients $\bar{\rho}_{\{k\}}(\{p\})$ in the rhs of (II. 2. 14) because of the collisions occuring in the system.

II. 3. Formal solution of the Liouville equation. Resolvent operator

The formal solution of the Liouville equation is of course very easily written :

$$\rho(t) = e^{-iLt} \rho(0) \qquad \text{(II. 3. 1)}$$

From this, we obtain for the various Fourier coefficients of the distribution function :

$$\rho_{\{\underset{\sim}{k}\}}(t) = \sum_{\{\underset{\sim}{k}'\}} \langle\{\underset{\sim}{k}\}|\, e^{-iLt}\, |\{\underset{\sim}{k}'\}\rangle\, \rho_{\{\underset{\sim}{k}'\}}(0)$$

(II.3.2)
$$= \Omega^{-N} \sum_{\{\underset{\sim}{k}'\}} \int \{d\underset{\sim}{q}\}^{N} \exp\left[-i \sum_j \underset{\sim}{k}_j \cdot \underset{\sim}{q}_j\right] e^{-iLt} \times$$

$$\times \exp\left[i \sum_j \underset{\sim}{k}'_j \cdot \underset{\sim}{q}_j\right] \rho_{\{\underset{\sim}{k}'\}}(0)$$

As L is an operator in the complete phase space, the matrix elements in the rhs of (II.3.2) are still operators in velocity space.

The operator e^{-iLt} can be expanded formally in a power series of the interaction :

(II.3.3)
$$e^{-iLt} = e^{-iL_o t} - i\lambda \int_o^t dt_1\, e^{-iL_o(t-t_1)}\, \delta L\, e^{-iL_o t_1}$$
$$+ (-i\lambda)^2 \int_o^t dt_1 \int_o^{t_1} dt_2\, e^{-iL_o(t-t_1)}\, \delta L\, e^{-iL_o(t_1-t_2)}\, \delta L\, e^{-iL_o t_2}$$
$$+ \ldots$$

This equality is most easily verified if one takes the time derivative of both sides of (II.3.3) .

However, the behavior of the system can be discussed much more easily if, rather than the operator e^{-iLt}, one considers its Laplace transform, the resolvent operator R(z) :

(II.3.4)
$$R(z) = -i\int_o^{\infty} dt\, e^{izt}\, e^{-iLt} = \frac{1}{z-L} \quad (z \in S^+)$$

From (II.3.2) and (II.3.4) , we obtain after an inverse Laplace transform :

(II.3.5)
$$\rho_{\{\underset{\sim}{k}\}}(t) = -\frac{1}{2\pi i} \int_C dz\, e^{-izt} \langle\{\underset{\sim}{k}\}|R(z)|\{\underset{\sim}{k}'\}\rangle \rho_{\{\underset{\sim}{k}'\}}(0)$$

where the contour C is a parallel to the real axis in the upper half plane, above all singularities of the integrand. (II. 3. 5) can be easily verified for a finite system. Indeed, then, the operator L is a hermitian operator and all its eigenvalues, although unknown, are real. For an infinite system ($\Omega \rightarrow \infty$), the properties of the Liouville operator are not known. We shall however assume that (II. 3. 5) remains valid when we perform the limiting procedure (II. 1. 10).

The resolvent operator can be expanded in a power series of the perturbation

(II. 3. 6)
$$R(z) = \sum_{n=0}^{\infty} \frac{1}{z - L_o} \left(\delta L \frac{1}{z - L_o}\right)^n$$

This result can also be obtained from (II. 3. 3) through a Laplace transform, using the convolution theorem.

(II. 3. 5) and (II. 3. 6) will be our basic equations for the following discussion. Of course this means that we assume that perturbation theory up to an infinite order is valid. Whether this is true or not is an unanswered question and we shall not discuss it.

The unperturbed resolvent operator is diagonal in the $|\{\underset{\sim}{k}\}\rangle$ representation. Its matrix elements are very simple :

(II. 3. 7)
$$\langle \{\underset{\sim}{k}\} | R_o(z) | \{\underset{\sim}{k}\}\rangle = \langle \{\underset{\sim}{k}\} | \frac{1}{z - L_o} | \{\underset{\sim}{k}\}\rangle = \frac{1}{z - \sum_j \underset{\sim}{k}_j \cdot \underset{\sim}{v}_j}$$

where

(II. 3. 8)
$$\underset{\sim}{v}_j = \underset{\sim}{p}_j / m_j$$

is the velocity of the jth particle.

As to the operator δL, we obtain from (II.1.7), (II.1.3) and (II.1.1):

$$(II.3.9) \qquad -i\delta L = \sum_{i<j} \left(\frac{\partial V_{ij}}{\partial \underset{\sim}{q}_i} \cdot \frac{\partial}{\partial \underset{\sim}{p}_i} + \frac{\partial V_{ij}}{\partial \underset{\sim}{q}_j} \cdot \frac{\partial}{\partial \underset{\sim}{p}_j} \right)$$

If we expand the potential in a Fourier series :

$$(II.3.10) \qquad V_{ij}(|\underset{\sim}{q}_i - \underset{\sim}{q}_j|) = (8\pi^3/\Omega) \sum_{\underset{\sim}{k}} V_k \, e^{i\underset{\sim}{k}\cdot(\underset{\sim}{q}_i - \underset{\sim}{q}_j)}$$

we have:

$$(II.3.11) \qquad -i\delta L = (8\pi^3 i/\Omega) \sum_{i<j} \sum_{\underset{\sim}{k}} V_k \, \underset{\sim}{k} \cdot \left(\frac{\partial}{\partial \underset{\sim}{p}_i} - \frac{\partial}{\partial \underset{\sim}{p}_j} \right) \times e^{i\underset{\sim}{k}\cdot(\underset{\sim}{q}_i - \underset{\sim}{q}_j)}$$

It is easy to verify that the only non vanishing matrix elements are those where the initial and final states have only two different wave vectors, the total wave vector being conserved :

$$(II.3.12) \qquad \langle \underset{\sim}{k}_1 \dots \underset{\sim}{k}_i \dots \underset{\sim}{k}_j \dots \underset{\sim}{k}_N | \delta L | \underset{\sim}{k}_1 \dots \underset{\sim}{k}'_i \dots \underset{\sim}{k}'_j \dots \underset{\sim}{k}_N \rangle = \delta_{\underset{\sim}{k}_i + \underset{\sim}{k}_j, \underset{\sim}{k}'_i + \underset{\sim}{k}'_j} (8\pi^3/\Omega) V_{|\underset{\sim}{k}_i - \underset{\sim}{k}'_i|} (\underset{\sim}{k}_i - \underset{\sim}{k}'_i) \cdot \left(\frac{\partial}{\partial \underset{\sim}{p}_i} - \frac{\partial}{\partial \underset{\sim}{p}_j} \right)$$

The fact that only two wave vectors are modified and that the total wave vector is conserved is of course due to our choice of binary central forces. An interesting consequence of the condition of conservation of the total wave vector is that the Fourier coefficients are divided into subsets corresponding to the different values of the total wave vector. Each subset evolves in time independently of all the others. Therefore, for instance, a system which is initially homogeneous in space will remain so in the course of time.

F. Henin

II.4 - Diagram representation of the formal solution of the Liouville equation.

The classification of the various terms in the series (II.3.6) is best performed if one uses a diagram technique . Let us associate with each state $|\{\mathbf{k}\}\rangle = |\mathbf{k}_1 \ldots \mathbf{k}_N\rangle$ with n non vanishing wave vectors a set of n lines running from the right to the left. Each line is labelled with an index corresponding to the particle; when necessary, we shall also indicate the wave vector An example is given in fig. II.4.1.

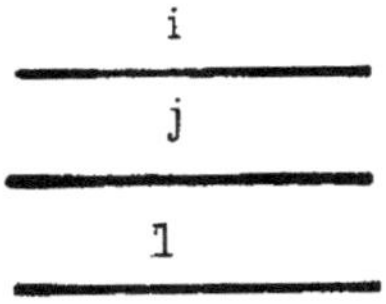

Diagrammatic representation of the state

$|\{0\}\mathbf{k}_i\,\mathbf{k}_j\,\mathbf{k}_l\rangle$

Fig. II.4.1

The matrix elements of δL provoke a modification of two wave vectors $\mathbf{k}_i\,\mathbf{k}_j \rightarrow \mathbf{k}'_i\;\mathbf{k}'_j$. Taking into account the fact that among these, none, one or two may correspond to the wave vector $\mathbf{0}$, we have 6 basic diagrams (see fig. II.4.2) .

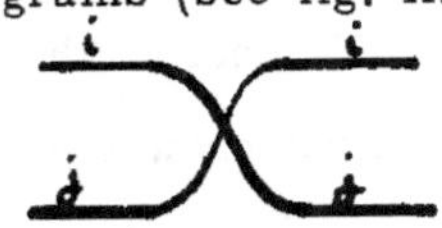

$\langle\ldots\mathbf{k}_i\ldots\mathbf{k}_j\,|\,\delta L\,|\ldots\mathbf{k}'_i\ldots\mathbf{k}'_j\ldots\rangle$

$\mathbf{k}_i+\mathbf{k}_j = \mathbf{k}'_i+\mathbf{k}'_j$

(a)

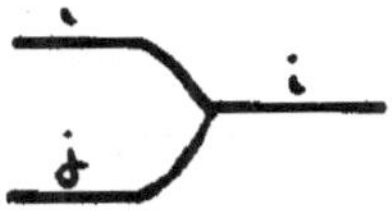

$\langle\ldots\mathbf{k}_i\ldots\mathbf{k}_j\ldots|\,\delta L\,|\ldots\mathbf{k}'_i\ldots\mathbf{0}\ldots\rangle$

$\mathbf{k}_i+\mathbf{k}_j = \mathbf{k}'_i$

(b)

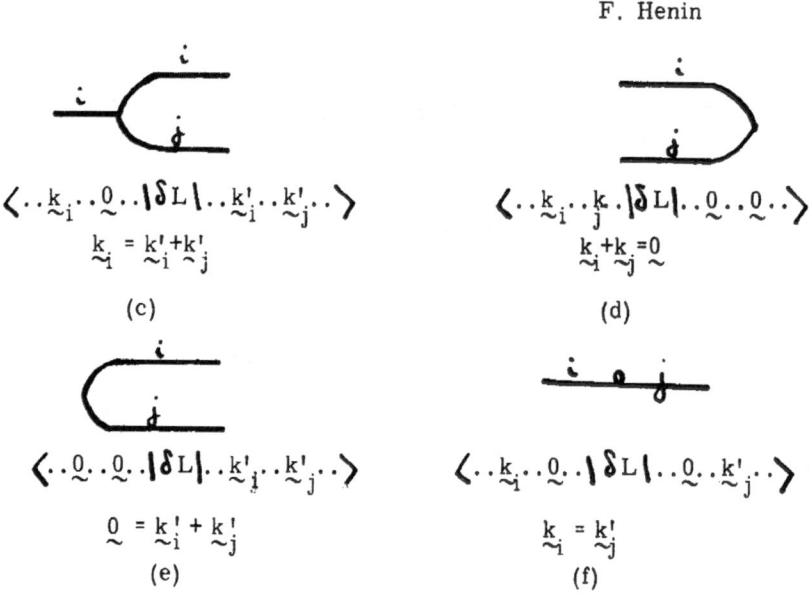

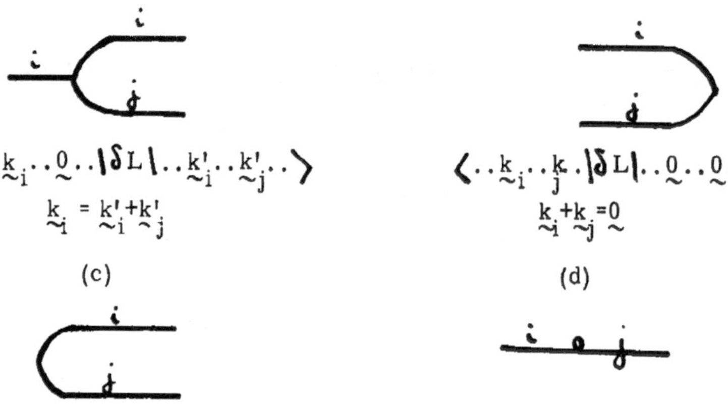

$$\langle ..\underset{\sim}{k}_i..\underset{\sim}{0}..|\delta L|..\underset{\sim}{k}'_i..\underset{\sim}{k}'_j..\rangle \qquad \underset{\sim}{k}_i = \underset{\sim}{k}'_i + \underset{\sim}{k}'_j$$

(c)

$$\langle ..\underset{\sim}{k}_i..\underset{\sim}{k}_j.|\delta L|..\underset{\sim}{0}..\underset{\sim}{0}..\rangle \qquad \underset{\sim}{k}_i + \underset{\sim}{k}_j = \underset{\sim}{0}$$

(d)

$$\langle ..\underset{\sim}{0}..\underset{\sim}{0}..|\delta L|..\underset{\sim}{k}'_i..\underset{\sim}{k}'_j..\rangle \qquad \underset{\sim}{0} = \underset{\sim}{k}'_i + \underset{\sim}{k}'_j$$

(e)

$$\langle ..\underset{\sim}{k}_i..\underset{\sim}{0}..|\delta L|..\underset{\sim}{0}..\underset{\sim}{k}'_j..\rangle \qquad \underset{\sim}{k}_i = \underset{\sim}{k}'_j$$

(f)

Basic interaction vertices .

Fig. II. 4. 2.

Taking into account the fact that the states $|\{\underset{\sim}{k}\}\rangle$ describe well defined correlations in the system, the diagrams indicate very clearly what changes in these correlations occur as a consequence of the interactions. The present formalism thus appear as a description of mechanics in terms of a dynamics of correlations.

With the diagrams, it is easy to represent any contribution to the formal solution of the Liouville equation. To obtain the nth order contribution to the evolution of $\rho_{\{\underset{\sim}{k}\}}(t)$, we first draw the final state $|\{\underset{\sim}{k}\}\rangle$. Then, we go to the right through n vertices, using all possible combinations of the six basic vertices which conserve the total wave vector. As an example , the second order contributions to the evolution of $\rho_{\underset{\sim}{k}_i}(t)$ are given in fig. II. 4. 3).

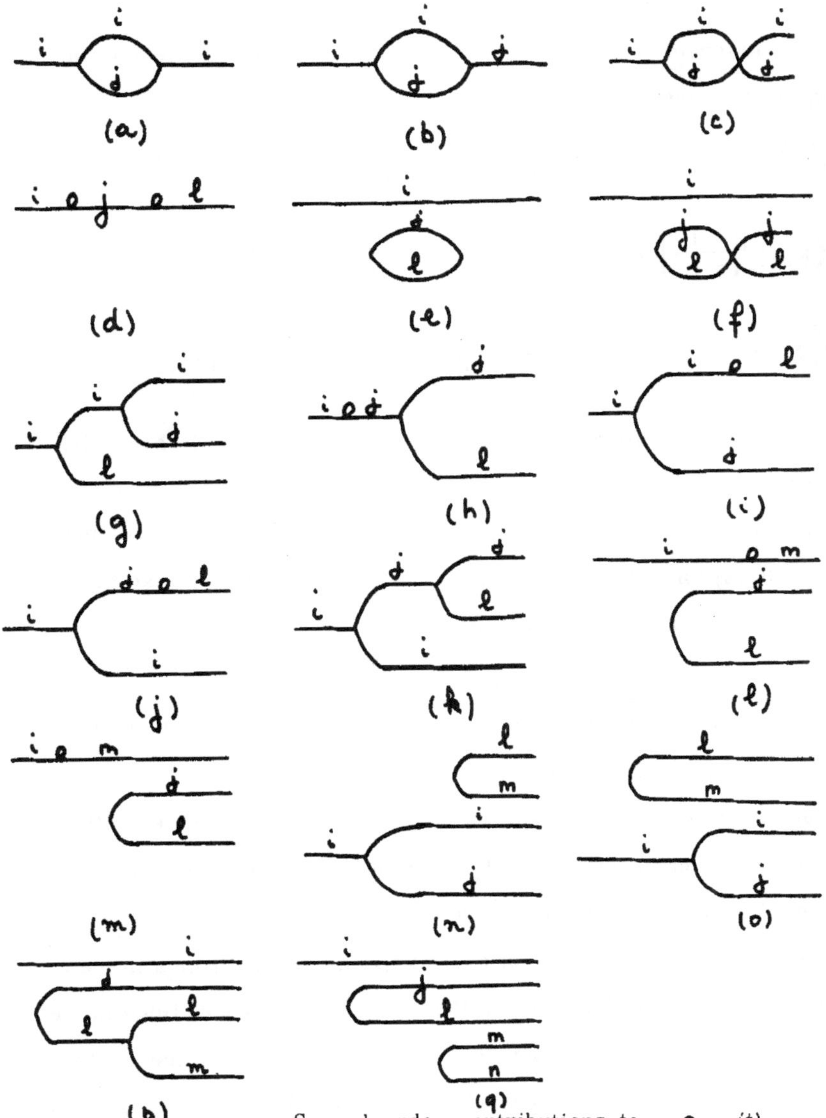

Second order contributions to $\rho_{\underset{\sim}{k}_i}(t)$

Fig. 4.3

Also, it is very easy, once we have a given diagram, to write down its analytic contribution . Let us for instance consider diagram (g) in fig. II. 4.3.

Reading the diagram from the left to the right, we obtain (we do not write explicitly wave vector equal to $\underset{\sim}{0}$) :

$$\langle\{\underset{\sim}{k}\}|\frac{1}{z-L_0}(\delta L\frac{1}{z-L_0})^2|\{\underset{\sim}{k}'\}\rangle_{(g)} = \langle\underset{\sim}{k}_i|\frac{1}{z-L_0}|\underset{\sim}{k}_i\rangle\langle\underset{\sim}{k}_i|\delta L|\underset{\sim}{k}_i',\underset{\sim}{k}_l'\rangle \times$$

$$\text{(II.4.1)}\quad\langle\underset{\sim}{k}_i',\underset{\sim}{k}_l'|\frac{1}{z-L_0}|\underset{\sim}{k}_i',\underset{\sim}{k}_l'\rangle\langle\underset{\sim}{k}_i',\underset{\sim}{k}_l'|\delta L|\underset{\sim}{k}_i'',\underset{\sim}{k}_j'',\underset{\sim}{k}_l'\rangle \times$$

$$\times\langle\underset{\sim}{k}_i'',\underset{\sim}{k}_j'',\underset{\sim}{k}_l'|\frac{1}{z-L_0}|\underset{\sim}{k}_i'',\underset{\sim}{k}_j'',\underset{\sim}{k}_l'\rangle\,\delta_{\underset{\sim}{k}_i,\underset{\sim}{k}_i'+\underset{\sim}{k}_l'}\,\delta_{\underset{\sim}{k}_i,\underset{\sim}{k}_i''+\underset{\sim}{k}_j''+\underset{\sim}{k}_l'}$$

In other words, we write a sequence of matrix elements of δL, each corresponding in a well defined order to the vertices; in between these matrix elements, we sandwich propagators (see II.3.7), which are matrix elements of $R_0(z)$ for the corresponding intermediate state; we also have such a propagator for the initial and final states.

II.5. Classification of diagrams.

We shall now discuss the topological structure of the diagrams which appear in the solution of the Liouville equation.

In the most general diagram, we may distinguish three different regions. Let us denote by $|\{\underset{\sim}{k}''\}\rangle$ the initial state of correlation (at the right), by $|\{\underset{\sim}{k}'\}\rangle$ the intermediate state where we have the minimum number of lines.(which may of course appear several times in the diagram and by $|\{\underset{\sim}{k}\}\rangle$ the final state. If s is the minimum number of lines in the diagram, it may of course happen that we have several different states with that number of lines. We then choose as $|\{\underset{\sim}{k}'\}\rangle$ the last one starting from the right. As an example, in fig. II.5.1b we have two different states with one line : $|\underset{\sim}{k}_\alpha,\{\underset{\sim}{0}\}\rangle$ and $|\underset{\sim}{k}_j,\{\underset{\sim}{0}\}\rangle$. As the latter is the second one when we start from the right, we choose it as our state $|\{\underset{\sim}{k}'\}\rangle$. Another example (with two lines) can be found in fig. II.5.1e.

In the most general case where $|\{\underset{\sim}{k}''\}\rangle \neq |\{\underset{\sim}{k}'\}\rangle \neq |\{\underset{\sim}{k}\}\rangle$ we have :

1. <u>a destruction region</u>, i.e. a region where we go from the state $|\{\underset{\sim}{k}''\}\rangle$ to the state $|\{\underset{\sim}{k}'\}\rangle$ in such a way that no intermediate state is identical to $|\{\underset{\sim}{k}'\}\rangle$; in such a region, we go from a state with given correlations to a state where we have less correlations.
2. <u>a diagonal region</u>, i.e. a region where we go from the state $|\{\underset{\sim}{k}'\}\rangle$ back to the state $|\{\underset{\sim}{k}'\}\rangle$; in general such a region contains a succession of irreducible diagonal fragments. By definition, an irreducible diagonal fragment is such that we go from a given state back to that state through a path such that no intermediate state is identical to the initial state.
3. <u>a creation region</u>, , i.e. a region where we go from the state $|\{\underset{\sim}{k}'\}\rangle$ to the final state $|\{\underset{\sim}{k}\}\rangle$ in such a way that no intermediate state is identical to $|\{\underset{\sim}{k}'\}\rangle$; in this region, we go from the state of correlations $|\{\underset{\sim}{k}'\}\rangle$ to a state of higher correlations.

Examples of this decomposition are given in fig. II.5.1 (diagrams (a) and (b) contain the three different types of regions while diagrams (c) , (d) and (e) have only one or two of them) .

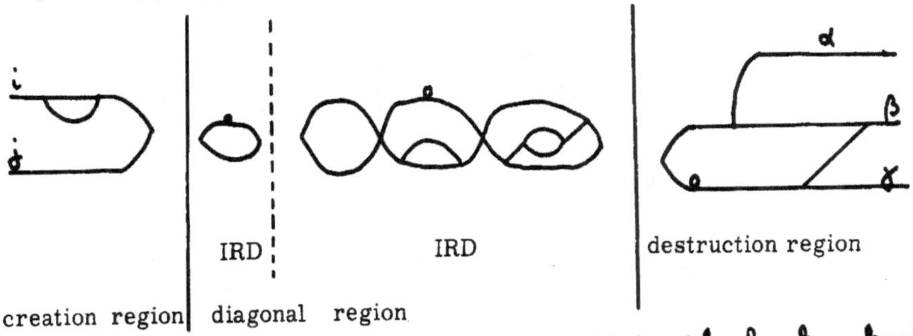

$|\{\underset{\sim}{k}\}\rangle = |\underset{\sim}{k}_i, \underset{\sim}{k}_j = -\underset{\sim}{k}_i\rangle$ $\quad |\{\underset{\sim}{k}'\}\rangle = |\{\underset{\sim}{0}\}\rangle$ $\quad |\{\underset{\sim}{k}''\}\rangle = |\underset{\sim}{k}_\alpha, \underset{\sim}{k}_\beta, \underset{\sim}{k}_\gamma = -\underset{\sim}{k}_\alpha - \underset{\sim}{k}_\beta\rangle$

(a)

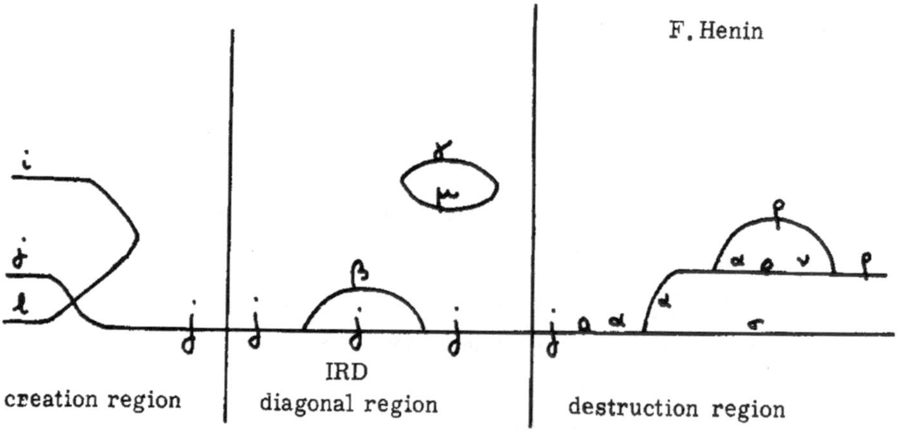

$$|\{k\}\rangle = |k_i, k_j, k_\ell = k'_j - k_j - k_i\rangle ; \; |\{k'\}\rangle = |k'_j\rangle \; ; \; |\{k''\}\rangle = |k_\sigma, k_\rho = k'_j - k_\sigma\rangle$$

(b)

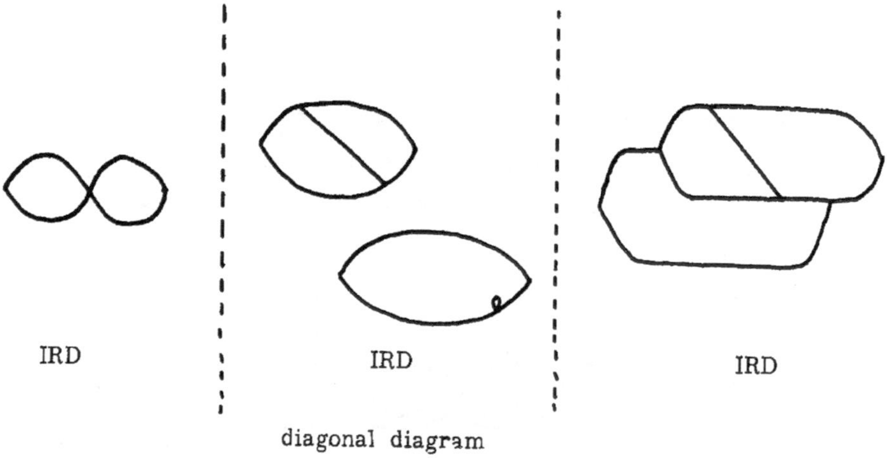

$$|\{k\}\rangle \equiv |\{k'\}\rangle \equiv |\{k''\}\rangle \equiv |\{0\}\rangle$$

(c)

F. Henin

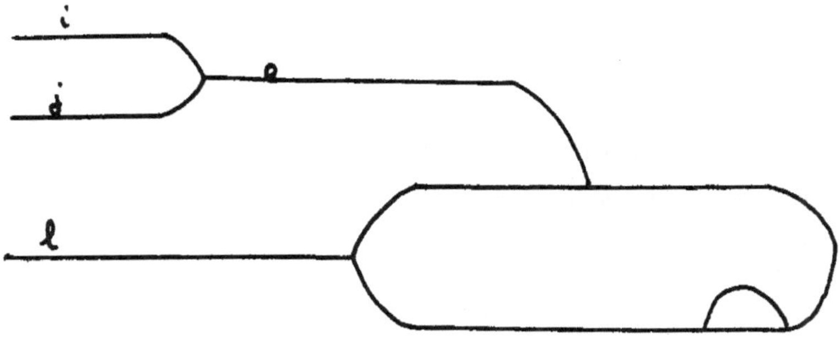

creation diagram

$$|\{\mathbf{k}\}\rangle = |\mathbf{k}_i, \mathbf{k}_j, \mathbf{k}_e = -\mathbf{k}_i - \mathbf{k}_j\rangle \qquad |\{\mathbf{k}'\}\rangle \equiv |\{\mathbf{k}''\}\rangle \equiv |\{\mathbf{0}\}\rangle$$

(d)

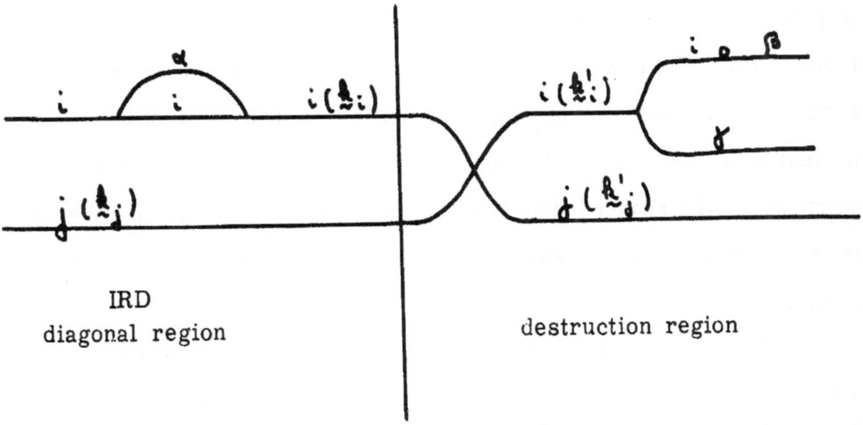

$$|\{\mathbf{k}\}\rangle \equiv |\{\mathbf{k}'\}\rangle = |\mathbf{k}_i, \mathbf{k}_j\rangle \qquad |\{\mathbf{k}''\}\rangle = |\mathbf{k}'_j, \mathbf{k}_\beta, \mathbf{k}_\gamma = \mathbf{k}_i + \mathbf{k}_j - \mathbf{k}'_j - \mathbf{k}_\beta\rangle$$

(e)

IRD = irreducible diagonal fragment

Examples of decomposition of diagrams in creation, diagonal and destruction regions.

Fig. II.5.1

As we shall see later on, the time dependence of the various contributions will be closely related to this decomposition.

II.6 - Initial conditions.

We shall always consider initial conditions such that macroscopic properties like the pressure, density etc.. are finite at every point of the system, even when the limiting case of an infinite system is considered. The interest of this class of initial conditions is obvious from the physical point of view; it can be shown that once the existence and finiteness of the reduced distribution functions for a finite number of degrees of freedom is imposed at t=0 , it will remain so at an arbitrary later time.

This choice of initial conditions introduces mathematical restrictions on the class of functions ρ we consider. It can be shown that this initial condition requires the following volume dependence for the various Fourier coefficients :

(II.6.1) $$\rho_{\underset{\sim}{k}_1 \ldots \underset{\sim}{k}_N} = (8\pi^3/\Omega)^{\nu} \tilde{\rho}_{\underset{\sim}{k}_1 \ldots \underset{\sim}{k}_N}$$

where ν is the number of independent non vanishing wave vectors which appear in the set $\underset{\sim}{k}_1 \ldots \underset{\sim}{k}_N$. (By this, we mean the total number of non vanishing wave vectors minus the number of relations of the form $\underset{\sim}{k}_i + \ldots + \underset{\sim}{k}_j = \underset{\sim}{0}$ which they satisfy). For instance :

$$\rho_{\underset{\sim}{k}_1, \underset{\sim}{k}_2} = (8\pi^3/\Omega)^2 \, \tilde{\rho}_{\underset{\sim}{k}_1, \underset{\sim}{k}_2} \qquad (\underset{\sim}{k}_1 \neq -\underset{\sim}{k}_2)$$

$$\rho_{\underset{\sim}{k}, -\underset{\sim}{k}} = (8\pi^3/\Omega) \, \tilde{\rho}_{\underset{\sim}{k}, -\underset{\sim}{k}} \tag{II.6.2}$$

The coefficients $\tilde{\rho}_{\underset{\sim}{k}_1 \ldots \underset{\sim}{k}_N}$ do no longer depend explicitly on Ω or N, although they might still depend on the ratio N/Ω.

With these assumptions, although in the formal solution of the Liouville equation, we find terms growing more and more rapidly $(N, N^2 \ldots)$, all contributions to the reduced distribution functions for a finite number of degrees of freedom remain finite. The proof of these theorems is rather lengthy and cannot be given here. However, we shall illustrate them with two examples. We shall consider the contribution of the two diagrams of fig. II.6.1 to the one particle velocity distribution function :

$$\varphi_1(\underset{\sim}{v}, t) = \int d\underset{\sim}{p}_2 \ldots d\underset{\sim}{p}_N \, \rho_0(\{\underset{\sim}{p}\}, t) \tag{II.6.3}$$

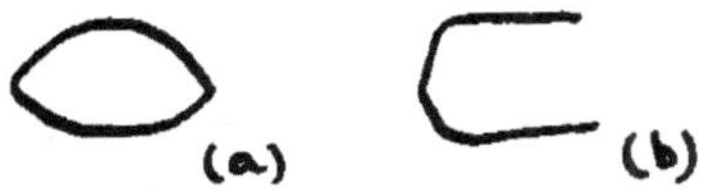

Lowest order diagonal and destruction contributions to $\rho_0(t)$

Fig. II.6.1

The contribution of the cycle (fig. II.6.1a) to the evolution of $\rho_0(t)$ is (see equ. (II.3.5), (II.3.7), (II.3.12) and (II.4.1))

$$[\rho_o(t)]_0 = -\frac{1}{2\pi i}\lambda^2\int_C dz\, e^{-izt}\sum_{i<j}\sum_{\underset{\sim}{k}}\langle 0|\frac{1}{z-L_o}|0\rangle\, \times$$

$$\times \langle 0|\delta L|\underset{\sim}{k}_i=\underset{\sim}{k}, \underset{\sim}{k}_j=-\underset{\sim}{k}\rangle\langle \underset{\sim}{k}_i=\underset{\sim}{k}, \underset{\sim}{k}_j=-\underset{\sim}{k}|\frac{1}{z-L_o}|\underset{\sim}{k}_i=\underset{\sim}{k}, \underset{\sim}{k}_j=-\underset{\sim}{k}\rangle\, \times$$

(II. 6. 4)
$$\times \langle \underset{\sim}{k}_i=\underset{\sim}{k}, \underset{\sim}{k}_j=-\underset{\sim}{k}|\delta L|0\rangle\langle 0|\frac{1}{z-L_o}|0\rangle \rho_o(0)$$

$$= \lambda^2(8\pi^3/\Omega)^2\sum_{i<j}\sum_{\underset{\sim}{k}}(1/2\pi i)\int_C dz\,\frac{e^{-izt}}{z^2}(V_k)^2\, \times$$

$$\times\, \underset{\sim}{k}\cdot\left(\frac{\partial}{\partial \underset{\sim}{p}_i}-\frac{\partial}{\partial \underset{\sim}{p}_j}\right)\frac{1}{z-\underset{\sim}{k}\cdot(\underset{\sim}{v}_i-\underset{\sim}{v}_j)}\,\underset{\sim}{k}\cdot\left(\frac{\partial}{\partial \underset{\sim}{p}_i}-\frac{\partial}{\partial \underset{\sim}{p}_j}\right)\rho_o(0)$$

In the limit of a large system (see II. 1. 10), the summation over $\underset{\sim}{k}$ becomes an integral :

(II. 6. 5)
$$(8\pi^3/\Omega)\sum_{\underset{\sim}{k}} \to \int d^3k$$

Hence

$$[\rho_o(t)]_0 = \lambda^2(8\pi^3/\Omega)\sum_{i<j}\int d^3k\,(1/2\pi i)\int_C dz\,\frac{e^{-izt}}{z^2}\, \times$$

(II. 6. 6)
$$\times\, |V_k|^2\,\underset{\sim}{k}\cdot\left(\frac{\partial}{\partial \underset{\sim}{p}_i}-\frac{\partial}{\partial \underset{\sim}{p}_j}\right)\frac{1}{z-\underset{\sim}{k}\cdot(\underset{\sim}{v}_i-\underset{\sim}{v}_j)}\,\underset{\sim}{k}\cdot\left(\frac{\partial}{\partial \underset{\sim}{p}_i}-\frac{\partial}{\partial \underset{\sim}{p}_j}\right)\rho_o(0)$$

$$= 0(NC)$$

if we take into account the fact that the sum over the particles contains N^2 terms.

Similarily, using (II. 3. 11), the contribution of the destruction diagram (fig. II. 6. 1b) is :

$$[\rho_o(t)]_C = -\lambda(8\pi^3/\Omega)^2 \sum_{i<j} \sum_{\underset{\sim}{k}} \int_C dz\, e^{-izt} \quad (2\pi i)^{-1} \times$$

$$\times \langle 0 | \frac{1}{z - L_o} | 0 \rangle \langle 0 | \delta L | \underset{\sim}{k}_i = \underset{\sim}{k},\ \underset{\sim}{k}_j = -\underset{\sim}{k} \rangle \times$$

$$\text{(II.6.7)} \quad \times \langle \underset{\sim}{k}_i = \underset{\sim}{k},\ \underset{\sim}{k}_j = -\underset{\sim}{k} | \frac{1}{z - L_o} | \underset{\sim}{k}_i = \underset{\sim}{k}, \underset{\sim}{k}_j = -\underset{\sim}{k} \rangle \tilde{\rho}_{\underset{\sim}{k}_i = \underset{\sim}{k}, \underset{\sim}{k}_j = -\underset{\sim}{k}}^{(0)}$$

$$= \lambda(8\pi^3/\Omega)^2 \sum_{i<j} \sum_{\underset{\sim}{k}} (1/2\pi i) \int_C dz \frac{e^{-izt}}{z} V_k \underset{\sim}{k} \cdot \left(\frac{\partial}{\partial \underset{\sim}{p}_i} - \frac{\partial}{\partial \underset{\sim}{p}_j}\right) \times$$

$$\times \frac{1}{z - \underset{\sim}{k} \cdot (\underset{\sim}{v}_i - \underset{\sim}{v}_j)} \tilde{\rho}_{\underset{\sim}{k}_i = \underset{\sim}{k},\ \underset{\sim}{k}_j = -\underset{\sim}{k}}^{(0)}$$

For large systems, this becomes :

$$[\rho_o(t)]_C = \lambda(8\pi^3/\Omega) \sum_{i<j} \int d^3k\, (1/2\pi i) \int_C dz \frac{e^{-izt}}{z} V_k$$

$$\text{(II.6.8)} \quad \underset{\sim}{k} \cdot \left(\frac{\partial}{\partial \underset{\sim}{p}_i} - \frac{\partial}{\partial \underset{\sim}{p}_j}\right) \frac{1}{z - \underset{\sim}{k} \cdot (\underset{\sim}{v}_i - \underset{\sim}{v}_j)} \tilde{\rho}_{\underset{\sim}{k}_i = \underset{\sim}{k},\ \underset{\sim}{k}_j = -\underset{\sim}{k}}^{(0)}$$

$$= 0(NC)$$

Let us now introduce these two results in (II.6.3) . Because of the integrations over the velocities, all contributions vanish except if $i = 1$. This means, that, among the $N(N-1)$ diagrams a or b of fig. II.6.1, we only keep the $(N-1)$ diagrams such that $i = 1$. Therefore we obtain :

$$[\varphi_1(\underset{\sim}{v}_1, t)]_O = \lambda^2(8\pi^3/\Omega) \sum_{j>1} \int d^3k\, (1/2\pi i) \int_C dz \frac{e^{-izt}}{z^2} \times$$

$$\times |V_k|^2\, \underset{\sim}{k} \cdot \frac{\partial}{\partial \underset{\sim}{p}_1} \int d\underset{\sim}{p}_2 \ldots d\underset{\sim}{p}_N \frac{1}{z - \underset{\sim}{k} \cdot (\underset{\sim}{v}_1 - \underset{\sim}{v}_j)} \times$$

$$\times \; \underset{\sim}{k} \cdot \left(\frac{\partial}{\partial p_1} - \frac{\partial}{\partial p_j}\right) \rho_o^{(0)} \tag{II.6.9}$$

$$= 0(C)$$

$$\left[\varphi_1 (\underset{\sim}{v}_1, t)\right]_C = \lambda \, (8\pi^3/\Omega) \sum_{j>1} \int d^3k \, (1/2\pi i) \int_C dz \; \frac{e^{-izt}}{z} \; \times$$

$$\times \; V_k \, \underset{\sim}{k} \cdot \frac{\partial}{\partial p_1} \int dp_2 \ldots dp_N \; \frac{1}{z - \underset{\sim}{k} \cdot (\underset{\sim}{v}_1 - \underset{\sim}{v}_j)} \; \tilde{\rho}_{\underset{\sim}{k}_1 = \underset{\sim}{k}, \, \underset{\sim}{k}_j = -\underset{\sim}{k}}^{(0)} \tag{II.6.10}$$

$$= 0(C)$$

The general mechanism which insures convergence at an arbitrary time for the reduced distribution function of a finite number of degrees of freedom is thus twofold ; first , our assumption (II.6.1) (which for instance introduced a factor Ω^{-1} in (II.6.7)), then the suppression of the contributions of many diagrams once we perform the integrations over all but a finite number of degrees of freedom.

In many problems we shall further reduce the class of initial conditions we consider . For instance we shall often restrict ourselves to the class of initial conditions where the correlations are over distances of the order of molecular distances. This will be discussed when necessary.

The property of finiteness of the reduced distribution functions plays a very important role in the obtention of irreversible equations for the macroscopic quantities. Indeed, once finiteness is ascertained with respect to N and Ω, we can further look at the time behavior of the system and find out that in the long time limit, some terms may become negligible. This is not the case for the complet distribution function because of the divergences with respect to N in the limit of a large system. However, for the sake of simplicity it appears often convenient not to worry about this N divergence and to write down asymptotic equations for the complete distribution function (kinetic equations, see § 9),. This procedure is perfectly legitimate provided we keep in mind the fact that all asymptotic equations we shall derive are valid only when they are used for the computation of average quantities which depend on a finite number of degrees of freedom.

II.7 - Time dependence

In order to get some feeling about the simplifications which may arise when we discuss the long time behavior of the system, let us consider in detail some simple and typical contributions which we meet in the evolution of homogeneous systems. To make things even clearer, let us choose a special form of a repulsive potential which will enable us to perform all calculations completely :

$$V(r) = V_o \; e^{-\kappa r} \tag{II.7.1}$$

κ^{-1} is here the range of the intermolecular force. The Fourier transform of this potential is :

$$V_k = V_o \frac{8 \pi \kappa}{(k^2+\kappa^2)^2} \tag{II.7.2}$$

Let us now first investigate the time dependence of the contribution of the simplest diagonal fragment to the evolution of the velocity distribution function : the cycle (fig. II.6.1a) . Using (II.3.6) ,(II.3.11) and (II.7.2), we have :

$$\left[\rho_o(t)\right]_{cycle} = -(1/2\pi i)\int_C dz\, \frac{e^{-izt}}{z^2}\, \psi_2(z)\, \rho_o(0) \tag{II.7.3}$$

with :

$$\psi_2(z) = \lambda^2 \langle 0 | \delta L \frac{1}{z - L_o} \delta L | 0 \rangle \tag{II.7.4}$$

$$= -\lambda^2 (8\pi \kappa V_o)^2 (8\pi^3/\Omega) \sum_{i<j} \int d^3k\, (k^2 + \kappa^2)^{-4}$$

$$\underset{\sim}{k} \cdot \left(\frac{\partial}{\partial \underset{\sim}{p}_i} - \frac{\partial}{\partial \underset{\sim}{p}_j}\right) \frac{1}{z - \underset{\sim}{k} \cdot (\underset{\sim}{v}_i - \underset{\sim}{v}_j)}\, \underset{\sim}{k} \cdot \left(\frac{\partial}{\partial \underset{\sim}{p}_i} - \frac{\partial}{\partial \underset{\sim}{p}_j}\right)$$

$$= -\lambda^2 \sum_{\alpha = x,y,z} \sum_{\beta = x,y,z} (8\pi^3/\Omega) \sum_{ij} \left(\frac{\partial}{\partial p_{i\alpha}} - \frac{\partial}{\partial p_{j\alpha}}\right)$$

$$I^+_{\alpha\beta}(z, \underset{\sim}{p}_i, \underset{\sim}{p}_j) \left(\frac{\partial}{\partial p_{i\beta}} - \frac{\partial}{\partial p_{j\beta}}\right) \tag{II.7.5}$$

where

$$I^+_{\alpha\beta}(z, \underset{\sim}{p}_i, \underset{\sim}{p}_j, k) = (8\pi \kappa V_o)^2 \int d^3k\, \frac{k_\alpha k_\beta}{(k^2 + \kappa^2)^4 \left[z - \underset{\sim}{k} \cdot (\underset{\sim}{v}_i - \underset{\sim}{v}_j)\right]}$$

$$(z \in S^+) \tag{II.7.6}$$

Using cylindrical coordinates with the z axis along the relative velocity :

$$\underset{\sim}{g} = \underset{\sim}{v}_i - \underset{\sim}{v}_j \tag{II.7.7}$$

one obtains easily :

$$I^+_{\alpha\beta}(z, \underset{\sim}{p}_i, \underset{\sim}{p}_j, \kappa) = (\pi/3)(8\pi \kappa V_o)^2\, \delta_{\alpha,\beta}\ \times$$

$$\times\Bigg[(\delta_{\alpha,x}+\delta_{\alpha,y})\ (1/4)\int_{-\infty}^{+\infty} dk_{\parallel}\,\frac{1}{(z-k_{\parallel}g)(k_{\parallel}^2+\kappa^2)^2}$$

$$+\delta_{\alpha,z}\int_{-\infty}^{+\infty} dk_{\parallel}\,\frac{k_{\parallel}^2}{(z-k_{\parallel}g)(k_{\parallel}^2+\kappa^2)^3}\Bigg] \qquad \text{(II.7.8)}$$

As can be easily seen when performing the $k_{\parallel}$ integration, $I^{+}_{\alpha\beta}(z)$ is regular in S^{+} and has poles in S^{-} at

$$z=-i\kappa g \qquad \text{(II.7.9)}$$

The quantity $(\kappa g)^{-1}$ represents the time during which the two particles i and j are interacting, i.e. the collision time τ_{coll}. Using this result, we can easily perform the z integration in (II.7.3). We obtain :

$$\left[\rho_0(t)\right]_{cycle}=-it\,\psi_2(0)+\psi_2'(0)+\mathrm{Res}\left[\frac{e^{-izt}\psi_2(z)}{z^2}\right]_{z=-ikg} \qquad \text{(II.7.10)}$$

Therefore, we have three types of contributions: one proportional to t, a second one which is of order τ_{coll}/t when compared to the first and finally an exponentially decaying contribution proportional to $\exp(-t/\tau_{coll})$. This last term becomes quite negligible for times much longer than the collision time.

If we do not make a special choice of the intermolecular potential, it is easily verified (see an example in chapter III, § 2) that the function $I^{+}_{\alpha\beta}$ has the form of a Cauchy integral :

$$I^{+}_{\alpha\beta}(z)=\int_{-\infty}^{+\infty} d\omega\,\frac{f(\omega)}{z-\omega} \qquad (z\in S^{+}) \qquad \text{(II.7.11)}$$

Provided $f(\omega)$ satisfies some general conditions [4], this function is regular in S^+ and can be continued analytically in S^-. Its singularities in S^- are at a finite distance of the real axis, of the order of τ_{coll}^{-1}. If we assume these singularities to be simple poles (although other types of singularities can also be discussed [3]), the general result (I.7.10) is valid (instead of a single residue, we must take a sum over the residues at all poles in S^-).

Let us now consider the contribution of the simplest destruction fragment to the evolution of $\rho_o(t)$ (fig. II. 6. 1b). In the limit of an infinite system, we have (see (II. 6. 8)):

$$[\rho_o(t)]_C = -(1/2\pi i)\int_C dz \, \frac{e^{-izt}}{z} \mathcal{D}_1(z) \tag{II.7.12}$$

with

$$\mathcal{D}_1(z) = (8\pi^3/\Omega) \sum_{i<j} \int d^3k \, V_k \, \underset{\sim}{k} \cdot \left(\frac{\partial}{\partial \underset{\sim}{p}_i} - \frac{\partial}{\partial \underset{\sim}{p}_j}\right) \times$$

$$\times \frac{1}{z - \underset{\sim}{k}\cdot(\underset{\sim}{v}_i - \underset{\sim}{v}_j)} \tilde{\rho}_{\underset{\sim}{k}_i = \underset{\sim}{k}, \underset{\sim}{k}_j = -\underset{\sim}{k}}(0) \tag{II.7.13}$$

In contrast with the operator $\psi_2(z)$, the singularities of the destruction operator $\mathcal{D}_1(z)$ depend not only on the type of intermolecular potential we choose, but also on the $\underset{\sim}{k}$ dependence of the function $\tilde{\rho}_{\underset{\sim}{k}, -\underset{\sim}{k}}$, i.e: on the initial conditions. Let us denote by κ_{corr}^{-1} the range of the initial correlations. We may for instance suppose that the binary correlation function $g(\underset{\sim}{x}, \underset{\sim}{x}', 0)$ (see II. 2. 9) is of the form :

$$g(\underset{\sim}{x}, \underset{\sim}{x}', 0) = g(|\underset{\sim}{x} - \underset{\sim}{x}'|, 0) = e^{-\kappa_{corr}(|\underset{\sim}{x} - \underset{\sim}{x}'|)} \tag{II.7.14}$$

Then we have :

(II.7.15) $$\tilde{\rho}_{\underset{\sim}{k},-\underset{\sim}{k}}(0) \sim \frac{\kappa_{corr}}{(k^2+\kappa^2_{corr})^2}$$

Therefore, besides the pole at $z = -i\,\kappa g$ due to the Fourier coefficient of the potential, we now have poles at :

(II.7.16) $$z = -i\,\kappa_{corr}\,g$$

in the rhs of (II.7.12) . We thus obtain :

(II.7.17) $$\left[\rho_o(t)\right]_C = \mathfrak{D}_1(0) + \mathrm{Res}\left[\frac{e^{-izt}}{z}\mathfrak{D}_1(z)\right]_{z=-i\kappa g} + \mathrm{Res}\left[\frac{e^{-izt}}{z}\mathfrak{D}_1(z)\right]_{z=-i\,\kappa_{corr}\,g}$$

As for the diagonal fragment, taking into account the fact that $\mathfrak{D}_1(z)$ is of the form of a Cauchy integral, one generalizes this result very easily for the case where one does not assume a particular form of the interaction and the correlation function.

For times t which are much longer than both the collision time and the characteristic time $(\kappa_{corr}\,g)^{-1}$, only the first term remain in the rhs of (II.7.17) . In what follows we shall restrict ourseivses situations where the initial correlations are due to molecular interaction tions. Then the range of the correlations is of the order of the range the interaction and both characteristic times are identical .

Let us now consider the simplest creation fragment:

(fig. II.7.1)

Simplest creation fragment

Fig. II.7.1

It is a contribution to the evolution of $\rho_{\underset{\sim}{k}, -\underset{\sim}{k}}(t)$. Analytically, we have:

(II.7.18) $$\left[\rho_{\underset{\sim}{k}, -\underset{\sim}{k}}(t)\right]_{\supset} = -(1/2\pi i)\int_C dz \frac{e^{-izt}}{z} C_1(z) \rho_o(0)$$

with

(II.7.19) $$C_1(z) = (8\pi^3/\Omega) V_k \frac{1}{z-\underset{\sim}{k}\cdot(\underset{\sim}{v}_i-\underset{\sim}{v}_j)} \underset{\sim}{k}\cdot\left(\frac{\partial}{\partial \underset{\sim}{p}_i} - \frac{\partial}{\partial \underset{\sim}{p}_j}\right)$$

The main difference with the two preceding cases is that we have no longer a summation over the wave vector. However, our aim is to compute average values of dynamical quantities in phase space. If we compute the contribution of (II.7.17) to the complete phase space distribution function $\rho(\{\underset{\sim}{p}\},\{\underset{\sim}{q}\},t)$, we have :

(II.7.20) $$\left[\rho(\{\underset{\sim}{p}\},\{\underset{\sim}{q}\}, t)\right]_{\supset} = -(1/2\pi i)\int_C dz \frac{e^{-izt}}{z} \Gamma_1(z) \rho_o(0)$$

where

(II.7.21) $$\Gamma_1(z) = \int d^3k \exp\left[i\underset{\sim}{k}\cdot(\underset{\sim}{q}_i-\underset{\sim}{q}_j)\right] C_1(z)$$

With our assumption (II.7.1) for the potential , we obtain:

(II.7.22) $$\left[\rho(\{\underset{\sim}{p}\},\{\underset{\sim}{q}\}, t)\right]_{\supset} = \Gamma_1(0) + \text{Res}\left[\frac{e^{-izt}\Gamma_1(z)}{z}\right]_{z=-i\kappa g}$$

The last term is proportional to $\exp\left[-\kappa(|\underset{\sim}{r} - \underset{\sim}{g}t|)\right]$ where $\underset{\sim}{r} = \underset{\sim}{q}_i-\underset{\sim}{q}_j$. It will become negligible for times t such that :

(II.7.23) $$t \gg r/g$$

Later on, we shall only be interested in the value of the distribution function for relative distances of the order of the range of the intermolecular forces. Then the characteristic time r/g will be of the order of the collision time and for $t \gg \tau_{coll}$, the asymptotic contribution will reduce to $\Gamma_1(0)$. This means that, for the computation of the quantities defined above, we may take the asymptotic expression :

$$\left[\rho_{\underset{\sim}{k}, -\underset{\sim}{k}}(t)\right]_{\supset} = C_1(0)\,\rho_o(0) \tag{II.7.24}$$

The contribution of other types of diagrams (diagonal fragments inserted on a line, free propagating lines, destruction diagrams involving exchange vertices (fig. II, 4.2 a, f) can be discussed in a similar way but shall not be considered here (see ref. [1]))

II.8 - Evolution of the velocity distribution function.

From (II.3.5) and (II.3.6), we have :

$$\rho_o(t) = -(1/2\pi i)\int_C dz\; e^{-izt} \sum_{\{\underset{\sim}{k}\}} \sum_{n=0}^{\infty} \langle 0 \, \frac{1}{z-L_o}\left(\delta L \frac{1}{z-L_o}\right)^n |\{\underset{\sim}{k}\}\rangle \times \rho_{\{\underset{\sim}{k}\}}(0) \tag{II.8.1}$$

If we separate out the diagonal part, we obtain :

$$\begin{aligned}\rho_o(t) = & -(1/2\pi i)\int_C dz\; e^{-izt} \sum_{n=0}^{\infty} \langle 0 \, \frac{1}{z-L_o}\left(\delta L \frac{1}{z-L_o}\right)^n |0\rangle \rho_o(0) \\ & -(1/2\pi i)\int_C dz\; e^{-izt} \sum_{\{\underset{\sim}{k}\}\neq\{\underset{\sim}{0}\}} \sum_{n=1}^{\infty} \langle 0 | \frac{1}{z-L_o}\left(\delta L \frac{1}{z-L_o}\right)^n |\{\underset{\sim}{k}\}\rangle \rho_{\{\underset{\sim}{k}\}}(0)\end{aligned} \tag{II.8.2}$$

It is quite obvious that all contributions to the first term in the rhs of (II.8.2) will be successions of irreducible diagonal fragments (see fig. II.8.1). As to the second term, we shall start from the right with a destruction region until we reach the vacuum of correlations $\{0\}$. Then we can again go on towards the left with a succession of irreducible diagonal fragments (see fig. II. 8. 2)

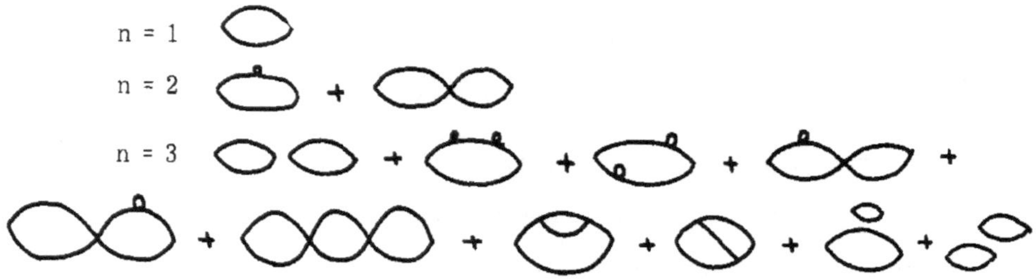

First diagonal contributions to $\rho_0(t)$

Fig. 11.8.1

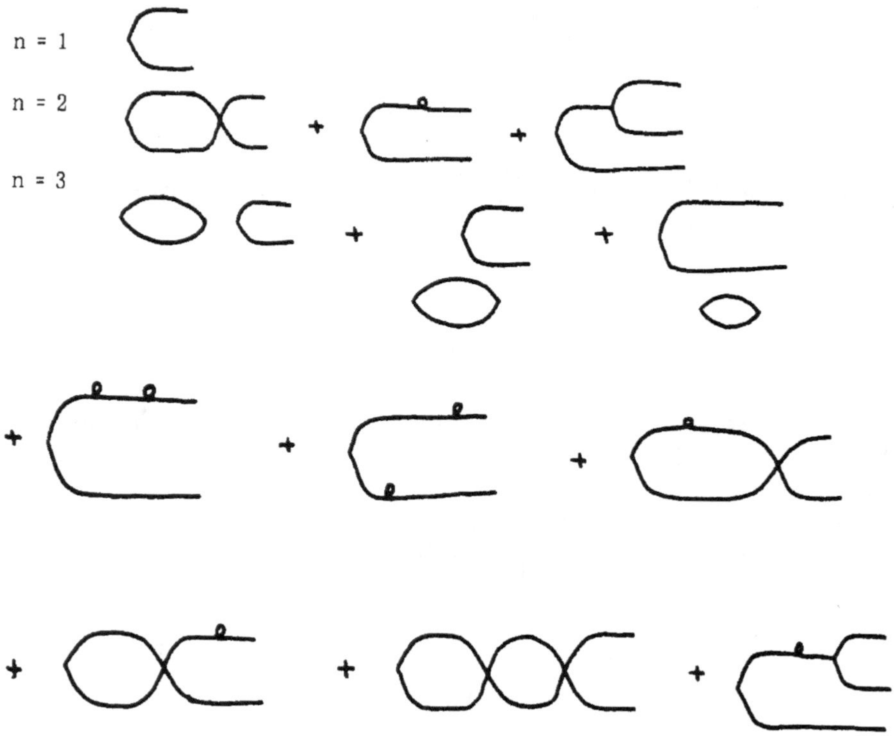

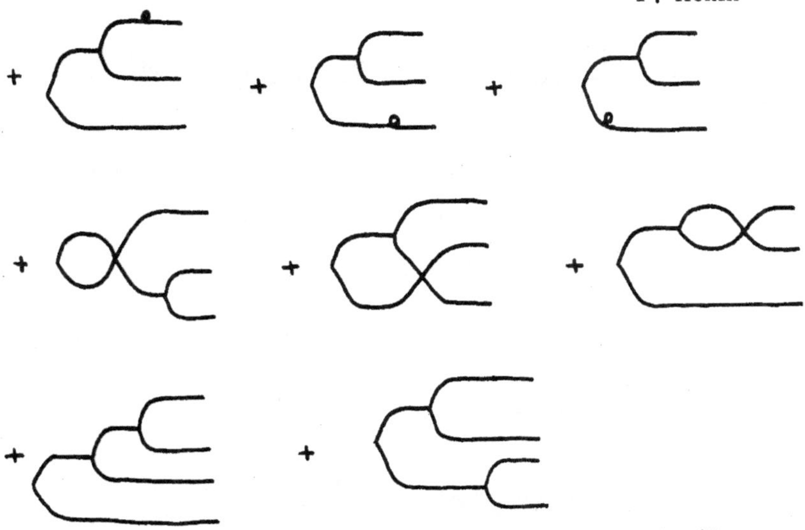

First non diagonal contributions to $\rho_0(t)$

Fig. II.8.2

If we use the diagrams of fig.(II.8.3) to denote

(a) the sum of all irreducible diagonal fragments whose initial (and final state) is the vacuum of correlations

(b) the sum af all destruction fragments whose final state is the vacuum of correlations

we easily obtain a regrouping of all terms in the rhs of (II.8.2) in terms of diagrams (fig. II.8.4)

(a)

(b)

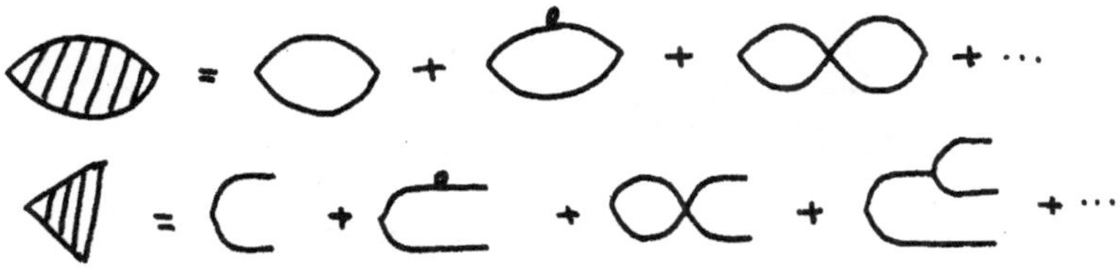

Diagrammatic representations of diagonal and destruction operators.

Fig. II.8.3

$$\rho_{\{0\}}^{(t)} \;\rightarrow\; \sum_{n=0}^{\infty} (\,\oslash\,)^n \;+\; \sum_{n=0}^{\infty} (\,\oslash\,)^n \,\triangleleft$$

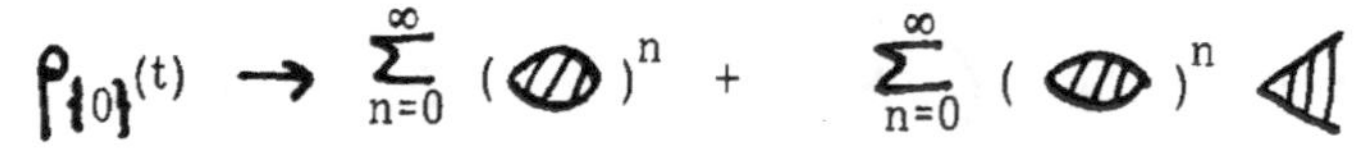

Diagrammatic representation of the evolution equation

for the velocity distribution function

Fig. II. 8. 4

Let us introduce the following operators :

$$\frac{1}{z}\Psi(z) \;=\; \sum_{m=2}^{\infty} \langle 0 \mid (\delta L \,\frac{1}{z-L_0})^m \mid 0 \rangle_{irr}$$

(II. 8. 3)
$$=\; \frac{1}{z} \sum_{m=1}^{\infty} \langle 0 \mid (\delta L \,\frac{1}{z-L_0})^m \,\delta L \mid 0 \rangle_{irr}$$

(II. 8. 4)
$$D_{\{\underset{\sim}{k}\}}(z) \;=\; \sum_{m=1}^{\infty} \langle 0 \mid (\delta L \,\frac{1}{z-L_0})^m \mid \{\underset{\sim}{k}\} \rangle_{irr}$$

where the index irr. means that only terms such that all intermediate states are different from the vacuum of correlations $|\{0\}\rangle$ must be taken into account. This condition means that all propagators in $\Psi(z)$ and $D_{\{\underset{\sim}{k}\}}(z)$ are different from z.

With these operators (II. 8. 2) may now easily be written as :

$$\rho_0(t) = -(1/2\pi i) \int_C dz \,\frac{e^{-izt}}{z} \sum_{n=0}^{\infty} \left[\frac{1}{z}\Psi(z)\right]^n \{\rho_0(0)$$

(II. 8. 5)
$$+ \sum_{\{\underset{\sim}{k}\} \neq \{\underset{\sim}{0}\}} D_{\{\underset{\sim}{k}\}}(z) \,\rho_{\{\underset{\sim}{k}\}}(0) \}$$

Differentiating with respect to time , we obtain :

$$\frac{\partial \rho_0(t)}{\partial t} = (1/2\pi) \int_C dz \, e^{-izt} \sum_{n=0}^{\infty} \left[\frac{1}{z}\Psi(z)\right]^n \{\rho_0(0)$$

$$+ i \sum_{\{\underset{\sim}{k}\} \neq \{\underset{\sim}{0}\}} D_{\{\underset{\sim}{k}\}}(z) \rho_{\{\underset{\sim}{k}\}}(0) \Big\}$$

$$= (1/2\pi) \int_C dz \frac{e^{-izt}}{z} \psi(z) \sum_{n=0}^{\infty} \left[\frac{1}{z}\psi(z)\right]^n \Big\{ \rho_o(0)$$

$$+ \sum_{\{\underset{\sim}{k}\} \neq \{\underset{\sim}{0}\}} D_{\{\underset{\sim}{k}\}}(z) \rho_{\{\underset{\sim}{k}\}}(0) \Big\}$$

$$+ (1/2\pi) \int_C dz\, e^{-izt} \rho_o(0) + (1/2\pi) \int_C dz\, e^{-izt} \times$$

$$\times \sum_{\{\underset{\sim}{k}\} \neq \{\underset{\sim}{0}\}} D_{\{\underset{\sim}{k}\}}(z) \rho_{\{\underset{\sim}{k}\}}(0) \tag{II.8.6}$$

The second term vanishes because its integrand has no singularities. As to the remaining terms, let us perform the z integration. Introducing time dependent operators $G(t)$ and $\mathcal{D}_{\{\underset{\sim}{k}\}}(t)$ which are respectively the inverse Laplace transforms of $\psi(z)$ and $D_{\{\underset{\sim}{k}\}}(z)$:

$$G(t) = - (1/2\pi i) \int_C dz\, e^{-izt} \psi(z) \tag{II.8.7}$$

$$\mathcal{D}_{\{\underset{\sim}{k}\}}(t) = - (1/2\pi i) \int_C dz\, e^{-izt} D_{\{\underset{\sim}{k}\}}(z) \tag{II.8.8}$$

and using the convolution theorem as well as (II.8.5), we obtain :

$$i \frac{\partial \rho_o(t)}{\partial t} = \int_o^t d\tau\, G(t-\tau) \rho_o(\tau) + \sum_{\{\underset{\sim}{k}\} \neq \{\underset{\sim}{0}\}} \mathcal{D}_{\{\underset{\sim}{k}\}}(t) \rho_{\{\underset{\sim}{k}\}}(0) \tag{II.8.9}$$

This generalized "master equation", which has been obtained in a straightforward way from a rearrangement of the terms in the formal solution of the Liouville equation describes the exact behavior of the velocity distribution function for any time. It may seem much more complicated than the original Liouville equation. However, as it will appear below, it has the great advantage that for a very wide class of initial states, it has simple properties in the long time limit.

Let us first notice that we have decomposed the time variation of $\rho_0(t)$ in two contributions of a very different kind. First of all, we have a non-markovian contribution which is expressed in terms of ρ_0 only ; this contribution describes scattering processes; the integration over the past corresponds to the physical fact that the scattering processes have a finite duration (collision time) . On contrast with the first term, the second term in the rhs of (II. 8. 9) does not depend on ρ_0 but on the initial correlations present in the system. This term describes the destruction of these initial correlations.

II. 9 - Kinetic equation.

Let us now consider the case of systems interacting through short range forces and such that the initial correlations are over a molecular range . For such systems, the duration of a collision is very short and many simplifying features appear if we consider the asymptotic behavior of the system, i. e. its behavior for times t such that

(II. 9. 1) $$t \gg \tau_{coll}$$

Generalizing our discussion of § 7 , we shall assume that the operators $\psi(z)$ and $\sum_{\{\underset{\sim}{k}\}} D_{\{\underset{\sim}{k}\}}(z)\, \rho_{\{\underset{\sim}{k}\}}(0)$ have the following properties :

1. they are analytical functions of z in the whole complex plane except

for a finite discontinuity along the real axis . These operators are analytical in S^{+} and can be continued analytically in the lower half plane.

These properties are consequences of the definition of these operators provided perturbation calculus converges. They have been verified in detail for the lowest order contribution to $\psi(z)$ and $\sum_{\{k\}} D_{\{k\}}(z)\rho_{\{k\}}(z)$ in § 7 ; there it has been shown that both these contributions are Cauchy integrals. A detailled discussion of the operator $\psi(z)$ at higher orders has been recently done for the problem of anharmonic solids and some quantum field theory problems [5)].

2. The singularities of the analytical continuation in S^{-} are poles at a finite distance from the real axis. This assumptinn must be considered as a sufficient condition for the validity of the kinetic equation we shall derive. We have seen how it can be realized for a simple type of interaction potential and a simple initial condition in §7. For more complicated interactions or initial conditions, singularities other than poles could appear and the following proofs must be amended but we shall not consider such cases here.

With these properties of the diagonal and destruction operator in the z plane, our results of § 7 can be easily generalized for the discussion of the two kinds of contributions in (II. 8. 9) .

Let us first consider the destruction term. Using (II. 8. 8) , we obtain :

$$\lim_{t \gg \tau_{coll}} \sum_{\{k\}} \mathcal{D}_{k}(t)\, \rho_{\{k\}}(0) = \lim_{t \gg \tau_{coll}} \sum_{j} \mathrm{Res} \left[\sum_{\{k\}} D_{\{k\}}(z)\, e^{-izt} \rho_{\{k\}}(0) \right]_{\zeta_j = z}$$

$$= 0 \tag{II. 9. 2}$$

where the ζ_j's are the poles in the lower half plane of the function

$\sum_{\{k\}} D_{\{k\}}(z)\, \rho_{\{k\}}(0)$. (II. 9. 2) corresponds to our result of 7 that the lowest order destruction contribution to the evolution of $\rho_o(t)$ is asymptotically constant; upon time differentiation, we obtain zero. Therefore, once (II. 9. 1) is fulfilled and the initial correlations are short range, the master equation becomes :

(II. 9. 3) $$i \frac{\partial \rho_o}{\partial t} = \lim_{t \gg \tau_{coll}} \int_0^t d\tau\, G(t-\tau)\rho_o(\tau)$$

while (II. 8. 5) reduces to :

$$\rho_o(t) = \mathrm{Res}\left[\frac{e^{-izt}}{z} \sum_{n=1}^{\infty} \left(\frac{1}{z}\psi(z)\right)^n\right]_{z=0} \bar{\rho}_o(0) + \rho_o(0)$$

(II. 9. 4) $$= \sum_{p=0}^{\infty} \sum_{q=0}^{\infty} (1/p!\, q!)\, (-it)^p \left\{\frac{d^p}{dz^p}\left[\psi(z)\right]^{p+q}\right\}_{z=0} \bar{\rho}_o(0) + \rho_o(0)$$

where the function $\psi(z)$ which has to be used for $z \to 0$ is the analytical continuation of the function defined in S^+ .[1)]

It can be shown , through some lengthy algebraic manipulations that this gives rise to the kinetic equation :

(II. 9. 5) $$i \frac{\partial \rho_o(t)}{\partial t} = \Omega\, \psi(0)\, \rho_o(t)$$

where Ω is a complicated functional of ψ and its derivatives for $z \to 0$:

(II. 9. 6) $$\Omega = \sum_{\alpha=0}^{\infty} \Omega_\alpha$$

(II. 9. 7) $$\Omega_o = 1$$

(II. 9. 8) $$\Omega_\alpha = \lim_{z \to 0} \Omega_\alpha(z) \qquad \alpha > 1$$

1) $\rho_o(0)$ is a modified initial creation and is given by the expression between in the rhs of (II. 8. 5)

The $\Omega_\alpha(z)$ are given by a recursion formula :

$$\Omega_\alpha(z) = (1/\alpha) \sum_{\beta=0}^{\alpha-1} \left\{ \frac{\partial}{\partial z} \Omega_{\alpha-1-\beta}(z)\psi(z) \right\} \Omega_\beta(z) \tag{II.9.8}$$

The operator Ω takes into account the finite duration of the collision. We shall not give this derivation here but rather use some simple considerations which will emphasize the meaning of both the operator ψ and Ω

The operator ψ has the dimension of the inverse of a time (see II.8.3) . As this operator describes the collisions occuring in the system, this time is of the order of the relaxation time.(for instance, in dilute gases (Boltzmann equation), the relaxation time is connected with binary collisions, i.e. those terms in ψ which involve only two particles). Derivation with respect to z of ψ increases by one the power of one of the unperturbed propagators. In our simple example of § 7, we have seen that this amounts to bring an extra factor τ_{coll}. Therefore, any contribution to rhe rhs of (II.9.4) corresponding to a given value of p and q is of the order :

$$(t/\tau_{rel})^p (\tau_{coll}/\tau_{rel})^q \tag{II.9.9}$$

Let us first neglect τ_{coll}/τ_{rel}, i.e. let us consider the collisions as instantaneous events. Then, we may restrict ourselves in (II.9.4) to the term q = 0 and we obtain :

$$\rho_o(t) = \sum_{p=0}^{\infty} (1/p!) (-it\psi(0))^p \rho_o(0) + 0 (\tau_{coll}/\tau_{rel}) \tag{II.9.10}$$

This leads us to :

$$\frac{\partial \rho_o(t)}{\partial t} = -i\psi(0) \rho_o(t) + 0 (\tau_{coll}/\tau_{rel}) \tag{II.9.11}$$

This equation is a very simple generalization of Boltzmann's equation; besides two-body collisions, it includes collisions between an arbitrary number of particles. However, if we do not neglect collisions between more than two particles , it is not consistent. Indeed, for a dilute gas of hard spheres for instance, where we can restrict ourselves to two-body collisions, it can be shown that the relaxation time is given by :

(II.9.12) $$\tau_{rel}^{-1} = C\, a^2 \bar{v}$$

where a is the diameter of the particles and $\bar{v}$ their average velocity. As the only dimensionless parameter we have is $a^3 C$, we must expect that, when we take into account higher order collision processes, we shall have an expansion analogous to the virial expansion :

(II.9.13) $$\tau_{rel}^{-1} = Ca^2\bar{v}\left[1 + \alpha\, a^3 C + \beta (a^3 C)^2 + \ldots\right]$$

Now, we also have :

(II.9.14) $$\tau_{coll} = a/\bar{v}$$

and thus

(II.9.15) $$\tau_{coll}/\tau_{rel} = 0\,(a^3 C)$$

Therefore, the procedure we have followed is certainly not consistent: we cannot keep higher order collision processes (i.e. corrections of order a^3C in τ_{rel}) and neglect terms of the order of τ_{coll}/τ_{rel} in the rhs of (II.9.4) . In order to understand the general evolution equation (II.9.5) , let us keep in (II.9.4) the first correction, that which is proportional to τ_{coll}/τ_{rel} (i.e. the contribution q =1) . Then we have :

$$\rho_0(t) = \sum_{p=0}^{\infty} (1/p!)\,(-it\,\psi(0))^p\,\rho_0(0)$$

$$+ \sum_{p=0}^{\infty} \sum_{q=0}^{\infty} (1/(p+q)!)(-it\psi(0))^p \psi'(0)(-it\psi(0))^q \rho_o(0)$$

(II.9.16)

$$+ 0(\tau_{coll}/\tau_{rel})^2$$

Upon differentiation with respect to t, it is easy to show that one obtains :

$$\text{(II.9.17)} \quad \frac{\partial \rho_o(t)}{\partial t} = -i(1 + \psi'(0))\psi(0)\rho_o(t) + 0(\tau_{coll}/\tau_{rel})^2$$

The operator $\psi'(0)$, which takes into account, in first order, the finite duration of the collision, is precisely identical to the operator Ω_1 in (II.9.6) .

II.10-Evolution of the correlations in an homogeneous system.

As all derivations are very similar to the derivation of the generalized master equation for the velocity distribution function, we shall only indicate how they proceed and what are the final results .

The most general diagram contributing to the evolution of a given correlation contains all three types of regions defined in § 5. We shall now write :

$$\text{(II.10.1)} \qquad \rho_{\{\underset{\sim}{k}\}}(t) = \rho'_{\{\underset{\sim}{k}\}}(t) + \rho''_{\{\underset{\sim}{k}\}}(t)$$

where, by definition :

$\rho'_{\{\underset{\sim}{k}\}}(t)$ contains all diagrams without creation region

$\rho''_{\{\underset{\sim}{k}\}}(t)$ contains all diagrams which end by a creation region

This decomposition is performed in detail for all second order diagrams contributing to the evolution of $\rho_{\underset{\sim}{k},-\underset{\sim}{k}}$ in fig;II.10.1.

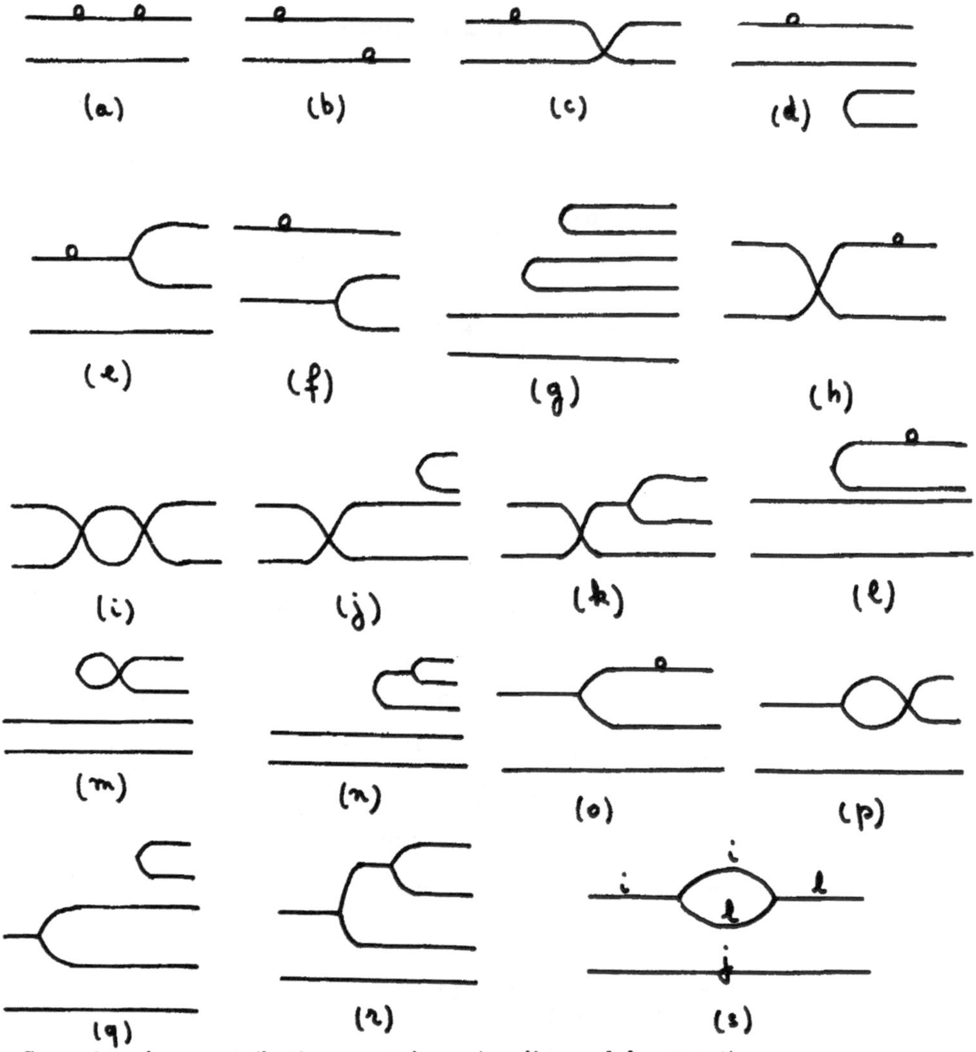

Second order contributions $\rho'_{\underset{\sim}{k}, -\underset{\sim}{k}}$ (no diagonal fragment)

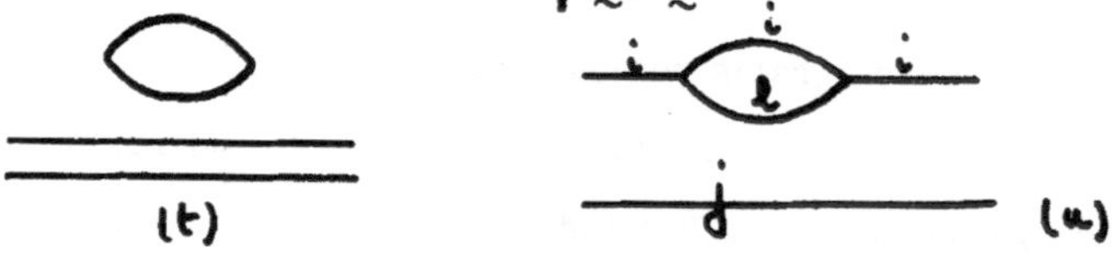

Second order contributions to $\rho'_{\underset{\sim}{k}', -\underset{\sim}{k}}$ (one diagonal fragment)

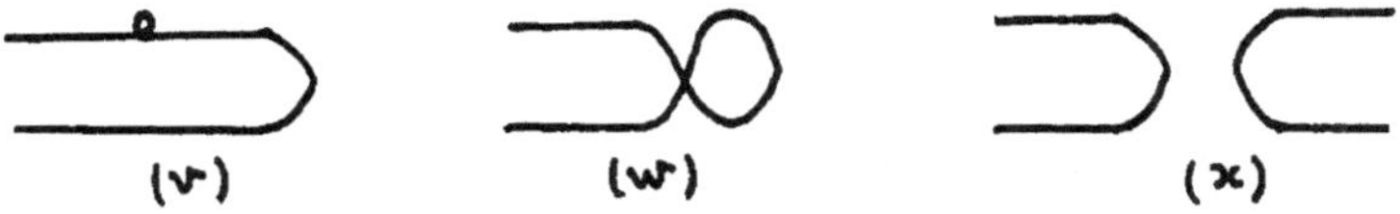

Second order contributions to $\rho''_{\underset{\sim}{k},-\underset{\sim}{k}}$

Second order contributions to the evolution of the binary correlation $\rho_{\underset{\sim}{k},-\underset{\sim}{k}}$

Fig. II. 10. 1

To discuss the evolution of $\rho'_{\{\underset{\sim}{k}\}}(t)$, one decomposes the relevant diagrams into those which are diagonal and those which contain a destruction region (see fig. II. 10. 1). In this way, one verifies easily that these functions obey an evolution equation very similar to the general master equation for the velocity distribution function :

$$i\frac{\partial \rho'_{\{\underset{\sim}{k}\}}(t)}{\partial t} = \int_0^t d\tau\, G_{\{\underset{\sim}{k}\}}(t-\tau)\rho'_{\{\underset{\sim}{k}\}}(\tau)$$

$$\text{(II. 10. 2)} \qquad + \sum_{\{\underset{\sim}{k}'\}}{}' \mathcal{D}_{\{\underset{\sim}{k}\}\{\underset{\sim}{k}'\}}(t, \rho_{\{\underset{\sim}{k}'\}}(0))$$

where $G_{\{\underset{\sim}{k}\}}(t)$ is the inverse Laplace transform of the diagonal operator:

$$\text{(II. 10. 3)} \qquad \psi_{\{\underset{\sim}{k}\}}(z) = \sum_{n=1}^{\infty} \langle\{\underset{\sim}{k}\}|\delta L\left(\frac{1}{z-L_0}\delta L\right)^n|\{\underset{\sim}{k}\}\rangle_{irr}$$

while $\mathcal{D}_{\{\underset{\sim}{k}\}\{\underset{\sim}{k}'\}}(t, \rho_{\{\underset{\sim}{k}'\}}(0))$ is the inverse Laplace transform of the destruction operator :

$$\text{(II. 10. 4)} \qquad D_{\{\underset{\sim}{k}\}\{\underset{\sim}{k}'\}}(z)\,\rho_{\{\underset{\sim}{k}'\}}(0) = \sum_{m=1}^{\infty}\langle\{\underset{\sim}{k}\}|\left(\delta L\frac{1}{z-L_0}\right)^m|\{\underset{\sim}{k}'\}\rangle_{dest.}\,\rho_{\{\underset{\sim}{k}'\}}(0)$$

The dash on the summation over $\{\underset{\sim}{k}'\}$ in (II. 10. 2) means that only those states $|\{\underset{\sim}{k}'\}\rangle$ which are such that the transition $\{\underset{\sim}{k}\}\leftarrow\{\underset{\sim}{k}'\}$ describes a destruction of correlations must be taken into account.

The evolution of $\rho'_{\{\underset{\sim}{k}\}}(t)$ is due to the dissipation of the initial correlations through the collision processes. For long times, a pseudomarkovian equation similar to (II.9.5) can also be derived from (II.10.2) .

As to the evolution of $\rho''_{\{\underset{\sim}{k}\}}(t)$, the main point is to notice that if we have at the left a given creation diagram, corresponding to a transition $\{\underset{\sim}{k}\} \leftarrow \{\underset{\sim}{k}'\}$, we may have at the right of this creation diagram any of the diagrams which contribute to the evolution of $\rho'_{\{\underset{\sim}{k}'\}}(\tau)$ (if $\{\underset{\sim}{k}'\} = \{\underset{\sim}{0}\}$, we may have all the diagrams which contribute to the evolution of the velocity distribution function), if τ is the time corresponding to the first creation vertex. This remark makes it possible to show rigorously that one has :

$$\text{(II.10.5)} \qquad \rho''_{\{\underset{\sim}{k}\}}(t) = \sum_{\{\underset{\sim}{k}'\}}{}' \int_0^t C_{\{\underset{\sim}{k}\}\{\underset{\sim}{k}'\}}(t-\tau)\, \rho'_{\{\underset{\sim}{k}'\}}(\tau)\, d\tau$$

where the dash on the summation over $\{\underset{\sim}{k}'\}$ means that only those states $|\{\underset{\sim}{k}'\}\rangle$ corresponding to a lower state of correlations than $|\{\underset{\sim}{k}\}\rangle$ must be taken into account.

$C_{\{\underset{\sim}{k}\}\{\underset{\sim}{k}'\}}(t)$ is the inverse Laplace transform of the creation operator :

$$\text{(II.10.6)} \qquad C_{\{\underset{\sim}{k}\}\{\underset{\sim}{k}'\}}(z) = \sum_{m=1}^{\infty} \langle\{\underset{\sim}{k}\}| \left(\frac{1}{z-L_0}\, \delta L \right)^m |\{\underset{\sim}{k}'\}\rangle$$

Equation (II.10.5) describes the continuous creation of fresh correlations by direct mechanical interactions from less excited states.

F. Henin

II.11 - Approach to equilibrium of the velocity distribution function.

For weakly coupled or dilute systems, the approach to equilibrium is usually discussed by means of an $\mathcal{H}$-theorem. More precisely, one shows that the quantity

$$\mathcal{H} = \int \{dp\}^N \rho_o \ln \rho_o \tag{II.11.1}$$

decreases monotonically in time and that the stationary solution (which is unique) corresponds to the equilibrium distribution.

Unfortunately, this theorem cannot be completely generalized when higher order contributions are taken into account. We shall only consider the case of systems where there exists a parameter such that a perturbation expansion in powers of that parameter has a meaning (coupling constant λ for weakly coupled systems, concentration C for dilute systems). As an example, we shall consider the case where an expansion in powers of λ has a meaning. Then, with the following expansions:

$$\rho_o(t) = \rho_o^{(0)}(t) + \lambda \rho_o^{(1)}(t) + \lambda^2 \rho_o^{(2)}(t) + \dots \tag{II.11.2}$$

$$\psi(0) = \lambda^2 \psi_2(0) + \lambda^3 \psi_3(0) + \lambda^4 \psi_4(0) + \dots \tag{II.11.3}$$

$$\Omega = 1 + \lambda^2 \psi_2'(0) + \dots \tag{II.11.4}$$

the kinetic equation (II.9.5) gives us a set of equations:

$$\frac{\partial \rho_o^{(0)}(t)}{\partial \lambda^2 t} = - i \psi_2(0) \rho_o^{(0)}(t) \tag{II.11.5}$$

$$\frac{\partial \rho_o^{(1)}(t)}{\partial \lambda^2 t} = - i \psi_2(0) \rho_o^{(1)}(t) - i \psi_3(0) \rho_o^{(0)}(t) \tag{II.11.6}$$

$$\text{(II.11.7)} \quad \frac{\partial \rho_o^{(2)}(t)}{\partial \lambda^2 t} = -i\,\psi_2(0)\,\rho_o^{(2)}(t) - i\,\psi_3(0)\,\rho_o^{(1)}(t) - i\,\psi_4(0)\,\rho_o^{(0)}(t)$$

$$- i\,\psi_2'(0)\,\psi_2(0)\,\rho_o^{(0)}(t)$$

etc...

The $\mathcal{H}$ -theorem for the lowest order approximation (for weakly coupled systems, see chapter III, §§ 2 and 3) shows us that $\rho_o^{(0)}(t)$ decreases monotonically towards its equilibrium value :

$$\text{(II.11.8)} \qquad \rho_o^{(0)}(t \rightarrow \infty) = f^{(0)}(H_o)$$

For times much longer than the relaxation time for $\rho_o^{(0)}(t)$, the next approximation is then given by :

$$\text{(II.11.9)} \quad \frac{\partial \rho_o^{(1)}(t)}{\partial \lambda^2 t} = -i\,\psi_2(0)\,\rho_o^{(1)}(t) - i\,\psi_3(0)\,f^{(0)}(H_o)$$

The interesting feature [*] is now that one can show that :

$$\text{(II.11.10)} \qquad \psi_n(0)\; g(H_o) = 0$$

where g is an arbitrary function of the unperturbed hamiltonian. A general method to verify this property can be found in [3)]. This method is based on the discussion of an integral equation and rather formal. A more cumbersome method [1) 6)] consists in the splitting of each ψ_n in a number of operators according to the number of particles which appear in the diagram . For instance, in the operator $\psi_3(0)$, we have

---

[*] This is valid for gases where the interaction is velocity independent. For anharmonic solids for instance, the situation is more complicated because of the action dependence of the potential and this property is not valid. This makes it very difficult to study the approach to equilibrium at higher orders than λ^4. [1) 5) 7)]

two diagrams (fig. II. 11. 1) : one with three particles (a) which we call $\psi_3^{(3)}$, the next one (b) with two particles which we call $\psi_3^{(2)}$.

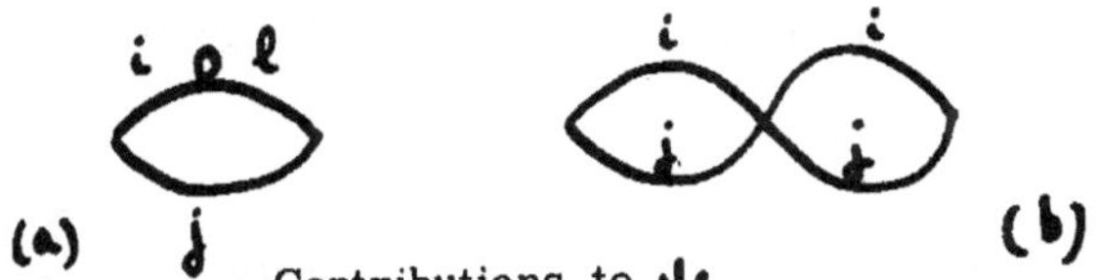

Contributions to ψ_3

Fig. II. 11. 1

One then shows that :

(II. 11. 11) $$\psi_n^{(\nu)}(0)\ g(H_o) = 0$$

As an example, let us verify this for $n=3$, $\nu = 3$ (diagram a, fig. II. 11. 1)

We have :

$$\psi_3^{(3)}(0)\, g(H_o) = \lim_{z\to 0} \sum_{ijl} \sum_{\mathbf{k}} \langle 0|\delta L\,|\,\mathbf{k}_i=\mathbf{k}, \mathbf{k}_j = -\mathbf{k}\rangle \times$$

(II. 11. 12) $$\langle \mathbf{k}_i=\mathbf{k}, \mathbf{k}_j = -\mathbf{k} \,|\, \frac{1}{z-L_o} \,|\, \mathbf{k}_i=\mathbf{k}, \mathbf{k}_j=-\mathbf{k}\rangle\langle \mathbf{k}_i=\mathbf{k}, \mathbf{k}_j = -\mathbf{k}\,|\delta L|\,\mathbf{k}_l=\mathbf{k}, \mathbf{k}_j=-\mathbf{k}\rangle \times$$

$$\times \langle \mathbf{k}_l=\mathbf{k}, \mathbf{k}_j=-\mathbf{k} \,|\, \frac{1}{z-L_o} \,|\, \mathbf{k}_l=\mathbf{k}, \mathbf{k}_j=-\mathbf{k}\rangle\langle \mathbf{k}_l=\mathbf{k}, \mathbf{k}_j=-\mathbf{k}\,|\delta L\,|\,0\rangle\, g(H_o)$$

Using (II. 1. 1) , (II. 2. 12) and (II. 3. 11) as well as :

(II. 11. 13) $$\lim_{z\to 0}\ a/(z-a) = -1$$

we easily obtain :

$$\psi_3^{(3)}(0)\, g(H_o) \sim \lim_{z\to 0} \sum_{ijl} \int d^3k\, |V_k|^2\, V_k\, \mathbf{k}\cdot\left(\frac{\partial}{\partial \mathbf{p}_i} - \frac{\partial}{\partial \mathbf{p}_j}\right) \times$$

(II. 11. 14) $$\times\ \frac{\mathbf{k}\cdot(\mathbf{v}_i-\mathbf{v}_l)}{z-\mathbf{k}\cdot(\mathbf{v}_i-\mathbf{v}_j)}\ \frac{\partial^2 g(H_o)}{\partial H_o^2}$$

Now, let : $i \to j$, $j \to i$, $\underset{\sim}{k} \to -\underset{\sim}{k}$ and take half sum of the rhs of (II. 11. 14) and the term obtained through this interchange of dummy variables. We obtain :

$$\psi_3^{(3)}(0)\, g(H_o) \sim \lim_{z \to 0} \sum_{ijl} \int d^3k \, |V_k|^2 \, V_k \, \underset{\sim}{k} \cdot (\underset{\sim}{v}_i - \underset{\sim}{v}_j) \, \frac{\partial^3 g(H_o)}{\partial H_o^3} \tag{II. 11. 15}$$

$$= 0$$

because the integrand is an odd function of $\underset{\sim}{k}$.

Let us now go back to the discussion of the set of equations (II. 11. 5) , (II. 11. 6) , (II. 11. 7) , etc... for long times. Using (II. 11. 10), we notice that (II. 11. 9) reduces to :

$$\frac{\partial \rho_o^{(1)}(t)}{\partial \lambda^2 t} = - i \psi_2(0) \rho_o^{(1)}(t) \tag{II. 11. 16}$$

As only the lowest order operator remains in this equation , we again obtain the result :

$$\rho_o^{(1)}(t \to \infty) = f^{(1)}(H_o) \tag{II. 11. 17}$$

It is then trivial , using again (II. 11. 10) to show by a recurrence procedure that :

$$\rho_o^{(n)}(t \to \infty) = f^{(n)}(H_o) \tag{II. 11.18}$$

Therefore, we obtain :

$$\rho_o(t \to \infty) = f(H_o) \tag{II. 11. 19}$$

which is the equilibrium distribution . The function $f(H_o)$ is arbitrary (normalized to unity) as far as this proof is concerned but is completely determined from the initial condition (see ref. [1]) .

Let us stress the fact that this generalized $\mathcal{H}$-theorem is not so powerful as the $\mathcal{H}$-theorem for the lowest order approximation. Indeed, the $\mathcal{H}$-theorem for the lowest order approximation actually amounts to proof that all eigenvalues of the hermitian operator $-i\psi_2$ are either negative or zero and that there is a unique eigenfunction (the equilibrium distribution) corresponding to the eigenvalue zero. In the case of stronger coupling, what we have actually done here is to verify that there exists one zero eigenvalue for the complete evolution operator, with the equilibrium distribution as eigenfunction. A true $\mathcal{H}$-theorem would require a proof that the eigenvalue zero is unique and that all other eigenvalues are negative. This is of course very likely, at least for systems where a perturbation expansion has a meaning, i.e. when the lowest order terms give the dominant features of the behaviour of the system.

II.12 - Approach to equilibrium of the correlations in an homogeneous system.

The asymptotic solution for the equation (II.10.2) for the part $\rho'_{\{\underset{\sim}{k}\}}$ of the correlation can be discussed in a way similar to the above discussion for the equation for the velocity distribution function. The main result is :

(II.12.1) $$\rho'_{\{\underset{\sim}{k}\}}(t \to \infty) = 0$$

As a result, in the equation for $\rho''_{\{\underset{\sim}{k}\}}(t)$, we only keep those creation fragments which start from the vacuum of correlations :

(II.12.2) $$\rho''_{\{\underset{\sim}{k}\}}(t) = \int_0^t d\tau \, C_{\{\underset{\sim}{k}\}\{0\}}(t-\tau)\, \rho_0(\tau)$$

These results mean that the initial correlations tend to dissipate. In the long time limit, only the fresh correlations which are continuously created from the velocity distribution function remain.

For times such that the velocity distribution function has reached its equilibrium value, we have ;

(II. 12. 3) $$\rho_{\{\underset{\sim}{k}\}}(t) = \rho''_{\{\underset{\sim}{k}\}}(t) = \int_0^t d\tau\, C_{\{\underset{\sim}{k}\}\{\underset{\sim}{0}\}}(\tau)\, f(H_o)$$

or, using the Laplace transform of the creation operator (see II. 10. 6):

(II. 12. 4)
$$\begin{aligned}\rho_{\{\underset{\sim}{k}\}}(t) = \rho''_{\{\underset{\sim}{k}\}}(t) &= -(1/2\pi)\int_0^t d\tau \int_C dz\; e^{-izt}\, C_{\{\underset{\sim}{k}\}\{\underset{\sim}{0}\}}(z)\, f(H_o) \\ &= (1/2\pi i)\int_C dz\,(e^{-izt}-1)\, z^{-1}\, C_{\{\underset{\sim}{k}\}\{\underset{\sim}{0}\}}(z)\, f(H_o)\end{aligned}$$

Now , as in our discussion of § 7, we take into account the fact that we shall always be interested in average values of dynamical quantities, i.e. in expressions which involve a sum over the wave vector. In such quantities, the operator $C_{\{\underset{\sim}{k}\}\{\underset{\sim}{0}\}}(z)$ is replaced by an operator $\Gamma(z)$ which is a Cauchy integral (see for instance II. 7. 21) . We then have :

(II. 12. 5)
$$\begin{aligned}&(1/2\pi i)\int_{\mathcal{C}} dz\,(e^{-izt}-1)\, z^{-1}\,\Gamma(z) \\ &= \sum_j \operatorname{res}\left[(e^{-izt}-1)\, z^{-1}\,\Gamma(z)\right]_{\zeta_j = z}\end{aligned}$$

where the ζ'_j s are the poles in S^- of $\Gamma(z)$.

For long times, this becomes :

(II. 12. 6) $$\lim_{t\to\infty} \sum_j \operatorname{res}\left[(e^{-izt}-1)\, z^{-1}\,\Gamma(z)\right]_{z=\zeta_j} = -\sum_j \operatorname{res}\left(z^{-1}\,\Gamma(z)\right)_{\zeta_j = z} = \Gamma(0)$$

This result allows us take [x] :

(II. 12.7)

$$\rho_{\{\underset{\sim}{k}\}}(\infty) = \rho''_{\{\underset{\sim}{k}\}}(\infty) = C_{\{\underset{\sim}{k}\}\{\underset{\sim}{0}\}}(0)\, f(H_o)$$

$$= \lim_{z\to 0} \langle\{\underset{\sim}{k}\}| \sum_{m=1} \left(\frac{1}{z-L_o}\delta L\right)^m |0\rangle f(H_o)$$

If we denote by

$$\text{(II. 12.8)} \qquad \rho^{(m)}_{\{\underset{\sim}{k}\}} = \lim_{z\to 0} \langle\{\underset{\sim}{k}\}| \left(\frac{1}{z-L_o}\delta L\right)^m |0\rangle f(H_o)$$

the set of all to $\rho_{\{\underset{\sim}{k}\}}$, contributions involving m vertices we have:

$$\text{(II. 12.9)} \qquad \rho^{(m)}_{\{\underset{\sim}{k}\}} = \langle\{\underset{\sim}{k}\}| \frac{\partial^m f(H_o)}{\partial H_o^m} \frac{V^m}{m!} |0\rangle$$

where V is the intermolecular potential (see (II. 1. 1)) . (II. 12. 9) can be proven , using a recurrence procedure. For m = 1, we have, using (II. 3. 9) :

(II. 12. 10)

$$\rho^{(1)}_{\underset{\sim}{k}_i = \underset{\sim}{k}, \underset{\sim}{k}_j = -\underset{\sim}{k}} = \lim_{z\to 0} \langle \underset{\sim}{k}_i = \underset{\sim}{k}, \underset{\sim}{k}_j = -\underset{\sim}{k} | \frac{1}{z-L_o} \delta L |0\rangle f(H_o)$$

$$= (8\pi^3/\Omega) \lim_{z\to 0} \frac{1}{z-\underset{\sim}{k}\cdot(\underset{\sim}{v}_i - \underset{\sim}{v}_j)} V_k \underset{\sim}{k}\cdot\left(\frac{\partial}{\partial \underset{\sim}{p}_i} - \frac{\partial}{\partial \underset{\sim}{p}_j}\right) f(H_o)$$

$$= (8\pi^3/\Omega)\, V_k \frac{\partial f(H_o)}{\partial H_o}$$

$$= \langle \underset{\sim}{k}_i = \underset{\sim}{k}, \underset{\sim}{k}_j = -\underset{\sim}{k} | V \frac{\partial f(H_o)}{\partial H_o} |0\rangle$$

---

[x] Some care has always to be taken in the use of (II. 12.7). The use of such an expression does not lead to difficulties when one is interested in average quan tities which are linear functionals of the correlations. When non linear functio nals must be considered, one must go back to (II. 12. 4) as has been shown in recent work on anharmonic solids [7].

Assuming (II. 12.9) to hold for a given value of m, we have :

$$\rho^{(m+1)}_{\{\underset{\sim}{k}\}} = \lim_{z\to 0} \langle\{\underset{\sim}{k}\}|\left(\frac{1}{z-L_o}\delta L\right)^{m+1}|0\rangle f(H_o)$$

(II. 12. 11)

$$= \lim_{z\to 0} \sum_{\{\underset{\sim}{k}'\}} \langle\{\underset{\sim}{k}\}| \frac{1}{z-L_o}\delta L|\{\underset{\sim}{k}'\}\rangle \rho^{(m)}_{\{\underset{\sim}{k}'\}}$$

$$= \lim_{z\to 0} \sum_{\{\underset{\sim}{k}'\}} \langle\{\underset{\sim}{k}\}| \frac{1}{z-L_o}\delta L|\{\underset{\sim}{k}'\}\rangle\langle\{\underset{\sim}{k}'\}| \frac{\partial^m f(H_o)}{\partial H_o^m}\frac{V^m}{m!}|0\rangle$$

$$= \lim_{z\to 0} \langle\{\underset{\sim}{k}\}|\frac{1}{z-L_o}\delta L\frac{V^m}{m}\frac{\partial^m f(H_o)}{\partial H_o^m}|0\rangle$$

Now, if $\{\underset{\sim}{k}\}$ contains ν non vanishing wave vectors $\underset{\sim}{k}_1 \dots \underset{\sim}{k}_\nu$, we obtain :

$$\rho^{(m+1)}_{\underset{\sim}{k}_1\dots\underset{\sim}{k}_\nu} = \lim_{z\to 0} \frac{1}{z-\sum_{i=1}^{\nu}\underset{\sim}{k}_i\cdot\underset{\sim}{v}_i}\langle\{\underset{\sim}{k}\}|\delta L\frac{V^m}{m!}\frac{\partial^m f(H_o)}{\partial H_o^m}|0\rangle$$

$$= -i(8\pi^3/\Omega)^{N+1}\lim_{z\to 0}\frac{1}{z-\sum_{i=1}^{\nu}\underset{\sim}{k}_i\cdot\underset{\sim}{v}_i}\times$$

$$\times\int\{d\underset{\sim}{q}\}^N\exp\left[-i\sum_{i=1}^{\nu}\underset{\sim}{k}_i\cdot\underset{\sim}{q}_i\right]\sum_{l=1}^{N}\frac{\partial V}{\partial\underset{\sim}{q}_l}\cdot\frac{\partial}{\partial\underset{\sim}{p}_l}\frac{V^m}{m!}\frac{\partial^m f(H_o)}{\partial H_o^m}$$

$$= i(8\pi^3/\Omega)^{N+1}\lim_{z\to 0}\frac{1}{z-\sum_{i=1}^{\nu}\underset{\sim}{k}_i\cdot\underset{\sim}{v}_i}\int\{d\underset{\sim}{q}\}^N\exp\left[-i\sum_{i=1}^{\nu}\underset{\sim}{k}_i\cdot\underset{\sim}{q}_i\right]\times$$

$$\times\sum_{l=1}^{N}\underset{\sim}{v}_l\cdot\frac{\partial V}{\partial\underset{\sim}{q}_l}\frac{V^m}{m!}\frac{\partial^{m+1}f(H_o)}{\partial H_o^{m+1}}$$

$$= i(8\pi^3/\Omega)^{N+1}\lim_{z\to 0}\frac{1}{z-\sum_{i=1}\underset{\sim}{k}_i\cdot\underset{\sim}{v}_i}\times$$

$$\times \int \{d\underset{\sim}{q}\}^N \exp\left[-i \sum_{i=1}^{\nu} \underset{\sim}{k}_i \cdot \underset{\sim}{q}_i\right] \sum_{l=1}^{N} \underset{\sim}{v}_l \cdot \frac{\partial V^{m+1}}{\partial \underset{\sim}{q}_1} \frac{1}{(m+1)!} \frac{\partial^{m+1} f(H_0)}{\partial H_0^{m+1}}$$

$$= i(8\pi^3/\Omega)^{N+1} \lim_{z\to 0} \frac{1}{z - \sum_{i=1}^{\nu} \underset{\sim}{k}_i \cdot \underset{\sim}{v}_i} \times$$

(II. 12. 12)

$$\times \int \{d\underset{\sim}{q}\}^N \exp\left[-i \sum_{i=1}^{\nu} \underset{\sim}{k}_i \cdot \underset{\sim}{q}_i\right] \sum_{l=1}^{\nu} \underset{\sim}{v}_l \cdot \frac{\partial V^{m+1}}{\partial \underset{\sim}{q}_1} \frac{1}{(m+1)!} \frac{\partial^{m+1} f(H_0)}{\partial H_0^{m+1}}$$

Integrating by parts over $\underset{\sim}{q}_1$, one obtains easily (II. 12. 9) with m replaced by m+1 .

Combining (II. 12. 9) and (II. 12. 7) , we easily obtain :

$$\rho_{\{\underset{\sim}{k}\}}^{equ.} = \sum_{m=1}^{\infty} \langle \{\underset{\sim}{k}\} \,|\, \frac{\partial^m f(H_0)}{\partial H_0^m} \frac{V^m}{m!} \,|\, 0 \rangle = \langle \{\underset{\sim}{k}\} \,|\, f(H) \,|\, 0 \rangle \qquad \text{(II. 12. 13)}$$

which is the correct value of the equilibrium correlation.

Therefore, once the velocity distribution function has reached its equilibrium value, the fresh correlations which are continuously created from ρ_0 are the equilibrium correlations. (II. 12. 7) gives us a dynamical description of the equilibrium correlations. A comparison with Mayer's cluster formalism can be done but will not be oonsidered here.

II. 13. Response to an external constraint.

As an example, let us consider a system of charged particles which is at equilibrium at t=0: at t=0, we switch on a spatially homogeneous externl electrical field $\underset{\sim}{E}$ (t) .

To the hamiltonian (II, 1. 1) , we now have to add a term describing the effect of the external field :

(II. 13. 1) $$H_E = \sum_i e_i \, \underset{\sim}{E}(t) \cdot \underset{\sim}{q}_i$$

where e_i is the the carge of the ith particle.

The Liouville operator corresponding to this problem is then:

(II. 13. 2) $$iL = iL_o + i\,\delta L + iL_E$$

with

(II. 13. 3) $$iL_E = \sum_i e_i \, \underset{\sim}{E} \cdot \frac{\partial}{\partial \underset{\sim}{p}_i}$$

As we have assumed the system to be at equilibrium at t=0, the initial condition is :

(II. 13. 4) $$\rho(0) = \rho_{equ.} = \frac{\exp\left[-(H_o + \lambda V)/kT\right]}{\int\{d\underset{\sim}{p}\,d\underset{\sim}{q}\}^N \exp\left[-(H_o + \lambda V)/kT\right]}$$

If the external field is sufficiently weak, we can restrict ourselves to a linear theory in E; therefore, we have:

(II. 13. 5) $$\rho(t) = \rho_{equ.} + \Delta\rho(t)$$

where $\Delta\rho$ is linear in E. Using :

(II. 13. 6) $$(L_o + \delta L)\,\rho_{equ.} = 0$$

(II. 13. 7) $$\partial \rho_{equ} / \partial t = 0$$

the Liouville equation reduces to :

(II. 13. 8) $$\frac{\partial \Delta\rho}{\partial t} = -i\,(L_o + \delta L)\Delta\rho - iL_E\,\rho_{equ.}$$

when terms of order E^2 and higher are neglected.

This equation can be solved formally very easily :

$$\text{(II. 13. 9)} \quad \Delta\rho(t) = -i\int_0^t dt' \exp\left[-i(L_o+\delta L)(t-t')\right] L_E(t')\, \rho_{equ.}$$

This solution takes into account the initial condition(II. 13. 9)) ($\Delta\rho(0)=0$).

Let us again expand the distribution function in a Fourier series of the position variables. Then, for the velocity distribution function $\rho_o(t)$ (which is the only coefficient we require if we want, for instance, to compute the current in the system), we obtain :

$$\text{(II. 13. 10)} \qquad \rho_o(t) = \rho_o^{equ.}(t) + \Delta\rho_o(t)$$

Taking into account the fact that $\underset{\sim}{E}$ is spatially constant, we have:

$$\text{(II. 13. 11)} \qquad \langle\{\underset{\sim}{k}\}|L_E(t)|\{\underset{\sim}{k}'\}\rangle = L_E(t)\,\delta_{\{\underset{\sim}{k}\}\{\underset{\sim}{k}'\}}$$

and therefore :

$$\text{(II. 13. 12)} \quad \Delta\rho_o(t) = -i\sum_{\{\underset{\sim}{k}\}}\int_0^t dt' \langle o|\exp\left[-i(L_o+\delta L)(t-t')\right]|\{\underset{\sim}{k}\}\rangle \times$$

$$\times\, L_E(t')\,\rho_{\{\underset{\sim}{k}\}}^{equ.}$$

Any time dependent field can be represented by a superposition of oscillating fields with various frequencies. Therefore, we shall restrict ourselves to the case of an external oscillating field :

$$\text{(II. 13. 12)} \qquad \underset{\sim}{E}(t) = \underset{\sim}{E}_o\, e^{-i\omega t}$$

Using the convolution theorem and (II. 13. 13) , we easily can write :

(II. 13. 14) $$\Delta \rho_o(t) = -(1/2\pi i) \sum_{\{\underset{\sim}{k}\}} \int_C dz\, e^{-izt} \langle 0|R(z)|\{\underset{\sim}{k}\}\rangle \frac{1}{z-\omega} \times \sum_i e_i\, \underset{\sim}{E}_0 \cdot \frac{\partial}{\partial \underset{\sim}{p}_i} \rho^{equ.}_{\{\underset{\sim}{k}\}}$$

where R(z) is the resolvent operator defined by (II. 3. 6) , in the absence of external field. The matrix elements of this operator have been discussed in detail in § 8, where we established the generalized master equation for the evolution of the velocity distribution function in the absence of external field. Using these results, we obtain :

(II. 13. 15) $$\Delta \rho_o(t) = -(1/2\pi i) \int_C dz \frac{e^{-izt}}{z} \sum_{n=0} \left[\frac{1}{z}\psi(z)\right]^n \left\{ \frac{1}{z-\omega} \sum_i e_i\, \underset{\sim}{E}_o \cdot \frac{\partial}{\partial \underset{\sim}{p}_i} \rho^{equ.}_o + \sum_{\{\underset{\sim}{k}\}} D_{\{\underset{\sim}{k}\}}(z) \frac{1}{z-\omega} \sum_i e_i\, \underset{\sim}{E}_o \cdot \frac{\partial}{\partial \underset{\sim}{p}_i} \rho^{equ.}_{\{\underset{\sim}{k}\}} \right\}$$

where the operators $\psi(z)$ and $D_{\{\underset{\sim}{k}\}}(z)$ are given by (II. 8. 3) a nd (II. 8. 4) respectively.

Differentiating with respect to t and using (II. 8. 7) , (II. 8. 8) , we obtain :

(II. 13. 6) $$\frac{\partial \Delta \rho_o(t)}{\partial t} + iL_E \rho^{equ.}_o + i \sum_{\{k\}} \int_0^t d\tau\, \mathcal{D}_{\{\underset{\sim}{k}\}}(t-\tau) L_E(\tau) \rho^{equ.}_{\{\underset{\sim}{k}\}} = \int_0^t d\tau\, G(t-\tau) \Delta \rho_o(\tau)$$

This transport equation is valid at any order in the coupling constant and the concentration.

The term $iL_E \rho^{equ.}_o$ in the lhs is the usual flow term which describes the effect of the field on the unperturbed particles (i. e. between successive collisions) .

As to the non markovian term in the lhs of (II. 13. 16) it describes the effect of the external field on the particles during the collisions. To see this, let us consider the static case ($\omega=0$) . Then, for times long with respect to the collision time, we have (see (II. 8. 8.)) :

(II. 13. 17) $$\sum_{\{k\}} \int_0^t d\tau\, \mathcal{D}_{\{k\}}(\tau)\, L_E \rho^{equ.}_{\{k\}} = \sum_{\{k\}} D_{\{k\}}(0) L_E \rho^{equ.}_{\{k\}}$$

As we have seen, the equilibrium correlations are created from the equilibrium velocity distribution (see (II. 12. 7)) and we can write :

$$\sum_{\{k\}} D_{\{k\}}(0)\ L_E \rho^{equ.}_{\{k\}} = \sum_{\{k\}} D_{\{k\}}(0)\ L_E\, C_{\{k\}}(0)\, \rho^{equ.}_0$$

$$= \lim_{z\to 0} \sum_{\{k\}} \sum_{m=1}^{\infty} \sum_{n=1}^{\infty} \langle 0 | \left(\delta L \frac{1}{z-L_0}\right)^m |\{k\}\rangle L_E \langle \{k\}| \left(\frac{1}{z-L_0}\delta L\right)^n |0\rangle \rho^{equ}_0$$

(II. 13. 18)

$$= \lim_{z\to 0} \sum_{\{k\}} \sum_{p=0}^{\infty} \sum_{q=0}^{\infty} \langle 0 | \delta L \left(\frac{1}{z-L_0}\delta L\right)^p \frac{1}{z-L_0} L_E \frac{1}{z-L_0} \left(\delta L \frac{1}{z-L_0}\right)^q \delta L | 0\rangle \times$$
$$\times\ \rho^{equ.}_0$$

If we compare the operator in the rhs with the operator $\psi(z)$ given by (II. 8. 3) , we notice that they differ only through the replacement of one the unperturbed propagator S $1/(z-L_0)$ in $\psi(z)$ by :

(II. 13. 19) $$\frac{1}{z-L_0}\ L_E\ \frac{1}{z-L_0} = \frac{1}{z-L_0-L_E} - \frac{1}{z-L_0} + 0(E^2)$$

In other words, the operator in the lhs of (II. 13. 16) describes the correction to a collision process which one obtains when one takes into account (to first order in E) the effect of the external field on one of the intermediate state.

II.14 - Stationary transport equation in the static case.

After a long time, we may expect that a system submitted to an external, spatially homogeneous, constant, electrical field will reach a stationary state :

$$\Delta\rho_o(t) \rightarrow \Delta\rho_o^{st} \tag{II.14.1}$$

where $\Delta\rho_o^{st.}$ is time independent. As the contributions to the rhs of (II.13.16) come from times τ such that :

$$t - \tau \lesssim \tau_{coll} \tag{II.14.2}$$

we also have, for t very long :

$$\Delta\rho_o(\tau) = \Delta\rho_o(t) = \Delta\rho_o^{st.} \tag{II.14.3}$$

Taking also into account (II.8.8), the general transport equation becomes :

$$iL_E\rho_o^{equ.} + \sum_{\{\underset{\sim}{k}\}} D_{\{\underset{\sim}{k}\}}(0) iL_E \rho_{\{\underset{\sim}{k}\}}^{equ.} = \lim_{t\to\infty} \int_o^t d\tau\, G(t-\tau)\Delta\rho_o^{st.} \tag{II.14.4}$$

or

$$iL_E\rho_o^{equ.} + \sum_{\{\underset{\sim}{k}\}} D_{\{\underset{\sim}{k}\}}(0)\, iL_E\, \rho_{\{\underset{\sim}{k}\}}^{equ.} = -i\psi(0)\Delta\rho_o^{st} \tag{II.14.5}$$

In chapter V, we shall use this equation as a starting point in a discussion of brownian motion of a heavy charged particle submitted to the action of an external constant electrical field and moving in a medium of light particles.

F. Henin

REFERENCES.

1. I. Prigogine, Non Equilibrium Statistical Mechanics, Inter science, London - New-York (1962)

1. R. Balescu, Statistical Mechanics of Charged Particles Interscience, London - New-York (1963)

3. P. Résibois, in "Many - Particle Physics" ed. by Meeron, Gordon and Breach (to appear, 1966)

4. C. George , Physica 30 , 1513 (1964)

5. I. Prigogine and F. Henin, J. Math; Phys. 1 , 349 (1960)

6. F. Henin , P. Résibois and F. Andrews, J. Math. Phys. 2 , 68 (1961)

7. I. Prigogine, F. Henin and C. George, Physica (32, 1873 (1966))

III. BROWNIAN MOTION IN AN HOMOGENEOUS, WEAKLY COUPLED SYSTEM

III. 1 - Introduction

We shall consider here a weakly coupled gas, i.e. a system such that the coupling constant is very small :

(III.1.1) $$\lambda \rightarrow 0$$

The relaxation time of such a system is proportional to λ^{-2} (see the Born approximation) :

(III.1.2) $$\tau_{rel} \sim \lambda^{-2}$$

We shall consider the evolution of such a system for times of the order of the relaxation time, i.e. we shall take the following limit :

(II.1.3) $$\lambda \rightarrow 0 \quad , \quad t \rightarrow \infty \, , \; \lambda^2 t \quad \text{finite}$$

As the collision time is independent of the strength of the interaction, we clearly have :

(II.I.4) $$\tau_{coll} / \tau_{rel} \rightarrow 0$$

and the evolution of the system will be described by a markovian equation.

The weak coupling condition implies that we exclude all forces with a strong repulsive core. Strictly speaking, there are no known intermolecular forces for which the theory of weakly coupled systems may be applied.

In all physical cases, the interaction becomes too strong at very short distances to be handled in a weak coupling theory . Nevertheless, we shall consider it here because it is the simplest example where the brownian motion problem can be discussed starting from a microscopic basis. This model has been discussed by Prigogine and Balescu [1)2)] and will already show up interesting features when compared with the phenomenological theory of chapter I, and in any case, we can expect that it will give us a good description of the effect of the collisions which are not too close.

We shall first find the equation for the reduced distribution function for one particle assuming that initially there are no long range correlations. Then, we shall specialize to the problem of brownian motion where the particle moves in a fluid at equilibrium. The equation so obtained for the one particle velocity distribution function will be of the Fokker-Planck type. However, in contrast with the assumptions of the stochastic theory, the friction coefficient will appear as velocity dependent. In fact, this dependence will be important only for velocities equal to or higher than the mean thermal velocity of the fluid.

III.2 - Equation of evolution of the velocity distribution function for weakly oaupled systems.

From the discussion of chapter II. § 9, it is quite clear that if we take into account (III.1.4), we must neglect all non markovian corrections to the kinetic equation, i.e. take

$$\Omega = 1 \qquad \text{(III.2.1)}$$

in (II.9.5).

When we do this, to be consistent, we must keep in the operator ψ only the lowest order contribution, of order λ^2. The only diagram which we have therefore to keep in the operator ψ is the cycle (fig. II.6 1a). The evolution equation then becomes :

$$\frac{\partial \rho_o(t)}{\partial t} = -i \lambda^2 \psi_2 \rho_o(t) \tag{III.2.2}$$

where ψ_2 is the operator associated with the cycle. We have already discussed this operator in ch. II. § 7, using the simple case of an exponential interaction law. If we do not make any special choice of the potential, we have (see II.7.4) ;

$$-i\psi_2(z) = (8\pi^3/\Omega)^2 \; i \sum_{i<j} \sum_{\underset{\sim}{k}} |V_k|^2 \underset{\sim}{k} \cdot \left(\frac{\partial}{\partial \underset{\sim}{p}_i} - \frac{\partial}{\partial \underset{\sim}{p}_j}\right) \times$$

$$\times \frac{1}{z - \underset{\sim}{k} \cdot (\underset{\sim}{v}_i - \underset{\sim}{v}_j)} \underset{\sim}{k} \cdot \left(\frac{\partial}{\partial \underset{\sim}{p}_i} - \frac{\partial}{\partial \underset{\sim}{p}_j}\right) \quad \cdot (z \in S^+) \tag{II.2.3}$$

We have now to find out the limiting value of $\psi_2(z)$ when $z \to 0^+$, i.e. the analytical continuation of this operator when z approaches the real axis. As we have already said this can be done easily, in the limit of a large system , using the theory of Cauchy integrals. Indeed, when the limit

$$N \to \infty \;, \quad \Omega \to \infty, \qquad N/\Omega = C \text{ finite} \tag{III.2.4}$$

in taken , the spectrum of values of $\underset{\sim}{k}$ becomes continuous and the summation over $\underset{\sim}{k}$ in (III.2.3) becomes an integral :

$$(8\pi^3/\Omega) \sum_{\underset{\sim}{k}} \to \int d^3k \tag{III2.5}$$

Then we have :

$$-i\psi_2(z) = (8\pi^3/\Omega)\; i \sum_{i<j} \int d^3k\, |V_k|^2 \underset{\sim}{k} \cdot \left(\frac{\partial}{\partial \underset{\sim}{p}_i} - \frac{\partial}{\partial \underset{\sim}{p}_j}\right) \times$$

$$\times \frac{1}{z - \underset{\sim}{k} \cdot (\underset{\sim}{v}_i - \underset{\sim}{v}_j)} \underset{\sim}{k} \cdot \left(\frac{\partial}{\partial \underset{\sim}{p}_i} - \frac{\partial}{\partial \underset{\sim}{p}_j}\right) \tag{III.2.6}$$

If we take as one of the integration variable the variable :

(II.2.7) $$x = \underset{\sim}{k} \cdot (\underset{\sim}{v}_i - \underset{\sim}{v}_j)$$

the dependence on z is of the form :

(III.2.8) $$F(z) = \int_{-\infty}^{+\infty} dx \; \frac{f(x)}{z - x} \qquad \left(z \in S^+\right)$$

This is precisely a Cauchy integral. Its analytical continuation is :

(III.2.9) $$F(0^+) = - \int_{-\infty}^{+\infty} dx \; \mathcal{P}(1/x) f(x) - \pi i \int_{-\infty}^{+\infty} dx \; \delta(x) \, f(x)$$

The rhs of (III.2.6) is an even function of $\underset{\sim}{k}$. Therefore, the contribution involving the principal part vanishes and we are left with the following kinetic equation for a weakly coupled gas :

(III.2.10) $$\frac{\partial \rho_o(t)}{\partial t} = (8\pi^4 \lambda^2/\Omega) \sum_{i<j} \int d^3k \, |V_k|^2 \, \underset{\sim}{k} \cdot \left(\frac{\partial}{\partial \underset{\sim}{p}_i} - \frac{\partial}{\partial \underset{\sim}{p}_j}\right) \times$$
$$\times \, \delta\left[\underset{\sim}{k} \cdot (\underset{\sim}{v}_i - \underset{\sim}{v}_j)\right] \, \underset{\sim}{k} \cdot \left(\frac{\partial}{\partial \underset{\sim}{p}_i} - \frac{\partial}{\partial \underset{\sim}{p}_j}\right) \rho_o(t)$$

Let us notice that with this equation it is very easy to verify Boltzmann's $\mathcal{H}$-theorem. Indeed, with Boltzmann's $\mathcal{H}$-quantity defined as :

(III.2.11) $$\mathcal{H} = \int \{d\underset{\sim}{p}\}^N \rho_o \ln \rho_o$$

the kinetic equation allows us to write :

(III.2.12) $$\frac{\partial \mathcal{H}}{\partial t} = - (8\pi^4 \lambda^2/\Omega) \sum_{i<j} \int d^3k \, |V_k|^2 \, \delta\left[\underset{\sim}{k} \cdot (\underset{\sim}{v}_i - \underset{\sim}{v}_j)\right] \frac{1}{\rho_o(t)} \times$$
$$\times \left[\underset{\sim}{k} \cdot \left(\frac{\partial}{\partial \underset{\sim}{p}_i} - \frac{\partial}{\partial \underset{\sim}{p}_j}\right) \rho_o(t)\right]^2 \leqslant 0$$

The function $\mathcal{H}$ (which is related to the entropy) decreases monotonically to its equilibrium value. The rhs of (III.2.12) vanishes once ρ_o has reached its equilibrium value :

$$\rho_o^{equ.} = f(H_o) \tag{III.2.13}$$

The function f cannot be determined by (III.2.12) but is determined by the initial conditions. We shall see in the next paragraph that the assumption of initial molecular chaos leads to the Maxwell-Boltzmann distribution.

III.3 - Kinetic equation for the one particle velocity distribution function.

Let us call φ_s the reduced distribution function for s momenta:

$$\varphi_s(\underset{\sim}{v}_1, \ldots \underset{\sim}{v}_s, t) = \int d\underset{\sim}{p}_{s+1} \ldots d\underset{\sim}{p}_N \; \rho_o(\{\underset{\sim}{p}\}, t) \tag{III.3.1}$$

Integrating (III.2.10) with respect to all momenta except $\underset{\sim}{v}_1$ and taking into account the fact that the distribution function vanishes at infinity, we obtain :

$$\begin{aligned} \frac{\partial \varphi_1(\underset{\sim}{v}_1, t)}{\partial t} &= (8\pi^4 \lambda^2/\Omega\, m_1^2) \sum_{j>1} \int d\underset{\sim}{v}_j \int d^3k \, |V_k|^2 \, \underset{\sim}{k} \cdot \frac{\partial}{\partial \underset{\sim}{v}_1} \times \\ &\times \delta\left[\underset{\sim}{k} \cdot (\underset{\sim}{v}_1 - \underset{\sim}{v}_j)\right] \underset{\sim}{k} \cdot \left(\frac{\partial}{\partial \underset{\sim}{v}_1} - \frac{m_1}{m_j} \frac{\partial}{\partial \underset{\sim}{v}_j}\right) \varphi_2(\underset{\sim}{v}_1, \underset{\sim}{v}_j, t) \end{aligned} \tag{III.3.2}$$

This equation gives us the evolution of the one-particle velocity distribution function in terms of the two-particle distribution function. Therefore , it is not a closed equation and if we do not make any further assumptions, we have actually to deal with an infinite hierarchy of equations.

According to our basic assumption (Ch. II. , § 9) , we consider a system where all correlations are of finite extension. Now, the velocity distribution function is an average over the positions of all particles of the complete distribution function. When we do this, if we consider two given particles, it is quite clear that the contribution of those configurations where the particles are correlated is much smaller than that of those configurations where they are uncorrelated. This allows us to neglect the effect of these correlations on the velocity distribution function, i.e. to make the assumption of molecular chaos at $t = 0$:

$$\rho_o(t=0) = \prod_j \varphi_1(\underset{\sim}{v}_j, t) \qquad \text{(III. 3. 3)}$$

Once molecular chaos is taken as an initial condition, it can be shown to persist for all times in the limit of infinite systems [1) 3)]. We shall not give the proof here.

With the initial condition (III. 3. 3) , we may write in the rhs of (III. 3. 2) :

$$\varphi_2(\underset{\sim}{v}_1, \underset{\sim}{v}_j, t) = \varphi_1(\underset{\sim}{v}_1, t)\varphi_1(\underset{\sim}{v}_j, t) \qquad \text{(III. 3. 4)}$$

and we obtain :

$$\frac{\partial \varphi_1(\underset{\sim}{v}_1, t)}{\partial t} = (8\pi^4 \lambda^2/\Omega\, m_1^2) \sum_j \int d\underset{\sim}{v}_j \int d^3k \, |V_k|^2 \, \underset{\sim}{k} \cdot \frac{\partial}{\partial \underset{\sim}{v}_1} \times$$

$$\times\, \delta\left[\underset{\sim}{k} \cdot (\underset{\sim}{v}_1 - \underset{\sim}{v}_j)\right] \underset{\sim}{k} \cdot \left(\frac{\partial}{\partial \underset{\sim}{v}_1} - \frac{m_1}{m_j} \frac{\partial}{\partial \underset{\sim}{v}_j}\right) \varphi_1(\underset{\sim}{v}_1, t)\varphi_1(\underset{\sim}{v}_j, t) \qquad \text{(III. 3. 5)}$$

A similar equation can of course be easily obtained for the reduced distribution function of any particle of the system. Therefore, we are now dealing with a closed set of equations.

Let us now verify that the equilibrium distribution corresponding to the initial condition (III.3.3) is indeed the Maxwell-Boltzmann distribution. Using (III.2.12) and the assumption of molecular chaos, at equilibrium, we must have :

(III.3.6) $$\underset{\sim}{k} \cdot \left(\frac{\partial}{\partial \underset{\sim}{p}_i} - \frac{\partial}{\partial \underset{\sim}{p}_j}\right) \varphi_1(\underset{\sim}{v}_i, t)\, \varphi_1(\underset{\sim}{v}_j, t) = 0$$

whenever $\underset{\sim}{k}$ is such that :

(III.3.7) $$\underset{\sim}{k} \cdot (\underset{\sim}{v}_i - \underset{\sim}{v}_j) = 0$$

We verify easily that (III.3.6) implies :

(III.3.8) $$\underset{\sim}{k} \cdot \left(\frac{\partial \ln \varphi_1(\underset{\sim}{v}_i, t)}{\partial \underset{\sim}{p}_i} - \frac{\partial \ln \varphi_1(\underset{\sim}{v}_j, t)}{\partial \underset{\sim}{p}_j}\right) = 0$$

Whenever (III.3.8) and (III.3.7) are simultaneously satisfied, we must have :

$$\frac{\dfrac{\partial \ln \varphi_1(\underset{\sim}{v}_i)}{\partial \underset{\sim}{p}_i}}{\dfrac{\underset{\sim}{p}_i - \underset{\sim}{p}_o}{m_i}} = \frac{\dfrac{\partial \ln \varphi_1(\underset{\sim}{v}_j)}{\partial \underset{\sim}{p}_j}}{\dfrac{\underset{\sim}{p}_j - \underset{\sim}{p}_o}{m_j}} = \alpha$$

where α and $\underset{\sim}{p}_o$ are constants .

Integrating (III.3.9) , we obtain :

(III.3.10) $$\ln \varphi_1(\underset{\sim}{v}_i) = \alpha(\underset{\sim}{p}_i - \underset{\sim}{p}_o)^2 / 2m_i + \ln \gamma_i$$

where γ_i is a constant .

This gives :

(III.3.11) $$\varphi_1(\underset{\sim}{v}_i) = \gamma_i \exp\left[\alpha \,|\underset{\sim}{p}_i - \underset{\sim}{p}_o|^2 / 2m_i\right]$$

The normalization condition requires :

(III. 3. 12) $$\alpha < 0 \qquad \gamma_i = 4\pi(-m_i\alpha/2\pi)^{3/2}$$

We also consider systems where the average velocity is zero ; hence

(III. 3. 13) $$\underset{\sim}{p}_o = \underset{\sim}{0}$$

Defining the temperature through :

(III. 3. 14) $$kT = \langle m_i v^2/2\rangle = \int d\underset{\sim}{v}_i\,(v_i^2/2)\,\varphi_1(\underset{\sim}{v}_i)\,m_i$$

one obtains easily the usual Maxwell - Boltzmann law ;

(III. 3. 15) $$\varphi_1(\underset{\sim}{v}) = 4\pi(m/2\pi kT)^{3/2}\,\exp(-mv^2/2\,kT)$$

In this way, we have verified our statement at the end of the previous paragraph.

III.4 - Brownian motion in a fluid at equilibrium.

We shall now consider the simple case where the particle 1 moves in a fluid at equilibrium . We then have :

(III. 4. 1) $$\varphi_1(\underset{\sim}{v}_j, t) = 4\pi(m_j/2\pi kT)^{3/2}\,\exp(-m_j v_j^2/2kT) \qquad j \neq 1$$

Then, equ. (III. 3. 5) becomes (assuming the masses of all fluid particles to be equal to m) :

(III. 4. 2) $$\frac{\partial\varphi_1(\underset{\sim}{v}_1, t)}{\partial t} = (32\pi^5\lambda^2 C/m_1^2)(m/2\pi KT)^{3/2}\int d\underset{\sim}{v}\exp(-mv^2/2kT)\times$$
$$\times\int d^3k\,|V_k|^2\,\underset{\sim}{k}\cdot\frac{\partial}{\partial\underset{\sim}{v}_1}\delta[\underset{\sim}{k}\cdot(\underset{\sim}{v}_1-\underset{\sim}{v})]\,\underset{\sim}{k}\left(\frac{\partial}{\partial\underset{\sim}{v}_1}+\frac{m_1\underset{\sim}{v}_1}{kT}\right)\varphi_1(\underset{\sim}{v}_1, t)$$

In this way, we have obtained a closed equation for $\varphi_1(\underset{\sim}{v}_1, t)$.
The $\underset{\sim}{k}$ integration is easily performed in a reference frame where the relative velocity

(III. 4. 3) $$\underset{\sim}{g} = \underset{\sim}{v}_1 - \underset{\sim}{v}$$

is along the z axis :

(III. 4. 4)
$$a_i \int d^3k \left|V_k\right|^2 k_i k_j \delta(\underset{\sim}{k} . \underset{\sim}{g})\, b_j = \pi B(a_x \frac{1}{g} b_x + a_y \frac{1}{g} b_y)$$
$$= \pi B(\underset{\sim}{a} . \frac{1}{g} \underset{\sim}{b} - \underset{\sim}{a} . \underset{\sim}{g} \frac{1}{g^3} \underset{\sim}{g} . \underset{\sim}{b})$$

where

(III. 4. 5) $$B = \int_0^\infty dk \;\; k^3 \left|V_k\right|^2$$

depends only on the intermolecular potential .
The last expression in (III. 4. 4) is valid in an arbitrary reference frame . Using dimensionless quantities :

(III. 4. 6) $$\underset{\sim}{u} = \sqrt{\frac{m}{2kT}}\, \underset{\sim}{v}_1 \qquad \underset{\sim}{w} = \sqrt{\frac{m}{2kT}}\, \underset{\sim}{g}$$

we obtain :

$$\frac{\partial \varphi_1(\underset{\sim}{u}, t)}{\partial t} = (8 \sqrt{2}\, \pi^{9/2} \lambda^2 CB / m_1^2)\, (m/kT)^{3/2} \times$$

(III 4. 7)
$$\times \left\{ \frac{\partial}{\partial u_i} \int d\underset{\sim}{w}\; e^{-(\underset{\sim}{w} - \underset{\sim}{u})^2} \frac{1}{w} \left(\frac{\partial}{\partial u_i} + 2 \frac{m_1}{m} u_i\right) - \frac{\partial}{\partial u_i} \int d\underset{\sim}{w}\; e^{-(\underset{\sim}{w} - \underset{\sim}{u})^2} \times \right.$$
$$\left. \times \; w_i w_j \frac{1}{w^3} \left(\frac{\partial}{\partial u_j} + 2 \frac{m_1}{m} u_j\right) \right\} \varphi_1(\underset{\sim}{u}, t)$$

In a reference frame where $\underset{\sim}{u}$ is along the z axis, we have :

$$I_1 = \int d\underset{\sim}{w}\, e^{-(\underset{\sim}{w}-\underset{\sim}{u})^2} (1/w) = 2\pi \int_0^\infty dw\, w \int_{-1}^{+1} d\cos\theta\, e^{-(w^2+u^2-2wu\cos\theta)}$$

(III. 4. 8)
$$= (\pi/u) \int_0^\infty dw \left\{ e^{-(w-u)^2} - e^{-(w+u)^2} \right\} = (\pi^{3/2}/u)\, \Phi(u)$$

where

(III. 4. 9)
$$\Phi(u) = (2/\sqrt{\pi}) \int_0^u dx\, e^{-x^2}$$

is the error function .

We also have :

$$a_i \int d\underset{\sim}{w}\, e^{-(\underset{\sim}{w}-\underset{\sim}{u})^2} w_i w_j w^{-3} b_j = (1/2)(\underset{\sim}{a}\cdot I_1 \underset{\sim}{b} - \underset{\sim}{a}\cdot\underset{\sim}{u} \frac{1}{u^2} I_1 \underset{\sim}{u}\cdot\underset{\sim}{b})$$

(III. 4. 10)
$$- (1/2)(\underset{\sim}{a}\cdot I_2 \underset{\sim}{b} - 3\, \underset{\sim}{a}\cdot\underset{\sim}{u} \frac{1}{u^2} I_2 \underset{\sim}{u}\cdot\underset{\sim}{b})$$

$$I_2 = 2\pi \int_0^\infty dw\, w \int_{-1}^{+1} d\cos\theta \cos^2\theta\; e^{-(w^2 + u^2\, wu\cos\theta)}$$

$$= (\pi/2u^2) \int_0^\infty dw\, w\, e^{-(w^2+u^2)} \frac{d^2}{dw^2} \int_{-1}^{+1} d\cos\theta\, e^{2wu\cos\theta}$$

(III. 4. 11)
$$= (\pi/2u^2) \left\{ \int_0^\infty dw \left[\frac{d^2}{dw^2} w\, e^{-(w^2+u^2)} \right] \int_{-1}^{+1} d\cos\theta\, e^{2wu\cos\theta} + 2e^{-u^2} \right\}$$

$$= \pi^{3/2} \left\{ u^{-2}\, \Phi'(u) - u^{-3} (1-u^2) \Phi(u) \right\}$$

where

(II. 4. 12)
$$\Phi'(u) = \frac{d\Phi(u)}{du} = (2/\sqrt{\pi})\, e^{-u^2}$$

Introducing (III. 4. 8) and (III. 4. 10) into (III. 4. 7) , one can finally write (III. 4. 7) as :

$$\frac{\partial \varphi_1(\underset{\sim}{u},t)}{\partial t} = \tau^{-1} \left\{ a(u) \frac{\partial^2}{\partial u_i^2} + b(u) \left[(u_i \frac{\partial}{\partial u_i})^2 - u_i \frac{\partial}{\partial u_i} \right] \right.$$

(III. 4. 13) $-2(1-\frac{m_1}{m})\left[u^{-3}\phi(u)-u^{-2}\phi'(u)\right]u_i\frac{\partial}{\partial u_i}+4\frac{m_1}{m}\phi'(u)\Big\}\varphi_1(\underset{\sim}{u},t)$,

where

(III. 4. 14) $$a(u)=\frac{1}{2u^3}\left[u\phi'(u)+(2u^2-1)\phi(u)\right]$$

(III.4. 15) $$b(u)=u^{-1}\frac{da(u)}{du}$$

and where τ has the dimension of a time :

(III. 4. 16) $$\tau^{-1}=(32\pi^6 C\lambda^2 B/m_1^2)(m/2kT)^{3/2}$$

III. 5. Link with stochastic theory .

Equation (III. 4. 13) can be easily written in the form of the general Fokker - Planck equation (I. 5. 7) :

$$\frac{\partial\varphi}{\partial t}=\frac{\partial}{\partial u_i}\left[-\frac{\langle\Delta u_i\rangle}{\Delta t}\varphi\right]+\frac{1}{2}\frac{\partial^2}{\partial u_i\partial u_j}\left[\frac{\langle\Delta u_i\Delta u_j\rangle}{\Delta t}\varphi\right]$$

(III. 5. 1)

with

(III. 5. 2) $$\frac{\langle\Delta u_i\rangle}{\Delta t}=-4\frac{u_i}{u}\left(1+\frac{m_1}{m}\right)g(u)\tau^{-1}$$

(III. 5. 3) $$\frac{\langle\Delta u_i\Delta u_j\rangle}{\Delta t}=\left\{6\frac{u_iu_j}{u^3}\left[g(u)-\frac{1}{3}\phi(u)\right]+\frac{2}{u}\delta_{i,j}\left[\phi(u)-g(u)\right]\right\}\tau^{-1}$$

where

(III. 5. 4) $$g(u)=\frac{1}{2}u^{-1}\left[u^{-1}\phi(u)-\phi'(u)\right]$$

This equation has been obtained from first principles as an asymptotic equation describing the motion of a particle of mass m_1 in a gas at thermal equilibrium , with the assumption of weak coupling. The averarage values $\langle \Delta u_i \rangle, \langle \Delta u_i \Delta u_j \rangle$ are such that the velocity distribution function reaches monotonically the Maxwell-Boltzmann distribution after a long time. No other assumption than the hypothesis of initial molecular chaos has been necessary to obtain this result.

The relaxation time, i.e. the characteristic time for the evolution of $\varphi(\underset{\sim}{u}, t)$ is given by (III. 4. 16) :

(III. 5. 5) $$\tau_{rel}^{-1} = (32\pi^6 \lambda^2 BC) / m_1^2)(m/2kT)^{3/2}$$

As expected, it decreases when the concentration increases. It is also a function of the temperature and the intermolecular forces (see III. 4.5 for B) ; it depends upon the ratio of the interaction energy and the mean thermal energy of the particles of the fluid. It decreases whenever this ratio increases.

If we compare (III. 5. 2) with the corresponding expression derived from the Langevin equation, we notice that the microscopic theory introduces a coefficient of dynamical friction η which is velocity dependent :

(III. 5. 6) $$\eta = 4\tau^{-1} \left(1 + \frac{m_1}{m}\right) \frac{g(u)}{u}$$

Let us introduce the following dimensionless quantities :

(III. 5. 7) $$\gamma = (m/m_1)^{1/2}$$

(III. 5. 8) $$x = (m_1/2kT)^{1/2} v_1 = u/\gamma$$

γ is the ratio of the masses of the fluid particle and the brownian particle ; x is the ratio of the velocity of the brownian particle and its thermal velocity. With these quantities, we have :

(III.5.9) $$\eta = (4(1+\gamma^2)\,\frac{g(\gamma x)}{\gamma x}\,\frac{\tau^{-1}}{\gamma^2}$$

For $x \ll 1$, (and $\gamma \leq 1$) we have

(III.5.10) $$\eta \simeq (8/3\sqrt{\pi})\,(1+\gamma^2)\,\tau^{-1}\gamma^{-2}$$

In this case, the dynamical friction coefficient is approximately constant: If $\gamma x \gg 1$, i.e. if the particle has a high velocity, the dynamical friction coefficient is very small :

(III.5.12) $$\eta \simeq 4(1+\gamma^2)\,\frac{1}{\gamma^3 x^3}\,\tau^{-1}\gamma^{-2} \simeq \frac{4\tau^{-1}}{\gamma^3 x^3}$$

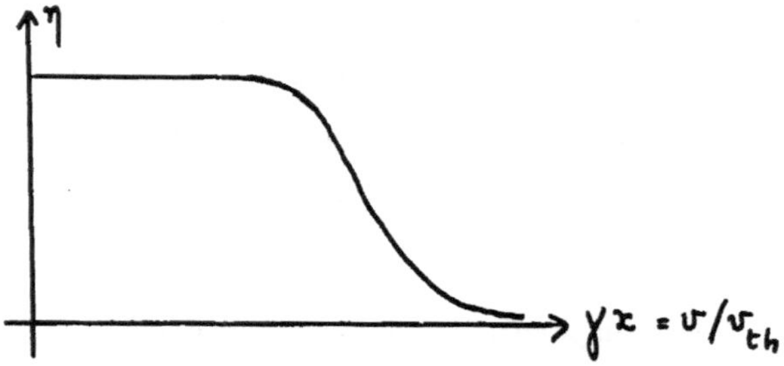

Dependence of dynamical friction coefficient on velocity

Fig. III.5.1

For $\gamma x \ll 1$, we also have :

(III.5.13) $$\frac{\langle \Delta u_i\, \Delta u_j \rangle}{\langle \Delta t \rangle} = \delta_{i,j}\,(8/3\sqrt{\pi})\,\tau^{-1}$$

For $\gamma \ll 1$ and $x \simeq 1$, i.e. for a heavy particle moving with thermal velocity in a medium of light particles at equilibrium , the Fokker Planck equation takes the simple form :

(III.6.14) $$\frac{\partial \varphi}{\partial t} = (4/3\sqrt{\pi})\,(1/\tau\gamma^2)\,\frac{\partial}{\partial x_i}\left(\frac{\partial}{\partial x_i} + 2x_i\right)\varphi$$

analogous to (I.5.11) .

The expressions (III.5.2) and (III.5.3) for the average value of the velocity and square of the velocity have been obtained first by Chandrasekhar [4].

In an analysis of the dynamical friction in systems of stars, Chandrasekhar first considered single stellar encounters idealized as two-body problems. If a star of mass m_1 and velocity $\underset{\sim}{v}_1$ collides with a star of mass m and velocity $\underset{\sim}{v}$, the increments parallel ($\delta \underset{\sim}{v}_{1\parallel}$) and perpendicular ($\delta \underset{\sim}{v}_{1\perp}$) to its direction of motion can be easily written. The net increments $\Delta \underset{\sim}{v}_{1\parallel}$ and $\Delta \underset{\sim}{v}_{1\perp}$, due to a large number of successive encounters with field stars during a time interval Δt such that $\underset{\sim}{v}_1$ does not change appreciably, are easily computed. Assuming the velocity distribution for field stars to be a gaussian, one obtains :

(III.6.15) $$\Delta \underset{\sim}{v}_{1\parallel} = \eta\, v_1$$

(III.6.16) $$\Delta \underset{\sim}{v}_{1\perp} = 0$$

(for more details, see Prof. Ferraro's notes in this volume)

III.6. Application.

These results have been used to discuss transport processes in fully ionized gases. A good account of this can be found in Spitzer's book [5]. However, the application is not straightforward. Indeed , the interaction law in this case is the Coulomb potential :

(III.6.1) $$V(r) = e^2/r$$

which Foûrier transform is :

(III. 6. 2) $$V_k = e^2/2\pi^2 k^2$$

Therefore the coefficient B which appears in the Fokker-Planck equation is :

(III. 6. 3) $$B = (e^4/4\pi^4) \int_0^\infty dk(1/k)$$

We notice that B diverges logarithmically both at the upper and lower limits of integration. This is due to the fact that the Coulomb potential has an infinite repulsive core at small distances (hence the upper limit divergence) and has a long range (hence the lower limit divergence). The long distance divergence is well known and appears also in the equilibrium properties. In fact, because of the long range of the potential, the interactions in such a medium have a collective character : configurations involving many particles play a dominant role. Both in equilibrium [6)7)8)] and non equilibrium properties, this problem can be settled by a summation over a well defined class of diagrams. The result of this summation is to introduce a screening effect : in simple cases, the effective interaction vanishes exponentially for distances greater than the Debye radius κ^{-1} :

(III. 6. 4) $$V_{eff} = e^2 e^{-\kappa r}/r$$

with

(III. 6. 5) $$\kappa^2 = 4\pi e^2 C/kT$$

One way to take into account these effects semi-empirically is to introduce a cut-off at both limits of integration :

(III. 6. 6) $$B = (e^4/4\pi^4) \int_\kappa^{1/a} dk(1/k)$$

The lower cut-off κ takes into account the screening effect while the upper one eliminates the effect of the very close collisions. The theory of weakly coupled gases may then be used.

However, this approach is not very satisfactory and the true way to solve this problem, although we shall not discuss it here, is, within a perturbation theory, to sum first all relevant contributions. From (III. 6. 4), it is quite apparent that the adequate procedure is not to limit ourselves to first power in the coupling constant e^2 or the concentration C, but to retain all terms proportional to any power of κ, i.e. of $e^2 C$, in the expansions. This summation introduces a dynamical screening effect, i.e. a screening which depends on the velocity of the brownian particle [9]. However, if the velocity of the particle is such that $(kT/m)^{1/2} \ll 1$, the dynamical effects may be neglected and the Debye potential is a good approximation. For more rapid particles, one can still write a Fokker-Planck equation but there appears a further velocity dependence of the coefficients due to the collective effects (excitation of plasma oscillations).

III. 7 - Brownian motion in a fluid which is not at equilibrium.

In this case, we must relax assumption (III. 4. 1) and use equation (III. 3. 5). The main feature is that, whereas, in the equilibrium case, we have a single closed equation for the distribution function of the brownian particle, in the non equilibrium case, we have a whole set of equations for the velocity distribution functions of the Brownian particle and the fluid particles.

Following the same procedure as above for the integration over the wave vector, we easily obtain :

$$\frac{\partial \varphi_1(\underset{\sim}{v}_1, t)}{\partial t} = (8\pi^5 \lambda^2 C B/m_1^2) \int d\underset{\sim}{v} \left\{ \frac{\partial}{\partial \underset{\sim}{v}_1} \quad \frac{1}{g} \quad \left(\frac{\partial}{\partial \underset{\sim}{v}_1} - \frac{\partial}{\partial \underset{\sim}{v}}\right)\right.$$

$$- \frac{\partial}{\partial \underset{\sim}{v}_1} \cdot \underset{\sim}{g} \frac{1}{g^3} \underset{\sim}{g} \cdot \left(\frac{\partial}{\partial v_1} - \frac{\partial}{\partial v} \right) \Big\} \varphi_1(\underset{\sim}{v}_1, t)\, \varphi(\underset{\sim}{v}, t) \tag{III.7.1}$$

This second order differential equation can be easily recast in the form of a generalized Fokker-Planck equation :

$$\frac{\partial \varphi_1(\underset{\sim}{v}_1, t)}{\partial t} = \left\{ \frac{\partial}{\partial v_{1_i}} \left(-\overline{\Delta v_{1_i} / \Delta \tau}\right) + \frac{1}{2} \frac{\partial^2}{\partial v_{1_i} \partial v_{1_j}} \left(\overline{\Delta v_{1_i} \Delta v_{1_j} / \Delta \tau}\right) \right\} \times \varphi_1(\underset{\sim}{v}_1, t) \tag{III.7.2}$$

where the transition moments are given by:

$$\overline{\Delta v_{1_i} / \Delta \tau} = (8\pi^5 \lambda^2 CB/m_1^2) \int d\underset{\sim}{v} \left\{ g^{-1} \frac{\partial \varphi(\underset{\sim}{v}, t)}{\partial v_i} - 2 g_i g^{-3} \varphi(\underset{\sim}{v}, t) - g_i g_j g^{-3} \frac{\partial \varphi(\underset{\sim}{v}, t)}{\partial v_j} \right\}$$

$$= - (16\pi^5 \lambda^2 CB/m_1^2) \int d\underset{\sim}{v}\, g_i\, g^{-3}\, \varphi(\underset{\sim}{v}, t) \tag{III.7.3}$$

$$\overline{\Delta v_{1_i} \Delta v_{1_j} / \Delta \tau} = -(16\pi^5 \lambda^2 CB/m_1^2) \int d\underset{\sim}{v} \left\{ g_i g_j g^{-3} - \delta_{ij} g^{-1} \right\} \varphi(\underset{\sim}{v}, t) \tag{III.7;4}$$

The transition moments are now functionals of the state of the fluid .

F. Henin

References

1. I. Prigogine, Non Equilibrium Statistical Mechanics Interscience, New-York London (1962)

2. I. Prigogine and R. Balescu, Physica 23 , 555 (1957)

3. M. Kac, Lecture on Probability Theory, Interscience, New-York (1959)

4. S. Chandrasekhar, Astrophys. J. 97 , 255 (1943)

5. L. Spitzer, Physics of Fully Ionized Gases, Interscience New-York (1956)

6. J. E. Mayer, J. Chem . Phys. 18 , 1426 (1950)

7. E. W. Montroll and J. C. Ward, Phys, of Fluids, 1 , 55 (1958)

8. M. Gell-Mann and K. A. Bruekner, Phys. Rev 106 , 364 (1957)

9. R. Balescu, Statistical Mechanics of Charged Particles, Interscience, New-York (1963)

F. Henin

IV. MICROSCOPIC THEORY OF BROWNIAN MOTION OF A HEAVY PARTICLE IN THE ABSENCE OF EXTERNAL FORCES.

IV.1 - Introduction

In the preceding chapter, we have discussed the brownian motion of a particle which is weakly coupled to a fluid at thermal equilibrium. However, as we have seen, there are no known intermolecular forces corresponding to the weak coupling approximation. Therefore, that problem was rather academic and we shall now consider a more realistic situation. The discussion which will follow will be valid for all cases where the forces are of short range (Coulomb forces can also be included provided the screening effects are taken into account in a phenomenological way). The case of long range forces with no screening (gravitational forces) will be dealt with in chapter VII.

For this model we can use as a starting point the kinetic equation (II.9.3) or (II.9.5) . We shall consider the case where the brownian particle is much heavier than the fluid particles. For the case of a brownian particle moving with thermal velocity in a fluid at equilibrium at temperature T, we shall show that an equation of the Fokker -Planck type is indeed obtained for the velocity distribution function of that particle if one retains only the lowest order terms in the expansion of the kinetic equation in the ratio of the masses of the light and heavy particles . The method we shall follow enables us also to compute the corrections to the Fokker-Planck equation. However , we shall not consider this problem here but we shall rather discuss it in the next chapter where we consider the same problem but with an external force acting on the particle .

IV.2 - Equation for the reduced velocity distribution function of the brownian particle.

Let us start with the kinetic equation in the form (II.9.3). Assuming the fluid to be at equilibrium, we have

$$\rho_0(t) = \varphi(\underset{\sim}{V},t)\,(\rho_0^f)\,\text{equ.}$$

(IV.2;1)

$$= \varphi(\underset{\sim}{V},t)\prod_{i=1}^{N}(2\pi kT/m)^{3/2}\,4\pi\, e^{-mv_i^2/2kT}$$

where $\underset{\sim}{V}$ is the velocity of the brownian particle (mass: M) while $\underset{\sim}{v}_i$ is the velocity of the ith fluid particle (all fluid particles have the same mass m).

Integrating both sides of (IV.2.1) with respect to the velocities of all the fluid particles, we readily obtain an equation for the reduced velocity distribution function $\varphi(\underset{\sim}{V},t)$ of the brownian particle:

(IV.2.2) $$\frac{\partial\varphi(\underset{\sim}{V},t)}{\partial t} = \int_0^t d\tau \int \{d\underset{\sim}{v}\}^N G(t-\tau)\,\varphi(\underset{\sim}{V},\tau)\,(\rho_0^f)\,\text{equ.}$$

which is valid only asymptotically.

Let us introduce the operator Γ:

(IV.2.3) $$\Gamma(t-\tau) = \int \{d\underset{\sim}{v}\}^N G(t-\tau)\,(\rho_0^f)_{\text{equ.}}$$

This operator is of course a differential operator with respect to the velocity $\underset{\sim}{V}$ of the brownian particle. (IV.2.2) now becomes:

(IV.2.4) $$\frac{\partial\varphi(\underset{\sim}{V},t)}{\partial t} = \int_0^t d\tau\,\Gamma(t-\tau)\,\varphi(\underset{\sim}{V},\tau)$$

From the very definition of G(t) and (IV.2.3), the Laplace transform of the operator $\Gamma(t)$ is obviously :

(IV.2.5) $$\phi(z) = \int \{ d\underset{\sim}{v} \}^N \psi(z) \, (\rho_o^f)_{equ.}$$

Following the same procedure as that used to go from (II.9.3) to (II.9.5) , the equation (IV.2.4) may be written in the pseudomarkovian form :

(IV.2.6) $$\frac{\partial \varphi(\underset{\sim}{V}, t)}{\partial t} = -i \varpi \, \phi(0) \, \varphi(\underset{\sim}{V}, t)$$

where the operator ϖ is given in terms of $\phi(0)$ and its derivatives $\phi'(0)$ by the same relation that holds between Ω and ψ (see (II.9.6) to (II.9.8)) :

(IV.2.7) $$\varpi = 1 + \phi'(0) + (1/2) \, \phi''(0) \phi(0) + [\phi'(0)]^2 + \dots$$

The equation (IV.2.6) will be our basic equation for a dynamical study of brownian motion. We shall now show how it reduces to an equation of the Fokker-Planck type when only the lowest order terms in the mass ratio m/M are retained .

IV.3 - Expansion in powers of the mass ratio.

As the fluid is at equilibrium at temperature T, we have:

(IV.3.1) $$\langle v_i \rangle = (2kT/m)^{1/2}$$

(IV.3.2) $$\langle p_i \rangle = (2mkT)^{1/2}$$

If the brownian particle moves with thermal velocity , we have:

(IV.3.3) $$V = 0(2kT/M)^{1/2} = 0(\gamma \langle v_i \rangle)$$

(IV.3.4) $$P = 0(2MkT)^{1/2} = 0(\gamma^{-1} \langle v_i \rangle)$$

with

(IV.3.5) $$\gamma = (m/M)^{1/2} \ll 1$$

The unperturbed and perturbed Liouville operators may be written in a way which exhibits their dependence on γ . We have :

(IV.3.6) $$L_o = L_o^f + \gamma L_o^A$$

where L_o^f is the unperturbed Liouville operator for the fluid :

(IV.3.7) $$L_o^f = \sum_{i=1}^{N} \underset{\sim}{v}_i \cdot \frac{\partial}{\partial \underset{\sim}{r}_i}$$

while γL_o^A is that for the brownian particle A :

(IV.3.8) $$\gamma L_o^A = \underset{\sim}{V} \cdot \frac{\partial}{\partial \underset{\sim}{R}}$$

Similarily, we may write :

(IV.3.9) $$\delta L = \delta L^f + \gamma \delta L^A$$

with

$$\delta L^f = \sum_{i<j} \frac{\partial V_{ij}(|\underset{\sim}{r}_i - \underset{\sim}{r}_j|)}{\partial \underset{\sim}{r}_i} \cdot \left(\frac{\partial}{\partial \underset{\sim}{p}_i} - \frac{\partial}{\partial \underset{\sim}{p}_j}\right)$$

(IV.3.10)

$$+ \sum_i \frac{\partial V_{iA}(|\underset{\sim}{r}_i - \underset{\sim}{R}|)}{\partial \underset{\sim}{r}_i} \cdot \frac{\partial}{\partial \underset{\sim}{p}_i}$$

(IV.3.11) $$\gamma \delta L^A = \sum_i \frac{\partial V_{iA}(|\underset{\sim}{r}_i - \underset{\sim}{R}|)}{\partial \underset{\sim}{R}} \cdot \frac{\partial}{\partial \underset{\sim}{P}}$$

With these expressions, we can easily expand the rhs of (IV.2.6) in powers of γ . Using (II.8.7) , we have :

(IV.3.12) $$\phi(z) = \int \{d\underset{\sim}{v}\}^N \langle 0 | \delta L \sum_{n=1}^{\infty} \left(\frac{1}{z - L_o} \delta L\right)^n | 0 \rangle_{irr} (\rho_o^f)_{equ.}$$

Taking into account the fact that δL^f is a differential operator with respect to the velocities of the fluid particles, we have :

$$\phi(z) = \int \{d\underset{\sim}{v}\}^N \langle 0 | (\delta L^f + \gamma \delta L^A) \sum_{n=1}^{\infty} (\frac{1}{z-L_o} \delta L)^n | 0 \rangle_{irr} (\rho_o^f)_{equ.}$$

$$(IV.3.13) \quad = \gamma \sum_{i=1}^{N} \sum_{\underset{\sim}{K}} \langle 0 | \delta L^A | \underset{\sim}{K}, \underset{\sim}{k}_i = -\underset{\sim}{K} \rangle \int \{d\underset{\sim}{v}\}^N \langle \underset{\sim}{K}, \underset{\sim}{k}_i = -\underset{\sim}{K} | \sum_{n=1}^{\infty} (\frac{1}{z-L_o} \delta L)^n | 0 \rangle_{irr} \times$$

$$\times (\rho_o^f)_{equ.}$$

Using (IV.3.6) and (IV.3.9) , we then immediately obtain the expansion:

$$(IV.3.14) \qquad \phi(z) = \gamma \phi_1(z) + \gamma^2 \phi_2(z) + \dots$$

with

$$\gamma \phi_1(z) = \sum_{i=1}^{N} \sum_{\underset{\sim}{K}} \{ \langle 0 | \delta L^A | \underset{\sim}{K}, \underset{\sim}{k}_i = -\underset{\sim}{K} \rangle \int \{d\underset{\sim}{v}\}^N \langle \underset{\sim}{K}, \underset{\sim}{k}_i = -\underset{\sim}{K} |$$

$$(IV.3.15) \qquad \sum_{n=1}^{\infty} (\frac{1}{z-L_o} \delta L^f)^n | 0 \rangle \}_{irr} (\rho_o^f)_{equ.}$$

$$\gamma^2 \phi_2(z) = \sum_{i=1}^{N} \sum_{\underset{\sim}{K}} \langle 0 | \delta L^A | \underset{\sim}{k}_i = -\underset{\sim}{K}, \underset{\sim}{K} \rangle \int \{d\underset{\sim}{v}\}^N \langle \underset{\sim}{k}_i = -\underset{\sim}{K}, \underset{\sim}{K} |$$

$$\sum_{n=0}^{\infty} (\frac{1}{z-L_o^f} \delta L^f)^n \frac{1}{z-L_o^f} L_o^A \sum_{m=0}^{\infty} (\frac{1}{z-L_o^f} \delta L^f)^m | 0 \rangle_{irr} (\rho_o^f)_{equ.}$$

$$(IV.3.16) \quad + \sum_{i=1}^{N} \sum_{\underset{\sim}{K}} \langle 0 | \delta L^A | \underset{\sim}{k}_i = -\underset{\sim}{K}, \underset{\sim}{K} \rangle \int \{d\underset{\sim}{v}\}^N \langle \underset{\sim}{k}_i = -\underset{\sim}{K}, \underset{\sim}{K} \sum_{n=0}^{\infty} (\frac{1}{z-L_o^f} \delta L^f)^n \times$$

$$\times \frac{1}{z-L_o^f} \delta L^A \sum_{m=0}^{\infty} (\frac{1}{z-L_o^f} \delta L^f)^m | 0 \rangle_{irr} (\rho_o^f)_{equ.}$$

In principle, in the first term in the rhs of (IV.3.16), the sum over m shuld go from 1 to infinity instead of from zero to infinity. However, taking into account the fact that L_o^A has no non diagonal elements and irreducible contributions only have to be kept (no intermediate state equal to the vacuum of correlations), we may add the term $m = 0$.

To the expansion (IV.3.14) of $\phi(z)$ corresponds the following expansion of the operator ϖ as given by (IV.2.7):

$$\varpi(z) = 1 + \gamma \phi_1'(z) + \gamma^2 \left[\phi_2'(z) + \frac{1}{2}\phi_1''(z)\phi_1(z) + \{\phi_1'(z)\}^2\right] + \dots \tag{IV.3.17}$$

Hence, if we do not retain terms of higher order than γ^2, we have:

$$\frac{\partial \varphi(\underset{\sim}{V}, t)}{\partial t} = - i\gamma\phi_1(0)\varphi(\underset{\sim}{V}, t) - i\gamma^2\left[\phi_1'(0)\,\phi_1(0) + \phi_2(0)\right]\varphi(\underset{\sim}{V}, t) \tag{IV.3.18}$$

We shall now discuss in details the various terms which appear in the rhs of (IV.3.18) and show how this equation reduces to a Fokker-Planck equation.

IV.4 - Study of the operator $\phi_1(0)$

This operator is the analytic continuation of the operator $\phi_1(z)$ given by (IV.3.1) for $z \to 0$. Using (II.12.14), for the canonical distribution for the fluid, we have :

$$\lim_{z \to 0} \sum_{m=1}^{\infty} \langle \{\underset{\sim}{k}_i\} \underset{\sim}{K} | \left(\frac{1}{z - L_o^f}\,\delta L^f\right)^m | 0 \rangle (\rho_o^f)_{equ.}$$

$$= (8\pi^3/\Omega)\int d\underset{\sim}{R}\; e^{-i\underset{\sim}{K}\cdot\underset{\sim}{R}} \lim_{z\to 0} \sum_{m=1}^{\infty} \langle \{\underset{\sim}{k}_i\} | \left(\frac{1}{z - L_o^f}\,\delta L^f\right)^m | 0 \rangle (\rho_o^f)_{equ.} \tag{IV.4.1}$$

$$= (8\pi^3/\Omega) \int d\underset{\sim}{R}\; e^{-i\underset{\sim}{K}.\underset{\sim}{R}}\; \rho^{equ.}_{\{\underset{\sim}{k}_i\}}$$

where $\rho^{equ}_{\{\underset{\sim}{k}_i\}}$ is the Fourier coefficient of the complete equilibrium distribution function for the fluid :

$$\rho^f_{equ.} = \exp\left[-\frac{1}{kT}\left\{\sum_i \frac{mv_i^2}{2} + \sum_{i<j} V_{ij}\,(|\underset{\sim}{r}_i - \underset{\sim}{r}_j|) + \sum_i V_{iA}(|\underset{\sim}{R} - \underset{\sim}{r}_i|)\right\}\right] \times$$

(IV. 4. 2)

$$\left[\int\{d\underset{\sim}{r}\,d\underset{\sim}{p}\}^N \exp\left[-\frac{1}{kT}\left\{\sum_i (mv_i^2/2) + \sum_{i<j} V_{ij}(|\underset{\sim}{r}_i - \underset{\sim}{r}_j|) + \sum_i V_{iA}(|\underset{\sim}{r}_i - \underset{\sim}{R}|)\right\}\right]\right]^{-1}$$

$$= \exp\left[-\frac{1}{kT}\left(\sum_i (mv_i^2/2) + V^f\right)\right]\left\{\int\{d\underset{\sim}{r}\,d\underset{\sim}{p}\}^N \exp\left[-\frac{1}{kT}\left(\sum_i \frac{mv_i^2}{2} + V^f\right)\right]\right\}^{-1}$$

(IV. 4. 3)

$$\rho^{equ.}_{\{\underset{\sim}{k}_i\}} = \int\{d\underset{\sim}{r}\}^N \exp\left[-i\sum_i \underset{\sim}{k}_i . \underset{\sim}{r}_i\right]\rho^f_{equ.}$$

Introducing (IV. 4. 1) into (IV. 3. 6) for $z \to 0$, we obtain :

(IV. 4. 4)

$$\gamma\,\phi_{\{0\}} \sim \sum_{i=1}\sum_{\underset{\sim}{K}}\int\{d\underset{\sim}{v}\}^N \langle 0|\delta L^A|\, \underset{\sim}{k}_i = -\underset{\sim}{K}, \underset{\sim}{K}\rangle \times$$

$$\times \int d\underset{\sim}{R}\; e^{-i\underset{\sim}{K}.\underset{\sim}{R}}\; \rho^{equ.}_{\underset{\sim}{k}_i = -\underset{\sim}{K}}$$

Now, we have :

(IV. 4. 5)

$$\langle 0|\delta L^A|\, \underset{\sim}{k}_i = -\underset{\sim}{K}, \underset{\sim}{K}\rangle \int d\underset{\sim}{R}\; e^{-i\underset{\sim}{K}.\underset{\sim}{R}}\; \rho^{equ.}_{\underset{\sim}{k}_i}$$

$$= V_K\,\underset{\sim}{K} \cdot \frac{\partial}{\partial \underset{\sim}{P}} \int d\underset{\sim}{R}\; e^{-i\underset{\sim}{K}.\underset{\sim}{R}} \int\{d\underset{\sim}{r}\}^N\; e^{i\underset{\sim}{K}.\underset{\sim}{r}_i}\rho^f_{equ.}$$

Taking into account the fact that V_K is the Fourier transform of V_{iA}, we have :

(IV. 4. 6)

$$\sum_{\underset{\sim}{K}} V_K\,\underset{\sim}{K}\; e^{i\underset{\sim}{K}.(\underset{\sim}{R} - \underset{\sim}{r}_j)} \sim \frac{\partial V_{iA}(|\underset{\sim}{R} - \underset{\sim}{r}_i|)}{\partial \underset{\sim}{R}}$$

Therefore, we obtain :

$$\gamma \phi_1(0) \sim \int \{d\underset{\sim}{v}\, d\underset{\sim}{r}\}^N d\underset{\sim}{R} \frac{\partial \sum_i V_{iA}(|\underset{\sim}{R} - \underset{\sim}{r}_i|)}{\partial \underset{\sim}{R}} \rho^f_{equ} \cdot \frac{\partial}{\partial \underset{\sim}{P}}$$

$$\text{(IV.4.7)} \qquad \sim \int d\underset{\sim}{R} \frac{\partial \int \{d\underset{\sim}{v} d\underset{\sim}{r}\}^N \rho^{equ.}_f}{\partial \underset{\sim}{R}} \cdot \frac{\partial}{\partial \underset{\sim}{P}} = 0$$

and the equation of evolution for the velocity distribution function of the brownian particle (IV.3.19) takes the simple form :

$$\text{(IV.4.8)} \qquad \frac{\partial \varphi(\underset{\sim}{V}, t)}{\partial t} = - i \gamma^2 \phi_2(0)\, \varphi(\underset{\sim}{V}, t)$$

The first non vanishing effects are proportional to the mass ratio m/M .

IV.5 - Study of the operator $\phi_2(0)$

From (IV.3.1) and (IV.4.2) , we obtain :

$$\gamma^2 \phi_2(0) = \gamma^2 \lim_{z \to 0} \sum_{i=1}^{N} \sum_{\underset{\sim}{K}} \sum_{\underset{\sim}{K'}} \sum_{\{\underset{\sim}{k'}\}} \langle 0 | \delta L^A | \underset{\sim}{k}_i = -\underset{\sim}{K}, \underset{\sim}{K} \rangle \times$$

$$\text{(IV.5.1)} \int \{d\underset{\sim}{v}\}^N \langle \underset{\sim}{k}_i = -\underset{\sim}{K}, \underset{\sim}{K} | \sum_{n=0}^{\infty} \left(\frac{1}{z - L^f_o} \delta L^f \right)^n \frac{1}{z - L^f_o} (L^A_o + \delta L^A) | \{\underset{\sim}{k'}\} \underset{\sim}{K'} \rangle \times$$

$$\times \frac{1}{\Omega} \int d\underset{\sim}{R}\; e^{-i \underset{\sim}{K'} . \underset{\sim}{R}} \rho^{equ}_{\{\underset{\sim}{k'}\}}$$

Now, using :

$$\text{(IV.5.2)} \int d\underset{\sim}{R}\; e^{-i\underset{\sim}{K'}.\underset{\sim}{R}} \rho^{equ.}_{\{\underset{\sim}{k'}\}} = \langle \{\underset{\sim}{k'}\} \underset{\sim}{K'} | \rho^{equ}_f | 0 \rangle \, \Omega^{N+1} \times$$

$$\text{(IV.5.3)} \times \sum_{n=0}^{\infty} \left(\frac{1}{z - L^f_o} \delta L^f \right)^n \frac{1}{z - L^f_o} = \frac{1}{z - L^f_o - \delta L^f}$$

and performing the summations over $\{\underset{\sim}{k'}\}$, $\underset{\sim}{K'}$, $\underset{\sim}{K}$, we obtain :

$$\gamma^2 \phi_2(0) = \gamma^2 \lim_{z \to 0} (\Omega)^{-1} \int \{d\underset{\sim}{r}\}^N d\underset{\sim}{R} \{\delta L^A \int \{d\underset{\sim}{v}\}^N \frac{1}{z - L^f_o - \delta L^f} \times$$

(IV.5.4) $$\times (L_o^A + \delta L^A)\, \rho^f_{equ.}$$

Using the explicit forms of L_o^A and δL^A, this becomes :

$$\gamma^2 \phi_2(0) = (\Omega)^{-1} \lim_{z \to 0} \int \{d\underset{\sim}{r}\}^N d\underset{\sim}{R}\, \lambda \frac{\partial \sum_i V_{iA}(|\underset{\sim}{R}-\underset{\sim}{r}_i|)}{\partial \underset{\sim}{R}} \cdot \frac{\partial}{\partial \underset{\sim}{P}} \times$$

(IV.5.5)
$$\times \int \{d\underset{\sim}{v}\}^N \frac{1}{z-L_o^f-\delta L^f} \left[-\underset{\sim}{V} \cdot \frac{\partial}{\partial \underset{\sim}{R}} + \lambda \frac{\partial \sum_i V_{iA}(|\underset{\sim}{R}-\underset{\sim}{r}_i|)}{\partial \underset{\sim}{R}} \cdot \frac{\partial}{\partial \underset{\sim}{P}} \right] \rho^f_{equ.}$$

Using the explicit form of $\rho^f_{equ.}$ (see IV .2.2) , we obtain easily :

$$\gamma^2 \phi_2(0) = \Omega^{-1} \lim_{z \to 0} \int \{d\underset{\sim}{r}\}^N d\underset{\sim}{R}\, \frac{\partial \sum_i V_{iA}(|\underset{\sim}{R}-\underset{\sim}{r}_i|)}{\partial \underset{\sim}{R}} \cdot \frac{\partial}{\partial \underset{\sim}{P}} \times$$

(IV.5.6)
$$\times \int \{d\underset{\sim}{v}\}^N \frac{1}{z-L_o^f-\delta L^f} \frac{\partial \sum_i V_{iA}(|\underset{\sim}{R}-\underset{\sim}{r}_i|)}{\partial \underset{\sim}{R}} \cdot \left(\frac{\partial}{\partial \underset{\sim}{P}} + \frac{1}{kT} \underset{\sim}{V}\right) \rho^f_{equ.}$$

Let us introducing the diffusion coefficient :

$$D_{ij} = \Omega^{-1} \int \{d\underset{\sim}{v}\, d\underset{\sim}{r}\}^N d\underset{\sim}{R}\; F_i(\underset{\sim}{R}, \{\underset{\sim}{r}\}) \times$$

(IV.5.7)
$$\times \lim_{z \to 0} \frac{1}{z-L_o^f-\delta L^f} F_j(R, \{\underset{\sim}{r}\})\, \rho^f_{equ.}$$

where

(IV.5.8) $$F_1(\underset{\sim}{R}, \{\underset{\sim}{r}\}) = \lambda \frac{\partial \sum_i V_{iA}(|\underset{\sim}{R}-\underset{\sim}{r}_i|)}{\partial R_1}$$

is the total force exerted by the fluid on the Brownian particle; the operator $\phi_2(0)$ becomes :

(IV.5.9) $$\gamma^2 \phi_2(0) = D_{ij} \frac{\partial}{\partial P_i} \left(\frac{\partial}{\partial P_j} + \frac{1}{kT} V_j\right)$$

The evolution equation (IV.4.8) may thus be written :

$$\frac{\partial \varphi(\underset{\sim}{V}, t)}{\partial t} = D_{ij} \frac{\partial}{\partial P_i} \left(\frac{\partial}{\partial P_j} + \frac{1}{kT} V_j \right) \varphi(\underset{\sim}{V}, t) \qquad \text{(IV.5.10)}$$

This is indeed an equation of the Fokker-Planck type. An explicit expression in terms of microscopic quantities is now obtained for the diffusion coefficient.

IV.6 - Diffusion coefficient.

Taking into account :

$$-i \lim_{z \to 0} \frac{1}{z - L_o^f - \delta L^f} = \lim_{z \to 0} \int_0^\infty dt \exp\left[-i (L_o^f + L^f - z) t\right] \qquad \text{(IV.6.1)}$$

the diffusion coefficient may be written as :

$$\begin{aligned} D_{ij} &= \lim_{z \to 0} \int_0^\infty dt\, (\Omega)^{-1} \int \{d\underset{\sim}{r}\, d\underset{\sim}{v}\}^N d\underset{\sim}{R}\; F_i(\underset{\sim}{R}, \{\underset{\sim}{r}\}) \times \\ &\times \exp\left[-i (L_o^f + \delta L^f - z) t\right] F_j(\underset{\sim}{R}, \{\underset{\sim}{r}\})\, \rho^f_{equ.} \end{aligned} \qquad \text{(IV.6.2)}$$

Taking into account :

$$L^f \rho^f_{equ.} = (L_o^f + \delta L^f) \rho^f_{equ.} = 0 \qquad \text{(IV.6.3)}$$

and the fact that the integrand is a function of relative distances only, we obtain :

$$\begin{aligned} D_{\alpha\beta} &= \lim_{z \to 0} \int_0^\infty dt \int \{d\underset{\sim}{v}\, d\underset{\sim}{r}\}^N \rho^f_{equ.}\; F_\alpha(\underset{\sim}{R}, \{\underset{\sim}{r}\}) \times \\ &\times \exp\left[-i (L^f - z) t\right] F_\beta(\underset{\sim}{R}, \{\underset{\sim}{r}\}) \\ &= \lim_{z \to 0} \int_0^\infty dt \left\langle F_\alpha(\underset{\sim}{R}, \{\underset{\sim}{r}\}) \exp\left[-i (L^f - z) t\right] F_\beta(\underset{\sim}{R}, \{\underset{\sim}{r}\}) \right\rangle \end{aligned} \qquad \text{(IV.6.4)}$$

The diffusion coefficient is thus the average value over the fluid equilibrium distribution of the time autocorrelation function of the force acting on the brownian particle for a fixed position of the brownian particle.

As an example, let us compute this coefficient in the case of a weakly coupled system. Then we have :

(IV.6.5) $$L^f \rightarrow L_0^f$$

(IV.6.6) $$\rho^f_{equ.} = (m/2\pi kT)^{3N/2} (4\pi)^{3N} \exp\left[-\sum_1 \frac{mv_1^2}{2kT}\right] \Omega^{-N}$$

Expanding the interaction potential in a Fourier series, we obtain :

$$D_{\alpha\beta} = -\lambda^2 (\Omega)^{-N} \sum_{i,j=1}^{N} \lim_{z\to 0} \int_0^\infty dt \int \{d\underset{\sim}{r} d\underset{\sim}{v}\}^N (m/2\pi kT)^{3N/2}$$

(IV.6.7) $$(4\pi)^{3N} \exp\left[-\sum_1 mv_1^2/2kT\right] \int d^3k \int d^3k' \, V_k V_{k'} k_\alpha k'_\beta \, e^{i\underset{\sim}{k}\cdot(\underset{\sim}{r}_i-\underset{\sim}{R})} \, e^{-i(L_0^f - z)t} \, e^{i\underset{\sim}{k}'\cdot(\underset{\sim}{r}_j-\underset{\sim}{R})}$$

Now, we have :

(IV.6.8) $$e^{-iL_0^f t} \, e^{i\underset{\sim}{k}'\cdot\underset{\sim}{r}_j} = e^{i\underset{\sim}{k}'\cdot(\underset{\sim}{r}_j-\underset{\sim}{v}_j t)}$$

With :

(IV.6.9) $$\int d\underset{\sim}{r}_i \, e^{i\underset{\sim}{k}\cdot\underset{\sim}{r}_i} = 8\pi^3 \delta(\underset{\sim}{k})$$

we verify easily that the only terms which contribute are those for which $i = j$. This result is of course quite obvious. Indeed, the only contributions to the evolution equation are those of diagonal diagrams, in which each particle must appear at least at two vertices. In the weakly coupled case, we only have to consider the cycle and have therefore only one fluid particle involved.

Therefore, with $\int d\underset{\sim}{r}_i \; e^{i(\underset{\sim}{k}+\underset{\sim}{k}')\cdot \underset{\sim}{r}_i} = 8\pi^3 \delta(\underset{\sim}{k}+\underset{\sim}{k}')$ (IV. ω. 10)

we have :

$$D_{\alpha\beta} = 32\pi^4 \lambda^2 C(m/2\pi kT)^{3/2} \lim_{z\to 0} \int_0^\infty dt \int d\underset{\sim}{v} \exp(-mv^2/2kT) \times \tag{IV. 6. 11}$$

$$\times \int d^3k \, |V_k|^2 \, k_\alpha k_\beta \, e^{i\underset{\sim}{k}\cdot\underset{\sim}{v}t} \, e^{izt}$$

We also have :

$$\lim_{z\to 0} \int_0^\infty dt \;\; e^{i(\underset{\sim}{k}\cdot\underset{\sim}{v} + z)t} = \pi\delta(\underset{\sim}{k}\cdot\underset{\sim}{v}) + i \; P(1/\underset{\sim}{k}\cdot\underset{\sim}{v}) \tag{IV. 6. 12}$$

As the remaining part of the integrand in (IV. 6. 11) is an even function of $\underset{\sim}{k}$, the contribution involving the principal part vanishes and we are left with :

$$D_{\alpha\beta} = 32\pi^5 \lambda^2 C(m/2\pi kT)^{3/2} \int d\underset{\sim}{v} \exp(-mv^2/2kT) \int d^3k \, |V_k|^2 \, k_\alpha k_\beta \times \tag{IV. 6. 13}$$

$$\times \; \delta(\underset{\sim}{k}\cdot\underset{\sim}{v})$$

One verifies easily that one has :

$$\int d^3k \, |V_k|^2 \; k_\alpha \, k_\beta \delta(\underset{\sim}{k}\cdot\underset{\sim}{v}) = \delta_{\alpha,\beta} \; (\pi B/v)\left[1 - \frac{(\underset{\sim}{1}\cdot\underset{\sim}{v})^2}{v^2}\right] \tag{IV. 6. 14}$$

where $\underset{\sim}{1} = (1_x, 1_y, 1_z)$ is the unit vector and B is given by (III. 4. 5) .

Performing the $\underset{\sim}{v}$ integration, one obtains :

$$D_{\alpha\beta} = \delta_{\alpha,\beta} \, 32\pi^6 \lambda^2 C(m/2kT)^{1/2} (4/3\sqrt{\pi}) \tag{IV. 6. 15}$$

One verifies easily that with this value of the diffusion coefficient, the evolution equation (IV. 5. 10) is indeed identical with the particular Fokker-Planck equation (III. 6. 14) we obtained for a heavy particle moving with thermal velocity and weakly coupled to the fluid.

F. Henin

V. BROWNIAN MOTION OF A HEAVY PARTICLE IN AN EXTERNAL FIELD

V.1 - Introduction .

We shall again consider the problem of a heavy particle moving in a fluid at thermal equilibrium. However, now we shall assume that the brownian particle is charged and that at $t = 0$, we switch on an external constant electrical field. The fluid particles are neutral and are not influenced by that field. After a long time, we shall reach a stationary state for the velocity distribution function of the brownian particle, corresponding to a balance between the effect of the external acceleration and the scattering by the fluid particles.

Our starting point for the discussion of this stationary state will be the transport equation (II. 14. 5). Here again, we shall show that, when only lowest order terms in the mass ratio are kept, the equation for the stationary state is in agreément with that of the stochastic theory. The calculations which have been performed originally by Résibois and Davis[1] will be closely parallel to that of the preceding chapter and we shall go over them very briefly and rather concentrate ourselves on a discussion of higher order corrections. First, we have corrections to the collision terms which are independent of the external field and introduce fourth order differential operators in the equation of evolution. Then , we also have corrections which take into account the effect of the field during a collision. We shall show that these corrections may formally be incorporated in the Fokker-Planck collision operator.

These results are in agreement with those obtained through a rather different method by Lebowitz and Rubin[2] . In order to make connection with this work , we shall show how the transport equation for this particular problem can be recast in another form, which is precisely that used as a starting point for the m/M expansion in Lebowitz and Rubin. This point has been discussed in great detail in a paper by Lebowitz and Résibois[3].

V.2 - Steady state equation for the velocity distribution function of the brownian particle.

In chapter II, § 14, we have obtained the linearized steady state equation (II.14.5) for the velocity distribution function of a system of charged particles submitted to the action of an external electrical field. We have assumed that at t=0 the system was in equilibrium and that the field was switched on only at t=0. Keeping only terms linear in E and restricting ourselves to the static case, we obtain :

$$\text{(V.2.1)} \quad iL_E \rho_o^{equ.} + \sum_{\{\underset{\sim}{k}\}} D_{\{\underset{\sim}{k}\}}(0)\, iL_E\, C_{\{\underset{\sim}{k}\}}(0)\, \rho_o^{equ} = i\psi(0)\Delta\rho_o^{st.}$$

where we have used (II.13.18) to rewrite the second term in the lhs of (II.14.5) as an operator acting on ρ_o.

As the brownian particle is the only charged particle we have here :

$$\text{(V.2.2)} \qquad iL_E = e\,\underset{\sim}{E} \cdot \frac{\partial}{\partial \underset{\sim}{P}}$$

instead of (II.13.3).

In (V.2.1), ρ_o^{equ} is the velocity equilibrium distribution for the whole system :fluid and brownian particle

$$\text{(V.2.3)} \qquad \rho_o^{equ} = \varphi^{equ}(\underset{\sim}{V}) \prod_i^N \varphi^{equ}(\underset{\sim}{v}_i)$$

(V.2.4) $$\varphi^{equ}(\underset{\sim}{v}_i) = 4\pi(m/2\pi kT)^{3/2} \exp(-mv_i^2/2kT)$$

(V.2.5) $$\varphi^{equ}(\underset{\sim}{V}) = 4\pi(M/2\pi kT)^{3/2} \exp(-MV^2/2kT)$$

$\Delta\rho_o^{st}$ is the linear (in E) correction to ρ_o^{equ} . As previously, we assume molecular chaos for the velocity distribution function. Therefore, we have :

(V.2.6) $$\Delta\rho_o^{st} = \delta\varphi(\underset{\sim}{V}) \prod_i^N \varphi^{equ}(\underset{\sim}{v}_i) + \sum_{i=1}^N \delta\varphi(\underset{\sim}{v}_i) \prod_{j\neq i}^N \varphi^{equ}(\underset{\sim}{v}_j)\, \varphi^{equ}(\underset{\sim}{V})$$

Therefore, (V.2.1) becomes :

(V.2.7) $$\prod_i^N \varphi^{equ}(\underset{\sim}{v}_i)\, e\underset{\sim}{E}.\frac{\partial\varphi^{equ}(\underset{\sim}{V})}{\partial \underset{\sim}{P}} + \sum_{\{\underset{\sim}{k}\}} D_{\{\underset{\sim}{k}\}}(0) iL_E\, C_{\{\underset{\sim}{k}\}}(0) \prod_i^N \varphi^{equ}(\underset{\sim}{v}_i)\, \varphi^{equ}(\underset{\sim}{V})$$
$$= i\psi(0)\delta\varphi(\underset{\sim}{V}) \prod_i^N \varphi^{equ}(\underset{\sim}{v}_i) + \sum_{i=1}^N i\psi(0)\delta\varphi(\underset{\sim}{v}_i) \prod_{j\neq i}^N \varphi^{equ}(\underset{\sim}{v}_j)\, \varphi^{equ}(\underset{\sim}{V})$$

Integrating this equation, first over the velocities of all the fluid particles, secondly over the velocity of the brownian particle and the velocities of all but one of the fluid particles, one obtains easily a set of coupled equations for the two unknown functions $\delta\varphi(\underset{\sim}{V})$ and $\delta\varphi(\underset{\sim}{v}_i)$. However, it can be proved [1] that, once terms of order $1/N$ are neglected the velocity distribution function of a fluid particle remains at equilibrium in the stationary case :

(V.2.8) $$\delta\varphi(\underset{\sim}{v}_i) = 0(1/N) \rightarrow 0$$

We shall not prove this here but it is a consequence of the fact that the probability of a given particle to interact with the single heavy par-cle is quite negligible in the limit of an infinite system. This allows us to neglect all vertices involving A and a fluid particle in the operators $D_{\{k\}}$, $C_{\{k\}}$ and ψ once we integrate over $\underset{\sim}{V}, \underset{\sim}{v}_2 \ldots \underset{\sim}{v}_N$. The equation for $\delta\varphi(\underset{\sim}{v}_1)$ then simplifies a great deal and using arguments similar to those which allowed us to establish the $\mathcal{H}$-theorem in chapter II, § 11 it is easy to show that (V.2.8) is its only solution.

Taking into account (V.2.8) , the evolution equation (V.2.7) becomes :

$$\prod_i^N \varphi^{equ}(\underset{\sim}{v}_i)\, e\underset{\sim}{E} \cdot \frac{\partial \varphi^{equ}(\underset{\sim}{V})}{\partial \underset{\sim}{P}}$$

$$(V.2.9) \qquad + \sum_{\{\underset{\sim}{k}\}} D_{\{\underset{\sim}{k}\}}(0)\; e\underset{\sim}{E} \cdot \frac{\partial}{\partial \underset{\sim}{P}}\, C_{\{\underset{\sim}{k}\}}(0) \prod_i^N \varphi^{equ}(\underset{\sim}{v}_i)\varphi^{equ}(\underset{\sim}{V})$$

$$= i\,\psi(0)\,\delta\varphi(\underset{\sim}{V}) \prod_i^N \varphi^{equ}(\underset{\sim}{v}_i)$$

After integration over the velocities of the fluid particles, we obtain :

$$(V.2.10) \qquad e\underset{\sim}{E} \cdot \frac{\partial \varphi^{equ}(\underset{\sim}{V})}{\partial P} + \Xi(0)\, \varphi^{equ}(\underset{\sim}{V}) = i\phi(0)\delta\varphi(\underset{\sim}{V})$$

where

$$(V.2.11) \qquad \phi(0) = \int \{d\underset{\sim}{v}\}^N \psi(0) \prod_i^N \varphi^{equ}(\underset{\sim}{v}_i)$$

$$(V.2.12) \qquad \Xi(0) = \sum_{\{\underset{\sim}{k}\}} \int \{d\underset{\sim}{v}\}^N D_{\{\underset{\sim}{k}\}}(0)\, e\underset{\sim}{E} \cdot \frac{\partial}{\partial \underset{\sim}{P}}\, C_{\{\underset{\sim}{k}\}}(0) \prod_i^N \varphi^{equ}(\underset{\sim}{v}_i)$$

The operator $\phi(0)$ is identical to the operator $\phi(0)$ given by (IV.2.5)

V.3 - Expansion in the mass ratio .

Here again, we write :

$$L_o = L_o^f + \gamma L_o^A \tag{V.3.1}$$

$$\delta L = \delta L^f + \gamma \delta L^A \tag{V.3.2}$$

$$e\underset{\sim}{E} . \frac{\partial}{\partial \underset{\sim}{P}} = \gamma L_E \tag{V.3.3}$$

where the fluid operators $L_o^f, \delta L^f$ are given by (IV.3.7) and (IV.3.10) respectively while the particle operators γL_o^A and $\gamma \delta L^A$ are given by (IV.3.8) and (IV.3.11) .

We also expand the operators $\phi(0)$ and $\Xi(0)$:

$$\phi(0) = \gamma \phi^{(1)}(0) + \gamma^2 \phi^{(2)}(0) + \dots \tag{V.3.4}$$

$$\Xi(0) = \gamma \Xi^{(1)}(0) + \gamma^2 \Xi^{(2)}(0) + \dots \tag{V.3.5}$$

The zero order term (γ^o) of ϕ vanishes as we have seen in chapter IV. The zero order term of $\Xi(0)$ vanishes because this operator involves the external field operator which is of order γ .

Up to order γ^4, we therefore obtain the following equation of evolution :

$$i L_E \varphi^{equ}(\underset{\sim}{V}) + \left[\gamma \Xi^{(1)}(0) + \gamma^2 \Xi^{(2)}(0) + \gamma^3 \Xi^{(3)}(0) + \gamma^4 \Xi^{(4)}(0)\right] \varphi^{equ}(V) = i \left[\gamma \phi^{(1)}(0) + \gamma^2 \phi^{(2)}(0) + \gamma^3 \phi^{(3)}(0) + \gamma^4 \phi^{(4)}(0)\right] \delta\varphi(\underset{\sim}{V}) . \tag{V.6}$$

In chapter IV, §§ 4 and 5, we have shown :

(V.3.7) $$\phi^{(1)}(0) = 0$$

(V.3.8) $$i\gamma^2 \phi^{(2)}(0) = D_{ij} \frac{\partial}{\partial P_i} \left(\frac{\partial}{\partial P_j} + \frac{1}{kT} V_j \right)$$

where the diffusion coefficient is given by (IV.5.7) . An alternative expression in terms of the average autocorrelation function of the force exerted by the fluid over the fixed brownian particle is given in (IV.6.4) . As to the interference term between the flow and the collision, it is again easy to show that :

(V.3.9) $$\Xi^{(1)}(0) = \Xi^{(2)}(0) = 0$$

Indeed the term of order γ in the rhs of (V.2.12) is

(V.3.10) $$\gamma \Xi^{(1)}(0) = \gamma \sum_{\{k\}} \int \{d\underset{\sim}{v}\}^N D^{(0)}_{\{\underset{\sim}{k}\}}(0) \; iL_E \; C^{(0)}_{\{\underset{\sim}{k}\}}(0) \prod_i^N \varphi^{equ}(\underset{\sim}{v}_i)$$

where $D^{(0)}_{\{\underset{\sim}{k}\}}(0)$ and $C^{(0)}_{\{\underset{\sim}{k}\}}(0)$ are the terms independent of γ of the destruction and creation operators :

(V.3.11) $$D^{(0)}_{\{\underset{\sim}{k}\}}(z) = \sum_{n=1}^{\infty} \langle 0 | \left(\delta L^f \frac{1}{z - L_o^f} \right)^n | \{\underset{\sim}{k}\} \rangle$$

(V.3.12) $$C^{(0)}_{\{k\}}(z) = \sum_{n=1}^{\infty} \langle \{\underset{\sim}{k}\} | \left(\frac{1}{z - L_o^f} \delta L^f \right)^n | 0 \rangle$$

Taking into account :

(V.3.13) $$\int \{d\underset{\sim}{v}\}^N \delta L^f \ldots = 0$$

we easily verify (V.3.9) for the operator $\Xi^{(1)}(0)$.

As to the operator $\Xi^{(2)}(0)$, we have :

$$\gamma^2 \; \Xi^{(2)}(0) = \gamma^2 \sum_{\{k\}} \int \{dv\}^N \; D^{(1)}_{\{k\}}(0) \; iL_E C^{(0)}_{\{k\}}(0) \prod_i \varphi^{equ}(v_i)$$

(V.3.14)
$$+ \gamma^2 \sum_{\{k\}} \int \{dv\}^N \; D^{(0)}_{\{k\}}(0) \; iL_E C^{(1)}_{\{k\}}(0) \prod_i \varphi^{equ}(v_i)$$

Using the same argument as above on the structure of $D^{(0)}_{\{k\}}(0)$, it is easy to show that the second term in the rhs vanishes . As to the first term, we have :

$$\gamma D^{(1)}_{\{k\}}(z) = \sum_{n=0}^{\infty} \langle 0 | \delta L^A \frac{1}{z-L_0^f} \left(\delta L^f \frac{1}{z-L_0^f} \right)^n | \{k\} \rangle$$

(V.3.15)
$$+ \sum_{n=1}^{\infty} \sum_{m=0}^{\infty} \langle 0 | (\delta L^f \frac{1}{z-L_0^f})^n \; (\delta L^A + L_0^A) \frac{1}{z-L_0^f} \; (\delta L^f \frac{1}{z-L_0^f})^m | \{k\} \rangle$$

Again the second term gives a vanishing contribution when we integrate over the velocities. Therefore, we have :

$$\gamma^2 \Xi^{(2)}(z) = \gamma^2 \sum_{n=0}^{\infty} \sum_{m=1}^{\infty} \sum_{\{k\}} \int \{dv\}^N \langle 0 | \delta L^A \frac{1}{z-L_0^f} \times$$

(V.3.16)
$$\times \left(\delta L^f \frac{1}{z-L_0^f} \right)^n | \{k\} \rangle \, iL_E \langle \{k\} | \left(\frac{1}{z-L_0^f} \delta L^f \right)^m | 0 \rangle \prod_i \varphi^{equ}(v_i)$$

Now, using (IV.4.1) , (IV.5.2) and (IV.5.3) , we easily obtain :

(V.3.17)
$$\gamma^2 \Xi^{(2)}(z) = \gamma^2 \int \{dv\}^N \langle 0 | \delta L^A \frac{1}{z-L_0^f - \delta L^f} \rho^f_{equ} | 0 \rangle \, iL_E$$

From (IV.3.11) , (IV.6.3) and (IV.4.2) , we easily obtain :

$$\gamma^2 \; \Xi^{(2)}(z) = \gamma^2 (1/z) \int \{d\underset{\sim}{r}d\underset{\sim}{v}\}^N \frac{\partial \sum_i V_{iA}(|\underset{\sim}{r}_i - \underset{\sim}{R}|)}{\partial \underset{\sim}{R}} \cdot \frac{\partial}{\partial \underset{\sim}{P}} \rho^f_{equ} \tag{V.3.18}$$

$$= 0$$

(see (IV.4.7)).

Therefore, if we restrict ourselves to terms of order m/M, we easily recover the stationary Fokker-Planck equation in presence of an external field acting on the heavy particle :

$$e\underset{\sim}{E} \cdot \frac{\partial}{\partial \underset{\sim}{P}} (\underset{\sim}{V}) = D_{ij} \frac{\partial}{\partial P_i} \left(\frac{\partial}{\partial P_j} + \frac{1}{kT} V_j\right) \delta\varphi(\underset{\sim}{V}) \tag{V.3.19}$$

The diffusion coefficient is not affected by the presence of the external field.

V.4 - Higher order corrections to the collision operator. Role of the irreducibility condition.

The γ^3 and γ^4 contributions are respectively :

$$\gamma^3 \phi^{(3)}(0) = \lim_{z \to 0} \gamma^3 \sum_{n=0}^{\infty} \sum_{p=0}^{\infty} \sum_{q=0}^{\infty} \int \{d\underset{\sim}{v}\}^N \langle 0 | \delta L^A$$

$$\frac{1}{z-L_o^f} \left(L^f \frac{1}{z-L_o^f} \right)^n (L^o_A + \delta L^A) \frac{1}{z-L^f_o} \left(\delta L^f \frac{1}{z-L_o^f}\right)^p (L_o^A + \delta L^A) \tag{V.4.1}$$

$$\left(\frac{1}{z-L_o^f} \delta L^f\right)^q | 0 \rangle_{irr} \prod_i \varphi^{equ}(\underset{\sim}{v}_i)$$

$$\gamma^4 \phi^{(4)}(0) = \lim_{z \to 0} \gamma^4 \sum_{n=0}^{\infty} \sum_{m=0}^{\infty} \sum_{p=0}^{\infty} \sum_{q=0}^{\infty} \int \{d\underset{\sim}{v}\}^N \langle 0 | \delta L^A$$

$$\frac{1}{z-L_o^f} \left(\delta L^f \frac{1}{z-L_o^f}\right)^n (L_o^A + \delta L^A) \quad \frac{1}{z-L_o^f} \left(\delta L^f \frac{1}{z-L_o^f}\right)^m (L_o^A + \delta L^A) \times$$

$$\text{(V.4.2)} \times \frac{1}{z-L_0^f} \left(\delta L^f \frac{1}{z-L_0^f}\right)^p \left(L_0^A + \delta L^A\right) \left(\frac{1}{z-L_0^f} \delta L^f\right)^q |0\rangle_{irr} \times$$

$$\times \prod_i \varphi^{equ}(\underset{\sim}{v}_i)$$

The first point we want to discuss here is the role played by the irreducibility condition. To do this, we will need a theorem first established by Balescu [4]:

If we have a succession of diagonal fragments, the only contributions to the reduced distribution function of particle α which do not vanish at the limit of an infinite system are those where the diagonal fragments are semi-connected, i.e. where they have a single particle in common with the preceding diagonal fragments.

We shall not prove this theorem in full generality but illustrate it on an example. Let us consider a succession of two cycles: we have three cases (see fig. V.4.1)

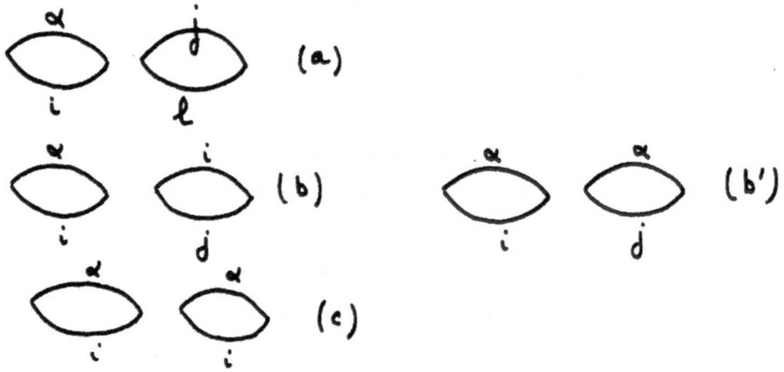

Possible connections in a succession of two cycles.

Fig. V.4.1

1) they are disconnected : no particle in common (a)
2) they are semi-connected : one particle in common (b), (b').
3) they have two particles in common (c).

Because of the integration over the velocities of all particles but α , α must necessarily appear at the first vertex on the left. We obtain :

$$(a) = \sum_{ijl} \int \{d\underset{\sim}{v}\}^{N-1} \langle 0 | \delta L^{\alpha i} | \underset{\sim}{k}_\alpha, \underset{\sim}{k}_i = -\underset{\sim}{k}_\alpha \rangle \langle \underset{\sim}{k}_\alpha, \underset{\sim}{k}_i = -\underset{\sim}{k}_\alpha | \delta L^{\alpha i} | 0 \rangle$$

$$\langle 0 | \delta L^{jl} | \underset{\sim}{k}_j, \underset{\sim}{k}_l = -\underset{\sim}{k}_j \rangle \langle \underset{\sim}{k}_j, \underset{\sim}{k}_l = -\underset{\sim}{k}_j | \delta L^{jl} | 0 \rangle \rho_o(0) \quad \text{(V.4.1)}$$

Now, the integrations over $\underset{\sim}{v}_j$ and $\underset{\sim}{v}_l$ commute with the first two matrix elements. Therefore, we have :

$$(a) \sim \int d\underset{\sim}{v}_j \, d\underset{\sim}{v}_l \, \delta L^{jl} \ldots = 0$$

This argument may easily be generalized to more complicated fragments and successions of more than two fragments; whenever there is at least two disconnected fragments, the contribution to the reduced distribution function vanishes.

Let us now consider the case of the semi-connected diagrams (b) and (b') . We have :

$$(b) + (b') = \sum_{ij} \int \{d\underset{\sim}{v}\}^{N-1} \langle 0 | \delta L^{i\alpha} | \underset{\sim}{k}_i = -\underset{\sim}{k}_\alpha, \underset{\sim}{k}_\alpha \rangle$$

$$\langle \underset{\sim}{k}_i = -\underset{\sim}{k}_\alpha, \underset{\sim}{k}_\alpha | \delta L^{i\alpha} | 0 \rangle \{ \langle 0 | \delta L^{ij} | \underset{\sim}{k}'_i, \underset{\sim}{k}_j = -\underset{\sim}{k}'_i \rangle$$

$$\langle \underset{\sim}{k}'_i, \underset{\sim}{k}_j = -\underset{\sim}{k}'_i | \delta L^{ij} | 0 \rangle + \langle 0 | \delta L^{\alpha j} | \underset{\sim}{k}'_\alpha, \underset{\sim}{k}_j = -\underset{\sim}{k}'_\alpha \rangle \quad \text{(V.4.3)}$$

$$\langle \underset{\sim}{k}'_\alpha, \underset{\sim}{k}_j = -\underset{\sim}{k}'_\alpha | \delta L^{\alpha j} | 0 \rangle$$

It is easily verified that none of these contributions vanishes; indeed, in the first case, the integral over $\underset{\sim}{v}_i$ does no longer commute with the first two matrix elements; in the second case the contribution proportional to $\partial / \partial \underset{\sim}{v}_\alpha$ of $\delta L^{\alpha j}$ in the third matrix element is non vanishing. As we have already seen ((II. 6. 6)), the contribution of each cycle is proportional to Ω^{-1}; because of the double summation over i and j, we obtain a contribution proportional to C^2. Such diagrams correspond indeed to those described by the above theorem.

As to (c), we may repeat the arguments for (b) or (b'); each cycle is proportional to Ω^{-1}; however, we only have one summation over i; hence we obtain a contribution proportional to C/Ω , which vanishes for $\Omega \to \infty$.

Let us now suppress the irreducibility condition in the operator $\Phi(0)$ and discuss the error introduced in this way. We consider a given contribution to $\Phi(0)$ with m vertices :

$$\Phi^{(m)} = \int \{d\underset{\sim}{v}\}^N \langle 0|\delta L^A (\frac{1}{z-L_o} \delta L)^m |0\rangle_{irr} \prod_i \varphi^{equ}(\underset{\sim}{v}_i) \tag{V.4.4}$$

and suppress the irreducibility condition on a given intermediate state (say after r vertices). In this way, we add to $\Phi^{(m)}$ the following contribution :

$$\begin{aligned} \alpha = \gamma \int \{d\underset{\sim}{v}\}^N \langle 0|\delta L^A (\frac{1}{z-L_o} \delta L)^{m-r} |0\rangle_{irr} \times \\ \times \langle 0|(\frac{1}{z-L_o} \delta L)^r |0\rangle_{irr} \prod_i \varphi^{equ}(v_i) \end{aligned} \tag{V.4.5}$$

As we have seen such a contribution will be different from zero in the limit of a large system only if the two diagonal fragments are semi-connected. We have two cases :

a) the semi-connection is through one fluid particle; but, then, A does not appear in the last fragment on the right and we have :

$$\begin{aligned} &\langle 0|(\frac{1}{z-L_o} \delta L)^r |0\rangle \prod_i \varphi^{equ}(v_i) \\ &= \langle 0|(\frac{1}{z-L_o^f} \delta L^f)^r |0\rangle \prod_i \varphi^{equ}(\underset{\sim}{v}_i) = 0 \end{aligned} \tag{V.4.6}$$

if we take into account (II.11.10).

b) the semi connection is through particle A. Then, all the fluid particles in the first fragment are different from those in the last and because of the integration over the velocities of the fluid particles, the first vertex in the second fragment must necessarily involve A ; hence :

$$\alpha = \gamma \int \{dv\}^{N-s} \langle 0 | \delta L^A (\frac{1}{z-L_o} \delta L)^{m-r} | 0 \rangle_{irr} \frac{1}{z} \times$$

(V.4.7)
$$\times \int \{dv\}^s \langle 0 | \delta L^A (\frac{1}{z-L_o} \delta L)^{r-1} | 0 \rangle_{irr} \prod_i \varphi^{equ}(v_i)$$

if $\int \{dv\}^s = \int dv_{i_1} \dots dv_{i_s}$, where $i_1 \dots i_s$ are the fluid particles involved in the second fragment . We easily recognize that the second contribution is a contribution to the operator ϕ given by (IV.3.13) . As we have seen , the first non-vanishing contribution is of order γ^2 and we have:

$$\alpha = \gamma \int \{dv\}^{N-s} \langle 0 | \delta L^A (\frac{1}{z-L_o} \delta L)^{m-r} | 0 \rangle_{irr} [\varphi^{equ}(v)]^{N-s} \times$$

(V.4.8)
$$\times \gamma^2 \frac{1}{z} \int \{dv\}^s \langle 0 | \delta L^A (\frac{1}{z-L_o} \delta L)^{r-1} | 0 \rangle_{irr} [\varphi^{equ}(v)]^s$$

With the same kind of arguments, it is again easy to convince oneself that the fragement on the left is also of order γ^2 at least. Thus, if we supress the irreducibility condition in the operator $\phi(0)$, we add a contribution of at least order four in γ . The contribution of order γ^4 which we add is :

$$\alpha = \gamma^4 \sum_{m=0}^{\infty} \sum_{n=0}^{\infty} \sum_{p=0}^{\infty} \sum_{q=0}^{\infty} \int \{dv\}^N \langle 0 | \delta L^A \frac{1}{z-L_f^o} \left(\delta L^f \frac{1}{z-L_f^o}\right)^m \times$$

(V.4.9)
$$\times (\delta L^A + L_o^A) \frac{1}{z-L_o^f} \left(\delta L^f \frac{1}{z-L_f^o}\right)^n | 0 \rangle_{irr} \langle 0 | \delta L^A \frac{1}{z-L_o^f} \left(\delta L^f \frac{1}{z-L_o^f}\right)^p \times$$

$$\times (\delta L^A + L_o^A) \left(\frac{1}{z-L_o^f} \delta L^f\right)^q | 0 \rangle_{irr} \prod_i \varphi^{equ}(v_i)$$

As a conclusion, we may suppress the irreducibility condition in $\phi^{(3)}(0)$. If we suppress it (which will appear convenient below) in $\phi^{(4)}(0)$, we have to subtract the contribution (V.4.9) . Because of the factor $1/z$, we easily notice that this contribution will diverge at the limit $z \to 0$.

This is easy to understand. Indeed, a product of two irreducible diagonal fragments brings a factor t^2 in the formal solution of the Liouville equation as compared with the single t factor brought by one irreducible fragment. In the long time limit, the first one diverges. The role of the irreducibility condition in $\psi(z)$ is precisely to suppress such contributions. This is well-known in the discussion of the three-body problem [5]. The operator $\psi(z)$ brings in the evolution equation only the contributions of genuine three-body collisions, i.e. of those collisions where the three particles interact almost simultaneously (i e. on a time scale of the order of the binary collision time) . The suppression of the irreducibility condition would amount to the inclusion of those three body processes which are a succession of 2 two binary collisions and would introduce a divergence. However, it is often convenient to write $\psi(z)$ as a difference between the reducible contribution (which includes all three-body processes, whatever the time ordering of events) and the reducible term (which describe those processes which are the result of succession of collisions). Both terms diverge but the difference is finite ; the cancellation occurs only for the diverging parts. We shall see an example of this procedure below.

V.5 - Higher order corrections to the collision operator. Explicit evaluation.

The above discussion shows us that we may forget the irreducibility condition in the third order operator ; it is then a simple matter, with the arguments we used in chapter IV, § 5 to compute $\phi^{(3)}(0)$. We obtain :

$$\gamma^3 \phi^{(3)}(0) = \lim_{z\to 0} \int \{d\underset{\sim}{v}\}^N \langle 0 | \delta L^A \frac{1}{z - L_o^f - \delta L^f} (\delta L^A + L_o^A) \times$$

$$\times \frac{1}{z - L_o^f - \delta L^f} (\delta L^A + L_o^A) \rho^f_{equ} | 0 \rangle$$

$$= \Omega^{-1} \lim_{z\to 0} \int_o^\infty dt_1 \int_o^{t_1} dt_2 \int \{d\underset{\sim}{v}\, d\underset{\sim}{r}\}^N d\underset{\sim}{R}\, \delta L^A\, e^{-i(L_o^f + \delta L^f - z)(t_1 - t_2)} \times$$

$$\text{(V.5.1)} \quad \times \left[\underset{\sim}{V} \cdot \frac{\partial}{\partial \underset{\sim}{R}} + \underset{\sim}{F} \cdot \frac{\partial}{\partial \underset{\sim}{P}} \right] e^{-i(L_o^f + \delta L^f - z)t_2} \left[\underset{\sim}{V} \cdot \frac{\partial}{\partial \underset{\sim}{R}} + \underset{\sim}{F} \cdot \frac{\partial}{\partial \underset{\sim}{P}} \right] \rho^f_{equ}$$

$$= \lim_{z\to 0} \int_o^\infty dt_1 \int_o^{t_1} dt_2 \int \{d\underset{\sim}{v}\, d\underset{\sim}{r}\}^N F_i(\underset{\sim}{R}, \{\underset{\sim}{r}\}) \frac{\partial}{\partial P_i} e^{-i(L_o^f + \delta L^f - z)(t_1 - t_2)} \times$$

$$\times \left[V_j \frac{\partial}{\partial R_j} + F_j(\underset{\sim}{R}, \{\underset{\sim}{r}\}) \frac{\partial}{\partial P_j} \right] e^{-i(L_o^f + \delta L^f - z)t_2} F_l(\underset{\sim}{R}, \{\underset{\sim}{r}\}) \rho^f_{equ} \times$$

$$\times \left[\frac{\partial}{\partial P_l} + \frac{1}{kT} V_l \right]$$

where the force $\underset{\sim}{F}$ acting on the fixed brownian particle is given by (I V.5.8) .

As the potential is spherically symmetric, one verifies easily that all contributions to the rhs of (V.5.1) vanish for symmetry reasons. Therefore :

$$\text{(V.5.2)} \qquad \gamma^3 \phi^{(3)}(0) = 0$$

Let us now consider the fourth order contribution. If we denote by $\tilde{\phi}^{(4)}(0)$ the operator $\phi^{(4)}$ in which we suppress the irreducibility condition, we obtain :

$$\text{(V.5.3)} \qquad \gamma^4 \phi^{(4)}(0) = \gamma^4 \tilde{\phi}^{(4)}(0) - \alpha$$

where α is given by (V.4.9).

Following the same procedure as that used to compute $\phi^{(2)}$ and $\phi^{(3)}$, we have :

$$\gamma^4 \tilde{\phi}^{(4)}(0) = \gamma^4 \lim_{z \to 0} \int_0^\infty dt_1 \int_0^{t_1} dt_2 \int_0^{t_2} dt_3 \int \{d\underset{\sim}{v}\, d\underset{\sim}{r}\}^N F_i(\underset{\sim}{R}, \{\underset{\sim}{r}\}) \frac{\partial}{\partial P_i} \times$$

$$\times\, e^{-i(L_o^f + \delta L^f - z)(t_1 - t_2)} \left[V_j \frac{\partial}{\partial R_j} + F_j(\underset{\sim}{R}, \{\underset{\sim}{r}\}) \frac{\partial}{\partial P_j} \right] \times$$

(V.5.4)

$$\times\, e^{-i(L_o^f + \delta L^f - z)(t_2 - t_3)} \left[V_l \frac{\partial}{\partial R_l} + F_l(\underset{\sim}{R}, \{\underset{\sim}{r}\}) \frac{\partial}{\partial P_l} \right] \times$$

$$\times\, e^{-i(L_o^f + \delta L^f - z)t_3} F_k(\underset{\sim}{R}, \{\underset{\sim}{r}\}) \rho^f_{equ} \left[\frac{\partial}{\partial P_k} + \frac{1}{kT} V_k \right]$$

This is a fourth order differential operator with respect to the velocity of the brownian particle . It diverges as can be easily verified if one keeps only lowest order terms in the coupling constant.

As to the operator α , we easily obtain :

$$\alpha = \gamma^4 \lim_{z \to 0} \sum_{p=0}^{\infty} \sum_{q=0}^{\infty} \int \{d\underset{\sim}{v}\}^N \langle 0 | \delta L^A \frac{1}{z - L_o^f - \delta L^f} (\delta L^A + L_o^A) \times$$

(V.5.5)

$$\times \left(\frac{1}{z - L_o^f} \delta L^f \right)^p | 0 \rangle_{irr} \frac{1}{z} \langle 0 | \delta L^A \frac{1}{z - L_o^f - \delta L^f} (\delta L^A + L_A^o) \left(\frac{1}{z - L_o^f} \delta L^f \right)^q | 0 \rangle \times$$

$$\times \prod_i \varphi^{equ}(\underset{\sim}{v}_i)$$

which we can easily rewrite as :

$$\alpha = \gamma^4 \lim_{z \to 0} \int \{d\underset{\sim}{v}\}^N \langle 0 | \delta L^A \frac{1}{z - L_f^o - \delta L^f} (\delta L^A + L_o^A) \frac{1}{z - L_o^f - \delta L^f} | 0 \rangle_{irr} \times$$

$$\text{(V.5.6)} \times \prod_i \varphi^{equ}(\underset{\sim}{v}_i) \int \{d\underset{\sim}{v}\}^N \langle 0 | \delta L^A \frac{1}{z - L_o^f - \delta L^f} (\delta L^A + L_o^A) \rho_{equ}^f | 0 \rangle_{irr}$$

Using

$$\rho_{equ}^f | 0 \rangle = \sum_{\{\underset{\sim}{k}\} \underset{\sim}{K}} | \{\underset{\sim}{k}\}, \underset{\sim}{K} \rangle \langle \{\underset{\sim}{k}\}, \underset{\sim}{K} | \rho_f^{equ} | 0 \rangle$$

$$\text{(V.5.7)} \qquad = \sum_{\{\underset{\sim}{k}\}, \underset{\sim}{K} \neq \{\underset{\sim}{0}\}} | \{\underset{\sim}{k}\}, \underset{\sim}{K} \rangle \langle \{\underset{\sim}{k}\}, \underset{\sim}{K} | \rho_f^{equ} | 0 \rangle + | 0 \rangle \prod_i \varphi^{equ}(\underset{\sim}{v}_i)$$

we have :

$$\sum_{\{\underset{\sim}{k}'\} \underset{\sim}{K}' \neq \{\underset{\sim}{0}\}} \langle \{\underset{\sim}{k}'\}, \underset{\sim}{K}' | \frac{1}{z - L_f^o - \delta L^f} | 0 \rangle \prod_i \varphi^{equ}(v_i)$$

$$\text{(V.5.8)} = \sum_{\{\underset{\sim}{k}'\} \underset{\sim}{K}' \neq \{\underset{\sim}{0}\}} \langle \{\underset{\sim}{k}'\}, \underset{\sim}{K}' | \frac{1}{z - L_o^f - \delta L^f} \rho_{equ}^f | 0 \rangle$$

$$- \sum_{\{\underset{\sim}{k}'\} \underset{\sim}{K}' \neq \{0\}} \langle \{\underset{\sim}{k}'\}, \underset{\sim}{K}' | \frac{1}{z - L_o^f - \delta L^f} \rho_{equ}^f | 0 \rangle_{irr}$$

Therefore, we obtain :

$$\text{(V.5.9)} \qquad \alpha = \alpha_1 + \alpha_2$$

with

$$\alpha_1 = \gamma^4 \lim_{z \to 0} \int \{d\underset{\sim}{v}\}^N \langle 0 | \delta L^A \frac{1}{z - L_o^f - \delta L^f} (\delta L^A + L_o^A) \frac{1}{z - L_o^f - \delta L^f} \rho_{equ}^f | 0 \rangle :$$

$$\text{(V.5.10)} \times \int \{d\underset{\sim}{v}\}^N \langle 0 | \delta L^A \frac{1}{z - L_o^f - \delta L^f} (\delta L^A + L_o^A) \rho_{equ}^f | 0 \rangle_{irr}$$

$$\alpha_2 = -\gamma^4 \lim_{z \to 0} \int \{d\underset{\sim}{v}\}^N \langle 0 | \delta L^A \frac{1}{z - L_o^f - \delta L^f} (\delta L^A + L_o^A) \frac{1}{z - L_o^f - \delta L^f} \rho_{equ}^f | 0 \rangle_{irr}$$

$$\text{(V.5.11)} \quad \times \int \{d\underset{\sim}{v}\}^N \langle 0 | \delta L^A \frac{1}{z - L_o^f - \delta L^f} (\delta L^A + L_o^A) \rho_{equ}^f | 0 \rangle_{irr}$$

Using (IV.6.1) and the convolution theorem we easily obtain :

$$\alpha_1 = \gamma^4 \lim_{z\to 0} \int_0^\infty dt_1 \int_0^{t_1} dt_2 \int_0^{t_2} dt_3 \int \{d\underset{\sim}{v} d\underset{\sim}{r}\}^N F_i(\underset{\sim}{R}, \{\underset{\sim}{r}\}) \frac{\partial}{\partial P_i} \times$$

$$\times e^{-i(L_o^f + \delta L^f - z)(t_1 - t_2)} F_j(\underset{\sim}{R}, \{\underset{\sim}{r}\}) \left(\frac{\partial}{\partial P_j} + \frac{1}{kT} V_j \right) \rho^f_{equ} \times$$

$$\times \int \{d\underset{\sim}{v} d\underset{\sim}{r}\}^N F_l(\underset{\sim}{R}, \{\underset{\sim}{r}\}) \frac{\partial}{\partial P_l} e^{-i(L_o + \delta L^f - z)t_3} F_k(\underset{\sim}{R}, \{\underset{\sim}{r}\}) \left(\frac{\partial}{\partial P_k} + \frac{1}{kT} V_k \right) \times \tag{V.5.12}$$

$$\times \rho^f_{equ}$$

Again this is a fourth order differential operator with diverging coefficients (this is again easily verified if one takes into account only lowest order terms in the coupling constant).

In the contribution α_2, because of the irreducibility condition in the diagonal fragment on the left, none of the propagators is identical to z and this contribution is perfectly finite at the limit $z \to 0$.

Introducing the operator :

$$\gamma^2 \overline{\phi}^{(2)}(z) = \gamma^2 \int \{d\underset{\sim}{v}\}^N \langle 0 | \delta L^A \frac{1}{z - L_o^f - \delta L^f} (\delta L^A + L_o^A) \frac{1}{z - L_o^f - \delta L^f} \times$$

$$\times \rho^f_{equ} | 0 \rangle_{irr} \tag{V.5.13}$$

we have

$$\alpha_2 = -\gamma^4 \overline{\phi}^{(2)}(0) \phi^{(2)}(0) \tag{V.5.14}$$

if we take into account (IV.5.6)

The irreducible operator $\phi^{(4)}$, which does not diverge at the limit $z \to 0$, may thus be written (see (V.5.3)) :

$$\phi^{(4)}(0) = \widetilde{\phi}^{(4)} - \alpha_1 - \alpha_2 \tag{V.5.15}$$

The infinite parts in $\tilde{\phi}^{(4)}$ and α_1 cancel each other. Introducing the fourth order differential operator :

$$\bar{\phi}^{(4)}(0) = \tilde{\phi}^{(4)}(0) - \alpha_1$$

$$= \lim_{z\to 0}\int_o^\infty dt_1 \int_o^{t_1} dt_2 \int_o^{t_2} dt_3 \Big\{\int \{d\underset{\sim}{r}\, d\underset{\sim}{v}\}^N F_i(\underset{\sim}{R},\{\underset{\sim}{r}\}) \frac{\partial}{\partial P_i} e^{-i(L_o^f+\delta L^f-z)(t_1-t_2)} \times$$

$$\times \left[V_j \frac{\partial}{\partial R_j} + F_j(\underset{\sim}{R},\{\underset{\sim}{r}\}) \frac{\partial}{\partial P_j}\right] e^{-i(L_o^f+\delta L^f-z)(t_2-t_3)} \left[V_l \frac{\partial}{\partial R_l} + \right.$$

$$\left. F_l(\underset{\sim}{R},\{\underset{\sim}{r}\}) \frac{\partial}{\partial P_l}\right] e^{-i(L_o^f+\delta L^f-z)t_3} F_k(\underset{\sim}{R},\{\underset{\sim}{r}\})\, \rho^f_{equ} \left(\frac{\partial}{\partial P_k} + \frac{1}{kT} V_k\right)$$

(V.5.16)

$$- \int \{d\underset{\sim}{r}\, d\underset{\sim}{v}\}^N F_i(\underset{\sim}{R},\{\underset{\sim}{r}\}) \frac{\partial}{\partial P_i} e^{-i(L_o^f+\delta L^f-z)(t_1-t_2)} F_j(\underset{\sim}{R},\{\underset{\sim}{r}\}) \rho^f_{equ}$$

$$\left(\frac{\partial}{\partial P_j} + \frac{1}{kT} V_j\right) \int \{d\underset{\sim}{r}\, d\underset{\sim}{v}\}^N F_l(\underset{\sim}{R},\{\underset{\sim}{r}\}) \frac{\partial}{\partial P_l} e^{-i(L_o^f+\delta L^f-z)t_3} F_k(\underset{\sim}{R},\{\underset{\sim}{r}\}) \times$$

$$\times\ \rho^f_{equ} \left(\frac{\partial}{\partial P_k} + \frac{1}{kT} V_k\right)\Big\}$$

we see that the fourth order correction to the collision operator is a sum of two finite terms :

$$\text{(V.5.17)} \quad \phi^{(4)}(0)\,\delta\varphi(\underset{\sim}{V}) = \bar{\phi}^{(4)}(0)\,\delta\varphi(\underset{\sim}{V}) + \bar{\phi}^{(2)}(0)\,\phi^{(2)}(0)\,\delta\varphi(\underset{\sim}{V})$$

The advantage of this form will appear once we calculate the corrections to the flow term. Then shall see that these exactly compensate the effect of the contribution α_2.

One verifies easily that the operator $\bar{\phi}^{(4)}(0)$ may be written :

$$\overline{\phi}^{(4)}{}_{(0)} = \frac{\partial}{\partial P_i}\Big\{ a_{ijkl} \frac{\partial^2}{\partial P_j \partial P_l} + b_{ijkl} \frac{\partial}{\partial P_j} V_l + c_{ijkl} V_l \frac{\partial}{\partial P_j}$$

(V.5.18)

$$+ d_{ijkl} V_l V_j \Big\}\Big\{ \frac{\partial}{\partial P_k} + \frac{1}{kT} V_k \Big\}$$

where

$$a_{ijkl} = \lim_{z \to 0} \int_o^\infty dt_1 \int_o^{t_1} dt_2 \int_o^{t_2} dt_3 \Big\{ \Big\langle F_i \, e^{-i(L_o^f + \delta L^f - z)(t_1 - t_2)} F_j \times$$

(V.5.19)

$$\times \, e^{-i(L_o^f + \delta L^f - z)(t_2 - t_3)} F_l \, e^{-i(L_o^f + \delta L^f - z)t_3} F_k \Big\rangle$$

$$- \Big\langle F_i \, e^{-i(L_o^f + \delta L^f - z)(t_1 - t_2)} F_j \Big\rangle \Big\langle F_l \, e^{-i(L_o^f + \delta L^f - z)t_3} F_k \Big\rangle \Big\}$$

$$b_{ijkl} = \lim_{z \to 0} \int_o^\infty dt_1 \int_o^{t_1} dt_2 \int_o^{t_2} dt_3 \Big\langle F_i \, e^{-i(L_o^f + \delta L^f - z)(t_1 - t_2)} F_j \times$$

(V.5.20)

$$\times \, e^{-i(L_o^f + \delta L^f - z)(t_2 - t_3)} \frac{\partial}{\partial R_l} e^{-i(L_o^f + \delta L^f - z)t_3} F_k \Big\rangle$$

$$c_{ijkl} = \lim_{z \to 0} \int_o^\infty dt_1 \int_o^{t_1} dt_2 \int_o^{t_2} dt_3 \Big\{ \Big\langle F_i \, e^{-i(L_o^f + \delta L^f - z)(t_1 - t_2)} \frac{\partial}{\partial R_l} \times$$

(V.5.21)

$$\times \, e^{-i(L_o^f + \delta L^f - z)(t_2 - t_3)} F_j \, e^{-i(L_o^f + \delta L^f - z)t_3} F_k \Big\rangle$$

$$- \Big\langle F_i \, e^{-i(L_o^f + \delta L^f - z)(t_1 - t_2)} F_l (kT)^{-1} \Big\rangle \Big\langle F_j e^{-i(L_o^f + \delta L^f - z)t_3} F_k \Big\rangle \Big\}$$

$$d_{ijkl} = \lim_{z \to 0} \int_o^\infty dt_1 \int_o^{t_1} dt_2 \int_o^{t_2} dt_3 \Big\langle F_i \, e^{-i(L_o^f + \delta L^f - z)(t_1 - t_2)} \times$$

(V.5.22)

$$\times \frac{\partial}{\partial R_j} e^{-i(L_o^f + \delta L^f - z)(t_2 - t_3)} \frac{\partial}{\partial R_l} e^{-i(L_o^f + \delta L^f - z)t_3} F_k \Big\rangle$$

where

$$\langle A \rangle = \int \{d\underset{\sim}{r}\, d\underset{\sim}{v}\}^{N} \; A \rho^{f}_{equ} \tag{V.5.23}$$

The form (V.5.18) may be compared to the results of Lebowitz and Rubin ; it is in complete agreement with their only correction to the lowest order Fokker-Planck equation.

V.6 - Higher order corrections to the flow term.

The first correction due to the effect of the field during a collision is given by :

$$\gamma^{3} \Xi^{(3)}(0) \varphi^{equ}(\underset{\sim}{V}) = \gamma^{3} \int \{d\underset{\sim}{v}\}^{N} \sum_{\{\underset{\sim}{k}\}} D^{(2)}_{\{\underset{\sim}{k}\}}(0) iL_{E} C^{(0)}_{\{\underset{\sim}{k}\}}(0) (\rho^{f}_{o})_{equ} \varphi^{equ}(\underset{\sim}{V})$$

$$+ \gamma^{3} \int \{d\underset{\sim}{v}\}^{N} \sum_{\{\underset{\sim}{k}\}} D^{(1)}_{\{\underset{\sim}{k}\}}(0) iL_{E} C^{(1)}_{\{\underset{\sim}{k}\}}(0) (\rho^{f}_{o})_{equ} \varphi^{equ}(\underset{\sim}{V}) \tag{V.6.1}$$

$$+ \gamma^{3} \int \{d\underset{\sim}{v}\}^{N} \sum_{\{\underset{\sim}{k}\}} D^{(0)}_{\{\underset{\sim}{k}\}}(0) iL_{E} C^{(2)}_{\{\underset{\sim}{k}\}}(0) (\rho^{f}_{o})_{equ} \varphi^{equ}(\underset{\sim}{V})$$

The third term vanishes because of the integration over the velocities of the fluid particles.

As to the second term, one verifies easily that one has :

$$C^{(1)}_{\{\underset{\sim}{k}\}}(0) (\rho^{f}_{o})_{equ} \varphi^{equ}(\underset{\sim}{V}) = \langle \{\underset{\sim}{k}\} | \frac{1}{z - L^{o}_{f} - \delta L^{f}} (L^{A}_{o} + \delta L^{A}) \frac{1}{z - L^{f}_{o} - \delta L^{f}}$$

$$| 0 \rangle_{irr} (\rho^{f}_{o})_{equ} \varphi^{equ}(\underset{\sim}{V}) \tag{V.6.2}$$

$$= \langle \{\underset{\sim}{k}\} | \frac{1}{z - L^{f}_{o} - \delta L^{f}} \rho^{f}_{equ} \; \underset{\sim}{F} \cdot (\frac{\partial}{\partial \underset{\sim}{P}} + \frac{1}{kT} \underset{\sim}{V}) \; | \; 0 \rangle \varphi^{equ}(\underset{\sim}{V}) = 0$$

In the first term, we take into account the fact that L_E commutes with ρ^f_{equ} . We then obtain easily :

$$\gamma^3 \, \Xi^{(3)}(0) \, \varphi^{equ}(\underset{\sim}{V}) = \gamma^3 \int \{d\underset{\sim}{v}\}^N \langle 0| \delta L^A \frac{1}{z - L^f_0 - \delta L^f} (\delta L^A + L^o_A) \times$$

(V.6.3) $$\times \frac{1}{z - L^f_0 - \delta L^f} \; iL_E \, \rho^f_{equ} \, |0\rangle_{irr} \varphi^{equ}(\underset{\sim}{V})$$

$$= \gamma^3 \, \overline{\phi}^{(2)}(0) \; iL_E \, \varphi^{equ}(\underset{\sim}{V})$$

if we take (V.5.13) into account.

The fourth order correction vanishes for symmetry reasons.

V.7 - Stationary transport equation up to order γ^4.

Summarizing the results of the previous paragraphs, we have:

(V.7.1) $$\Xi(0) \varphi^{equ}(\underset{\sim}{V}) = \gamma^3 \overline{\phi}^{(2)}(0) \; iL_E \, \varphi^{equ}(\underset{\sim}{V})$$

(V.7.2) $$\phi(0) = \gamma^2 \phi^{(2)}(0) + \gamma^4 \overline{\phi}^{(4)}(o) \qquad + \gamma^4 \overline{\phi}^{(2)}(0) \phi^{(2)}(0)$$

where the Fokker-Planck operator $\phi^{(2)}(0)$ is given by (V.3.8) while the operators $\overline{\phi}^{(4)}(0)$ and $\overline{\phi}^{(2)}(0)$ are given by (V.5.18) and (V.5.13) respectively.

Therefore , up to order γ^4, the stationary transport equation becomes :

$$\gamma \left[1 + \gamma^2 \overline{\phi}^{(2)}(0) \right] iL_E \varphi^{equ}(\underset{\sim}{V}) = \gamma^2 \left[1 + \gamma^2 \overline{\phi}^{(2)}(0) \right] \phi^{(2)}(0) \delta\varphi(\underset{\sim}{V})$$

(V.(.3) $$+ \gamma^4 \overline{\phi}^{(4)}(0) \delta\varphi(\underset{\sim}{V})$$

Expanding $\delta\varphi(\underset{\sim}{V})$ in powers of γ , we have :

(V.7.4) $$\delta\varphi(\underset{\sim}{V}) = \gamma^{-1}\left[\delta\varphi(\underset{\sim}{V})\right]_o + \gamma\left[\delta\varphi(\underset{\sim}{V})\right]_1$$

which leads to :

(V.7.5) $$iL_E\varphi^{equ}(\underset{\sim}{V}) = \phi^{(2)}(0)\left[\delta\varphi(\underset{\sim}{V})\right]_o$$

(V.7.6) $$0 = \phi^{(2)}(0)\left[\delta\varphi(\underset{\sim}{V})\right]_1 + \overline{\phi}^{(4)}(0)\left[\delta\varphi(\underset{\sim}{V})\right]_o$$

which we may rewrite as :

(V.7.7) $$iL_E\varphi^{equ}(\underset{\sim}{V}) = \left[\gamma^2\phi^{(2)}(0) + \gamma^4\overline{\phi}^{(4)}(0)\right]\delta\varphi(V)$$

which shows that the corrections due to the action of the field during a collision can be formally incorporated in a modification of the collision operator. The corrections to the lowest order Fokker-Planck equation are thus entirely given by the operator $\overline{\phi}^{(4)}(0)$. Part of these corrections may of course be incorporated in a modification of the diffusion coefficient D which appears in the lowest order equation. Equation (V.7.7) agrees exactly with the transport equation derived by Lebowitz and Rubin.

V.8 - Alternative form of the transport equation for the brownian motion problem.

Through a rather different method, Lebowitz and Rubin have obtained the following transport equation for the velocity distribution function of the heavy particle :

(V.8.1) $$\frac{\partial\,\delta\varphi(\underset{\sim}{V},t)}{\partial t} + iL_E\varphi^{equ}(\underset{\sim}{V}) = \int_0^t d\tau\,\mathcal{K}(t-\tau)\,\delta\varphi(\underset{\sim}{V},\tau)$$

The collision operator $\mathcal{K}$ is given by :

(V.8.2) $$\mathcal{K}(t) = -(2\pi i)^{-1} \int dz \; e^{-izt} K(z)$$

(V.8.3) $$K(z) = -i\gamma \int \{d\underset{\sim}{r}\, d\underset{\sim}{v}\}^N \delta L^A \frac{1}{z-(1-\mathcal{P})(L_o+\delta L)} (1-\mathcal{P})(L_o+\delta L)\rho^f_{equ}$$

where $\mathcal{P}$ is a projection operator :

(V.8.4) $$\mathcal{P}\ldots = \rho^f_{equ} \int \{d\underset{\sim}{r}\; d\underset{\sim}{v}\}^N \ldots$$

The remarkable feature is that, except for the trivial flow term, all the dynamics of the problem has been incorporated in the collision operator $\mathcal{K}$; at first sight this seems to be in contradiction with our general result of chapter II. However, we have already seen that, up to fourth order in γ , in this brownian motion problem, the corrections to the flow term in (V.2.9) can indeed be taken formally as a modification of the collision operator.

We shall not give here the original derivation of (V.8.1) which can be found in ref. [2], but rather concentrate ourselves on the equivalence between the two transport equations for this problem.

First of all, we shall show that the solutionsof (V.8.1) are identical to the solutions of the following equation :

(V.8.5) $$\frac{\partial \delta\varphi(\underset{\sim}{V},t)}{\partial t} + iL_E\varphi^{equ}(\underset{\sim}{V}) + \int_0^t d\tau\, \Delta(t-\tau) iL_E \varphi^{equ}(\underset{\sim}{V}) = \int_0^t d\tau \chi(t-\tau)\delta\varphi(\underset{\sim}{V},\tau)$$

where Δ (t) and χ(t) are the inverse Laplace transforms of the operators:

F. Henin

$$(V.8.6)\quad \overline{\chi}(z) = -i\int\{d\underset{\sim}{r}\, d\underset{\sim}{v}\}^N \delta L^A \frac{1}{z-(1-I)(L_o+\delta L)}(1-I)(L_o+\delta L)I\,\rho^f_{equ}$$

$$(V.8.7)\quad \overline{\Delta}(z) = \int\{d\underset{\sim}{r}\, d\underset{\sim}{v}\}^N \delta L^A \frac{1}{z-(1-I)(L_o+\delta L)}(1-I)\rho^f_{equ}$$

where I is a projection operator :

$$(V.8.8)\qquad I\ldots = (\Omega)^{-N}(\rho^f_o)_{equ}\int\{d\underset{\sim}{r}\,d\underset{\sim}{v}\}^N\ldots$$

(I involves the fluid velocity equilibrium distribution function while ρ involves the complete fluid equilibrium distribution function for a fixed position of the brownian particle) .

Using Laplace transforms, one easily obtains the formal solutions of (V.8.1) :

$$(V.8.9)\qquad \overline{\delta\varphi}(z) = \left[-\ iz-K(z)\right]^{-1} i(z+\omega)^{-1}\ \underset{\sim}{E}.\underset{\sim}{\nabla}\varphi^{equ}(\underset{\sim}{V})$$

and of (V.8.5) :

$$(V.8.10)\qquad \overline{\delta\varphi}(z) = \left[-iz-\overline{\chi}(z)\right]^{-1}\left[1+\overline{\Delta}(z)\right] i(z+\omega)^{-1}\underset{\sim}{E}.\underset{\sim}{\nabla}\varphi^{equ}(\underset{\sim}{V})$$

(We consider here the case of an oscillating field as given by (II.13.12))

In order to establish the identity of these two functions, we have to show that the rhs are identical, or equivalently, that we have :

$$(V.8.11)\quad \left[\overline{\chi}(z) - iz\,\overline{\Delta}(z)\right]f(\underset{\sim}{V}) = \left[1\ +\overline{\Delta}(z)\right]K(z)\ f(\underset{\sim}{V})$$

where $f(\underset{\sim}{V})$ is an arbitrary function of $\underset{\sim}{V}$;

Now , using the identity, valid for an arbitrary function $A(\underset{\sim}{V},\{\underset{\sim}{r}\},\{\underset{\sim}{v}\})$:

$$(\mathcal{P}-I)A(\underset{\sim}{V},\{\underset{\sim}{r}\},\{\underset{\sim}{v}\}) = \left[\rho^f_{equ} - (\Omega)^{-N}(\rho^f_o)_{equ}\right]\int\{d\underset{\sim}{r}\, d\underset{\sim}{v}\}^N A$$

(V.8.12)

$$=(1-I)\,\rho^f_{equ}\int\{d\underset{\sim}{r}\, d\underset{\sim}{v}\}^N A$$

as well as :

$$\frac{1}{z-(1-\mathcal{P})(L_o+\delta L)} = \frac{1}{z-(1-I)(L_o+\delta L)+(\mathcal{P}-I)(L_o+\delta L)}$$

$$=\frac{1}{z-(1-I)(L_o+\delta L)}\left\{1 + (\mathcal{P}-I)(L_o+\delta L)\frac{1}{z-(1-I)(L_o+\delta L)+(\mathcal{P}-I)(L_o+\delta L)}\right\}$$

(V.8.13)

$$=\frac{1}{z-(1-I)(L_o+\delta L)}\left\{1 +(1-I)\rho^f_{equ}\int\{d\underset{\sim}{r}\, d\underset{\sim}{v}\}^N(L_o+\delta L)\,\frac{1}{z-(1-\mathcal{P})(L_o+\delta L)}\right\}$$

$$=\frac{1}{z-(1-I)(L_o+\delta L)}\left\{1+(1-I)\,\rho^f_{equ}\int\{d\underset{\sim}{r}\, d\underset{\sim}{v}\}^N(L^A_o+\delta L^A)\,\frac{1}{z-(1-\mathcal{P})(L_o+\delta L)}\right\}$$

we easily obtain :

$$\left[1 + \tilde{\Delta}(z)\right]K(z)f(\underset{\sim}{V}) = -\,i\gamma\int\{d\underset{\sim}{r}\, d\underset{\sim}{v}\}^N\delta L^A\,\frac{1}{z-(1-I)(L_o+\delta L)}\,(1-\mathcal{P})\times$$

$$\times(L_o+\delta L)\rho^f_{equ}\quad f(\underset{\sim}{V})$$

(V.8.14)

$$+\,i\gamma^2\int\{d\underset{\sim}{r}d\underset{\sim}{v}\}^N\delta L^A\,\frac{1}{z-(1-I)(L_o+\delta L)}\,(1-I)\rho^f_{equ}\quad L^A_o\quad\times$$

$$\times\int\{d\underset{\sim}{r}\, d\underset{\sim}{v}\}^N\,\frac{1}{z-(1-\mathcal{P})(L_o+\delta L)}\,(1-\mathcal{P})\,(L_o+\delta L)\,\rho^f_{equ}\quad f(\underset{\sim}{V})$$

Now, if :

(V.8.15)
$$\Gamma = \frac{1}{z-(1-\mathcal{P})(L_o+\delta L)}\,(1-\mathcal{P})(L_o+\delta L)\,\rho^f_{equ}$$

we also have :

(V.8.16) $$\left\{ z - (1-\mathcal{P})(L_o + \delta L) \right\} \Gamma = (1-\mathcal{P})(L_o + \delta L)\, \rho^f_{equ}$$

Applying the operator $\mathcal{P}$ to both sides , we have :

(V.8.17) $$z\mathcal{P}\Gamma \equiv z\rho^f_{equ} \int \left\{ d\underset{\sim}{r}\, d\underset{\sim}{v} \right\}^N \Gamma = 0$$

which shows us that the second term in the rhs of (V.8.14) vanishes. Hence ,

$$\left[1 + \bar{\Delta}(z)\right] K(z) f(\underset{\sim}{V}) = -i \gamma \int \left\{ d\underset{\sim}{r}\, d\underset{\sim}{v} \right\}^N \delta L^A \frac{1}{z-(1-I)(L_o+\delta L)} (1-\mathcal{P}) \times$$

(V.8.18) $$\times (L_o + \delta L) \rho^f_{equ}\, f(\underset{\sim}{V})$$

The lhs of (V.8.11) can be written , with the definitions (V.8.6) and (V.8.7) :

$$\left[\bar{\chi}(z) - iz\bar{\Delta}(z)\right] f(\underset{\sim}{V}) = -i\gamma \int \left\{ d\underset{\sim}{r}\, d\underset{\sim}{v} \right\}^N \delta L^A \frac{1}{z-(1-I)(L_o+\delta L)} (1-I) \times$$

(V.8.19) $$\times \left\{ z + (L_o + \delta L) I \right\} \rho^f_{equ}\, f(\underset{\sim}{V})$$

In order to prove identity (V.8.11) , we thus have to show that the quantity α given by :

$$\alpha = \int \left\{ d\underset{\sim}{v}\, d\underset{\sim}{r} \right\}^N \delta L^A \frac{1}{z-(1-I)(L_o+\delta L)} \left\{ (1-\mathcal{P})\ (L_o + \delta L) \right.$$

(V.8.20) $$\left. - (1-I)(L_o + \delta L)\, I - z(1-I) \right\} \rho^f_{equ}\ f(\underset{\sim}{V})$$

vanishes .

Now, using again (V.8.12) , we have :

$$\left\{ (1-\mathcal{P})(L_o + \delta L) - (1-I)(L_o + \delta L) I - z(1-I) \right\} \rho^f_{equ}\ f(\underset{\sim}{V}) =$$

(V.8.21) $$(1-I) \left\{ (L_o + \delta L)(1-I)\rho^f_{equ} - \rho^f_{equ} \int \left\{ d\underset{\sim}{r}\, d\underset{\sim}{v} \right\}^N (L_o + \delta L)\rho^f_{equ} - z\rho^f_{equ} \right\} f(\underset{\sim}{V})$$

We also have :

(V.8.22)
$$\int \{ d\underset{\sim}{r}\, d\underset{\sim}{v} \}^N (L_o + \delta L)\, \rho^f_{equ} = \gamma L_o^A \int \{ d\underset{\sim}{r}\, d\underset{\sim}{v} \}^N \rho^f_{equ} + \gamma \int \{ d\underset{\sim}{r}\, d\underset{\sim}{v} \}^N \delta L^A \rho^f_{equ}$$
$$= \gamma L_o^A + \gamma \int \{ d\underset{\sim}{r}\, d\underset{\sim}{v} \}^N \delta L^A \rho^f_{equ}$$

(V.8.23)
$$\chi L_o^A f(\underset{\sim}{V}) = 0$$

If we combine (V.8.20), (V.8.21), (V.8.22) and (V.8.23), we obtain

(V.8.24)
$$\alpha = \Big[- \int \{ d\underset{\sim}{r}\, d\underset{\sim}{v} \}^N \delta L^A (1-I) \rho^f_{equ} - \int \{ d\underset{\sim}{r}\, d\underset{\sim}{v} \}^N \delta L^A \frac{1}{z-(1-I)(L_o + \delta L)} \times$$
$$\times (1-I)\, \rho^f_{equ} \int \{ d\underset{\sim}{r}\, d\underset{\sim}{v} \}^N \delta L^A \rho^f_{equ} \Big] f(\underset{\sim}{V})$$

Now, we have :

$$\int \{ d\underset{\sim}{r}\, d\underset{\sim}{v} \}^N \delta L^A \rho^f_{equ} \; f(\underset{\sim}{V}) = 0$$

(see IV.4.7)

and:

(V.8.25)
$$\int \{ d\underset{\sim}{r}\, d\underset{\sim}{v} \}^N \delta L^A \; I \rho^f_{equ} = \int \{ d\underset{\sim}{r}\, d\underset{\sim}{v} \}^N \delta L^A (\rho^f_o)_{equ} (\Omega)^{-N} = 0$$

Therefore, we indeed obtain :

(V.8.26)
$$\alpha = 0$$

and the identity of the solutions of (V.8.1) and (V.8.5) is thus established. Equation (V.8.5) has the same structure as the transport equation we obtained in Chapter II, § 13. In the steady state, we obtain :

(V.8.27)
$$iL_E \varphi^{equ}(\underset{\sim}{V}) + \bar{\Delta}(0)\, iL_E \varphi^{equ}(\underset{\sim}{V}) = \bar{\chi}(0)\, \delta \varphi(\underset{\sim}{V})$$

To establish the equivalence with (V.2.10), we have to show :

(V.8.28)
$$\bar{\Delta}(0)\, iL_E \varphi^{equ}(\underset{\sim}{V}) = \sum_{\{\underset{\sim}{k}\}} D_{\{\underset{\sim}{k}\}}(0)\, iL_E C_{\{\underset{\sim}{k}\}}(0)\, \varphi^{equ}(\underset{\sim}{V})$$

(V.8.29) $$\bar{\chi}(0)\,\delta\varphi(\underset{\sim}{V}) = \phi(0)\,\delta\varphi(\underset{\sim}{V})$$

Let us first consider the lhs of (V.8.29) . We have :

(V.8.30) $$\bar{\chi}(z)\,\delta\varphi(\underset{\sim}{V}) = \int\{d\underset{\sim}{v}\}^N \langle 0|\,\delta L^A \frac{1}{z-L_o^f-(1-I)(L_o^A+\delta L)}\left[L_o^f+(1-I)(L_o^A+\delta L)\right] |0\rangle (\rho_o^f)_{equ}\,\delta\varphi(\underset{\sim}{V})$$

if we take into account :

V(8.31) $$IL_o^f = 0$$

(V.8.32) $$I\rho^f_{equ}\,\delta\varphi(V) = \Omega^{-N}(\rho_o^f)_{equ}\,\delta\varphi(\underset{\sim}{V})$$

Through a straightforward expansinn, we obtain :

$$\bar{\chi}(z)\,\delta\varphi(\underset{\sim}{V}) = \sum_{m=0}^{\infty}\int\{d\underset{\sim}{v}\}^N \langle 0|\,\delta L^A\,\frac{1}{z-L_o}\left\{\left[(1-I)\,\delta L-IL_o^A\right]\frac{1}{z-L_o}\right\}^m \times$$

(V.8.33) $$\times\left\{L_o^f+(1-I)(L_o^A+\delta L)\right\}|0\rangle\,(\rho_o^f)_{equ}\,\delta\varphi(\underset{\sim}{V})$$

However, we have :

$$\langle\{\underset{\sim}{k}\},\underset{\sim}{K}|\,I(L_o^A+\delta L)|\{\underset{\sim}{k}'\},\underset{\sim}{K}'\rangle F(\{\underset{\sim}{v}\},\underset{\sim}{V})\,\delta_{\sum_i \underset{\sim}{k}_i+\underset{\sim}{K},0}\,\delta_{\sum_i \underset{\sim}{k}'_i+\underset{\sim}{K}',\underset{\sim}{0}}$$

$$=\int\{d\underset{\sim}{r}\}^N d\underset{\sim}{R}\,\exp\left[-i\sum_i \underset{\sim}{k}_i\cdot\underset{\sim}{r}_i\right]e^{-i\underset{\sim}{K}\cdot\underset{\sim}{R}}\,\Omega^{-N}(\rho_o^f)_{equ}\int\{d\underset{\sim}{r}'d\underset{\sim}{v}'\}^N\left[L_o^A(\underset{\sim}{R},\underset{\sim}{V})\right.$$

$$\left.+\,\delta L(\underset{\sim}{R},\underset{\sim}{V},\{\underset{\sim}{r}'\},\{\underset{\sim}{v}'\})\right]\exp\left[i\sum_i \underset{\sim}{k}'_i\cdot\underset{\sim}{r}'_i\right]e^{i\underset{\sim}{K}'\cdot\underset{\sim}{R}}\,F(\{\underset{\sim}{v}'\},\underset{\sim}{V})\,\delta_{\sum_i \underset{\sim}{k}_i+\underset{\sim}{K},\underset{\sim}{0}}\,\times$$

(V.8.34) $$\times\,\delta_{\sum_i \underset{\sim}{k}'_i+\underset{\sim}{K}',\underset{\sim}{0}}$$

$$=\prod_i \delta_{\underset{\sim}{k}_i,\underset{\sim}{0}}\,\delta_{\underset{\sim}{K},\underset{\sim}{0}}\,\delta_{\sum_i \underset{\sim}{k}'_i+\underset{\sim}{K}',\underset{\sim}{0}}\,(\rho_o^f)_{equ}\int\{d\underset{\sim}{v}'\}^N\langle 0|\,L_o^A(\underset{\sim}{V})+\delta L(\underset{\sim}{V},\{\underset{\sim}{v}'\})$$

$|\{k'\}, \underset{\sim}{K}'\rangle F\ (\{\underset{\sim}{v}'\}, \underset{\sim}{V})$

As a consequence all irreducible contributions involving IL_o^A or $I\delta L$ in (V.8.33) vanish (see the product of Kronecker's deltas in the rhs of (V.8.34)).

Let us now consider a reducible contribution made of a product of two irreducible contributions :

$$\beta = \int \{d\underset{\sim}{v}\}^N \langle 0|\delta L^A \frac{1}{z-L_o}\left\{\left[(1-I)\delta L - IL_o^A\right]\frac{1}{z-L_o}\right\}^p |0\rangle \times$$

(V.8.35)

$$\times \langle 0|\left\{\left[(1-I)\delta L - IL_o^A\right]\frac{1}{z-L_o}\right\}^q \{L_o^f + (1-I)(L_o^A + \delta L)\}|0\rangle_{irr} (\rho_o^f)_{equ}\, \delta\varphi(\underset{\sim}{V})$$

(V.8.34) shows us that we may drop the operator I every where in the second fragment except at the left. (combine the Kronecker's deltas in (V.8.34) and the irreducibility condition). Let us first consider the case q=0 . Then , using (V.8.34), we have :

$$\langle 0|L_o^f + (1-I)(L_o^A + \delta L)|0\rangle (\rho_o^f)_{equ}\, \delta\varphi(\underset{\sim}{V}) = 0 \tag{V.8.36}$$

If q is different from zero, we have :

$$\langle 0|\left\{\left[(1-I)\delta L - IL_o^A\right]\frac{1}{z-L_o}\right\}^q \{L_o^f + (1-I)(L_o^A + \delta L)\}|0\rangle_{irr} (\rho_o^f)_{equ}\, \delta\varphi(\underset{\sim}{V})$$

$$= \sum_{\{\underset{\sim}{k}\}\underset{\sim}{K} \neq \{0\}} \langle 0|(1-I)\delta L - IL_o^A|\{\underset{\sim}{k}\}, \underset{\sim}{K}\rangle\langle\{\underset{\sim}{k}\}, \underset{\sim}{K}|\left(\frac{1}{z-L_o}\delta L\right)^q|0\rangle_{irr} (\rho_o^f)_{equ}\ (\underset{\sim}{V})$$

(V.8.37)

$$= \langle 0|\delta L\ \left(\frac{1}{z-L_o}\delta L\right)^q|0\rangle_{irr} (\rho_o^f)_{equ}\, \delta\varphi(\underset{\sim}{V})$$

$$- (\rho_o^f)_{equ} \int \{d\underset{\sim}{v}\}^N \langle 0|\delta L\ \left(\frac{1}{z-L_o}\delta L\right)^q|0\rangle_{irr} (\rho_o^f)_{equ}\, \delta\varphi(\underset{\sim}{V})$$

and we obtain :

$$\beta = \int \{d\underset{\sim}{v}\}^N \langle 0|\delta L^A \frac{1}{z-L_o}\{[(1-I)\delta L - IL_o^A]\frac{1}{z-L_o}\}^p |0\rangle\langle 0|\delta L(\frac{1}{z-L_o}\delta L)^q$$

$$|0\rangle_{irr} (\rho_o^f)_{equ} \delta\varphi(\underset{\sim}{V})$$

(V.8.38)
$$-\int \{d\underset{\sim}{v}\}^N \langle 0|\delta L^A \frac{1}{z-L_o}\{[(1-I)\delta L - IL_o^A]\frac{1}{z-L_o}\}^p |0\rangle (\rho_o^f)_{equ} \times$$

$$\times \int \{d\underset{\sim}{v}\}^N \langle 0|\delta L (\frac{1}{z-L_o}\delta L)^q |0\rangle_{irr} (\rho_o^f)_{equ} \delta\varphi(\underset{\sim}{V})$$

If no fluid particle is common to both fragments, the rhs vanishes trivially; now, in our discussion of the role of the irreducibility condition, we have seen that when such a product of fragments has a single semi-connection, it is through particle A; if we have more than a semiconnection, we obtain a contribution of order N^{-1}. Therefore, at the limit of a large system :

(V.8.39)
$$\beta = 0$$

This means that all reducible contributions to (V.8.33) vanish. Taking also into account the remark following (V.8.34) , we obtain :

$$\bar{\chi}(z)\,\delta\varphi(\underset{\sim}{V}) = \sum_{m=0}^{\infty} \int \{d\underset{\sim}{v}\}^N \langle 0|\delta L^A (\frac{1}{z-L_o}\delta L)^{m+1} |0\rangle_{irr} (\rho_o^f)_{equ} \times$$

(V.8.40)
$$\times\,\delta\varphi(\underset{\sim}{V}) = \phi(z)\,\delta\varphi(\underset{\sim}{V})$$

which establishes (V.8.29) .

(V.8.28) can be established in a similar way, if one takes into account:

(V.8.41)
$$\begin{aligned}\langle \{\underset{\sim}{k}\}, \underset{\sim}{K}|(1-I)\rho^f_{equ}|0\rangle iL_E &= 0 \quad \text{if } |\{\underset{\sim}{k}\}, \underset{\sim}{K}\rangle = |0\rangle \\ &= \langle \{\underset{\sim}{k}\}, \underset{\sim}{K}|\rho^f_{equ}|0\rangle iL_E \\ &= iL_E \langle \{\underset{\sim}{k}\}, \underset{\sim}{K}|\rho^f_{equ}|0\rangle\end{aligned}$$

$$= iL_E C_{\{\underset{\sim}{k}\}\underset{\sim}{K}} (\rho_o^f)_{equ} \quad \text{if} \, |\{\underset{\sim}{k}\}, \underset{\sim}{K}\rangle \neq |0\rangle$$

This completes the proof of the equivalence between our starting point and that of Lebowitz and Rubin. This equivalence shows us that quite generally all corrections due to the effect of the field during a collision can, in this brownian motion problem, be incorporated formally in a modification of the collision operator ($\phi \to \mathcal{K}$).

References.

1. P. Résibois and T. Davis , Physica 30 , 1077 (1964)

2. J. Lebowitz and E. Rubin, Phys. Rev. 131 , 2381 (1963)

3. J. Lebowitz and P. Résibois, Phys. Rev. 139 , A1101 (1965)

4. R. Balescu, Physica 27 , 693 (1961)

5. P. Résibois, J. Math. Phys. 4 , 166 (1963)

F. Henin

VI. BROWNIAN MOTION IN AN EXTERNAL FIELD. QUANTUM CASE

VI. - Introduction

We shall now consider the problem of brownian motion of of a heavy particle in a quantum system. Our starting point for a microscopic discussion of the evolution equation of quantum systems will be the Von Neuman equation for the density matrix. As we shall see, provided we choose suitable variables, this equation can be written in a form which is very similar to the Liouville equation. The main difference will be the replacement of differential operators by displacement operators. This corresponds to the physical fact that energy transfers are infinitesimal in the classical case while they are finite in the quantum case. The similarity between the quantum equation for the density matrix and the Liouville equation for the distribution function enables us to extend the whole formalism very easily to quantum systems.

However, the problem of brownian motion in quantum systems presents features which are quite different from those of the classical problem. In the classical case, we performed an expansion in powers of the mass ratio and showed that, to lowest order, the velocity distribution function of the heavy particle obeys a Fokker-Planck equation. At first sight, we might expect this to be true also for the quantum case, the quantum effects appearing in the diffusion coefficient. However, if we go back for a while to the classical problem, we easily notice that, what we did, was to assume that we were dealing with a particle moving with thermal velocity (i. e. a velocity of the order of its equilibrium velocity) in a fluid at equilibrium. We then had:

$$\langle p \rangle / P = 0(m/M)^{1/2} = 0(\gamma) \ll 1 \qquad \text{(VI.1.1)}$$

Our expansion in a power series of γ was actually an expansion in $\langle p \rangle /P$.

It is easy to convince oneself that (VI. 1. 1.) does not necessarily hold in the quantum case. Let us for instance consider the case of a heavy particle moving in a weakly coupled Fermi fluid. At very low temperature, the particle collides with fermions whose energy is very close to the Fermi energy ε_F. Then we have:

$$\langle p \rangle /P = O(m\varepsilon_F/MkT)^{1/2} = O(\gamma \xi) \qquad \text{(VI. 1. 2.)}$$

where

$$\xi = (\epsilon_F/kT)^{1/2} \qquad \text{(VI. 3. 3)}$$

and we may expect to find a Fokker-plank equation only in the region where

$$\gamma \xi \ll 1 \qquad \text{(VI. 1. 4)}$$

This condition is much more restrictive than the condition we met in the classical problem.

With this example, one might think that such difficulties will appear only if we consider fermions, because of the exclusion principle and the existence of the Fermi energy. However, we shall see that it is not the case and that the difficulty is more general. At very low temperatures, we must always expect that the Fokker-Planck equation will not be valid.

We shall first show how the Von Neumann equation may be written in a form very similar to the Liouville equation[1]. Then, assuming that an expansion in $\langle p \rangle /P$ is valid, we shall easily obtain a Fokker-Planck equation. We shall discuss the quantum corrections to the diffusion coefficient in some simple cases [2)3)] and compare briefly the theoretical results with the results obtained in experiments on heavy ions moving in liquid helium. Finally, we shall discuss on somewhat more general grounds than above the validity of expansions in powers of $\langle p \rangle / P$. [4)]

VI. 2 - Von Neumann - Liouville equation.

Let us consider one heavy charged particle acted upon by a constant external field moving in a fluid of light neutral particles. If we use a second quantization representation for free fluid particles and a plane wave representation for the heavy particle, the hamiltonian operator is (we take $\hbar = 1$):

(VI. 2. 1) $$H = H_o + \lambda V + H_E$$

The unperturbed hamiltonian is a sum over the kinetic energies of the fluid particles and the brownian particles :

(VI. 2. 2) $$H_o = \sum_{\underset{\sim}{k}} (k^2/2m)\, a^+_{\underset{\sim}{k}}\, a_{\underset{\sim}{k}} + K^2/2M$$

where $a^+_{\underset{\sim}{k}}$, $a_{\underset{\sim}{k}}$ are the creation and destruction operators for a fluid particle of momentum $\underset{\sim}{k}$.

As in the classical case, the interaction is a sum of two terms : one which describes the interactions of the fluid particles among themselves and a second one which describes the interactions of the heavy particle with the fluid particles :

(VI. 2. 3)
$$V = (\lambda/2\Omega) \sum_{\underset{\sim}{k}\underset{\sim}{l}\underset{\sim}{p}\underset{\sim}{r}} v(\underset{\sim}{k},\underset{\sim}{l},\underset{\sim}{p},\underset{\sim}{r})\; a^+_{\underset{\sim}{k}}\, a^+_{\underset{\sim}{l}}\, a_{\underset{\sim}{p}}\, a_{\underset{\sim}{r}}\; \delta_{\underset{\sim}{k}+\underset{\sim}{l}-\underset{\sim}{p}-\underset{\sim}{r},\,\underset{\sim}{0}}$$
$$+ (\lambda/\Omega) \sum_{\underset{\sim}{k}\underset{\sim}{l}} u(\underset{\sim}{k}-\underset{\sim}{l})\, e^{-i(\underset{\sim}{k}-\underset{\sim}{l}).\underset{\sim}{R}}\; a^+_{\underset{\sim}{k}}\, a_{\underset{\sim}{l}}$$

The contribution due to the action of the external field on the heavy particle is :

(VI. 2. 4) $$H_E = e\, \underset{\sim}{E} . \underset{\sim}{R}$$

In the mixed representation $|\underset{\sim}{K}, \{n\}\rangle$ of the eigenstates of H_o :

(VI. 2. 4) $$H_o |\underset{\sim}{K}, \{\underset{\sim}{n}\}\rangle = \{K^2/2M + \sum_{\underset{\sim}{k}} (k^2/2m) n_{\underset{\sim}{k}}\} |\underset{\sim}{K}, \{n\}\rangle$$

the Von Neumann equation for the density matrix is :

(VI. 2. 5) $$i \frac{\partial \langle \underset{\sim}{K}, \{n\} | \rho | \underset{\sim}{K}', \{n'\}\rangle}{\partial t} = \langle \underset{\sim}{K}, \{n\} | [H, \rho] | \underset{\sim}{K}', \{n'\}\rangle$$

where $[H, \rho]$ is the commutator of the two operators H and ρ .

Let us now perform the following change of variables: for the heavy particle :

(VI. 2. 6) $$\underset{\sim}{\varkappa} = \underset{\sim}{K} - \underset{\sim}{K}' \qquad \underset{\sim}{P} = (\underset{\sim}{K} + \underset{\sim}{K}')/2$$

for the fluid particles :

(VI. 2. 7) $$\nu_{\underset{\sim}{k}} = n_{\underset{\sim}{k}} - n'_{\underset{\sim}{k}} \qquad N_{\underset{\sim}{k}} = (n_{\underset{\sim}{k}} + n'_{\underset{\sim}{k}})/2$$

We also write any matrix element of an operator A in the following form :

(VI. 2. 8) $$\langle \underset{\sim}{K}, \{n\} | A | \underset{\sim}{K}', \{n'\}\rangle = A_{\underset{\sim}{K}-\underset{\sim}{K}', \{n-n'\}}((\underset{\sim}{K}+\underset{\sim}{K}')/2, \{(n+n')/2\}) = A_{\underset{\sim}{\varkappa}, \{\nu\}}(\underset{\sim}{P}, \{N\})$$

To the operator A, we associate an operator $\mathcal{A}$, which we define through the relation :

$$\langle \underset{\sim}{\varkappa}, \{\nu\} | \mathcal{L}(\underset{\sim}{P}, \{N\}) | \underset{\sim}{\varkappa}', \{\nu'\} \rangle = \zeta^{\underset{\sim}{\varkappa}'} \eta^{\sum_{\underset{\sim}{k}} \nu'_{\underset{\sim}{k}}} A_{\underset{\sim}{\varkappa}, \{\nu\}}(\underset{\sim}{P}, \{N\}) \zeta^{-\varkappa}$$

$$\times \eta^{-\sum_{\underset{\sim}{k}} \nu_{\underset{\sim}{k}}} - \zeta^{-\underset{\sim}{\varkappa}'} \eta^{-\sum_{\underset{\sim}{k}} \nu'_{\underset{\sim}{k}}} A_{\underset{\sim}{\varkappa}, \{\nu\}}(\underset{\sim}{P}, \{N\}) \zeta^{\underset{\sim}{\varkappa}} \eta^{\sum_{\underset{\sim}{k}} \nu_{\underset{\sim}{k}}} \tag{VI.2.9}$$

where the ζ and η 's are displacement operators acting on the variables P and N respectively :

$$\zeta^{\pm \underset{\sim}{\varkappa}} f(\underset{\sim}{P}) = \exp\left[\pm \frac{\underset{\sim}{\varkappa}}{2} \cdot \frac{\partial}{\partial \underset{\sim}{P}} \right] f(\underset{\sim}{P}) = f\left(\underset{\sim}{P} \pm \frac{\underset{\sim}{\varkappa}}{2}\right) \tag{VI.2.10}$$

$$\eta^{\pm \sum_{\underset{\sim}{k}} \nu_{\underset{\sim}{k}}} f(\{N\}) = \exp\left[\pm \frac{1}{2} \sum_{\underset{\sim}{k}} \nu_{\underset{\sim}{k}} \frac{\partial}{\partial N_{\underset{\sim}{k}}} \right] f(\{N\}) = f\left(\left\{N \pm \frac{\nu}{2}\right\}\right) \tag{VI.2.11}$$

One can then easily write the Von Neumann equation as :

$$i \frac{\partial \rho_{\underset{\sim}{\varkappa}, \{\nu\}}(\underset{\sim}{P}, \{N\})}{\partial t} = \sum_{\underset{\sim}{\varkappa}', \{\nu'\}} \langle \underset{\sim}{\varkappa}, \{\nu\} | \mathcal{H}(\underset{\sim}{P}, \{N\}) | \underset{\sim}{\varkappa}' \{\nu'\} \rangle \times$$

$$\times \rho_{\underset{\sim}{\varkappa}', \{\nu'\}}(\underset{\sim}{P}, \{N\}) \tag{VI.2.12}$$

which is formally analogous to the classical Liouville equation. The Von Neumann-Liouville operator $\mathcal{H}$ can be split into three terms :

$$\mathcal{H} = \mathcal{H}_0 + \mathcal{W} + \mathcal{H}_E \tag{VI.2.13}$$

The unperturbed operator $\mathcal{H}_0$ is given by :

$$\langle \underset{\sim}{\varkappa}, \{\nu\} | \mathcal{H}_0 | \underset{\sim}{\varkappa}', \{\nu'\} \rangle = \delta_{\underset{\sim}{\varkappa}, \underset{\sim}{\varkappa}'} \prod_{\underset{\sim}{k}} \delta_{\nu_{\underset{\sim}{k}}, \nu'_{\underset{\sim}{k}}} \times$$

$$\times \left[\frac{\underset{\sim}{\varkappa} \cdot \underset{\sim}{P}}{M} + \sum_{\underset{\sim}{k}} (k^2/2m) \nu_k \right] \tag{VI.2.14}$$

The interaction operator is given by :

$$\langle \underset{\sim}{\varkappa}, \{\nu\} | \lambda \mathcal{W} | \underset{\sim}{\varkappa}', \{\nu'\} \rangle = (\lambda / 2\Omega) \sum_{\underset{\sim}{k} \underset{\sim}{l} \underset{\sim}{p} \underset{\sim}{r}} v(\underset{\sim}{k}, \underset{\sim}{l}, \underset{\sim}{p}, \underset{\sim}{r}) \delta_{\underset{\sim}{k} + \underset{\sim}{l} - \underset{\sim}{p} - \underset{\sim}{r}, 0} \times$$

$$\Big\{ \eta^{\nu'_k+\nu'_l+\nu'_p+\nu'_r} \Big[(N_k+\tfrac{1}{2})(N_l+\tfrac{1}{2})(N_p+\tfrac{1}{2})(N_r+\tfrac{1}{2})\Big]^{1/2} \times$$

$$\times\, \eta^{-\nu_k-\nu_l-\nu_p-\nu_r} - \eta^{-\nu'_k-\nu'_l-\nu'_r-\nu'_p} \Big[(N_k+\tfrac{1}{2})(N_l+\tfrac{1}{2})$$

$$(N_p+\tfrac{1}{2})(N_r+\tfrac{1}{2})\Big]^{1/2} \eta^{\nu_k+\nu_l+\nu_p+\nu_r}\Big\} \delta_{\nu'_k,\nu_k-1}\, \delta_{\nu'_l,\nu_l-1} \times$$

$$\text{(V.2.15)}\quad \delta_{\underset{\sim}{\varkappa}}\, \delta_{\nu'_p,\nu_p+1}\, \delta_{\nu'_r,\nu_r+1} \prod_{i\neq klpr} \delta_{\nu'_i,\nu_i}\, \delta_{\underset{\sim}{\varkappa}',\underset{\sim}{\varkappa}}$$

$$+(\lambda/\Omega) \sum_{kl} u(k-l) \Big\{ \eta^{\nu'_k+\nu'_l} \zeta^{\underset{\sim}{\varkappa}'} \Big[(N_k+\tfrac{1}{2})(N_l+\tfrac{1}{2})\Big]^{1/2} \times$$

$$\times\, \eta^{-\nu_k-\nu_l} \zeta^{-\underset{\sim}{\varkappa}} - \eta^{-\nu'_k-\nu'_l} \zeta^{-\underset{\sim}{\varkappa}'} \Big[(N_k+\tfrac{1}{2})(N_l+\tfrac{1}{2})\Big]^{1/2} \times$$

$$\times\, \eta^{\nu_k+\nu_l} \zeta^{\underset{\sim}{\varkappa}} \Big\} \delta_{\nu_k,\nu_k-1}\, \delta_{\nu'_l,\nu_l+1} \prod_{i\neq lk} \delta_{\nu'_i,\nu_i} \times$$

$$\times\, \delta_{\underset{\sim}{k}-\underset{\sim}{l},\underset{\sim}{\varkappa}-\underset{\sim}{\varkappa}'}$$

The external hamiltonian is given by :

$$\langle \underset{\sim}{\varkappa}, \{\nu\} | \mathcal{H}_E | \underset{\sim}{\varkappa}', \{\nu'\} \rangle = ie\underset{\sim}{E} \cdot \frac{\partial}{\partial \underset{\sim}{P}} \times$$

$$\text{(V.2.16)}\qquad \times\, \delta_{\underset{\sim}{\varkappa},\underset{\sim}{\varkappa}'} \prod_k \delta_{\nu_k,\nu'_k}$$

The algebra which leads to these results is very simple. We give examples in Appendix VI.1.

The equation (VI.2.12) can be treated in the same way as the Liouville equation. Whenever we had a matrix element of the classical Liouville operator, we must now replace it by a matrix element of the Von Neumann Liouville operator . If we compare those matrix elements, we notice that the main difference between quantum and classical matrix elements is the appearance of displacements operators in the former while we had differential operators in the latter. This is specially striking for the operators acting on the heavy particle variables: the classical operator $\partial / \partial P$ is now replaced by the displacement operator $\zeta^{\pm \underset{\sim}{\varkappa}}$; this is due to the fact that we used a plane wave representation in the quantum case which is the analog of the Fourier expansion in the position variables in the classical case. For the fluid particles, the similarity is also very striking when one compares the quantum equation to the Liouville equation for a system of oscillators in action (J) - angle (α) variables). The displacement operators acting on the N variables are the analog of the differential operators acting on the action variables in the classical problem (for more details see [5)6)]).

VI. 3 - Stationary transport equation.

The quantum analog of the stationary transport equation (II. 14.5) in a static field will be :

(VI. 3. 1)
(V. 3. 1)

$$i \mathcal{H}_E \rho_o^{equ}(P, \{N\}) + i \sum_{\underset{\sim}{\varkappa} \{\nu\}} D_{\underset{\sim}{\varkappa} \{\nu\}}(0) \mathcal{H}_E C_{\{\nu\} \underset{\sim}{\varkappa}}(0) \rho_o^{equ}(\underset{\sim}{P}, \{N\}) = i \psi(0) \Delta \rho_o(\underset{\sim}{P}, \{N\})$$

Here $\rho_o(P, \{N\})$ represents the diagonal elements of the Von Neumann matrix. At equilibrium, we have in the $|\{n\}, \underset{\sim}{K}\rangle$ representation:

(VI. 3. 2)

$$\langle \{n\}, \underset{\sim}{K} | \rho^{equ} | \{n\}, \underset{\sim}{K} \rangle = \langle \{n\}, \underset{\sim}{K} | e^{-(H_o + \lambda V)/kT} | \{n\}, \underset{\sim}{K} \rangle \times$$
$$\times \left[\sum_{\{n'\}, \underset{\sim}{K}'} \langle \{n'\}, \underset{\sim}{K}' | e^{-(H_o + \lambda V)/kT} | \{n'\}, \underset{\sim}{K}' \rangle \right]^{-1}$$

Because of the non commutativity of H_o and V, this expression is much less simple, if all orders in λ are kept, than the corresponding classical expression. It is only in the weak coupling case, where we can neglect the interaction that we obtain :

$$\langle \{n\} \underset{\sim}{K} | \rho^{equ}_{\lambda \to 0} | \{n\}, K \rangle = \exp\left\{ - \left[\sum_{\underset{\sim}{k}} \frac{k^2}{2m} n_{\underset{\sim}{k}} + \frac{K^2}{2M} \right] \Big/ k\, T \right\} \times \tag{VI.3.3}$$
$$\times \left[\sum_{\{n\} \underset{\sim}{K}} \exp\left\{ - \left[\sum_{\underset{\sim}{k}} \frac{k^2}{2m} n_{\underset{\sim}{k}} + \frac{K^2}{2M} \right] \Big/ k\, T \right\} \right]^{-1}$$

or

$$\rho^{equ}_{o\, \lambda \to o} (\underset{\sim}{P}, \{N\}) = \exp\left\{ - \left[\sum_{\underset{\sim}{k}} \frac{k^2}{2m} N_{\underset{\sim}{k}} + \frac{P^2}{2M} \right] \Big/ kT \right\} \times \tag{VI.3.4}$$
$$\times \left[\sum_{\{N\} \underset{\sim}{P}} \exp\left\{ - \left[\sum_{\underset{\sim}{k}} \frac{k^2}{2m} N_{\underset{\sim}{k}} + \frac{P^2}{2M} \right] \Big/ kT \right\} \right]^{-1}$$

This will play an important role in our discussion of the validity of the Fokker-Planck equation in the quantum case .

The collision operator ψ is given by :

$$\psi(0) = \lim_{z \to 0} \langle 0 | \mathcal{V} \sum_{n=1}^{\infty} \left(\frac{1}{z - \mathcal{H}_o} \mathcal{V} \right)^n | 0 \rangle_{irr} \tag{VI.3.5}$$

while for the operators of creation or destruction of correlations we have :

$$C_{\underset{\sim}{\varkappa} \{\nu\}}(0) = \lim_{z \to 0} \langle \underset{\sim}{\varkappa} \{\nu\} | \sum_{n=1}^{\infty} \left(\frac{1}{z - \mathcal{H}_o} \mathcal{V} \right)^n | 0 \rangle_{irr} \tag{VI.3.6}$$

$$D_{\underset{\sim}{\varkappa} \{\nu\}}(0) = \lim_{z \to 0} \langle 0 | \sum_{n=1}^{\infty} \left(\mathcal{V} \frac{1}{z - \mathcal{H}_o} \right)^n | \underset{\sim}{\varkappa} \{\nu\} \rangle_{irr} \tag{VI.3.7}$$

The index irr means that only irreducible contributions have to be kept, i.e. contributions such that no intermediate state is identical to the vacuum of correlations (diagonal elements of the Von Neumann matrix).

As in the classical problem, we assume molecular chaos ; hence :

(VI.3.8) $$\rho_o^{equ}(\underset{\sim}{P},\{N\}) = \varphi^{equ}(\underset{\sim}{P})\{\rho_o^f(N)\}_{equ}$$

(VI.3.9) $$\Delta\rho_o(\underset{\sim}{P},\{N\}) = \delta\varphi(\underset{\sim}{P})\{\rho_o^f(N)\}_{equ}$$

if we take into account the fact that the modification of the density matrix of a fluid particle is of order N^{-1}, hence negligible in the limit of a large system.

VI.4 - Expansion in the mass ratio.

Whenever the ratio $\langle p\rangle/P$ satisfies the relation

(VI.4.1) $$\langle p\rangle/P = 0\,(m/M)^{1/2} = 0(\gamma)$$

we easily obtain an expansion in the mass ratio if we follow the same procedure as in the classical case. We decompose the unperturbed operator $\mathcal{H}_o$:

(VI.4.2) $$\mathcal{H}_o = \mathcal{H}_o^f + \gamma\mathcal{H}_o^A$$

with (VI.4.3) $$\langle\{\nu\}\underset{\sim}{x}|\mathcal{H}_o^f|\{\nu'\}\underset{\sim}{x}'\rangle = \sum_{\underset{\sim}{k}}(k^2/2m)\nu_{\underset{\sim}{k}} \prod_{\underset{\sim}{k}}\delta_{\nu_{\underset{\sim}{k}},\nu'_{\underset{\sim}{k}}}\delta_{\underset{\sim}{x},\underset{\sim}{x}'}$$

(VI.4.4) $$\langle\{\nu\},\underset{\sim}{x}|\gamma\mathcal{H}_o^A|\{\nu'\},\underset{\sim}{x}'\rangle = (\underset{\sim}{x}.\underset{\sim}{P})/M \prod_{\underset{\sim}{k}}\delta_{\nu_{\underset{\sim}{k}},\nu'_{\underset{\sim}{k}}}\delta_{\underset{\sim}{x},\underset{\sim}{x}'}$$

Similarily :

(VI.4.5) $$\lambda\mathcal{W} = \mathcal{W}^f + \gamma\mathcal{V}_1^A + \gamma^2\mathcal{V}_2^A + \dots$$

where

(VI.4.6) $$\mathcal{W}^f = \mathcal{V} + \mathcal{V}_o^A$$

where $\mathcal{V}$ represents the fluid-fluid interaction (first term in the rhs of (VI.2.15). $\mathcal{V}^A$ is the fluid-particle interaction (second term in the rhs of (VI.2.15). If one expands the displacement operators $\zeta^{\pm \kappa}$ in a power series, one easily obtains an expansion of $\mathcal{V}^A$ in a power series of γ. For instance, $\mathcal{V}^A_o$ is obtained by the mere replacement of the displacement operators by unity:

$$\langle\{\nu\},\kappa|\mathcal{V}^A_o|\kappa',\{\nu'\}\rangle = \langle\{N+\frac{\nu}{2}\},\frac{\kappa}{2}|U|\{N-\frac{\nu}{2}+\nu'\},-\frac{\kappa}{2}+\kappa'\rangle\,\eta^{\sum_k(\nu'_k-\nu_k)}$$

(VI.4.7)

$$-\langle\{N+\frac{\nu}{2}-\nu'\},\frac{\kappa}{2}-\kappa'|U|\{N-\frac{\nu}{2}\},-\frac{\kappa}{2}\rangle\,\eta^{\sum_k(-\nu'_k+\nu_k)}$$

while

$$\langle\{\nu\},\kappa|\mathcal{V}^A_1|\kappa',\{\nu'\}\rangle = (\lambda/\Omega)\sum_{kl} u(k-l)\Big\{\eta^{\nu'_k+\nu'_l}\times$$

$$\times\left[(N_k+\frac{1}{2})(N_l+\frac{1}{2})\right]^{1/2}\eta^{-\nu_k-\nu_l}+\eta^{-\nu'_k-\nu'_l}\left[(N_k+\frac{1}{2})(N_l+\frac{1}{2})\right]^{1/2}\times$$

$$\eta^{\nu_k+\nu_l}\Big\}\,\delta_{\nu'_k,\nu_k-1}\,\delta_{\nu'_l,\nu_l+1}\prod_{i\neq kl}\delta_{\nu'_i,\nu_i}\,\delta_{k-l,\kappa-\kappa'}\times$$

(VI.4.8)

$$\times\ \frac{1}{2}(\kappa-\kappa')\cdot\frac{\partial}{\partial P}$$

$$=\Big\{\langle\{N+\frac{\nu}{2}\},\frac{\kappa}{2}|U|\{N-\frac{\nu}{2}+\nu'\},\kappa'-\frac{\kappa}{2}\rangle\,\eta^{\sum_k(\nu'_k-\nu_k)}$$

$$+\langle\{N+\frac{\nu}{2}-\nu'\},\frac{\kappa}{2}-\kappa'|U|\{N-\frac{\nu}{2}\},-\frac{\kappa}{2}\rangle\,\eta^{\sum_k(\nu_k-\nu'_k)}\Big\}\frac{1}{2}(\kappa'-\kappa)\cdot\frac{\partial}{\partial P}$$

The external field contribution is, as in the classical case, of order γ. The main difference with the classical case is the appearance of higher order terms in the expansion of $\mathcal{V}^A$. However, one verifies easily again, that, to lowest order in γ, the equation may be written;

$$i\gamma\langle 0|\mathcal{H}_E|0\rangle\varphi^{equ}(\underset{\sim}{P}) = \lim_{z\to 0}\gamma^2\int_0^\infty dt\sum_{\{N\}}\langle 0|\mathcal{V}_1^A\ e^{-i(\mathcal{H}^f-z)t}\ \times$$

(VI. 4. 9)

$$\times\ (\mathcal{V}_1^A+\mathcal{H}_o^A)\rho^f_{equ}|0\rangle\delta\varphi(\underset{\sim}{P})$$

where

(VI. 4. 10) $$\mathcal{H}^f = \mathcal{H}_o^f + \mathcal{W}^f$$

Moreover, at this order, we can restrict ourselves to the lowest order term $\varphi_o^{equ}(\underset{\sim}{P})$ in the equilibrium distribution function.

Hence , we have :

$$i\gamma\langle 0|\mathcal{H}_E|0\rangle\varphi_o^{equ}(\underset{\sim}{P}) = \lim_{z\to 0}\gamma^2\int_0^\infty dt\sum_{\underset{\sim}{\kappa}\{N\}\{\nu\}}\langle 0|\mathcal{V}_1^A\ \times$$

(VI. 4. 11)

$$\times\ e^{-i(\mathcal{H}^f-z)t}(\mathcal{V}_1^A+\mathcal{H}_o^A)|\{\nu\},\underset{\sim}{\kappa}\rangle(\rho^f_{\{\nu\}\underset{\sim}{\kappa}})_{equ}\delta\varphi(\underset{\sim}{P})$$

where

(VI. 4. 12) $$(\rho^f_{\{\nu\}\underset{\sim}{\kappa}})_{equ} = \frac{\int d\underset{\sim}{R}\ e^{-i\underset{\sim}{\kappa}\cdot\underset{\sim}{R}}\langle\{N+\nu/2\}|e^{-H^f/kT}|\{N-\nu/2\}\rangle}{\sum_{\{N\}}\langle\{N\}|e^{-H^f/kT}|\{N\}\rangle}$$

where

$$H^f = H_f^o + V$$

is the fluid hamiltonian operator

VI. 5 - Fokker-Planck equation.

Let us consider the rhs of (VI. 4. 11) at t=0 (the calculation of the contribution involving twice $\mathcal{V}_1^A$ at a time t different from zero is given in appendix VI. 2). For the contribution involving twice $\mathcal{V}_1^A$, we have*:

$$\alpha = \sum_{\{N\}}\sum_{\{\nu\}\{\nu'\}}\sum_{\kappa\kappa'}\langle 0|\mathcal{V}_1^A|\kappa,\{\nu\}\rangle\langle\{\nu\},\kappa|\mathcal{V}_1^A|\kappa',\{\nu'\}\rangle\ \times$$

(VI. 5. 1)

$$\times\left[\rho^f_{\kappa',\{\nu'\}}(\{N\})\right]_{equ}$$

---

*When no confusion is possible , we do no longer use a special notation for vectors.

Using (VI.4.8) , we obtain :

$$\sum_{\{\nu'\}\varkappa'} \langle \varkappa, \{\nu\} | \mathcal{V}_1^A | \varkappa' \{\nu'\} \rangle \left[\rho^f_{\varkappa', \{\nu'\}}(\{N\}) \right]_{equ}$$

$$= (\lambda/\Omega) \sum_{kl} u(k-l) \Big\{ \left[(N_k + \nu'_k/2)(N_l + 1 + \nu'_l/2) \right]^{1/2} \times$$

(VI.5.2) $$\times \left[\rho^f_{\{\nu'\}', \nu'_k - 1, \nu'_l + 1, \varkappa' - k + l} (\{N\}', N_k - 1, N_l + 1) \right]_{equ} + (N_k + 1 - \nu'_k/2) \times$$

$$\times (N_l - \nu'_l/2) \Big]^{1/2} \left[\rho^f_{\{\nu'\}', \nu'_k - 1, \nu'_l + 1, \varkappa' - k + l} (\{N\}', N_k + 1, N_l - 1) \right] \Big\} \times$$

$$(1/2)(l-k) \cdot \frac{\partial}{\partial P}$$

Going back to the occupation number-plane wave representation we obtain :

$$\sum_{\{\nu'\}\varkappa'} \langle \varkappa \{\nu\} | \mathcal{V}_1^A | \varkappa' \{\nu'\} \rangle \left[\rho^f_{\varkappa' \{\nu'\}} (\{N\}) \right]_{equ} =$$

(VI.5.3) $$\sum_{i=x,y,z} \sum_K \langle \{N + \nu/2\} | F_i(K) \rho^f_{\varkappa - K} + \rho^f_{\varkappa - K} F_i(K) | \{N - \nu/2\} \rangle \frac{\partial}{\partial P_i}$$

where the force operator is given by :

(VI.5.4) $$F_i(K) = \sum_k u(K) K_i \, a_k^+ \, a_{K-k}$$

Therefore, we have :

$$\alpha = \sum_{\{N\}} \sum_{\{\nu\}} \sum_{\varkappa} \langle 0 | \mathcal{V}_1^A | \varkappa \{\nu\} \rangle \sum_i \sum_K \langle \{N + \nu/2\} | F_i(K) \rho^f_{\varkappa - K}$$

$$+ \rho^f_{\varkappa - K} F_i(K) | \{N - \nu/2\} \rangle (1/2) \frac{\partial}{\partial P_i}$$

$$= \sum_{\{N\}} \sum_{ij} \sum_{K} \sum_{\varkappa} \{ \langle\{N\}|U|\{N+\nu\},\varkappa\rangle\langle\{N+\nu\}|F_i(K)\rho^f_{\varkappa-K} + \rho^f_{\varkappa-K} F_i(K)|\{N\}\rangle$$

$$+ \langle\{N-\nu\},-\varkappa|U|\{N\}\rangle\langle\{N\}|F_i(K)\rho^f_{\varkappa-K} + \rho^f_{\varkappa-K} F_i(K)|\{N-\tfrac{\nu}{2}\}\rangle\} \times$$

$$\times (1/4) \frac{\partial^2}{\partial P_i \partial P_j} \varkappa_j$$

(VI.5.5)

$$= \sum_{\{N\}} \sum_{ij} \sum_{\varkappa K} \langle\{N\}|F_j^+(\varkappa)F_i(K)\rho^f_{\varkappa-K} + F_j^+(\varkappa)\rho^f_{\varkappa-K}F_i(K)$$

$$+ F_i(K)\rho^f_{\varkappa-K} F_j^+(\varkappa) + \rho^f_{\varkappa-K} F_i(K)F_j^+(\varkappa)|\{N\}\rangle \frac{1}{4} \frac{\partial^2}{\partial P_i \partial P_j}$$

Taking into account the fact that we can interchange $\varkappa$, K and i and j, and that the trace of a product of operators the invariance for cyclic permutations of the operators , as well as the hermiticity properties of the operators, we obtain :

$$\alpha = \sum_{\{N\}} \sum_{ij} \sum_{K} \sum_{\varkappa} \langle\{N\}|F_i^+(K) F_j(\varkappa) \rho^f_{\varkappa-K}|\{N\}\rangle \frac{\partial^2}{\partial P_i \partial P_j}$$

(VI.5.6)

$$= \langle F_i^+ F_j \rangle \frac{\partial^2}{\partial P_i \partial P_j}$$

As in the classical case, we see that we have here the average over the fluid equilibrium distribution of the tensor operator FF.

As to the second term, at t = 0 , we have :

$$\beta = \sum_{\{N\}} \sum_{\{\nu\}} \sum_{\varkappa} \langle 0|U_1^A|\varkappa\{\nu\}\rangle\langle\varkappa\{\nu\}|\mathcal{X}_A^o|\varkappa\{\nu\}\rangle [\rho^f_{\{\varkappa\nu\}}]_{equ.}$$

(VI.5.7)

$$= \sum_{\{N\}} \sum_{ij} \sum_{\varkappa} \langle\{N\}|F_i^+(\varkappa)\varkappa_j \rho^f_{\varkappa} + \varkappa_j \rho^f_{\varkappa} F_i^+(\varkappa)|\{N\}\rangle \frac{\partial}{\partial P_i} (P_j/M)$$

Now we have , performing an integration by parts in the second step,

$$\langle \{n'\}| k_j\, \rho^f_k |\{n\}\rangle = \int d\underset{\sim}{R}\; k_j\, e^{-i\underset{\sim}{k}\underset{\sim}{R}} \langle \{n'\}|\, e^{-H^f/kT}\, |\{n\}\rangle \times$$

$$\times \left[\sum_n \langle \{n\}|\, e^{-H^f/kT}\, |\{n\}\rangle \right]^{-1}$$

$$(VI.5.8) = \sum_{\{n''\}} \sum_{\underset{\sim}{k'}} \int d\underset{\sim}{R}\; e^{-i(k+k')R} \langle \{n'\}|\, F_j(k') |\{n''\}\rangle (kT)^{-1} \langle \{n''\}|\, e^{-H^f/kT}\, |\{n\}\rangle \times$$

$$\times \left[\sum_{\{n\}} \langle \{n\}|\, e^{-H^f/kT}\, |\{n\}\rangle \right]^{-1}$$

$$= \sum_{k'} \langle \{n'\}| F_j(k')\, \rho^f_{k+k'} |\{n\}\rangle$$

and therefore :

$$(VI.5.9) \quad \sum_{\{N\}} \langle 0| \mathcal{V}^A_1\, \mathcal{X}^A_o\, \rho^f_{equ} |0\rangle = \langle F^+_i\; F_j \rangle \frac{\partial}{\partial P_i} (P_j/kTM)$$

Therefore, at $t = 0$, the integrand in the rhs of (VI.4.11) becomes :

$$(VI.5.10) \qquad \langle F^+_i\; F_j \rangle \frac{\partial}{\partial P_i} \left[\frac{\partial}{\partial P_j} + \frac{1}{kTM} P_j \right]$$

In appendix VI.2, we show that the first term, at a time t different from zero, is identical to (VI.5.6) but with F_i replaced by $F_i(t)$ where $F_i(t)$ is the Heisenberg representation of the force operator:

$$(VI.5.11) \qquad F_i(t) = e^{-iH^f t}\; F_i\; e^{iH^f t}$$

A similar proof can be given for the second term and we obtain finally the Fokker-Planck form of (VI.4.11) :

$$(VI.5.12) \quad i\gamma \langle 0| \mathcal{X}_E |0\rangle \varphi^{equ}_o(P) = \zeta_{ij} \frac{\partial}{\partial P_i} \left[\frac{\partial}{\partial P_j} + \frac{1}{MkT} P_j \right] \delta\varphi(P)$$

where the diffusion coefficient in terms of microscopic quantities is given by

(VI.5.13) $$\zeta_{ij} = \int_0^\infty dt \left\langle F_i(t)\, F_j^+ \right\rangle$$

As it can also be shown that, to first order in γ, we have*:

(VI.5.14) $$\varphi_o^{equ}(P) = (M/2\pi kT)^{3/2}\, 4\pi\, \exp(-P^2/2MkT)$$

we have :

(VI.5.15) $$-(1/MkT)\ \underset{\sim}{E}.\underset{\sim}{P}\, \varphi_o^{equ}(P) = \zeta_{ij} \frac{\partial}{\partial P_i}\left[\frac{\partial}{\partial P_j} + \frac{1}{MkT}\, P_j\right]\varphi(P)$$

This completes our outline of the derivation of the Fokker-Planck equation in the quantum case. Let us stress that this equation is valid whenever the condition :

(VI.5.16) $$\langle p\rangle/P \ll 1$$

is satisfied.

VI.6 - Diffusion coefficient for a heavy ion in a slightly imperfect Bose fluid.

In a weakly coupled Bose gas, the condition (VI.5.16) for the validity of the Fokker-Planck equation would always be fulfilled. When strong interactions are present, the zero point motion of both kind of particles starts to play a role and it is difficult to make general assertions.

We shall assume that the interaction between the fluid and the

*One can for instance use an expansion of φ^{equ} in powers of γ. As we must have : $\phi(0)\varphi^{equ} = 0$, we have $\phi^{(2)}(0)\,\varphi_o^{equ} = 0$. But as we have just seen, $\phi^{(2)}(0)$ is the Fokker - Planck operator : hence φ_o^{equ} must be the Boltzmann distribution.

heavy ion is very weak and that we can neglect it in H^f. We shall also restrict ourselves to a slightly imperfect Bose fluid and assume that the temperature is sufficiently low to insure that the Bose-Einstein condensation has already occurred to a large extend. Then most of the bosons are in the ground state and we have :

(VI. 6. 1) $$n_o \simeq N$$

where n_o is the number of particles in the ground state. Then we can apply the well-known assumption :

(VI. 6. 2) $$a_o \simeq a_o^+ \simeq (n_o)^{1/2}$$

The force operator becomes :

(VI. 6. 3) $$\begin{aligned} \underset{\sim}{F}(k) = {} & \Omega^{-1} \underset{\sim}{k} \; u(k) \sqrt{n_o} \; (a_k^+ - a_k) \\ & + \Omega^{-1} \sum_{l \neq 0} \underset{\sim}{k} \; u(k) \, a_l^+ \, a_{l-k} \end{aligned}$$

while , if we use a pseudopotential (see ref. [4]) with $U \simeq 4\pi a/m$ (a: radius of the particles) :

(VI. 6. 4) $$\begin{aligned} V = (U/2\Omega) \Big[n_o^2 & + 2n_o \sum_{p \neq 0} (a_{-p}^+ a_p + a_p^+ a_{-p}) \\ & + n_o \sum_{p=0} (a_p^+ a_{-p}^+ + a_p^+ a_{-p}) \Big] \end{aligned}$$

We shall use the Bogoliubov transformation to phonons operators b_k and b_k^+. (see for instance ref [4]) :

(VI. 6. 5) $$a_k = g_k b_k + f_k b_{-k}^+$$

(VI. 6. 6) $$a_k^+ = g_k b_k^+ + f_k b_{-k}$$

(VI. 6. 7) $$g_k = (1 - \alpha_k^2)^{1/2}$$

(VI. 7. 8) $$f_k = \alpha_k (1 - \alpha_k^2)^{-1/2}$$

(VI. 6. 9) $$\alpha_k = 1 + x^2 - x(x^2+2)$$

(VI. 6. 10) $$x^2 = k^2\, \Omega / 8\pi a n_o = k^2/c^2$$

This transformation does not change the commutation rules and we have

(VI. 6. 11) $$b_k^+ |n_k\rangle = \sqrt{n_k+1}\, |n_k+1\rangle \; ; \; b_k |n_k\rangle = \sqrt{n_k}\, |n_k-1\rangle$$

(VI. 6. 12) $$H^f = E_o + \sum_k \omega_k b_k^+ b_k$$

(VI. 6. 13) $$\omega_k = (k/2M)\,(k^2 + 16\pi a n_o\, \Omega^{-1})^{1/2}$$

Similarly, the force operator can be written in the phonon representation b_k, b_k^+. In this representation, it is a matter of algebra to compute the diffusion coefficient (see an example of computation in the next paragraph for Fermi systems). One obtains [2], if one restricts oneself to small wave numbers (i.e. to the linear term in the dispersion relation (VI. 6. 13)):

$$\zeta_{ij} = (\pi/\Omega^2) \sum_{kl} k_i k_j \,(f_l^2 f_{l-k}^2 + g_l^2 g_{l-k}^2)\, m_l^o (m_{l-k}^o + 1)$$

(VI. 6. 14) $$\delta(|l-k| - l)$$

m_l^o is the distribution function in a slightly imperfect Bose gas in the limit of small momentum:

(VI. 6. 15) $$m_l^o = \left[\exp(\omega_l/kT) - 1\right]^{-1} = \left[\exp(cl/kT) - 1\right]^{-1}$$

The integrations can be performed and one obtains, if one reintroduces explicitly Planck's constant :

(VI. 6. 16) $\zeta = (2\pi^3/45(kT)^5 \; a^2 \; \hbar^{-3} \; c^{-4}$

The mobility of a heavy ion is given by [5]:

(VI. 6. 18) $\mu = eD/kT$

where D is the diffusion coefficient in ordinary space. As we have seen in chapter I, we have :

(VI. 6. 19) $D = (kT/M\beta) = (kT/M)^2(M/kT\beta) = (kT/M)^2(M^2/\zeta)$

(note that D is the coefficient appearing in the diffusion equation in ordinary space, $(M/kT\beta)^{-1}$ that appearing in the Fokker-Planck equation in velocity space , ζ that appearing in the Fokker-Planck equation in momentum space) .

Therefore we obtain :

(VI. 6. 20) $\mu = (45/2\pi^3)\, e\, \hbar^3 \; c^4 \; a^{-2}(KT)^{-4} \sim T^{-4}$

In a measure of the static mobility of $(He^+)_n$ in liquid He^4 at low temperature, Meyer and Reif [5] have found :

(VI. 6. 21) $\mu \sim T^k$ with $k = - 3.3 \pm 0.3$

which is in good agreement with the above result if one takes into account the fact that our model is quite rough .

VI. 7 - Diffusion of a heavy ion in a weakly coupled Fermi fluid.

We have seen in the introduction that in a Fermi fluid, at low temperature, the Fokker-Planck equation is valid only if the condition (VI. 1. 4) is satisfied. We shall discuss again this problem in the next paragraph but presently we consider a situation where this condition is satisfied and compute the diffusion coefficient, assuming the fluid to be weakly coupled.

As well known, in a Fermi system, the creation and destruction operators anticommute

$$[a_k^+ , a_k]_+ = 1 \tag{VI.7.1}$$

and the only possible values of the occupation numbers are :

$$n_k = 0 \quad \text{or} \quad 1 \tag{VI.7.2}$$

In order to avoid difficulties with the subsidiary condition

$$\sum_k n_k = N \tag{VI.7.2}$$

we shall consider the grand canonical ensemble . Then the equilibrium distribution for a weakly coupled system is given by :

$$\langle\{n\}|\rho^f_{\varkappa-K}|\{n'\}\rangle = \prod_k \delta_{n_k,n'_k} \exp\left[-\sum_k n_k(k^2/2m-\varepsilon_F)/kT\right] \times$$

$$\times \prod_k \left\{1 + \exp\left[-(k^2/2m - \varepsilon_F)/kT\right]\right\}^{-1} \delta_{\varkappa,K} \tag{VI.7.4}$$

where ε_F is the Fermi energy .

Thus, going back to the occupation number representation we have :

$$\langle F_i(t) F_j^+\rangle = \sum_{\{n\}}\sum_{\{n'\}}\sum_{\{n''\}}\sum_{\{n'''\}}\sum_{\{n''''\}}\sum_k\sum_l\sum_{l'} |u(k)|^2 \times$$

$$\times \; k_i k_j \langle\{n\}|\exp\left[-i\sum_k (k^2/2m)\, a_k^+ a_k t\right]|\{n'\}\rangle\langle\{n'\}|a_k^+ a_{k-l}|\{n''\}\rangle \times$$

$$\text{(VI.7.5)} \times \langle \{n''\} | \exp\left[i \sum_k (k^2/2m) a_k^+ a_k t\right] | \{n'''\} \rangle \langle \{n'''\} | a_k a_{k-1'}^+ | \{n''''\} \rangle \times$$
$$\times \langle \{n''''\} | \rho_o^f | \{n\} \rangle$$

Using

$$\text{(VI.7.6)} \qquad a_k^+ | n_k \rangle = | n_k + 1 \rangle \delta_{n_k \, 0}$$

$$\text{(VI.7.7)} \qquad a_k | n_k \rangle = n_k - 1 \, \delta_{n_k, 1}$$

one easily verifies that one must have :

$$\text{(VI.7.8)} \qquad 1 = 1'$$

and one obtains :

$$\langle F_i(t) F_j^+ \rangle = (\lambda^2/\Omega^2) \sum_{\{n\}} \sum_k \sum_1 |u(k)|^2 \, k_i k_j \, \delta_{n_k, 1} \, \delta_{n_{k-1}, 0} \times$$
$$\times \exp\left[-i \{k^2 - (k-1)^2\} t/2m\right] \exp\left[- \sum_k n_k (k^2/2m - \epsilon_F)/kT\right] \times$$
$$\times \prod_k \left\{1 \exp\left[- (k^2/2m - \epsilon_F)/kT\right]\right\}^{-1}$$
$$\text{(VI.7.9)} \quad = (\lambda^2/\Omega^2) \sum_k \sum_{k'} |u(k)|^2 \, k_i k_j \exp\left[-i (k^2 - k'^2) t/2m\right] \times$$
$$\times \exp\left[- (k^2/2m - \epsilon_F)/kT\right] \left\{1 + \exp\left[- (k^2/2m - \epsilon_F)/kT\right]\right\}^{-1} \times$$
$$\times \left\{1 + \exp\left[-(k'^2/2m - \epsilon_F)/kT\right]\right\}^{-1}$$

At the limit of a large system, the summations over k and 1 become integrals. If we also perform the asymptotic time integration in (VI.5.13) , we obtain

$$\zeta_{ij} = 2m \pi \lambda^2 \int d^3k \int d^3k' \, |u(k)|^2 \, k_i k_j \, \delta(k^2 - k'^2) \exp\left[+ (k^2/2m - \epsilon_F)/kT\right] \times$$
$$\text{(VI.7.10)} \quad \times \left\{1 + \exp\left[(k^2/2m - \epsilon_F)/kT\right]\right\}^{-1} \left\{1 + \exp\left[(k'^2/2m - \epsilon_F)/kT\right]\right\}^{-1}$$

One easily verifies that this tensor is diagonal and that one has :

(VI.7.11) $$\underline{\underline{\zeta}} = \underline{\underline{1}}\,\zeta$$

with

$$\zeta = (16\pi^3 m/3)\int_0^\infty dk k^3 \int_0^\infty dk' k'^2 |u(k)|^2 \left[\delta(k-k') + \delta(k+k')\right] \times$$

(VI.7.12) $$\times \quad \exp\left[(k'^2/2m - \varepsilon_F)/kT\right]\left\{1 + \exp\left[(k^2/2m - \varepsilon_F)/kT\right]\right\}^{-1} \times$$

$$\times \left\{1 + \exp\left[(k'^2/2m - \varepsilon_F)/kT\right]\right\}^{-1}$$

We integrate over k' and approximate the potential by a constant; we then have :

$$\zeta = (128/3)\pi^3 (kT)^3 m^4 u^2 \int_0^\infty d\varepsilon\, \varepsilon^2 e^{(\varepsilon - \xi^2)}\left\{1 + e^{(\varepsilon - \xi^2)}\right\}^{-2}$$

(VI.7.13) $$= -(128/3)\pi^3 m^4 u^2 (kT)^3 \int_0^\infty d\varepsilon\, \varepsilon^2 \frac{d}{d\varepsilon}\left\{1 + e^{(\varepsilon - \xi^2)}\right\}^{-1}$$

$$= (256/3)\pi^3 m^4 (kT)^3 u^2 \int_0^\infty d\varepsilon\,\varepsilon\left[1 + e^{(\varepsilon - \xi^2)}\right]^{-1}$$

where

(VI.7.14) $$\xi = (\varepsilon_F/kT)^{1/2}$$

At sufficiently low T, we have :

(VI.7.15) $$\xi \gg 1$$

and we may write :

$$\int_0^\infty d\varepsilon\,\varepsilon\left[1 + e^{(\varepsilon - \xi^2)}\right]^{-1} = \int_{-\xi^2}^\infty dy\,(y + \xi^2)(1 + e^y)^{-1}$$

$$= \xi^2 \int_{-\xi^2}^\infty dy\,(1 + e^y)^{-1} + \int_{-\xi^2}^0 dy\, y + 2\int_0^\infty dy\; y\,(1 + e^y)^{-1}$$

(VI.7.16)

$$- \int_{\xi^2}^\infty dy\, y\,(1 + e^y)^{-1}$$

$$= \xi^4/2 + \pi^2/6$$

where we have neglected exponentially decreasing contributions in the first integral as well as the last one whose integrand is exponentially small.

Therefore, we obtain :

(VI.7.17) $$\zeta = (128/3)\,\pi^3\, m^4\, kT\, u^2 \left\{ \varepsilon_F^2 + (\pi^2/3)(kT)^2 \right\}$$

From this, we easily obtain the mobility (see VI.6.18). Introducing $\hbar$ and the collision cross section σ in the Born approximation ($\sigma = 2m\,\pi\,u^2$), we obtain :

(VI.7.18) $$\mu = \frac{3\,\pi\, e\hbar^3}{8m^2\sigma\left\{\varepsilon_F^2 + (\pi^2/3)\,(kT)^2\right\}}$$

Davis and Dagonnier[1] compared this result with the experimental mobility for a heavy ion in liquid He^3 at 1.2 °K measured by Meyer and al.[6]. However, the comparison is not very easy because of the lack of information about the collision cross section (radius of the ion) and the effective mass of the ion. With reasonable estimates for these quantities, they find a good agreement.

VI.8 - Validity of the Fokker-Planck equation for brownian motion in quantum systems.

Let us first consider again the case of motion in a weakly coupled Fermi fluid. We may distinguish three temperature regions:

a) the temperature is so high that both the fluid and the particle behave classically. Then, if the particle moves with thermal velocity, we have:

(VI.8.1) $$\langle p \rangle / P = 0(m/M)^{1/2} = 0(\gamma)$$

and the classical Fokker-Planck equation is valid.

F. Henin

At lower temperature, quantum effects become important for the fluid. Only those fluid particles near the Fermi surface interact with the particle and we obtain :

(VI. 8. 2) $\langle p \rangle / P = 0\ (m\,\varepsilon_F / MkT)^{1/2} = 0\ (\gamma \xi)$

We may now distinguish two cases :

b) an intermediate temperature range where the more restrictive condition :

(VI. 8. 3) $\gamma \xi << 1$

is satisfied and the quantum Fokker-Planck equation (VI. 4. 11) is valid .

c) the case of very low temperature where

$$\gamma \xi > 1$$

and where the Fokker-Planck description does no longer hold.

In a discussion of the possibility of convergence of the $\langle p \rangle / P$ development presented here, Résibois and Dagonnier [3)8)] have shown that, in general, one must not expect this convergence to be realized at very low temperature, whatever the statistics. A very simple argument is the fact that at very low temperature, the average momentum P is independent of the mass ratio. It is essentially determined by the interactions with the fluid molecules and momentum transfers can become very large; the Fokker-Planck description is then no longer valid.

Let us show briefly how this conclusion about the independence of the average momentum at very low temperature can be obtained. For simplicity, we consider the case of Boltzmann statistics. The equilibrium distribution function for the brownian particle is :

$$\varphi^{equ}(P) = \frac{Tr_{fluid}\, \rho_{equ}}{Tr_{fluid,A}\, \rho_{equ}}$$

(VI. 8. 5)
$$= \frac{tr_{fluid} \exp\left[-\beta(H_o^f + H_o^A + V + U)\right]}{tr_{fluid,A} \exp\left[-\beta(H_o^f + H_o^A + V + U)\right]}$$

where

(VI. 8. 6) $$\beta = (kT)^{-1}$$

Let us write :

(VI. 8. 7) $$\rho(\beta) = e^{-\beta H_o} \omega(\beta) = e^{-\beta H}$$

Then:

(VI. 8. 8) $$\omega(\beta) = \exp\left[\beta(H_o^f + H_o^A)\right] \exp\left[-\beta(H_o^A + H_o^f + V + U)\right]$$

One verifies easily that this quantity satisfies the Bloch equation :

(VI. 8. 9) $$\frac{\partial \omega(\beta)}{\partial \beta} = \left\{\exp\left[\beta(H_o^A + H_o^F)\right](V + U) \exp\left[-\beta(H_o^A + H_o^f)\right]\right\} \omega(\beta)$$

Taking into account the commutation relations

(VI. 8. 10) $$\left[V, e^{-\beta H_o^A}\right] = 0$$

(VI. 8. 11) $$\left[H_o^f, H_o^A\right] = 0$$

and introducing

(VI. 8. 12) $$\tilde{V} = \exp(\beta H_o^f)\, V \exp(-\beta H_o^f)$$

(VI. 8. 13) $$\tilde{U} = \exp(\beta H_o^f)\, U \exp(-\beta H_o^f)$$

this equation can be rewritten :

(VI. 8. 14) $$\frac{\partial \omega(\beta)}{\partial \beta} = \left\{\tilde{V} + \exp(\beta H_o^A)\, \tilde{U} \exp(-\beta H_o^A)\right\} \omega(\beta)$$

Expanding the rhs in powers of H_o^A , we obtain :

(VI. 8. 15) $$\frac{\partial \omega(\beta)}{\partial \beta} = \left\{ \tilde{V} + \tilde{U} + \beta [H_o^A , \tilde{U}] + \beta^2 [H_o^A, [H_o^A, \tilde{U}]] + \dots \right\} \omega(\beta)$$

After a rather lengthy calculation [8], one can finally obtain the γ^2 (=m/M) correction to the maxwellian φ_o^{equ} (P) :

(VI. 8. 16) $$\varphi^{equ}(P) = \varphi_o^{equ}(P) + \gamma^2 \left[(P^2/2MkT)(kT_Q/kT) + \alpha \right] \varphi_o^{equ}(P) + 0(\gamma^4)$$

where α is a factor which guarantees the normalization of φ^{equ} to unity. T_Q is a characteristic quantum temperature which expression is quite compicated but can be shown to have the following properties:

(VI. 8.17) $$\lim_{T \to 0} T_Q(\hbar, T) = \text{constant}$$

(VI. 8. 18) $$\lim_{\substack{\hbar \to 0 \\ T \to \infty}} T_Q(\hbar, T) = 0$$

From the above equilibrium distribution, we obtain easily :

(VI. 8. 19) $$\langle P^2 \rangle = 3m\, \gamma^{-2} \left[(kT + \gamma^2 kT_Q) + 0(\gamma^4) \right]$$

At sufficiently high temperature, we have

(VI. 8. 19) $$kT \gg \gamma^2 kT_Q$$

(VI. 8. 20) $$\langle P^2 \rangle \simeq \frac{3m}{\gamma^2} kT$$

However, at sufficiently low temperature, we may reach a regime where:

(VI. 8. 21) $$\gamma^2 kT_Q / kT \gg 1$$

In that case :

(VI. 8. 22) $$\langle P^2 \rangle \simeq 3m\, kT_Q$$

does no longer depend on γ .

A more rigorous mathematical analysis is of course quite difficult in general because of the complexity of dense systems. A more elaborate discussion can be found in [3)8)].

Appendix VI.1 - Examples of the algebra leading to the Von Neumann - Liouville equation (VI.2.12)

In the mixed occupation numbers-plane wave representation the Von Neumann equation is :

$$i\frac{\partial\langle K\{n\}|\rho|K'\{n'\}\rangle}{\partial t} = \sum_{K''}\sum_{\{n''\}}\{\langle K\{n\}|H|K''\{n''\}\rangle\langle K''\{n''\}|\rho|K'\{n'\}\rangle$$

(AVI.1.1)

$$-\langle K\{n\}|\rho|K''\{n''\}\rangle\langle K''\{n''\}|H|K'\{n'\}\rangle\}$$

Let us consider in the first term in the rhs one contribution to the fluid-fluid interaction :

(AVI.1.2) $$\alpha = \sum_{K''}\sum_{\{n''\}}\langle K\{n\}|\bar{V}|K''\{n''\}\rangle\langle K''\{n''\}|\rho|K'\{n'\}\rangle$$

with

(AV1.1.3) $$\bar{V} = v(k,l,p,r)\,\delta_{k+l-p-r,0}\,a_k^+\,a_l^+\,a_p\,a_r$$

We perform the following change of variables :

(AVI.1.4)

$$K-K' = \varkappa \qquad n_k-n'_k = \nu_k$$

$$K''-K' = \varkappa' \qquad n''_k-n'_k = \nu'_k$$

$$\frac{K+K'}{2} = P \qquad \frac{n_k+n'_k}{2} = N_k$$

Then, we have

$$\alpha = \sum_{\varkappa'} \sum_{\{\nu'_k\}} \langle P + \frac{\varkappa}{2}, \{N + \frac{\nu}{2}\} | \bar{V} | P - \frac{\varkappa}{2} + \frac{\varkappa'}{2}, \{N - \frac{\nu}{2} + \frac{\nu'}{2}\} \rangle \times$$

(AVI. 1.5)
$$\times \langle P - \frac{\varkappa}{2} + \frac{\varkappa'}{2}, \{N - \frac{\nu}{2} + \frac{\nu'}{2}\} | \rho | P - \frac{\varkappa}{2}, \{N - \frac{\nu}{2}\} \rangle$$

Using the notation (VI. 2. 8) , this becomes :

$$\alpha = \sum_{\varkappa'} \sum_{\{\nu'_k\}} \bar{V}_{\varkappa - \varkappa', \{\nu - \nu'\}} (P + \frac{\varkappa'}{2}, \{N + \frac{\nu'}{2}\}) \times$$

(AVI. 1.6)
$$\times \rho_{\varkappa' \{\nu'\}} (P - \frac{\varkappa}{2} + \frac{\varkappa'}{2}, \{N - \frac{\nu}{2} + \frac{\nu'}{2}\})$$

Introducing the displacement operators defined by (VI. 2. 10) and (VI. 2. 11) we may write :

$$\alpha = \sum_{\varkappa'} \sum_{\{\nu'_k\}} \zeta^{\varkappa'} \eta^{\sum_k \nu'_k} \bar{V}_{\varkappa - \varkappa', \{\nu - \nu'\}} (P, \{N\}) \zeta^{-\varkappa} \eta^{-\sum_k \nu_k} \times$$

(AVI. 1.7)
$$\times \rho_{\varkappa' \{\nu'\}} (P, \{N\})$$

(the displacement operators act on everything that stands at their right.)

Now, using again (VI. 2. 8) and (AVI. 1. 3) , we have :

$$\bar{V}_{\varkappa - \varkappa', \{\nu - \nu'\}} (P, \{N\}) = v(k, l, p, r) \delta_{k+l-p-r, 0} \times$$

$$\times \langle P + \frac{\varkappa - \varkappa'}{2}, \{N + \frac{\nu - \nu'}{2}\} | a_k^+ a_l^+ a_p a_r | P - \frac{\varkappa - \varkappa'}{2}, \{N + \frac{\nu - \nu'}{2}\} \rangle$$

(AVI. 1. 8)
$$= v(l, lp, r) \delta_{k+l-p-r, 0} \delta_{\varkappa, \varkappa'} \prod_{i \neq klpr} \delta_{\nu_i, \nu'_i} \delta_{\nu'_k, \nu_k - 1} \times$$

$$\times \delta_{\nu'_l, \nu_l - 1} \delta_{\nu'_p, \nu_p + 1} \delta_{\nu'_r, \nu_r + 1} \Big[(N_k + 1/2)(N_l + 1/2) \times$$

$$\times (N_p + 1/2)(N_r + 1/2) \Big]^{1/2}$$

If we combine (AVI.1.7) and (AVI.1.8), we readily obtain the first term in the rhs of (VI.2.12) .

As another example, let us consider the external field term. One can also very easily obtain the following contribution to the Von Neumann - Liouville equation :

$$\sum_{\varkappa'}\sum_{\{\nu'\}}\langle \varkappa\{\nu\}|\mathcal{H}_E|\{\nu'\}\varkappa'\rangle \rho_{\varkappa'\{\nu'\}}(P,\{N\})$$

$$= \sum_{\varkappa'}\sum_{\{\nu'\}}\Big\{\xi^{\varkappa'}\eta^{\sum_k \nu_k} H^E_{\varkappa-\varkappa',\{\nu-\nu'\}}(P,\{N\})\xi^{-\varkappa}\eta^{-\sum_k \nu_k}$$

$$-\xi^{-\varkappa'}\eta^{-\sum_k \nu_k} H^E_{\varkappa-\varkappa',\{\nu-\nu'\}}(P,\{N\})\xi^{\varkappa}\eta^{\sum_k \nu_k}\Big\}\rho_{\varkappa'\{\nu'\}}(P,\{N\})$$

$$= \sum_{\varkappa'}\sum_{\{\nu'\}}\Big\{\exp\left(\frac{\varkappa'}{2}\frac{\partial}{\partial P}\right)\exp\left[\frac{1}{2}\sum_k \nu'_k \frac{\partial}{\partial N_k}\right]\Big\langle P+\frac{\varkappa-\varkappa'}{2},\{N+\frac{\nu-\nu'}{2}\}\Big|$$

$$\underset{\sim}{E}.\underset{\sim}{R}\Big|P-\frac{\varkappa-\varkappa'}{2},\{N-\frac{\nu-\nu'}{2}\}\Big\rangle\exp\left(-\frac{\varkappa}{2}\frac{\partial}{\partial P}\right)\exp\left[-\frac{1}{2}\sum_k \nu_k \frac{\partial}{\partial N_k}\right]$$

(AVI.1.9)

$$-\exp\left(-\frac{\varkappa'}{2}\frac{\partial}{\partial P}\right)\exp\left[-\frac{1}{2}\sum_k \nu'_k \frac{\partial}{\partial N_k}\right]\Big\langle P+\frac{\varkappa-\varkappa'}{2},\{N+\frac{\nu-\nu'}{2}\}\Big|$$

$$\underset{\sim}{E}.\underset{\sim}{R}.\Big|P-\frac{\varkappa-\varkappa'}{2},\{N-\frac{\nu-\nu'}{2}\}\Big\rangle\exp\left(\frac{\varkappa}{2}\frac{\partial}{\partial P}\right)\exp\left[\frac{1}{2}\sum_k \nu_k \frac{\partial}{\partial N_k}\right]\Big\}\times$$

$$\times\rho_{\varkappa'\{\nu'\}}(P,\{N\})$$

$$= \sum_{\varkappa'}\sum_{\{\nu'\}}\prod_k \delta_{\nu_k,\nu'_k}\int d\underset{\sim}{R}\;\underset{\sim}{E}.\underset{\sim}{R}.\;e^{i(\varkappa'-\varkappa).R}\times$$

$$\times\Big\{\rho_{\varkappa'\{\nu'\}}\Big(P+\frac{\varkappa'-\varkappa}{2},\{N\}\Big)-\rho_{\varkappa'\{\nu'\}}\Big(P-\frac{\varkappa'-\varkappa}{2},\{N\}\Big)\Big\}$$

$$= \sum_{\varkappa'}\sum_{\{\nu'\}}\prod_k \delta_{\nu_k,\nu'_k}\,\delta_{\varkappa',\varkappa}\; iE.\frac{\partial}{\partial(\varkappa'-\varkappa)}\times$$

$$\times\Big\{\rho_{\varkappa'\{\nu'\}}\Big(P+\frac{\varkappa'-\varkappa}{2},\{N\}\Big)-\rho_{\varkappa'\{\nu'\}}\Big(P-\frac{\varkappa'-\varkappa}{2},\{N\}\Big)\Big\}$$

$$= i\,E.\ \frac{\partial}{\partial P} \sum_{\varkappa'} \sum_{\{\nu'\}} \prod_k \delta_{\nu_k,\nu'_k}\,\delta_{\varkappa',\varkappa}\,\rho_{\varkappa'\{\nu'\}}(P,\{N\})$$

which agrees with (VI. 2. 16)

Appendix VI. 2 - Equivalence between (VI. 4. 11) and (VI. 5. 6)

Let us discuss the contribution involving twice V_1^A :

$$\alpha(t) = \sum_{\{N\}\{\nu\}\varkappa} \sum_{\{\nu'\}\varkappa'} \sum_{\{\nu''\}\varkappa''} \langle 0|V_1^A|\{\nu\}\varkappa\rangle \times$$

$$(\text{AVI. 2. 1}) \times \langle\{\nu\}\varkappa|\exp\left[-i\,(\mathcal{H}^f - z)t\right]|\varkappa'\{\nu'\}\rangle\langle\varkappa'\{\nu'\}|V_1^A|\varkappa''\{\nu''\}\rangle \times$$

$$\times\ \rho^f_{\varkappa''\{\nu''\}}(\{N\})$$

Using (VI. 5. 6), we have (see § 5) :

$$\sum_{\{\nu''\}} \sum_{\varkappa''} \langle\varkappa'\{\nu'\}|V_1^A|\varkappa''\{\nu''\}\rangle\rho^f_{\varkappa''\{\nu''\}}(\{N\})$$

$$(\text{AVI. 2. 2}) \quad = \sum_i \sum_K \langle\{N + \frac{\nu'}{2}\}|F_i(K)\,\rho^f_{\varkappa'-K} + \rho^f_{\varkappa'-K}\,F_i(K)|\{N - \frac{\nu'}{2}\}\rangle\frac{\partial}{\partial P_i}$$

$$= \sum_i \sum_K \langle\{N + \frac{\nu'}{2}\}\varkappa'|\,F_i(K)\,\rho^f + \rho^f\,F_i(K)|K\{N - \frac{\nu'}{2}\}\rangle\frac{\partial}{\partial P_i}$$

where the force operator is given by (VI. 5. 4). Hence, we obtain :

$$\alpha(t) = \sum_i \sum_K \sum_{\{N\}} \sum_{\{\nu\}\varkappa} \sum_{\{\nu'\}\varkappa'} \langle 0|V_1^A|\{\nu\}\varkappa\rangle \times$$

$$(\text{AVI. 2. 3}) \times \langle\{\nu\}\varkappa|\exp(-i\mathcal{H}^f t)|\{\nu'\}\varkappa'\rangle\langle\{N + \frac{\nu'}{2}\},\varkappa'|\,F_i(K)\,\rho^f$$

$$+\,\rho^f\,F_i(K)|\{N - \frac{\nu'}{2}\},\ K\rangle\frac{\partial}{\partial P_i}$$

Now, let us differentiate with respect to t the quantity :

(AVI. 2. 4)
$$\beta_{\nu,\varkappa-K}(\{N\},t) = \sum_{\{\nu'\}}\sum_{\varkappa'} \langle\{\nu\}\varkappa|\exp(-i\mathcal{H}^f t)|\{\nu'\}\varkappa'\rangle \times$$
$$\times\langle\varkappa',\{N+\frac{\nu'}{2}\}|F_i(K)\rho^f|\{N-\frac{\nu'}{2}\},K\rangle$$

with

(AVI. 2. 5)
$$\beta_{\nu,\varkappa-K}(\{N\},0) = \langle\varkappa,\{N+\frac{\nu}{2}\}|F_i(K)\rho^f|\{N-\frac{\nu}{2}\},K\rangle$$

hence

(AVI. 2. 6)
$$\beta(0) = F_i(K)\rho^f$$

We have :

(AVI. 2. 7)
$$\frac{\partial\beta_{\nu,\varkappa-K}(\{N\},t)}{\partial t} = -i\sum_{\{\nu'\}}\sum_{\varkappa'}\langle\{\nu\}\varkappa|\mathcal{H}^f|\{\nu'\}\varkappa'\rangle\beta_{\nu',\varkappa'-K}(\{N\},t)$$

or , going back to the occupation numbers-plane wave representation;

(AVI. 2. 8)
$$\frac{\partial\langle\{N+\frac{\nu}{2}\},\varkappa|\beta(t)|\{N-\frac{\nu}{2}\},K\rangle}{\partial t} = -i\langle\{N+\frac{\nu}{2}\},\varkappa|[H_f,\beta(t)]|\{N-\frac{\nu}{2}\},K\rangle$$

where

(AVI. 2. 9)
$$H_f = H_o^f + V$$

(To obtain this result, we take into account the fact that, in the term involving V_o^A, we have :

(AVI. 2. 10)
$$\sum_{\varkappa}\int dR\, e^{-i(\varkappa-k+l-\varkappa').R}\int dR'\, e^{-i(\varkappa'-K).R'}\beta^f(R,t)$$
$$= \int dR\, e^{-i\varkappa R}\, e^{i(k-l).R}\beta^f(R,t)\, e^{iKR} = \langle\varkappa|e^{i(k-l).R}\beta^f|K\rangle$$

Therefore, we have

(AVI. 2. 11)
$$\beta(t) = e^{-iH^f t}\beta(K)\, e^{iH^f t} = e^{-iH^f t}F_i(K)\, e^{iH^f t}\rho^f = F_i(K,t)\rho^f$$

where $F_i(K,t)$ is the force operator in Heisenberg representation.

Therefore :

$$\alpha(t) = \sum_i \sum_K \sum_{\{N\}} \sum_{\{\nu\}\varkappa} \langle 0 | V_1^A | \{\nu\}\varkappa \rangle \langle \varkappa, \{N + \frac{\nu}{2}\} | F_i(K,t) \rho^f$$

$$+ \rho^f F_i(K,t) | \{N - \frac{\nu}{2}\}, K \rangle \frac{\partial}{\partial P_i} \qquad \text{(AVI. 2. 12)}$$

From now, on , the derivation goes on as in the text and leads to the second order differential contribution in (VI. 6. 12) .

References.

1. T. Davis and R. Dagonnier, J. Chem. Phys. 44 , 4030 (1966)
2. R. Dagonnier and P. Résibois, Bull. Acad. Roy. Belg. Cl Sci. 52 , 229 (1966)
3. P. Résibois and R. Dagonnier , Physics Letters (1966)
4. A. Abrikosov, L. Gorkov and I. Dzyaloshinski, Methods of Quantum Field Theory in Statistical Physics, Prentice Hall, New Jersey (1963)
5. L. Meyer, and F. Reif, Phys. Rev . 123 , 727 (1961)
6. L. Meyer, T. Davis, S. Rice and R. Donelly, Phys. Rev. 126 , 1927 (1962)
7. P. Résibois and R. Dagonnier, Bull. Acad. Roy . Belg. Cl. Sci. (to be published)

VII. GRAVITATIONAL PLASMAS.

VII.1 - Introduction.

In all the preceding chapters on brownian motion, we have used the kinetic equation (II.9.5) as the starting point . However, when we derived this equation, we stressed the importance of time scales. This equation fails to describe correctly the asymptotic behavior of a system where there is no net separation between the collision time and the relaxation time, or where the initial correlations are over long distances. Now, long collision times or long range correlations must obviously be expected in systems interacting through long range forces. As examples of such systems, we have immediatly in mind, on the one hand, systems interacting through electrostatic forces and, on the other hand, systems interacting through gravitational forces. In both cases, the interaction between two particles is inversely proportional to their relative distance.

A great deal of effort has been done to understand the situation in the case of electrostatic forces. If we consider a charged test particle moving in a plasma, it polarizes the medium : the charge distribution around the particle is no longer uniform. The medium screens the interaction between two particles and we are no longer dealing with a pure Coulomb force. This idea, in its simplest form leads to the Debye Huckel theory. Out of equilibrium, it has been shown that, if one sums in the operator ψ all contributions proportional to $(e^2 C)^m$, the result of this summation is to introduce a dynamical screening. This screening introduces a short time scale :

$$t_p = (m/e^2 c)^{1/2} = (k_D v)^{-1} \tag{VII.1.1}$$

and a long time scale

$$t_r = (e^4 c)^{-1} m^{1/2} (kT)^{3/2} \sim c^{-1} \tag{VII.1.2}$$

We are again in the conditions where an asymptotic kinetic equation may be written (Balescu-Lenard equation).

What is now the situation with gravitational forces? The essential feature is that the interaction is now purely, attractive, whereas when we deal with electrostatic forces we have a mixture of attractive and repulsive forces (with a condition of overall neutrality). In a discussion of equilibrium properties in the framework of Mayer's cluster integrals, one notices very easily the importance of this difference. Indeed, there it appears clearly that the most divergent contributions can be eliminated because of the electroneutrality condition; the next dominant terms (which diverge also) can then be summed and lead to the screening. This problem arises always as soon as one computes average quantities, but, in the derivation of the Balescu-Lenard equation, one can restrict oneself to an electron plasma with a positive background; one notices then that the background plays no role in the derivation. In this case , we are dealing with purely repulsive forces. The first idea is then to perform in the case of purely attractive gravitational forces, the same summation that worked for purely electrostatic repulsive forces, or for a mixture of both attractive and repulsive electrostatic forces. In simple models, which leads for electrostatic potentials to the Debye potential :

(VII. 1. 3) $$V_D = e^{-\kappa r}/r$$

one obtains, for gravitational systems an effective potential of the form :

(VII. 1. 4) $$V_G = e^{i\kappa r}/r$$

This partial summation does not lead to a screening; the "effective interaction " has the same range as the gravitational interaction.

The intuitive ideas which lead to the Debye potential for an electrostatic plasma can obviously not be extended to the gravitational systems. As we have seen, more elaborate techniques did not succeed to bring anything which ressembles whatsoever to an effective short range interaction. In view of the failure of all those attemps to justify the use of the kinetic equation for such systems, one may try to use a completely different starting point . If we cannot use the kinetic equation, we must go back to the exact equation, the generalized master equation (II. 8. 9) . One can then try to discuss the new features introduced by the long range character of the interaction. This is what Prigogine and Severne [1] attempted very recently. They considered a very idealized model of a gravitational system: a weakly coupled, homogeneous system, with no correlations at t=0. Although the first condition can be justified with the conditions prevailing at present in our galaxy [1][2], the other conditions of the model are certainly not realistic. A real gravitational system is an inhomogeneous system. However, the finding of a proper treatment for this idealized model , which is the simplest model of a gravitational system one can find, is certainly the first obvious step one has to take if one wants to achieve an understanding of the much more complex actual systems (or at least of realistic simplifications of the actual systems) .

We shall first show on a very simplified collision operator what is the basic difference between systems with a very short collision time (as compared to the relaxation time) and systems where the collision time is infinite . This will enable us to convince ourselves that, even if we require only an understanding of the asymptotic behavior of the latter systems, it is necessary to start with the generalized master equation . We shall then discuss the characteristics of the long time evolution as described by this equation in the limit of infinite collision times.

More details as well as general considerations about the following discussion can be found in the paper by Prigogine and Severne[1] as well as in a paper by Prigogine [5].

VII. 2 - Simple model for the collision operator.

In order to see what are the difficulties which arise when one deals with long range forces, i.e. long collision times, let us investigate a simple model for the collision operator [3)4)] :

(VII. 2. 1) $$\psi(z) = \frac{\alpha}{z + i\beta} \qquad (\alpha, \beta > 0)$$

The collision time in this model is given by β^{-1} while the relaxation time is given by $\alpha^{-1}\beta$.

The non markovian equation for this case, if we do not take into account the destruction term, is :

(VII. 2. 2)
$$\frac{\partial \rho_o}{\partial t} = \int_0^t d\tau\, G(t-\tau)\, \rho_o(\tau)$$
$$= (2\pi i)^{-1} \int_0^t d\tau \int_C dz\, e^{-iz(t-\tau)} \psi(z) \rho_o(\tau)$$

Provided that we can neglect terms of the form $\exp(-t/\tau_{coll})$, an expansion of (VII. 2. 3) in powers of $(\tau_{coll}/\tau_{rel}) = \alpha\beta^{-2}$, leads to the pseudo-markovian equation (see chapter II, § 9) :

(VII. 2. 3)
$$\frac{\partial \rho_o}{\partial t} = -i\, \Omega(0)\psi(0)\, \rho_o^{(t)} = -i \left\{ \psi(0) + \psi'(0)\psi(0) + \left[\tfrac{1}{2}\psi''(0)\psi(0) + \psi'^2(0)\right]\psi(0) + \dots \right\} \rho_o(t)$$
$$= -\left\{ 1 + \alpha\beta^{-2} + 2\,(\alpha\beta^{-2})^2 + \dots \right\} (\alpha/\beta)\, \rho_o(t)$$

This expansion has a simple meaning for small values of τ_{coll}/τ_{rel}. On the other hand, if we take the extreme case of infinite collision times ($\beta \to 0$), it is meaningless; the operator $\psi(0)$ is no longer defined. However, in any case the non markovian equation predicts a well defined behavior of the distribution function. Indeed, from (VII. 2. 1) and (VII. 2. 2) we have

$$\frac{\partial \rho_0}{\partial t} = -\alpha \int_0^t d\tau\, e^{-\beta(t-\tau)} \rho_0(\tau) \tag{VII. 2. 4}$$

and we easily obtain the second order differential equation :

$$\frac{\partial^2 \rho_0}{\partial t^2} + \beta \frac{\partial \rho_0}{\partial t} + \alpha \rho_0 = 0 \tag{VII. 2. 5}$$

The solution of this equation is very simple and perfectly well defined whatever β (note that $[\partial \rho_0 / \partial t]_{t=0} = 0$) :

$$\begin{aligned} \rho_0(t) &= (1/2)\rho_0(0)\left[1 - \beta(\beta^2-4\alpha)^{-1/2}\right]\exp\left[-\left(\frac{\beta}{2} + \frac{(\beta^2-4\alpha)^{1/2}}{2}\right)t\right] \\ &+ (1/2)\rho_0(0)\left[1 + \beta(\beta^2-4\alpha)^{-1/2}\right]\exp\left[-\left(\frac{\beta}{2} - \frac{(\beta^2-4\alpha)^{1/2}}{2}\right)t\right] \end{aligned} \tag{VII. 2. 6}$$

When

$$\tau_{coll} / \tau_{rel} = \alpha\beta^{-2} \leqslant 4 \tag{VII.2. 7}$$

we have a monotonic decay of $\rho_0(t)$.

If $\alpha\beta^{-2} \ll 1$, neglecting terms proportional to $e^{-t/\tau_{coll}}$ $(e^{-\beta t})$, we obtain :

$$\rho_0(t) \simeq \rho(0) \exp(-\alpha\beta^{-1} t) \tag{VII. 2. 8}$$

which is the behavior that is given by the pseudomarkovian equation if we restrict ourselves to the first term in the rhs of (VII. 2. 3)

For

(VII. 2. 9) $$\alpha \beta^{-2} > 4$$

i. e. for large collision times, we obtain damped oscillations . In the extreme case of infinite collision time, we obtain a purely oscillating behavior :

(VII. 2. 10) $$\rho_o(t) = (1/2)\, \rho_o(0)\left(e^{-i\sqrt{\alpha}\, t} + e^{i\sqrt{\alpha}\, t}\right)$$

In other words, for long range interactions (small β), the corrections due to the non markovian character of the equation are quite important.

VII. 3 - Non markovian equation in the weak coupling approximation.

Let us consider a gravitational plasma :

(VII. 3. 1) $$H = \sum_{i=1}^{N} (1/2) m v_i^2 + \lambda \sum_{i<j} \left| \underset{\sim}{r}_i - \underset{\sim}{r}_j \right|^{-1}$$

where

(VII. 3. 2) $$\lambda = - Gm^2$$

where G is the constant of gravitation .

The characteristic parameters of such a system are (numerical values correspond to conditions prevailing now in our galaxy and in the vicinity of the sun [2]) :

(VII. 3. 3)

$$G = 6{,}7 . 10^{-8} \text{ cgs} \qquad \text{(constant of gravitation)}$$

$$m = 10^{33} \text{ g} \qquad \text{(average mass)}$$

$$\bar{v} = 3 . 10^{6} \text{ cm/sec} \qquad \text{(mean star velocity)}$$

$$C = 3{,}4 . 10^{-57} \text{ cm}^{-3} \qquad \text{(number density)}$$

With these parameters we can form one non dimensional quantity proportional to G:

(VII.3.4) $$\Gamma = C^{1/3} \lambda (m\bar{v}^2)^{-1} = 0\,(10^{-6})$$

The smallness of this parameter justifies a weak coupling approximation.

In our discussion in chapter III, § 6, we have seen that the asymptotic weak coupling operator $\psi(0)$ for a potential in $1/r$ presents a double divergence. We shall show here that the non-markovian corrections, even in the weak coupling approximation, permits to remove the long distance divergence. The short distance divergence, due to the close collisions cannot be removed in this way; however as close encounters are not very frequent, we shall neglect them in the evolution equation; in other words, we shall cut-off the potential at some short distance R.

In the following analysis, we shall find-two time scales : : the "nominal" relaxation time :

(VII.3.5) $$\tau_r = C^{-1} \lambda^{-2} m^2 \bar{v}^3 = C^{-1/3} \Gamma^{-2} \bar{v}^{-1} \simeq 7.10^{16} \text{ years}$$

the duration of close encounters :

(VII.3.6) $$\tau_c = R\bar{v}^{-1} = C^{-1/3} \Gamma \bar{v}^{-1} \simeq 7.10^{-2} \text{ years}$$

if one takes for R the distance corresponding to a mean $90°$ deflection in the two body scattering problem.

We shall, as usual, consider the limiting case

(VII.3.7) $$N \rightarrow \infty, \quad \Omega \rightarrow \infty, \quad C \text{ finite}$$

We shall also neglect all time variations occuring on the close collisions time scale τ_c .

We shall assume that we are dealing with an homogeneous system in which there are no spatial correlations at t=0. Then , the destruction term in the non markovian equation vanishes identically and this equation has the form (VII. 2. 2) .

In the weak coupling approximation, the collision operator $\psi(z)$ is given by (see III. 2. 3) :

$$\psi(z) = (8i\pi^3\lambda^2 C/m^2)\int d^3k|V_k|^2\, \underset{\sim}{k}\cdot\left(\frac{\partial}{\partial \underset{\sim}{v}_1}-\frac{\partial}{\partial \underset{\sim}{v}_2}\right)\frac{1}{\underset{\sim}{k}\cdot(\underset{\sim}{v}_1-\underset{\sim}{v}_2)-z}\times$$

(VII. 3. 8)

$$\times\; \underset{\sim}{k}\cdot\left(\frac{\partial}{\partial \underset{\sim}{v}_1}-\frac{\partial}{\partial \underset{\sim}{v}_2}\right)$$

In order to treat in a proper way the divergence, we shall consider the gravific potential as the limit of a screened potential:

(VII. 3. 9) $$V_k = \lim_{K\to 0}(2\pi^2)^{-1}(k^2+K^2)^{-1}$$

Following the technique of chapter III, § 4 , one obtains easily :

(VII. 3. 10) $$\psi(z) = (32i\pi^3\lambda^2 C/m^2)\frac{\partial}{\partial g_r}T_{rs}\frac{\partial}{\partial g_s}$$

with

(VII. 3. 11) $$T_{rs} = T_{\parallel}\, g_r g_s g^{-2} + T_{\perp}(g^2\delta_{r,s} - g_r g_s)g^{-2}$$

where

(VII. 3. 12) $$\underset{\sim}{g} = \underset{\sim}{v}_1 - \underset{\sim}{v}_2$$

is the relative velocity. $T_{\parallel}$ and $T_{\perp}$ are the parallel and transverse components :

(VII. 3. 13) $$T_{\parallel} = \lim_{K\to 0}(4\pi^2)^{-1}\left\{\frac{K}{z+iKg} - \frac{\mu}{z+i\mu g}\right\}$$

(VII. 3. 14) $$T_{\perp} = \lim_{K\to 0} (4\pi^2)^{-1}\left\{\frac{i}{g}\ln\frac{z+iKg}{z+i\mu g} - \frac{1}{2}\,\frac{1}{\mu R^2}\,\frac{1}{z+i\mu g}\right\}$$

where R is the short distance cut-off and

(VII. 3. 15) $$\mu = (R^{-2} + K^2)^{1/2}$$

For instance, in a reference frame where the z axis is along g, using cylindrical coordinates, we have :

$$T_{\parallel} = (2\pi^3)^{-1} \lim_{K\to 0}\int_{-\infty}^{+\infty} dk_z \int_0^R dk_{\perp}\, k_{\perp}\, k_z^2\, (k_z^2 + k_{\perp}^2 + K^2)^{-2} (k_z g - z)^{-1}$$

(VII. 3. 16) $$= (4\pi^2)^{-1} \lim_{K\to 0}\int_{-\infty}^{+\infty} dk_z\; k_z^2\,(k_z g - z)^{-1}\left\{(k_z^2+K^2)^{-1} - (k_z^2+K^2+R^{-2})^{-1}\right\}$$

The last integration is easily performed by the method of residues ($z \in S^+$) and leads to (VII. 3. 13) .

Also :

$$T_{\perp} = (4\pi^3)^{-1} \lim_{K\to 0}\int_{-\infty}^{+\infty} dk_z \int_0^R dk_{\perp}\, k_{\perp}^3\, (k_z^2 + k_{\perp}^2 + K^2)^{-2} (k_z g - 1)^{-1}$$

(VII. 3. 17) $$= (8\pi^3)^{-1} \lim_{K\to 0}\int_{-\infty}^{+\infty} dk_z \left\{\ln\frac{k_z^2+\mu^2}{k_z^2+K^2} - \frac{1}{R^2(k_z^2+\mu^2)}\right\}(k_z g - z)^{-1}$$

The second term is easily computed, using, the method of residues and gives the second contribution in the rhs of (VII. 3. 14) . As to the first contribution, we complete the real axis with a half circle at infinity in S^- and avoid the two branch points at $-i\mu$ and $-iK$ by making a cut. (see fig. VII. 3. 1)

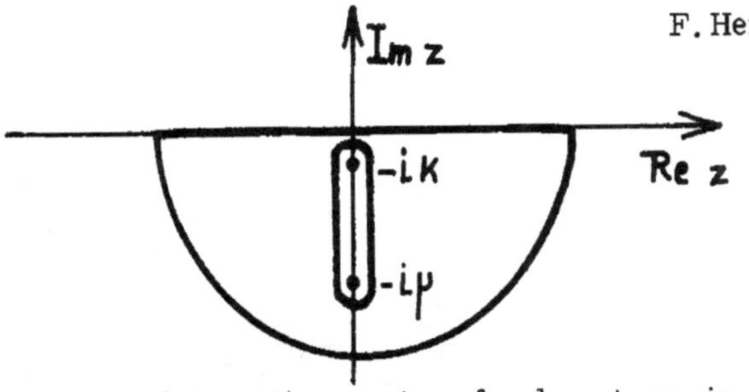

Integration contour for log. term in (VII. 3. 17)

Fig. VII. 3. 1

One can then easily obtain the first contribution in the rhs of (VII. 3. 14).

The most important feature of the operator $\psi(z)$ is that, at the limit $K \to 0$, the transverse part $T_{\perp}$ has a logarithmic singularity at $z = 0$. The collision operator may be rewritten :

(VII. 2. 18) $$-i\psi(z) = \phi(z) + \left\{ \frac{R^{-2}}{2i\mu} \frac{g}{(z+i\mu g)} - \ln \frac{z+iKg}{z+i\mu g} \right\} \hat{\psi}$$

with

(VII. 3. 19) $$\phi(z) = (8\pi\lambda^2 C/m^2) \frac{\partial}{\partial g_r} \frac{g_r g_s}{g^3} \left[\frac{Kg}{z+iKg} - \frac{\mu g}{z+i\mu g} \right] \frac{\partial}{\partial g_s}$$

(VII. 3. 20) $$\hat{\psi} = (8\pi\lambda^2 C/m^2) \frac{\partial}{\partial g_r} \frac{g^2 \delta_{r,s} - g_r g_s}{g^3} \frac{\partial}{\partial g_s}$$

if one takes into account the fact that any function of g commutes with the differential operator $\hat{\psi}$.

We have singularities both at $z = -iKg$ and $z = -i\mu g$. The singularities at $z = -i\mu g$ are related to the close collisions and are at a finite distance from the real axis. As the time scale for the close collisions is much shorter than any time scale we shall meet, we treat the close collisions as instantaneous. This means that we consider values of z such that :

(VII. 3. 21) $$z \ll -i\mu g$$

We then obtain the approximate form :

$$\text{(VII.3.22)} \qquad -i\,\psi(z) = \phi(z) + \left[\frac{R^{-2}}{2\mu^2} + \ln\left(\frac{K}{\mu} + \frac{z}{i\mu g}\right)\right]\hat{\psi}$$

$$\text{(VII.3.23)} \qquad \phi(z) = (8\pi\lambda^2 C/m^2)\,\frac{\partial}{\partial g_r}\,\frac{g_r g_s}{g^3}\left[\frac{Kg}{z+iKg} + i\right]\frac{\partial}{\partial g_s}$$

Now, in the case of electrostatic plasmas , K is finite and if we neglect effects proportional to $(Kgt)^{-1}$, we obtain the asymptotic form :

$$\text{(VII.3.24)} \qquad -i\psi(0) = -\left[\frac{1}{2(1+K^2R^2)} + \ln\frac{KR}{(1+K^2R^2)^{1/2}}\right]\hat{\psi}$$

$$\text{(VII.3.24)} \qquad = \int_0^{R^{-1}} dk\,k^3\,(k^2+K^2)^{-2}\,\hat{\psi}$$

which is easily verified to be identical to (III.3.5) when one takes as the interaction the screened Coulomb potential. However, in the limit $K \to 0$, this procedure is meaningless and we obtain :

$$\text{(VII.3.25)} \qquad -i\,\psi(z) = \phi - \left\{(1/2) + \ln(-izRg^{-1})\right\}\hat{\psi}$$

with

$$\text{(VII.3.26)} \qquad = (8\pi\lambda^2 C/m^2)\,\frac{\partial}{\partial g_r}\,\frac{g_r g_s}{g^3}\,\frac{\partial}{\partial g_s}$$

The solution of the non markovian equation (VII.2.2) may be written (see II.9.4) :

$$\text{(VII.3.27)} \qquad \begin{aligned} \rho_o(t) &= -(2\pi i)^{-1}\int dz\; e^{-izt}\sum_{n=0}^{\infty}\frac{1}{z^{n+1}}\left[\psi(z)\right]^n \rho_o(0) \\ &= -(2\pi i)^{-1}\int dz\; e^{-izt}\,\frac{1}{z-\psi(z)}\,\rho_o(0) \end{aligned}$$

VII.4 - Time evolution of the velocity distribution function.

The operators $\hat{\Psi}$ and Φ do not commute; hence they do not have a common set of eigenfunctions and the analysis of the time evolution is not very easy. However, the problem will be considerably simplified if one of the contributions turns out to be dominant. Let us first investigate the type of behavior determined by the second term.

The characteristic time involved in $\hat{\Psi}$ is the nominal relaxation time τ_r given by (VII.3.5). The operator $-i\hat{\Psi}$, as we have already seen (see the discussion of the $\mathcal{H}$-theorem, chapter III, § 2) has real, negative or zero eigenvalues which define a spectrum of relaxation times. We may in a qualitative discussion introduce the following approximation :

(VII.4.1) $$-i\hat{\Psi}\rho_o = -\rho_o/\tau_r$$

Also, we replace Rg^{-1} by its average value, the collision time for close encounters τ_c. Then, we obtain :

(VII.4.2) $$\rho_o(t) = -(2\pi i)^{-1}\int dz\ e^{-izt}\left[z-\chi(z)\right]^{-1}\rho_o(0)$$

where

(VII.4.3) $$\chi(z) = -\left\{(1/2) + \ln(-iz/\tau_c)\right\}(i\tau_r)^{-1}$$

In order to understand the evolution of $\rho_o(t)$, we have to discuss the singularities of the integrand, i.e. find out the roots of the equation :

(VII.4.4) $$z - \chi(z) = 0$$

Let us introduce :

(VII.4.5) $$z = -\omega - i\alpha$$

Then (VII. 4. 4) and (VII. 4. 4) give us :

(VII. 4. 6) $$-x+iy = (1/2) + \ln \Gamma^{-1}(-x+iy)$$

with

(VII. 4. 7) $$x = \alpha\tau_r \qquad y = \omega\tau_r \qquad \Gamma = \tau_r/\tau_c = 10^{18}$$

A detailed discussion of the dispersion equation (VII. 4. 6) can be found in ref.[1]. The main point is that the equation can be very much simplified if one takes into account the largeness of Γ. A whole spectrum of solutions is found. For consistency, the range of the spectrum is restricted to frequencies such that :

(VII. 4. 8) $$\omega \ll \omega_{max} \approx \tau_c^{-1}$$

The frequencies ω_n are essentially the odd harmonics of $\omega_o = \tau_r^{-1}$

(VII. 4. 9) $$\omega_n \simeq (2n+1)\pi/\tau_r$$

In the useful part of the spectrum (frequencies less than $10^{-2}\omega_{max}$), the damping is such that :

(VII. 4. 10) $$\alpha_n > 4\tau_r^{-1}$$

In fact, in the major part of the spectrum, the damping is found to be of order :

(VII. 4. 11) $$\alpha_o = \tau_r^{-1} \ln \Gamma^{-3} = 40\tau_r^{-1}$$

The time scale for the oscillations is of the order of the nominal relaxation time τ_r^{-1} while for the damping it is much shorter (at least by a factor 4). The essential feature is that we have now an oscillatory relaxation of $\rho_o(t)$ on a time scale much shorter than the nominal relaxation time and given by :

(VII. 4. 12) $$t_{rel} = \tau_r / \ln(\tau_r / \tau_c)$$

The consideration of the complete spectrum of eigenfunctions and eigenvalues of ψ, avoided in the approximation (VII. 4. 1), would only lead to further complications of detail.

However, we still have to examine the effect of the operator ϕ which appears in the complete equation (VII. 3. 27). The time scale for the effects due to ϕ is again the nominal relaxation time τ_r, while the term involving $\hat{\psi}$ has a much shorter time scale t_{rel}. Therefore, the system will first reach a quasiequilibrium distribution, which will be further modified by the action of the ϕ contribution. One verifies easily that, if the $\hat{\psi}$ contribution conserves the kinetic energy, this is not true for the ϕ contribution. One obtains easily for the variation of the kinetic energy per star :

$$\frac{\partial E_{kin}}{\partial t} = N^{-1} \int \{d\underset{\sim}{v}\}^N \sum_{i=1}^{N} (mv_i^2/2) (\partial \rho_o / \partial t)$$

(VII. 4. 13)
$$= -(2\pi i)^{-1} N^{-1} \int \{d\underset{\sim}{v}\}^N \sum_{i=1}^{N} (mv_i^2/2) \int dz\, e^{-izt} \psi(z) \frac{1}{z-\psi(z)} \rho_o(o)$$
$$= -(2\pi i)^{-1} N^{-1} \int \{d\underset{\sim}{v}\}^N \sum_{i=1}^{N} (mv_i^2/2) \phi\, \rho_o(t)$$
$$= 4\pi \lambda^2 C m^{-1} \int d\underset{\sim}{g}\, d\underset{\sim}{w}\; g^{-1} \rho_o(\underset{\sim}{v}_1, \underset{\sim}{v}_2; t) > 0$$

where

(VII. 4. 14) $$\underset{\sim}{g} = \underset{\sim}{v}_1 - \underset{\sim}{v}_2$$

(VII. 4. 15) $$\underset{\sim}{w} = (\underset{\sim}{v}_1 + \underset{\sim}{v}_2)/2$$

are respectively the relative and center of mass velocities.

Therefore, the ϕ contribution plays the role of a source term and leads to a continuous increase of the kinetic energy. In other words, we have the following picture of the time evolution of ρ_0: first , a quasiequilibrium distribution will be reached on a time scale t_{rel}; the aged system then remains in the quasiequilibrium state but with a time dependent temperature.

VII. 5 - Role of the initial correlations .

In this whole discussion, we have assumed that there were no initial correlations. Then, we have seen that there is a continuous increase of kinetic energy. Because of total energy conservation, we have at the same time a decrease of potential energy (the complete energy balance can be verified in detail but requires the evaluation of binary correlation Fourier coefficient and will not be considered here) . This continuous exchange between kinetic and potential energy of course occurs for any system when the non markovian description is retained . The particularity of the gravitational plasma is that it occurs at lowest order, which finally is due to the fact that there exists no approximation corresponding to instantaneous collisions. However, we may wonder whether this picture could not be affected if initial correlations were present. As we are dealing with long range forces, once initial correlations are present, there is no mechanism by which the system can loose the memory of these conditions in a short time as it happens for systems interacting through short range forces. The fact that the true collision time for such a system is very long on the time scale over which we discuss the behavior of the system, has the consequence that neither can we consider the collisions as complete nor can we assume that the system has forgotten its initial conditions. Therefore, we have to retain both the non markovian character of the collision term and the destruction term in the master equation.

We can easily see that the presence of initial correlations will indeed modify our results. Let us take as the initial condition a function of the hamiltonian :

$$\rho(0) = f(H) \tag{VII.5.1}$$

If one adds to (VII.2.2), the destruction term and computes it for this initial situation with the same assumptions as the collision term, one verifies easily that the increase of the kinetic energy which results from the ϕ contribution is completely cancelled by the destruction contribution.

This example clearly shows us the important role played by the initial correlations in the description of systems interacting through long range forces. It would therefore be of great importance to have realistic models of non equilibrium correlations.

References.

1. I. Prigogine and G. Severne, Physica (to appear, 1966)
2. S. Chandrasekhar, "Principles of Stellar Dynamics" U. Of Chicago Press (1942) , Dover, New York (1960)
3. J. Brocas, Bull. Acad. Roy. Belg. cl. Sci. 50 , 765 (1964)
4. K. Hauboldt, Physica 28 , 834 (1962)
5. I. Prigogine, Nature 209 , (1966)

CENTRO INTERNAZIONALE MATEMATICO ESTIVO

(C. I. M. E.)

T. KAHAN

"THEORIE DES REACTEURS NUCLEAIRES METHODES DE RESOLUTION PERTURBATIONNELLES, ITERACTIVES ET VARIATIONNELLES".

Corso tenuto a Varenna dal 19 al 27 settembre 1966.

THEORIE CINETIQUE DES REACTEURS NUCLEAIRES METHODES DE RESOLUTION PERTURBATIONNELLES ITERACTIVES ET VARIATIONNELLES .

par

T. KAHAN

Avant d'aborder la théorie mathèmatique de l'équation de transport qui régit l'évolution des phénomènes neutroniques dans un réacteur nucléaire, il me parait indispensable de définir et de préciser un certain nombre de notions majeures en physique nucléaire .

Introduction physique.

On admet en physique atomique que chaque atome est constitué par un centre ou noyau positivement chargé, contenant à lui seul presque toute la masse de l'atome, autour duquel gravitent un certain nombre d'électrons négatifs(ou négatons) .

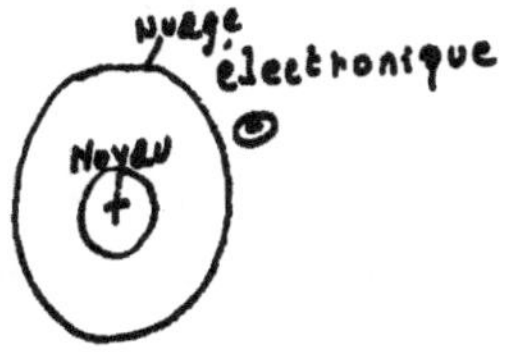

Dans un atome neutre, la charge positive du noyau est égale et de signe contraire à la somme des charges négatives des négatons : c'est un multiple autre Ze de la charge e.Le nombre Z est appelé le 'nombre atomique' de l'atome . (Z = 1, 2, 3, ...)

Tous les noyaux se composent de Z protons (noyau de l'atome d'H) et de N néutrons : particules de charge électric. que . Le nombre

$$A = Z + N$$

est le nombre de masse du noyau. Ainsi les nombres de masse du proton et du néutron sont l'un et l'autre égal à A = 1 . Ces deux particules élémentaires de masse à peu près égale, représentent, au fond, deux états quantiques différents d'une seule et même particule fondamentale, dite nucléon.

Une réaction nucléaire est un processus qui a lieu lorsqu'une particule nucléaire - nucléon, noyau, photon, - entre en choc avec une autre particule . Considérons une réacteur du type

$$a + C \longrightarrow R + b$$

qu'on écrit aussi

$$C(a, b)\, R\, .$$

Cette notation signifié qu'une particule a frappé le noyau cible C et produit le noyau résiduel R et une particule émergente b :

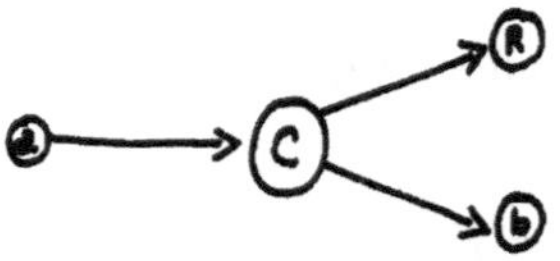

Ainsi par exemple un neutron (n) frappe le noyau de base $_5B^{10}$ pour donner lieu à un noyau de lillium avec émission d'une particule α

$$n + {}_5B^{10} \longrightarrow {}_3L_i^{7} + \alpha, \text{ soit } B(n, \alpha)\, L_i\, .$$

Le terme fission signifie rupture d'un noyau lourd en deux (plus rarement en plus de deux) fragments sensiblement egayx . C'est un phénomène exo énergétique , c'est à dire un réaction accompagnée d'une énorme libératioh d'énergie ΔE(suivant la relation d'Einstein

$$\Delta m = c^2 \Delta E$$

(diminution de masse (Energie libérée)
du noyau)

T. Kahan

La première fission découverte fut celle dell'isotope 236 de l'uranium obtenu en bombardant l'isotope rare (0,7 %) 235 de celui-ci avec des neutrons thermiques (c'est à dire avec des neutrons animés de la vitesse d'agitation thermique (E = 0,025 ev à T = 300°K)) suivant la réaction :

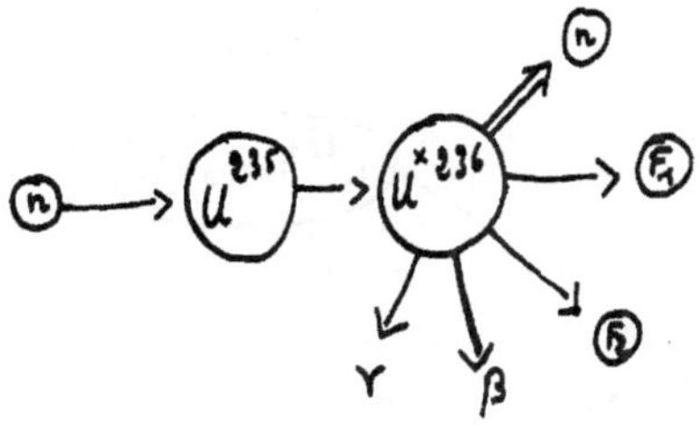

$$n + U^{235} \longrightarrow \underset{\text{(instable)}}{U^{*236}} \longrightarrow F_1 + F_2 \quad \text{neutrons+ rayons } \beta \text{ +rayor}$$

+énerg

en F_1 et F_2 étant deux fragments nucléaires tels que $_{52}T_e^{137}$ et $_{40}Z_2^{97}$ qui sont des nuclides instables qui se desintègrent en chaine jusqu'à ce qu'un nuclide stable soit atteint selon le schéma suivant

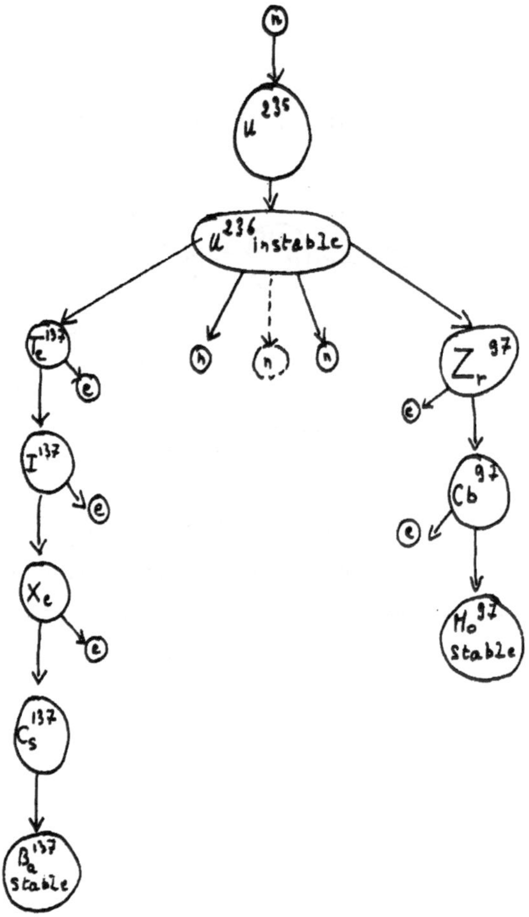

Fait très imputant , il apparait qu'un petit nombre de neutrons (en moyenne 2, 5 neutron) sont émis par les fragments de fission après que la fission a eu lieu ; ce on les néutrons prompts ; en très petit nombre de neutrons (N 1 % dits retardés sont émis par les fragments au bout d'un temps relativement long après la fission (Fraction de seconde a 55 se.) Ces neutrons retardés jouent un rôle majeur dans les réacteurs atomiques .

Réaction en chaine . Le fait qu'un neutron capturé par le noyau U^{235} provoque l'émission d'environ 3 neutron (2,5 en moyenne) avec une énergie cinétique de quelques M e v au total, entraine la possibilité que ces nouveaux neutrons aillent en dégager par la suite

$$3^2 = 9 ,$$

puis

$$3^3 = 27 ,$$

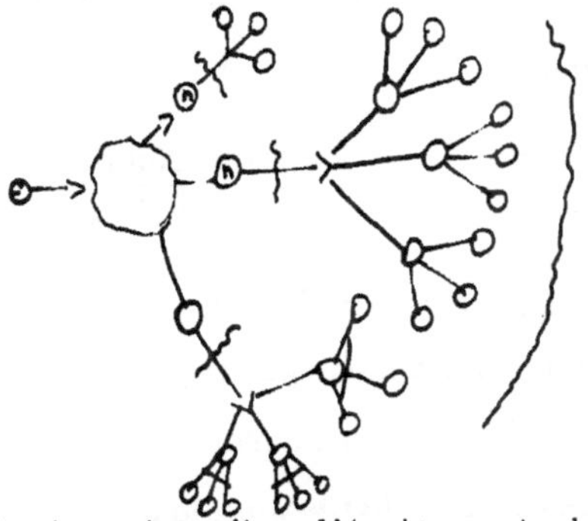

la chaine continuant ainsi en principe, jusqu'à l'épuisement des noyaux fissiles , à condition que chacun de ces neutrons induise de nouvelles fissions. En cinquante générations, on aurait de la sorte

$$3^{50} = 10^{25}$$

neutrons présents dans le système . Il se produit ainsi une chaine de réactions où e neutron initial agit comme une allumette appliquée à un corps combustible : la chaleur dégagée par la flamme de l'allumette met le feu à une partie du corps et la chaleur résultante provoque la combustion de proche en proche d'autres portions jusqu'à ce que le tout soit consumé et son enérgie chimique liberéé

Faisons quelques ordres de grandeur . Il se dégage environ 200 Mev au cours d'une seule fission par un seul neutron du 0,03 e v Ceci est equivalent à

$$\sim 3,2 \,.\, 10^{-11} \text{ watt.sec}$$

La fission complète de 1g d'u libère donc l'énorme quantité d'énergie de

$$8,2 \,.\, 10^{7} \text{ kw.sec.}$$

Comme chaque atome d'u possède à la température, ordinaire une enérgie thermique de ∼ 0, 0 3 ev on voit que l'énergie dégagée autour de la fission est d'énviron :

$$\frac{200 \text{ Mev}}{0,03 \text{ ev}} \cong 7.\ 10^{9} \text{ fois}$$

plus grande que l'energie calorifique de l'U . Si le temps nécessaire par la fission est supposé être de 10^{-8} sec, la 50° genération où 10^{25} neutrons auront été, serait atteinte en moins d'une micro-seconde —> bombe atomique . En realité, une potion des neutrons est absorbée par des processus de capture autres que la fission et une autre partie s'évade du système Néanmoins si la perte des neutrons n'est pas excessive la possibilité d'une réaction en chaine et d'une explosion subsiste. Si d'autre part , on part à maitriser la cadence de liberation de cette énérgie et à la plier à une utilisation contrôlable dans le temps, on obtient un réacteur nucléaire (ou pile atomique) .

Réacteurs nucléaires. Le réacteur nucléaire est un dispositif comprenant une matière fissile en quantité suffisante, disposée de façon a pouvoir entretenir une réaction en chaine contralée En principe, une telle machine est possible, car elle exige seulement que la vitesse de production des néutrons par fission ayant lieu dans le réacteur soit égale à la vitesse de disparition des neutrons due à toutes les causes de parte.

Cette condition minimale se traduit en disant que, pour chaque noyau subissant la fission, il faut qu'il se produise en moyenne au moins un neutron qui induise la fission dans un autre noyau. Cette condition s'exprime par un facteur de reproduction ou de multiplication k du réacteur défini comme la

$$C = \frac{\text{nombre de neutrons d'une génération quelconque}}{\text{nombre de neutrons de la génération immédiatement précedente.}}$$

Si C est rigoreusement égal à 1 ou quelque peu supérieur à 1, la réaction en chaine pourra s'amorcer, mais si $C < 1$, la chaine ne pourra pas s'entretenir. Si $C6 > 1$ il suffit d'un petit nombre de neutrons pour amorcer une chaine divergente de fissions. Pour empêcher qu'une telle chaine échappe au contrôle, on peut introduire un absorbeur de neutrons. Si $C < 1$, la chaine, au lieu de se propager, finirait par se désamorcer. .

Ceci étant, les neutrons libérés par fission possédent des energies cinétiques de l'ordre de 100 à 200 Mev. Bien que de tels neutrons rapides puissent induire la fission alla fois dans $l'U^{238}$ (abondance normale 99, 3 %) et dans $l'U^{235}$ (abondance 0,7 %), les probalités de fission (ou section efficace de fissinn) sont bien plus petites par les neutrons rapides (energie 1 à 200 MeV) que par les neutrons thermiques (énergie ~ 0,03 e v) Il est donc indispensable de ramener l'énergie de ces neutrons rapides à des énergies thermiques (v. thermique ~200m/sec)

. On est ainsi conduit à placer dans le réacteur un matériel dit <u>modérateur</u> qui sera d'autant meilleur qu'il amènera les neutrons rapides au prix du nombre minimal de chaos élastiques au niveau thermique . L'hydrogène serait idéal, n'était la réaction de capture $H^{1}(n,\gamma)^{2}H$, c. a.d .

$$n + H^{1} \longrightarrow H^{2} + \text{photon}$$. (donc neutron perdu définitivement)

On utilisé avec succès le carboné sous forme de graphite pur, leau lourde etc.

Une autre difficulté tient à ce que l'U^{238} est un fort absorant de neutrons , dans la région dite de résonance, à environ E = 100 e V

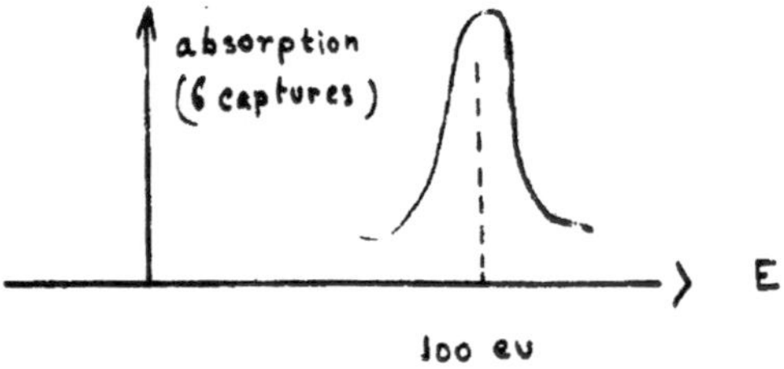

Un telle capture donne pas lieu à la fission. Si les neutrons rapides sont ralentis, leur énergie passera obligatoirement par cette région. Cette difficulté peut être tournée en plaçant l'uranium, par exemple, sous forme de barres au sein du modérateur, au lieu de le mélanger intimement.

Les neutrons qui naissent près de parois de réacteur peuvent s'en échapper et être perdus ainsi pour la réaction ramifiée .Comme cette fuite des neutrons est un phénomène de surface et que la production des neutrons est un effet de volume, on peut rendre le rapport d'espace volume aussi petit qu'on veut en augmentant la masse du réacteur. On entoure aussi le réacteur d'une enveloppe protectrice de matériau , dit réflecteur ayant un grande section efficace pour la diffusion des neutrons et une faible section efficace pour la capture et qui réfléchit les neutrons vers l'interieur du réacteur . Ce réflecteur peut être construit en graphite, en oxide de beryllium, etc. La figure suivante donne une représentation schématique de la fission d'U^{235} et de la capture des neutrons par l'U^{238} dans un Réacteur nucléaire .

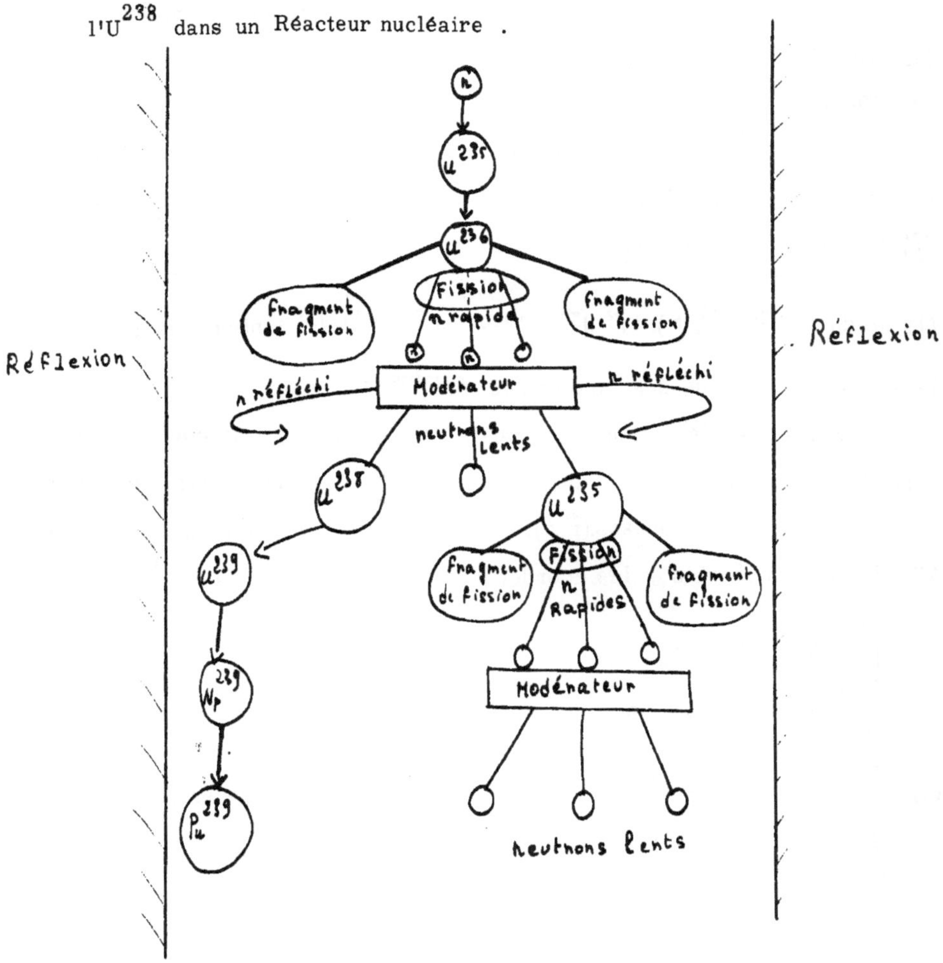

Chapt. I

L'equation du réacteur et les équations de transport cinétiques

L'état dynamique d'un réacteur est déterminé avant tout par son bilan ou budget neutronique. Le point de départ d'une analyse cinétique est par conséquent une relation mathématique qui exprime l'évolution de ce bilan au cours du temps. Cette relation mathématique ponte le nom d'équation du réaction.

La prototype de l'équation du réacteur est l'équation intégro-différentielle de transport de Boltzmann qui formule le bilan du nombre de molécules dans un élément de volume de l'espace des phases

Soit $f(\vec{r}, \vec{v}, t)$ la fonction de distribution d'un gaz moléculaire. Le nombre de molécules contenues dans l'élément de volume de phase $d\vec{r}, d\vec{v}$ est pour définition

(1) $dn = f(\vec{r}, \vec{v}, t)\, d\vec{r}\, d\vec{v}$, avec $n(\vec{r}) = \int f\, d\vec{v}$ $n(\vec{r})$ étant la densité numérique des molécules gazeuses.

Quelle est l'équation qui régit l'évolution de f au cours du temps sous l'action de diverses causes ? Pour l'établir, analysons l'évolution de f dans l'espace des phases $(\vec{r}, \vec{v})$. Dans l'intervalle de temps δ en l'absence de chaos entre molécules, les coordonnees des molécules $(\vec{r}, \vec{v})$ deviennent

$$(2)\quad \begin{cases} \vec{r}\,' = \vec{r} + \vec{v}\,\delta t \\ \vec{v}\,' = \vec{v} + \delta\vec{v} = \\ \quad = \vec{v} + \dfrac{\vec{F}}{m}\,\delta t \\ \dfrac{\delta v}{\delta t} = \dfrac{\vec{F}}{m} \end{cases}$$

$\vec{F}$ = forme extérieure

m = masse des molécules. Au bout de δt, les molécules initialement dans $d\vec{r}\, d\vec{v}$, se retrouveront dans l'élément de phase $(r + \vec{v}\,\delta t,\ \vec{v} + \frac{\vec{F}}{m}\,\delta t)$

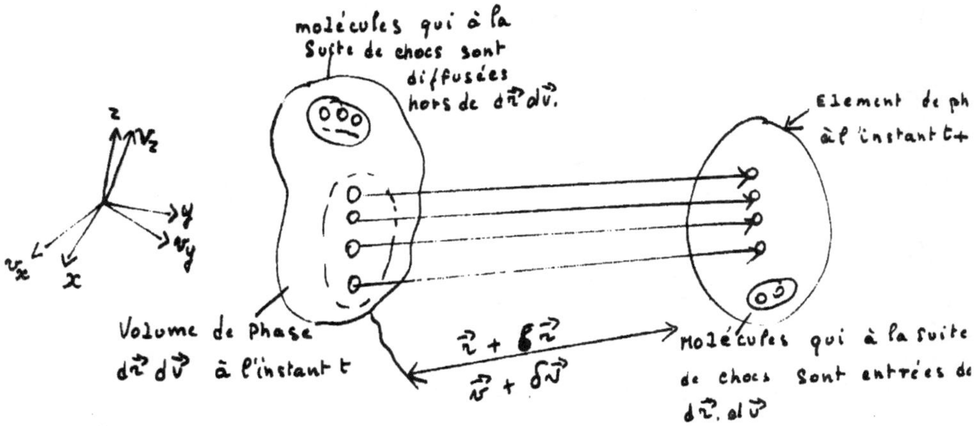

Par développement en série de Taylor

(3) $$f(\vec{r} + \delta\vec{r},\ \vec{v} + \frac{\vec{F}}{m}\delta t,\ \ + \delta t) =$$

$$= f(\vec{r},\ \vec{v},\ t) + \sum_i v_i \frac{\partial f}{\partial x_i} + \sum_i \frac{F_i}{m}\frac{\partial f}{\partial v_i} + \frac{\partial f}{\partial t}\,\delta t$$

$$= f + \frac{df}{dt}\,\delta t$$

La variation totale le long des lignes de flux sera ainsi

(4) $$\frac{df}{dt} = \frac{\partial f}{\partial t} + \vec{v}\cdot\frac{\partial f}{\partial \vec{r}}\cdot\nabla_r f \qquad + \frac{\vec{F}}{m}\cdot\frac{\partial f}{\partial \vec{v}}\cdot\nabla_v f$$

Admettons maintenant que δt soit beaucoup plus grand que la durée de choc. Alors plusieurs chocs auront lieu dans l'intervalle δt dont quelques uns vont faire entrer quelques molécules dans l'élément $d\vec{r}\,d\vec{v}$, et dont d'autres vont sortir des molécules hors de $d\vec{r}\,d\vec{v}$ (voir fig. pag. 22).

Si nous désignons la variation par unité de temps de f à la suite des chocs pour

(5) $$\left(\frac{\delta f}{\delta t}\right)_{\text{chocs}},$$

l'équa. (3) prend la forme

(6) $$f(\vec{r} + \delta \vec{r}, \vec{v} + \delta \vec{v}, t + \delta t) - f(\vec{r}, \vec{v}, t) = \left(\frac{\delta f}{\delta t}\right)_{\text{choc}} \delta t$$

Soit

(7) $$\frac{\partial f}{\partial t} + \vec{v} \cdot \frac{\partial f}{\partial \vec{r}} + \frac{\vec{F}}{m} \cdot \frac{\partial f}{\partial \vec{v}} = \left(\frac{\delta f}{\delta t}\right)_{\text{chocs}}$$

C'est là l'équation générale de Boltzmann .

Le premier nombre de (7) n'est autre chose que la dérivée totale de f le long de la trajectoire (2) . Reste à établir la forme fonctionnelle du second membre de (7) à savoir

$$\left(\frac{\delta f}{\delta t}\right)_{\text{ch.}}$$

Dans le budget neutronique d'un réacteur les forces extérieures $\vec{F}$, contrairement à ce qui se passent dans les gaz traités par Boltzmann ne jouent aucun rôle .Car pour les neutrons seule entrerait en ligne de compte la pesanteur dont l'effet, pour des vitesses moyennes des neutrons thermiques de 2200 m/ sec et des libres parcours moyens de quelques centimètres, est cependant négligeable .

De même, choc néutron-néutron est un événement fort rare: même dans le cas d'un flux neutronique de $10^{14}/\text{cm}^2 \cdot \text{sec}$. il ne se trouve que $4{,}5 \cdot 10^8$ neutrons dans un centimètre cube comparé à 10^{22} molécules gazeuses.

T. Kahan

D'autre part, le sort d'un neutron au cours du choc contre les matériaux du réacteur tels que modérateur, matière fissile, matériaux de structure, réfrigerant, poisons , etc. est très varié . Les chocs élastiques et inélastiques laissent le nombre de neutrons inchangé alors que la fission et les processus de capture provoquent un changement du nombre de particules et dont l'équation de Boltzmann ne tient pas compte.

Depuis la création de la physique neutronique on a formulé des équation des transport d'un degré de généralité variable pour la diffusion des neutrons.

Dans notre exposé, forcément limité, nous allons, établir une formulation mathématiquement aussi générale que possible mise sous une forme indépendante du temps et tenant compte des neutrons retardés.

Aprés avoir précisé quelques concepts de physique nucléaire je présenterai l'équation générale des réacteurs sous une formulation de transport .

Avant de procéder à l'établissement de l'équation des réacteurs, il est indispensable de préciser un certain nombre de concepts atomiques que nous avons introduits déjà .

Rappel de physique nucléaire.

Soit un faisceau de neutrons monomètrique de vitesse v qui pénètre dans une couche de substance contenant n particule , par cm^3 (cf . fig. 1) Ces neutrons subiront des chocs de diverses sortes.

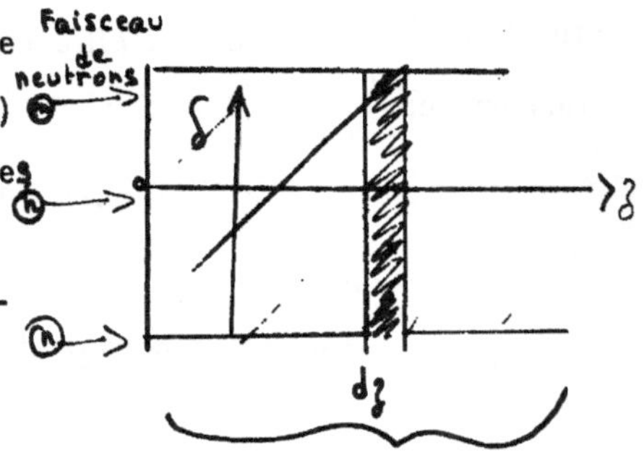

(Fig 1)

Soit P(z) la probabilité de séjour des neutrons dans le corps à la distance z. La perte de neutrons par absorption dans le faisceau incident peut être mise sous la forme (S = surface du corps)

$$(1) \quad - d\,P(z) = P(z) \times \frac{n\sigma S\, d z}{S} = P(z)\, n\sigma\, d z$$

où σ désigne la section efficace microscopique du noyau individuel absorbant qui est définie de la façon suivante. Soit N le nombre de neutrons par cm^3 voyageant dans la direction z dans notre faisceau et qui viennent frapper une couche mince substance (cible) . Le nombre de processus ou de réactions ou de chocs observés sera proportionnel au nombre de neutrons f venant frapper la cible et au nombre de noyaux par unité d'aire de la cible.

$$(2) \quad \frac{\text{Nombre de réactions (chocs)}}{cm^2 \cdot sec} = \sigma \times \frac{\text{nombre de noyaux cibles}}{cm^2} \times N v \frac{\text{neutrons}}{cm^2 \cdot sec}$$

Nv , le flux, est le nombre de neutrons venant frapper une surface unité de la cible par seconde. La constante de proportionalité σ est précisément la section efficace microscopique pour l'évènement donné. L'équation (2) montre que σ a la dimension d'une surface. Beaucoup de sections efficases de réactions nucléaires sont conprises entre 10^{-27} et 10^{24} cm^2 . Les sections efficaces sont souvent indiquées en "barns" au lieu de cm^2

$$1 \text{ barn} = 10^{-24}\ cm^2$$

Il résulte de (2)

$$\sigma = \frac{\text{fraction de noyaux cibles réagissant par sec.}}{N v}$$

ce qui peut s'interpréter en attribuant à chaque noyau cible une surface σ

perpendiculaire à la direction de mouvement du neutron incident . Si le neutron incident arrive à frapper cette surface, il réagira avec le noyau: cible correspondant.

σ dépend bien entendu de la vitesse (ou de l'énergie $E = mv^2/2$) du neutron incident et du produit de la réaction . Ainsi $\sigma(E)$ sera différent pour la réaction (n, γ) et (n, p) . L'un des objectifs majeurs de la théorie nucléaire est de donner une expression pour la grandeur de σ.

Revenons maintenant à (1) dont l'intégration conduit à

$$P(z) = e^{-n\sigma z} \tag{2}$$

compte tenude la relation de normalisation

$$\int_0^\infty P(z)\,dz = 1. \tag{3}$$

L'equ. (2) permet de définir un libre parcours moyen

$$\lambda = \int_0^\infty z\,P(z)\,dz = \frac{1}{n\sigma}\ [\text{en cm}] \tag{4}$$

ce qui conduit à son tour à l'introduction de la section efficace <u>macroscopique</u>

$$\Sigma = n\sigma \quad [\text{cm}^{-1}] \tag{5}$$

c'est la surface efficace non pas par noyau mais par cm^3. Définissons encore le <u>nombre de chocs moyens</u> d'un neutron par seconde par :

$$\gamma = \frac{v}{\lambda} = v\Sigma\ (\text{sec}^{-1}) \tag{6}$$

Si le réacteur renferme un mélange homogène de diverses espèces de noyaux , l'on a la règle de melange

$$\nu = \sum_i \nu_i = v \sum_i n_i \sigma_i = v \sum_i \Sigma_i \tag{7}$$

n_i étant la densité du i-ème constituant du réacteur.

Les considérations précèdentes se rapportaient à des processus d'absorption, mais les mêmes raisonnements valent pour touts les autres phènomènes nucléaires. Ainsi on parle de section efficace pour la diffusion (ou chocs) élastique (σ_{ea}), pour la diffusion (choc) inélastique) (σ_i) pour la fission (σ_f), pour la capture de neutrons (σ_c). et pour l'absorption ($\sigma_a = \sigma_c + \sigma_f$) . Tous les concepts introduits jusqu'à présent s'appliquent aux processus précédents: il suffit de mettre l'indice approprié correspondant.

En divisant grosse modo la bande d'énergie E des neutrons en trois régions

1) la région de résonance comprise entre 0 et 10^k e v ,

2) la region de vitesse moyenne de 10 M ev à 0, 5 Mev, et 3) la région des neutrons rapides de $E > 0,5$ Mev, on obtient l'allure suivante de la section efficace de capture $_c$ en fonction de $E = m\, v^2/2$

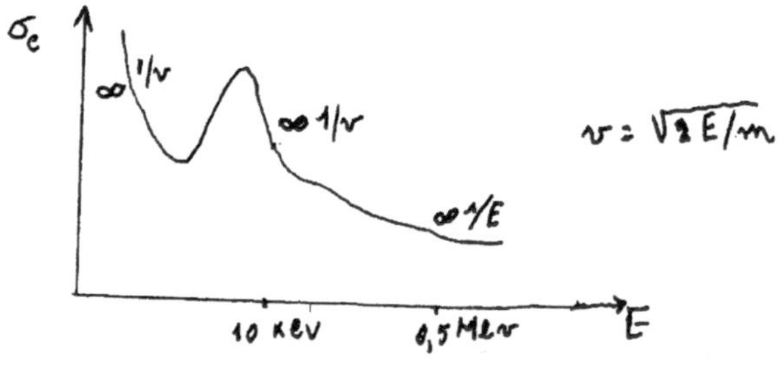

Le réacteur au sein duquel les neutrons évoluent possède une structure dont la composition et la densité varient d'un point à l'autre . En passant de la zone de fission au réflecteur, etc. En outre, il se produit au cours due temps des changements tels que :

- combustion, rechauffement, intervention des bazzes de régulation du fluc neutronique etc. c'est pourquoi les sections efficaces et les fréquences de choc sont fonctions à la fois de la position ($\vec{r}$) et du temps. A fin de ne pas surcharger l'écriture, ces variables r et t seront impicitement supposé dans nos formules sauf avis contraire.

Les divers processus de diffusion affichent des dépendances de l'énergie et de la direction (angle de diffusion. C'est pourquoi il est indispensable de préciser avec soin la nature du processus de diffusion. Pour dresser de façon simple le bilan neutronique, nous allons, à titre de convention, compter tout neutron (neutron dit primaire)qui participe à une réaction nucléaire comme absorbé par le noyau choqué . A sa place, un certain nombre n_s de neutrons secondaires vont appaitre:

neutron primaire

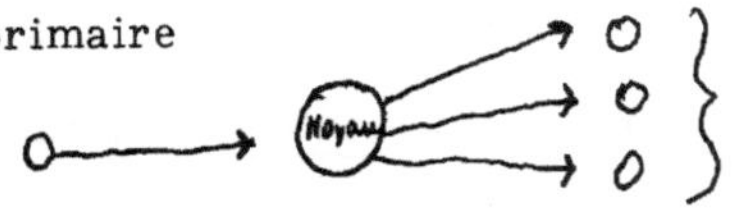

neutrons secondaires

Ce nombre n est nul (n_p = 0) dans le cas de l'absorption. n_p = 1 pour la diffusion et $n_s = \nu$ pour la fission .

Ceci posé, envisageons un neutron d'énergie cinétique E', se déplaçant dans la direction du vecteur unité $\vec{\Omega}$ vers le noyau supposé d'abord ou repos . La Mécanique quantique des processus nucléaires nous donne la probabilité.

$$\mathcal{P}(E', \vec{\Omega}' \to E, \vec{\Omega})\, d E\, d \vec{\Omega} / 4\pi$$

pour que le neutron secondaire engendré soit émis avec une énergie comprise entre E et E + d E dans un cône d $\vec{\Omega}$ autour de $\vec{\Omega}$ (fig. . . .)

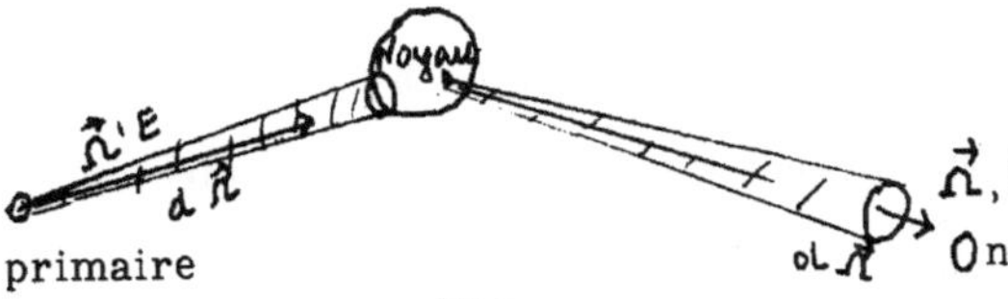

neutron primaire

neutron secondaire

Fig. . .

On peut ainsi définir pour le processus envisagé un <u>nombre de chocs différentiel</u>

$$(8) \qquad (E', \vec{\Omega}', \to E, \vec{\Omega})\, dE\, d\vec{\Omega} = \gamma \times \mathcal{P}(E', \vec{\Omega}' \to E, \vec{\Omega})^{1/}\, 4\pi dE d\vec{\Omega}$$
$$= v' \Sigma (E') \frac{1}{4\pi} \mathcal{P}(E', \vec{\Omega}' \to E, \vec{\Omega})\, dE\, d\vec{\Omega}$$

Comme probabilité, est normé à

(9) $$\frac{1}{4\pi}\iint \mathcal{P}\, d E\, d \vec{\Omega} = 1$$

Le nombre de chocs total, défini par (6), est donc

(10) $$\gamma = \iint \Gamma\, dE\, d\Omega = v'\Sigma\,(E')$$

Dans ce sens

(11) $$\Sigma . \mathcal{P}/4 = n\sigma\, \mathcal{P} / 4\pi$$

est la section efficace différentielle pour la diffusion dans l'angle solide d $\vec{\Omega}$ autour de $\vec{\Omega}$ et dans la bande d'énergie d E autour de E .

S'il s'agit d'une capture réelle de neutron , alors $\mathcal{P}$ ne peut être différent de zéro que si E = 0 , c'est à dire que si aucun neutron n'est émis. On doit donc poser

(12) $$\Gamma_c\, d E\, d\Omega = v'\, \Sigma_c\, (E') \cdot \frac{1}{4\pi}\, \delta(E)\, dE\, d\vec{\Omega}$$

ici δ est la fonction de Dirac.

Par intégration

(13) $$\gamma_c = \iint \Gamma_c\, d E\, d\vec{\Omega} = v'\, \Sigma_c(E')$$

La diffusion élastique est associée à une cession d'énergie par le neutron incident au noyau cible. On peut alors écrire :

(14) $$\Gamma_d\, d E\, d\vec{\Omega} = v'\, \Sigma_d\, (E')\, \frac{1}{4\pi}\, \delta\left[E - (E' - \Delta E')\right] \times d E\, d\vec{\Omega}$$

La valeur de la perte d'énergie $\Delta E'$ (qui dépend de l'angle de diffusion resp. de son cosinus $0 = \Omega . \Omega'$) se calcule par la mécanique du choc.

Pour la fission, on peut admettre une distribution uniforme des neutrons de fission sous toutes les directions spatiales. En outre , le spectre d'énergie de ces neutrons f(E) est, dans une large mesure, indépendant de l'énergie des neutrons, causes de la fission, de sorte que

(15) $\Gamma_f \, dE \, d\vec{\Omega} = v' \Sigma_f \, (E') \frac{1}{4\pi} \, f(E) \, dE \, d\vec{\Omega}$

Comme $\Sigma = \Sigma$ l'on a pour le nombre de chocs total par seconde pour tous les processus mis en jeu

(16) $\Upsilon = v\Sigma = v(\Sigma_d + \Sigma_c + \Sigma_f)$

et pour le nombre des neutrons secondaires produits par ces processus par seconde

(17) $\Gamma(E', \vec{\Omega}' \longrightarrow E, \vec{\Omega}) \, d\,E \, d\vec{\Omega} = (\Gamma_d + \nu \Gamma_f) \, dE \, d\vec{\Omega}$

on $\nu(E)$ est le nombre d'accroissement de neutrons par fission.

L'intégration de (17) donne

(18) $\iint \Gamma \, d\,E \, d\vec{\Omega} = v' \, (\Sigma'_d + \nu' \Sigma'_f)$

Problèmes majeurs en théorie de transport neutronique

Nous voilà maintenant en possession des concepts les plus importants pour l'établissement de l'équation des réacteurs

Avant d'aborder l'analyse mathématique du transport et de la migration des neutrons, je vais brosser un tableau rapide des problèmes majeurs dont la solution incombe à la théorie du transport.

En considérant la variation avec le temps du nombre total des neutrons N dans un système réactif, on voit que (*)

(*) Sans tenir compte de la (n, 2n)

(a) Production par des sources indépendantes + Production par fission - capture - évasion hors du système $= \frac{\partial N}{\partial t}$. ,

où le terme $\frac{\partial N}{\partial t}$, s'il est positif ($\frac{\partial N}{\partial t} > 0$) représente l'accroissement par unité de temps de la population en neutrons du réacteur, s'il est négatif ($\frac{\partial N}{\partial t} < 0$), est la diminution par unité de temps, prise avec le signe négatif .

"Production par fission" signifie l'éxcès du nombre de néutrons libérés par fission sur le nombre de neutrons absorbés en produisant la fission.

Des sources indépendantes telles que les neutrons, dûs aux rayons cosmiques, sont toujours présentes dans un réacteur mais elles sont normalement négligeables comparées aux sources artificielles introduites ou à la production par fission. Ainsi en l'absence de sources artificielles, le premier membre de (1) se réduit à la production par fission .

Supposons maintenant que les dimensions linéaires R du système soient augmentées dans le rapport $r = R' / R$ en devenant $R' = rR$, la composition chimique et la densité demeurant les mêmes.

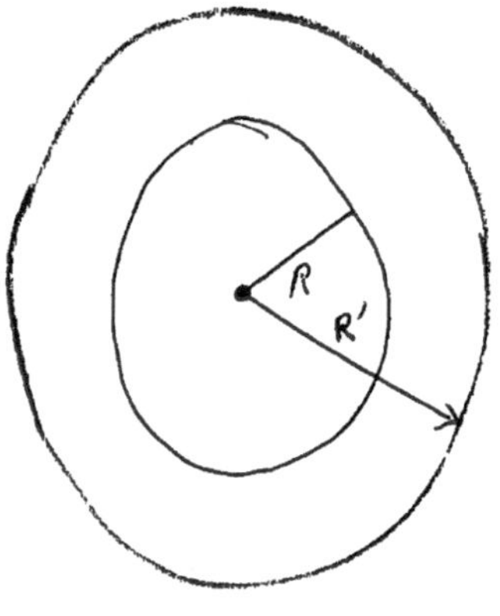

La production de neutrons sera sensiblement proportionnelle au volume occupé pour la matière fissile et variera ainsi par un facteur r^3 .

La capture va changer aussi par un facteur r^3, et comme elle est nécessairement inférieure à la production , elle représentera une fonction constante de cette dernière . L'évasion de neutrons hors du réacteur , sera toutefois sensiblement proportionnelle à la surface frontière, et croitra , ainsi par un facteur r^2. L'équ. (1) fournit alors

$$\text{(b)} \qquad \frac{\partial N}{\partial t} = N\left[F\,r^3 - C\,r^3 - E\,r^2\right] = N\left[A\,r^3 - E\,r^2\right]$$

où figure aussi le facteur N, car la production par fission, la capture et l'évasion, seront toutes proportionnelles à la population neutronique existante. F , C, E et A sont des quantités sensiblement constantes.

L'équ. (b) montre immédiatement qu'il existe une valeur de r soit r_o , telle que par

$$r < r_o ,$$

$$\frac{\partial N}{\partial t} \text{ est négatif } \left(A\,r_o^3 - E\,r_o^2 < 0\right), \quad r_o = E/A$$

en l'absence de sources indépendantes, la densité neutronique diminuera exponentiellement (*)

$$N = N_o e^{-t/T} \qquad T > 0$$

Pour $r > r_o$, toutefois ,

$$\frac{\partial N}{\partial t} \text{ est positif}$$

et la population neutronique va croitre exponentiellement

$$N = N_o e^{t/T} \qquad T > 0$$

(*) Dans les problèmes dépendant du temps , cette décroissance peut n'être qu'approximativement exponentielle ;puisque A et E peuvent dépendre eux mêmes du temps pur, exemple si des fractions de la population neutronique les diverses portions du réacteur avec le temps.

Si $r = r_o$, le réacteur est dit critique , et si r est une dimension linéaire caractéristique, τ_o porte le nom de valeur critique de cette dimention. Pour $r > \tau$ en effet la population croît indéfiniment et à moins que l'état du réacteur varie avec le temps, le résulta risque de solder par une catastrophe.

La détermination de la dimension critique est le problème le plus important de la théorie du transport neutronique . Il arrive aussi qu'on soit confronté, soit avec des problèmes fonctions du tèmps, soit avec des problèmes ou l'èxistence de sources, indépendante de neutrons entre en jeu .

Forme générale de l'équation des réacteurs(Bilan neutronique)

La grandeur fondamentale de la théorie de transport est la densité

$$N(\vec{r},\ t\ ,\ E\ ,\ \vec{\Omega})\ /d/E\ d/\vec{\Omega}/\quad (\text{ en cm}^{-3})\ ; \tag{19}$$

c'est la densité des électrons au point $\vec{r}$, à l'instant t , dont les vitesses $v = \sqrt{2\, e\, E/m}$ sont dans le cône d'ouverture $d\vec{\Omega}$ autour de $\vec{\Omega}$ et adont les énergies sont dans la bande d E autour de E (dont les vitesses sont dans la bande d v autour de v) . Le nombre probable de neutrons dans une élément de volume

$$d\,v = dE\ .\ d\vec{\Omega}\ .\ d\,V \qquad (\vec{v} = v\,\vec{\Omega}) \tag{20}$$

de l'espace des phases v est alors $N(\vec{r},\ t,\ E\ ,\ \vec{\Omega})\ d\,v$. Le bilan pour cet élément se met sous la forme :

$$d\,v\ \frac{\partial N}{\partial t} = \text{gains}\ - \text{pertes} \tag{21}$$

Commençons par le calcul des pertes neutroniques . Définissons le flux de vecteurs par $\vec{\mathcal{J}}$

$$\vec{\mathcal{J}} = \vec{\Omega}\ v\ .\ N(\vec{r}, t,\ E,\ \vec{\Omega}) = \vec{v}.\ N\ . \tag{22}$$

Pour conséquent, la perte nette par fuite hors de l'élément spatial d V se traduira pas

(23) $\nabla\cdot\hat{f}\, d\mathcal{V}z$ - div $\hat{f}$ d-$\mathcal{V}$ = $-\vec{v}\cdot\nabla$ N. d $\mathcal{V}$

En outre, d'après notre convention de p. 33, chaque nêutron qui aura subi un choc quelconque sera tout d'abord compté comme perte. Par (16), l'on a pour la "perte par choc"

(24) $- \Upsilon$ N d$\mathcal{V}$

Venons-en maintenant aux gains. Les neutrons secondaires dûs aux processus nucléaires qui ont lieu dans $d\mathcal{V}'$ = d V d E' d $\vec{\Omega}'$ peuvent parvenir dans ($d\mathcal{V}$ = dV d E d $\vec{\Omega}$). Le nombre de neutrons dans l'élément d $\mathcal{V}$ ' est

(25) $N(\vec{r}, t, E', \vec{\Omega}')$ d E' d Ω' d V. Par (17),

il faudra comptabiliser un gain de

(26) $$d\mathcal{V} \int_0^\infty \int_\Omega N(\vec{r}, t, E', \vec{\Omega}')\, \Gamma(E', \vec{\Omega}' \rightarrow E, \Omega)\, dE'\, d\vec{\Omega}' .$$

Enfin, il a lieu de tenir compte de sources extérieures $S(\vec{r}, t, E, \vec{\Omega})d\mathcal{V}$. Le bilan (21) prendra donc avec (22) à (26), la forme

(27) $$\frac{\partial N}{\partial t} = -\vec{\Omega}\, v\, .\nabla N - \Upsilon\, N + \int_0^\infty \int_\Omega N'\,\Gamma\, dE'\, d\Omega' + S .$$

qui est en principe, l'équation des réacteurs, où nous n'avons pas tenu compte des neutrons retardés.

Dans la fission il se crée entre autres des fragments avec excès de neutrons (*) qui sont des éléments négatogenes. Le noyau créé par

(*) Pour la stabilité d'un noyau, il ne faut pas que le nombre de neutrons N = A-Z exède une certaine limitea. Pour les noyaux légers $N \simeq Z$

fission peut être assez excité pour émettre un neutron immédiatement après sa naissance. Le fragment de fission en question, noyau mere du néutron, a une vie moyenne, déterminée . C'est pourquoi le neutron sera émis en moyenne avec un certain retard après la naissance du fragment de fission au cours de la fission. Les neutrons retardés, ergèndrés de cette façon, se distinguent ainsi des neutrons dits prompts, nés lors de le fission en moins d'énviron 10^{-14} sec. On distingue actuellement 6 groupes différents de neutrons retardés .

Comment tenir compte de ces neutrons retardés dans l'équation des réacteurs? Revenons pour cela à l'equ. (15)

$$\Gamma_f \, dE \, d\vec{\Omega} = v' \Sigma_f (E') \frac{1}{4\pi} f(E) \, dE \, d\vec{\Omega} \tag{15}$$

La fraction $[1 - \beta(E')]$ de ces neutrons sera émise prompteme , la fonction $\beta(E')$ de manière rétardées . L'on aura $\beta = \sum_i \beta_i$ où β_i désigne le nombre des neutrons retardés du i-éme groupe rapporté à 1 neutron prompt . Le spectre de fission $f_o(E)$ des neutrons prompts est représentable entre de larges limites, par la formule

$$f_o(E) = 0,48 \, e^{-E} \, \mathrm{sh} \sqrt{2E} \quad (E \text{ en Mev}) \tag{28}$$

avec la condition de normalisation

$$\int_o^\infty f_o(E) \, dE = 1 . \tag{29}$$

Le nombre de secondaires nés par seconde à la suite des divérs processus envisagés sera dès lors, au lieu de (17),

$$\Gamma(E', \vec{\Omega} \to E, \vec{\Omega}) \, dE \, d\vec{\Omega} = (\Gamma_d + \Gamma_i + \nu' \Gamma_f) \, dE \, d\vec{\Omega} \tag{30}$$

et

$$\iint \Gamma \, dE \, d\vec{\Omega} = v' (\Sigma'_d + \Sigma'_i + \nu' \Sigma'_f) . \tag{30-1}$$

D'après (30), on peut séparer le gain par entrée dans $d\mathcal{V}$

(31) $$d\mathcal{V} \int_0^\infty \int_\Omega N(\vec{r}, t\ E', \vec{\Omega})\ (\Gamma_d + \Gamma_i)\, dE'\, d\vec{\Omega}'$$

du gain par fission

(32) $$d\mathcal{V} \int_0^\infty \int_\Omega N(\vec{r}, t, \vec{E}', \vec{\Omega}')\, \nu(E')\, \Gamma_g\, dE'\, d\vec{\Omega}'.$$

La densité de probabilité Γ_f définie par (15) a pour expression complète

(33) $$\Gamma_f = v'\ \Sigma_f(E')\ \frac{1}{4\pi} \left\{ [1 - \beta(E')]\ f_o(E) + \right.$$
$$\left. + \sum_i \beta_i(E')\ f_i(E) \right\}$$

Pour le nombre des neutrons prompts, l'on tire de (32), avec (33),

(33-1) $$\frac{d\mathcal{V}}{4\pi}\ f_o(E) \int_0^\infty \int_{\vec{\Omega}} [1 - \beta(E')]\, \nu(E')\ \Sigma_f(E')\ v'\ N(\vec{r}, d, E', \vec{\Omega}')$$
$$\times\ dE'\, d\vec{\Omega}'$$

En portant le second terme du second membre de (33) pour Γ_g dans (32), on obtient les neutrons retardés. Il convient, toutefois, d'observer que les neutrons retardés, engendrés à l'instant t, sont dus aux noyaux mères qui sont nés des fissions ayant eu lieu à un instant antérieur $t' < t$. Envisageons, à titre d'exemple le i-ème groupe de noyaux mères, ayant la constante radioactive λ_i. Leur concentration a pour expression

(34) $$G_i(\vec{r}, t) = G_i\ (\vec{r}, t')\ e^{-\lambda_i(t-t')}$$

En admettant que les fragments de fission qui résultent de fissions ayant lieu dans dV, demeurent dans dV, le nombre $d\ C_i(\vec{r}, t')dV$

de noyaux mères engendrés pendant l'intervalle de temps dt' dans d V est déterminé par le nombre total de fissions dans d V dt' il faut mettre dans (32) t ' à la place de t et intégrer sur E et $\vec{\Omega}$. Il vient avec (30-1)

$$(35) \quad d\, C_i(\vec{r}, t')\, d\, V = d\, V\, dt' \int_0^{\infty} \int_{\Omega} \beta_i(E')\, \nu\, (E')\, \Sigma_f(E) \times v'\, N(\vec{r}, t', E', \vec{\Omega}')\, d\, E'\, d\, \vec{\Omega}' .$$

Dans l'intervalle dt , il se produit

$$(36) \quad \left(\frac{d\, G_i}{dt}\right. = \lambda_i\, G_i\, (\vec{r}, t')\, e^{-\lambda_i (t-t')}$$

des intégrations par sec dont chacune conduit à un neutron retardé. Par conséquent, le nombre de neutrons rétardés, nés à l'instant t, dans l'élément de volume dV par seconde, sera au total

$$(37) \quad d\, v .\, \lambda_i \int_{-\infty}^{t} e^{-\lambda_i (t-t')} \int_0^{\infty} \int_{\Omega} \beta_i(E')\, \Sigma_f(E') \times v'\, N\, (\vec{r}, t', E', \vec{\Omega}')\, d\, E'\, d\vec{\Omega}'\, dt'$$

Comme les neutrons retardés naissent avec la distribution d'énergie $f_i(E)$, et que leur distribution angulaire peut passer pour isotrope, c'est la fraction $f_i(E)\, d\, E\, d\, \vec{\Omega} / 4\pi$ qui pénètre dans l'élément de l'espace des phases . On obtient ainsi comme contribution de m groupes de neutrons retardés .

$$(38) \quad \frac{d\, v}{4\pi} \sum_{i=1}^{m} \lambda_i\, f_i(E) \int_{-\infty}^{t} e^{-\lambda_i (t-t')} \int_0^{\infty} \int_{\Omega} \beta_i(E')\, \nu\, (E')\, \nu \times \Sigma_g(E')\, v'\, N(\vec{r}, t, E', \vec{\Omega}')\, dE'\, d\vec{\Omega}'\, dt'$$

Dès lors, l'équ. (27) se mettra par (31), (33-1) et (38) , sous la forme.

$$\frac{\partial N}{\partial t} = -\vec{v}\,.\,\nabla N - \gamma N + \int_0^\infty \int_\Omega N' \, (\Gamma_d + \Gamma_i)\, dE'\, d\Omega'$$

$$(39) \qquad + \frac{1}{4\pi} \; f_o(E) \int_0^\infty \int_\Omega (1-\beta')\, \nu' \Sigma'_f \; v' \; N' \; dE' \, d\vec{\Omega}'$$

$$+ S + \frac{1}{4\pi} \sum_{i=1}^{m} \lambda_i \; f_i(E) \int_{-\infty}^{+} e^{-\lambda_i (t-t')} \int_0^\infty \int_\Omega \beta_i \Sigma_f v' N'(t')$$

$$\times \quad dE' \, d\vec{\Omega}' \; dt'$$

que j appelerai équation générale des réacteurs . Pour simplifier l'écriture, j'ai indiqué par des primes (') dans N, Σ, etc, les variables sur lesquelles il faut intégrer .

Si le réacteur contient divers matériaux fissiles, k=1, 2, . . . r , il faut poser dans (39) ν^k, β_i^k , λ_i^k , f_i et Σ_i^k et sommes par rapport à k.

I. Cas particuliers : Une seule matière fissile. et $\partial\beta/\partial E' = 0$

Supposons maintenant 1°) qu'il y a essentiellement une seule matière fissile pour entretenir la chaine de neutrons par exemple U^{235};
2°) Si , en outre , les fissions sont dues pour une part prépondérante à des neutrons d'une petite bande d'énergie (réacteur à neutrons thermiques respe réacteur à neutrons rapides) , alors la contribution β_i des fragments de fission peut être supposée comme étant largement indépendante de la distribution énergétique des neutrons. $\left(\frac{\partial \beta_i}{\partial E'} = \frac{\partial \beta(E')}{\partial E'} = 0\right)$

Pour la contribution des neutrons retardés dans (39) au nombre de neutrons contenus dans l'élément considéré de l'espace des phases, on peut poser en outre d'une façon abrégée

$$(40) \qquad \frac{1}{4\pi} \sum_{i=1}^{m} \lambda_i \; f_i \; (E) \,.\, C_i \,(\vec{r}_1 \; t)$$

où $C_i(\vec{r}, t)$ est la densité des noyaux mères correspondants . La dérivation de (37) par rapport au temps, fournit alors un système de m équations intégro-différentielles qui sont à joindre à (39) :

$$(41) \quad \frac{\partial C_i}{\partial t} = - \lambda_i C_i + \beta_i \int_0^\infty \int_\Omega \nu' \Sigma_f' N' \, dE' \, d\vec{\Omega}' , \quad (i = 1, 2, 3, \ldots m)$$

L'équ. (39) se met, dès lors, sous la forme

$$(42) \quad \frac{\partial N}{\partial t} = - \vec{v} . \nabla N - \gamma N + \int_0^\infty \int_\Omega N (\Gamma_d + \Gamma_i) \, dE' \, d\vec{\Omega}' + \delta$$

$$+ f_o(E) \, (1 - \beta) \int_0^\infty \int_\Omega \nu' \Sigma_f' v' N' \, dE' \, \frac{d\vec{\Omega}'}{4\pi} + \frac{1}{4\pi} \sum_{i=1}^m \lambda_i f_i C_i$$

Elle porte aussi le nom d'équation du transport ou d'équation de Boltzmann en théorie de transport des neutrons.

Pour abréger, on peut introduire l'opérateur de pertes et de diffusion par (*)

$$(43) \quad \hat{H} = \vec{v} . \nabla + \gamma - \int_0^\infty \int_\Omega dE \, d\vec{\Omega}' \, (\Gamma_d + \Gamma_i) \quad \text{en sec.}^{-1}$$

et l'opérateur de production .

$$(44) \quad \hat{K}_\ell = \frac{1}{4\pi} f_\ell(E) \int_0^\infty \int_\Omega dE' \, d\vec{\Omega}' \, \nu' \, \Sigma_f' \, v' \quad \text{en sec.}^{-1}$$

où il convient de poser $\ell = 0$ pour les neutrons prompts et $\ell = i = 1, 2, \ldots m$, pour les retardés. Dès lors, les équations (41) et (42) prennent la forme

$$(45) \quad (a) \quad \frac{\partial N}{\partial t} = (1 - \beta) \hat{K}_0 N - \hat{H} N + \frac{1}{4\pi} \sum_{i=1}^m \lambda_i f_i C_i + \delta$$

$$(b) \quad f_i \frac{\partial C_i}{\partial t} = 4\pi \beta_i \hat{K}_i N - \lambda_i f_i C_i$$

(*) Les opérateurs sont indiqués par un chapeau $\wedge$, par ex. $\hat{K}$

Les équations (39) respt. (45) déterminent univoquement la densité $N(\vec{r}, t, E, \vec{\Omega})$, compte tenu des conditions initiales et aux frontières données I- Sur une frontière délimitant le vide $(\vec{r} = \vec{r}_f)$ il ne peut pas y avoir de flux de neutrons venant du vide pour pénétrer dans l'intérieur du réacteur (direction $\vec{\Omega}_i$).

(46) $$N(\vec{r}_f, t, E, \vec{\Omega}_i) = 0$$

II- En outre, dans le cas dépendant du temps, il faut se donner les grandeurs :

(47) $N(\vec{r}, t = 0, \vec{E}, \vec{\Omega})$ et $C_i(\vec{r}, t = 0)$. III Enfin, N doit être d'après son sens physique, au sein du réacteur partout continu, fini et positif .

Si le réacteur est stationnaire (critique) les derivées temporelles sont nulles dans (45)

$$\frac{\partial N}{\partial t} = \frac{\partial C_i}{\partial t} = 0 .$$

En portant (45 b) dans (45-a) il vient alors

(48) $$(\hat{K} - \hat{H})\, N + S = 0 \quad \text{soit} \quad \hat{L}\, N = - S \quad \text{avec} \quad \hat{L} \equiv \hat{K} - \hat{H}$$

L'opérateur $\hat{K}$ doit, d'après (44), être formé alors à la place de f(E) avec le spectre f(E) global défini pour β_i = const, par

(49) $$f(E) = (1 - \beta)\, f_0 + \sum_{i=1}^{n} f_i\, \beta_i \;, \quad \text{avec} \quad \int_0^\infty f(E)\, dE = t$$

II. Cas général : plusieurs matières fissiles

Dans les équations qui précèdent, j'ai supposé, d'une part que les neutrons retardés dans le réacteur ne proviennent que de la fission d'une seule matière fissile et que d'autre part, la fraction β_i avec laquelle les neutrons retardés du groupe i apparaissent parmi les neutrons de fission est indépendante de l'énergie des neutrons de fission . La première hypothèse tombe en défaut lorsque le réacteur contient plusieurs matières fissiles en quantités comparables. Dans ce dernier cas, il faut, ainsi que je l'ai indiqué, mettre des indices de substance k dans l'intégrale de fission et sommer par rapport aux divérses matières fis siles. Dans le réacteur thermique à haute concéntraction de substances "breedes" ou sur régéneration il est dans certaines circonstances nécéssaire d'ajouter une intégrale de fission aussi pour ces substances fertiles .

Au cas où dans un réacteur tel que celui dont il vien d'étre question, les fissions aussi, bien thermiques que rapides, jouent un rôle, les valeurs des β_i sont à prendre pour l'énergie correspondante . Dans le passage de (39) (p.32) à (45) (p.36) il n'est plus légitime alors de faire sortir β et β_i de l'intégrale de fission. Si l'indice k sert à désigner les divérses matières , il faut remplacer dans (45) (p.31) les opérateurs $(1 - \beta)\, \hat{K}_o$ et $\beta_i \cdot \hat{K}_i$ par

$$(50) \qquad \hat{K}^{bt} = \frac{1}{4\pi} \sum_k f_o^{(k)}(E) \int_0^\infty \int_\Omega dE' \, d\vec{\Omega}' \, [1 - \beta^{(k)}(E')]$$

$$\times \nu^{(k)}(E') \, \Sigma_f^{(k)}(E') \, v' = \sum_k \hat{K}_o^{(k)}$$

et

$$(51) \qquad \hat{K}_i^{bt} = \frac{1}{4\pi} \sum_k f_i^{(k)}(E) \int_0^\infty \int_\Omega dE' \, d\vec{\Omega}' \, \beta_i^{(k)}(E')$$

$$\times \nu^{(k)}(E') \, \Sigma_f^{(k)}(E') \, v' = \sum_k \hat{J}_i^{(k)}$$

Les équations (45) prennent alors la forme

$$(52)\quad \begin{cases} \dfrac{\partial N}{\partial t} = \hat{J}_0^{tot} - \hat{H}\, N + \dfrac{1}{4\pi} \displaystyle\sum_{i=1}^{m} \sum_{k} \lambda_i^{(k)} f_i^{(k)} C_i^{(k)}\, t\, S, \\ f_i^{(k)} \dfrac{\partial C_i^{(k)}}{\partial t} = 4\pi\, \hat{K}_i^{(k)} N - \lambda_i^{(k)} f_i^{(k)} C_i^{(k)} \end{cases}$$

L'équation générale des réacteurs (39) (p.30) peut alors se mettre sous la forme abrégée

$$(53)\quad \frac{\partial N}{\partial t} = \hat{K}_0^{tot}\, N + \sum_{i=1}^{m} \sum_{k} \lambda_i^{(k)} \int_{-\infty}^{t} e^{-\lambda_i^{(k)}(t-t')}\, \hat{K}_i^{(k)} N'(t')\, dt' - \hat{H}\, N + S$$

L'équation (48) (p.32) pour le réacteur stationnaire s'écrit alors sous cette forme plus exacte

$$(54)\quad (\hat{K}^{tot} - H)\, N + S = 0,$$

où

$$(55)\quad \hat{K}^{tot} = \hat{K}_0^{tot} + \sum_{i=1} \hat{K}_i^{tot}$$

Pour une seule matière fissile et pour des β_i constants, on retombe à partir de (54 , avec (49), de nouveau sur (48) (p. 32)

Réduction de l'équation des réacteurs fonction du temps à l'équation indépendante du temps ou équation stationnaire.

L'équation des réacteurs (45) (p 31) peut se ramener à l'onde d'une transformation de Laplace à une forme analogue à celle de l'équation (48) (p32) à condition que les sections efficaces Σ soient indépendantes du temps et pourvu qu'aucun changement dans le temps ne se produise dans le milieu réactionnel . Pour cela soumettons, le système (45) (p31) à une transformation de Laplace en posant

$$(56)\quad \mathcal{L}\left\{ X(t) \right\} = \int_0^{\infty} X(t)\, e^{-pt}\, dt \equiv \tilde{X}(p)$$

Comme l'on a par ailleurs

$$\mathcal{L}\left\{\frac{dx}{dt}\right\} = -\chi(o) + p\tilde{\chi}(p) \quad , \tag{57}$$

il vient

$$-N_o + p\tilde{N} = (1-\beta)\hat{K}_o\tilde{N} - \hat{H}\tilde{N} +$$

$$+ \frac{1}{4\pi}\sum_{i=1}^{m}\lambda_i f_i \tilde{C}_i + \tilde{S} \, , \tag{58}$$

$$f_i\left[-C_{io} + p\tilde{C}_i = 4\pi\beta_i \hat{K}_i N - \lambda_i t_i \tilde{C}_i\right.$$

où $N_o = N(\vec{r}, t = 0, E, \vec{\Omega})$ et $N = N(\vec{r}, p, E, \Omega)$

$C_{io} = C_i(r, t = o)$ $\quad C_i = C_i(\vec{r}, p)$.

En éliminant $\hat{C}_i$, eu égard à (49) (p 32) , l'on a

$$\left[(\hat{K} - \sum_{i=1}^{m}\frac{p}{p+\lambda_i}\beta_i\hat{K}_i) - (\hat{H}+p)\right]\tilde{N} + (N_o + \frac{1}{4\pi}\sum_{i+1}^{m}\frac{\lambda_i}{p+\lambda_i}f_i C_{io} + \tilde{S}) = 0 \tag{59}$$

ou encore en désignant les grandeurs entre parenthèse par

$\mathcal{K}$, $\mathcal{H}$ et $\hat{\mathcal{S}}$

$$(\mathcal{K} - \mathcal{H})\tilde{N} + \hat{\mathcal{S}} = 0 \; ; \tag{60}$$

c'est une équation du type de (48) (p. 32) avec p comme paramètre (valeur propre) .

Chapter II <u>La fonction d'influence ou d'importance (importance fonctio</u>

1. Compte tenu des conditions aux limites et des conditions initial correspondantes , les solutions de l'équation générale des réacteurs (39) p. 30 ou (45) (p. 31) décrivent la distribution de la densité néutronique

$$N(r, t, E, \Omega)$$

comme fonction de la postion $\vec{r}$, du temps t, de l'énergie cinétique E et de la direction de , vol $\vec{\Omega}$. C'est là la description différentielle par des équations aux derivées partielles du budget neutronique. Or, en théorie cinétique, des réacteurs dont la mission est de livrer les fondements théorique pour la commande, le contrôle et la régulation de réacteurs, on ne s'intéresse guère à la répartition neutronique en detail. Ce qui importe davantage ce sont des relations intégrales qui decrivent le comportement du réacteur dans son ensemble. Un grandeur de ce genre qui caractérise cette allure globale serait par exemple le nombre de fission, par seconde rapporté au réacteur entièr.

Si le réacteur est homogène et nu (c'est à dire sans réflecteur), il est possible d'établir une cinématique simple des réacteurs en admettant simplement que l'allure de N en un endroit donné du réacteur, tel que le centre (ou coeur) du réacteur, est représentatif pour l'ensemble de l'allure du réacteur tout entièr .

En effet l'égalisation (ou le retour à l'équilibre) de perturbations de densité se fait si rapidement dans le sein du réacteur que la forme de la distribution du flux dans le réacteur reste presque inaltérée dans la plupart des cas pratiques tandis que sa grandeur où sou intensité subit de fortes variations. De ce point de vue, on peut imaginer la densité neutronique du réacteur comme composée d'un facteur dépendant du temps et d'un facteur indépendant de t.

$$N(\vec{r}, t, E, \vec{\Omega}) = T(t) \,.\, F(\vec{r}, E, \vec{\Omega}) \,.$$

On peut alors prendre la fonction T(t) pour mesure relative pour le comportement du réacteur qui serait par exemple proportionnelle à la puissance du réacteur.

d'indiquer n'est pas approprie. Il en est plus particulièrement ainsi pour les réacteurs à reflecteurs dans lesquels le budget dans le réflecteur diffère sènsiblement de celui qui règne dans le coeur du réacteur. Dans cette conception plus large de la cinétique des réacteurs la fonction d'influence (ou importante fonction des Anglo-Saxons) joue un rôle éminent et décisif.

2. Définition de la fonction d'influence

Le réacteur contient un certain nombre de matériaux doués de diverses propriétés nucléaires. Leur répartition est sujette entre autres à des variations temporelles qui se font sentir en partie pendant des periodes brevès(régulation, démarrage, arrêt), en partie sur des intervalles de temps plus longs (épuisement des matières fissiles) , accumulation des poisons, breeding). A cela viennent s'ajouter le rôle joué par les grandeurs d'état telles que la température , la pression, et la densité. Les diverses zones du réacteur jouent dans la réaction ramifiée, des rôles distincts. Des neutrons qui se trouvent dans les zones frontières se perdent plus aisément par évasion vers l'extérieur sans provoquer de fission, que les neutrons évoluant dans le coeur. En outre, l'effet d'une zone déterminée sur les neutrons varie avec l'énergie de ceux-ci car le rendement réactionnel dépend des noyaux qui s'y trouvent ainsi que de l'énergie des neutrons

En principe, on peut décrire l'importance d'un élément de volume déterminé dans le réacteur en donnant sa position, les matières présentes et les grandeurs d'état. Il est, toutefois, physiquement plus raisonnable d'exprimer cette importance par le budget neutronique. On peut caractériser l'endroit évisagé par exemple par le rôle d'un neutron qui s y trouve, et ayant une vitesse (énergie E) et une direction déterminée $\vec{v} = \vec{\Omega} v = \vec{\Omega}\sqrt{2E/m}$, en considération des réactions en chaîne à venir . On parvient de la sorte à une fonction définie pour tout le volume du réacteur, de la position , du temps, de l'énérgie et de la direction, fonction qui porte le nom de fonction d'influence (importance fonction des Anglo-Saxons). Il est à souligner des maintenant que cette fonction d'influence n'a pas pour mission de caractériser la distribution de neutrons régnant momentanément dans le réacteur qui dépend en effet des conditions initiales données , a mais plutôt l'état du réacteur tant géométrique que matériel (nature, etc. des matériaux)

La fonction d'importance a reçu ce nom parce que le nombre total de neutrons " filles" qu'un neutron initialement introduit fournira au total au réacteur en chaîne est une mesure de l'"importance" qu'a le neutron initial pour entretenir la réaction ramifiée. Un neutron introduit sur la frontière d'un réacteur n'a pas beaucoup de chance de laisser beaucoup de descendants dans le réacteur parce que lui et sa progéniture risquent de d'évader . C'est précisement ce que la fonction d'influence prévoit ; sur la frontière du réacteur le flux et par suite l'importance est très petite.

Il est utile pour les raisonnements suivants, de raisonner sur un réacteur critique , et cependant exempt de neutrons . Cette conception abstraite permet de se faire une représentation concrète car l'on n'est pas obligé, de la sorte, de distinguer dans le réacteur divers groupes de neutrons comme étant constamment distincts . Cette introduction des réacteurs initialement exempt de neutrons ne restreint pas la généralité de nos raisonnements et peut s'étendre aisément aussi à des réacteurs critique ayant un nouveau neutronique arbitraire. A vrai dire, cette friction est rendue possible par le fait que les neutrons présents dans le réacteur ne se gênent pas les uns les autres en raison même de leur faible densité et qu'on peut négliger les chors entre neutrons.

Afin de préciser le cocept d'influence , plaçons par la pensée, dans le réacteur critique excempt de neutrons au point $\vec{r}_o$, S_o neutrons au total qui appartiennent au princeau $(E_o, \vec{\Omega}_o)$:

$$\text{II.(1)} \qquad S(\vec{r}_o, E_o, \vec{\Omega}_o) = S_o \, \delta(\vec{r}-\vec{r}_o) \, \delta(E-E_o) \delta(\vec{\Omega} - \vec{\Omega}_o)$$

Notons que la fonction d'influence se rapporte à <u>un</u> neutron . Nous considérons pour l'instant un nombre Q_o suffisamment grand pour ne pas avoir à tenir compte des fluctuations. Je procederai par la suite a la normalisation à un neutron)

Ces neutrons se répartissent par diffusion à travers le réacteur et au bout d'une période assez longue seront consommés par evasion et par absorption. En même temps un certain pourcentage des absorptions est productif et il se produit un niveau de puissance déterminé, différént de zéro dans le réacteur. Ce niveau va dépendre de la position, de l'énergie et de la direction des neutrons de "démarrage" (1) et il définit la fonction d'influence (à une constante de normalisation près).

Suivons , pour préciser les idées, le destin des neutrons de démarrage introduits dans notre réacteur. Initialement le réacteur était exempt de neutrons et l'équation du réacteur stationnaire 1-48) p.32 , à savoir

$$\text{(I-(48)} \qquad (\hat{K} - \hat{H}) \, N + S = 0$$

était satisfaite en vertu de N = 0. Démarrons alors à l'instant t=0 avec (2-1) et désignons la densité de ces électrons par $N_o(\vec{r}, t, E, \vec{\Omega})$ avec

$$N_o(\vec{r}, t, = 0) \; E, \vec{\Omega}) = S$$

Abstraction faite des neutrons retardés, cette distribution obéit , par (I-45) (de pag. 31)

$$\text{(II-2)} \qquad \frac{\partial N_o}{\partial t} = - K N_o$$

Comme je n'envisagerai jusqu'ici que des neutrons de démarrage, le terme de fission est supprimé de même S = 0 pour t > 0). Au bout d'un temps suffisamment long , ces neutrons seront consommés par évasion et par absorption :

$$N(\vec{r},\ t. = \infty\ ,\ E\ ,\vec{\Omega}) = 0$$

Intégrons (II-2) par rapport à t et définissons par

(II. 3) $$\frac{1}{\ell} \int_0^\infty N_o \, dt = \bar{N}_o$$

une probabilité de séjour des neutrons de source (à la normalisation près), en désignant par ℓ la 'vie moyenne' des neutrons. En intégrant (II-2) d'après (II-3) par rapport au temps, il vient compte tenu des valeurs de N pour t = 0 et t = ∞ ,

$$\int_0^\infty \frac{\partial N_o}{\partial t} dt = N_o(\infty) - N_o(0) = - S = - \hat{H}\ \bar{N}_o\ .\ \ell$$

soit

(II-4) $$S = \hat{H}\ \bar{N}_o\ \ell\ .$$

Cette relation peut s'interprét de la façon suivante. La répartition $\bar{N}_o$ se trouve en quelque sorte "rassemblée" par l'opérateur $+\hat{K}$ de sorte qu'il résulte une distribution finale pronctuelle S. La signification de la relation (II-4) ressort plus particulièrement en intégrant (II-4) sur le volume du réacteur ainsi que par rapport à toutes les énergies et à toutes les directions [(par (7), (23) et (10)]

(II. 5) $$\iiint S \, d E \, d \vec{\Omega} \, d V = \iiint \hat{H}\ \bar{N}_o\ \ell \, d E \, d \Omega \, d V$$

$$= S_o = \ell \iiint_{V\,0\,\Omega}^{\infty} dE \, d\vec{\Omega} dV \left[v.\nabla + \gamma - \int_0^\infty \int_\Omega dE' d\Omega' (\Gamma_d + \Gamma_i)\ N \right]$$

Evasion

(II.5) $$S_o = \ell \frac{\overbrace{\int dE \int d\Omega \int (\vec{\mathcal{F}}_o \, d\vec{F})}^{\text{Evasion}} + \text{absorption}}{\ell \iiint v\,(\Sigma_e + \Sigma_f)\,\overline{N}_o \, dE\, d\vec{\Omega}\, dV}$$

Cette équation de bilan montre que les neutrons introduits, en partie s'évadent , à travers la frontière F du réacteur, en partie sont absorbés. Il résulte (II.4) que $\overline{N}_o$ est une fonction des conditions de démarrage où d'ignition".

$$\overline{N}_o = \overline{N}_o(\vec{r}, E, \Omega, \vec{r}_o, E_o, \vec{\Omega}_o)$$

Intégrée sur E et $\vec{\Omega}$, cette fonction permet d'indiquer le nombre des absorptions ayant eu lieu en $\vec{r}$:

(II-6) $$\ell d V \iint v\,(\Sigma_e + \Sigma_f)\,\overline{N}_o \, d E\, d\vec{\Omega}$$

Une partie de cas absorptions est productive et l'on a pour la densité des neutrons de fission produits à l'endroit de leur naissance :

(II.7) $$S_1(\vec{r}, E, \vec{\Omega}, \vec{r}_o, E_o, \vec{\Omega}_o) = \ell \nu f(E) \iint v\, \Sigma_f \,\overline{N}_o\, d E\, d\vec{\Omega}$$

Désignons la distribution de cette première génération par

$$N_1(\vec{r}, t, E, \vec{\Omega}, \vec{r}_o, E_o, \vec{\Omega}_o) ;$$

elle dépend aussi du temps car les fissions ont eu lieu à des instants différents et elle obéit à l'équation II-2 . L'intégration par rapport au temps d'après (II.3) conduit à une probabilité de présence $\overline{N}_1$. Le nombre de fissions produits par ces neutrons peut se calculer d'après (II.6) et finalèment on obtient de maniere analogue à (II-7) , la répartition de source de la deuxième génération S_2 .

On peut procéder de proche en proche de cette manière. A la distribution extrême de source (II-1) se substitue une distribution interessant tout le réacteur au cours des générations qui se relayent par egalisation de diffusion . Cette distribution ne peut toutefois être , lorsqu un nombre suffisant de générations ne sont succédé qu'une distribution de neutrons stationnaire, compatible avec (I-4) à savoir .

(I-48) $$(\hat{K} - \hat{H})\ N + S = 0$$

Cette équation est avec S = 0 , une équation homogène et par suite, par un théorème classique, la distribution de neutrons stationnaire dans le réacteur, n'est définie qu'à une constante près qui caractérise le niveau de puissance. Le niveau de puissance d'un réacteur critique est fonction de l' histoire antérieure, dans notre cas du processus de mise à feu ou de démarrage.

Il résulte, de ce que je viens dire, la distribution asymptotique N_∞ doit admettre la même <u>forme</u> que les fonctions propres du réacteur stationnaire $N(\vec{r}, E, \vec{\Omega})$. <u>L'amplitude</u>, par contre , est une fonction des conditions de source seule :

$$A = A(\vec{r}_0, E_0, \vec{\Omega}_0) .$$

On peut donc poser

(II-8) $$\overline{N}_\infty = A(\vec{r}_0, E_0, \vec{\Omega}_0)\quad N(\vec{r}, E, \vec{\Omega}) ,$$

et l'on obtient par intégration et division par S_0, une fonction sans dimension, normée à <u>un</u> neutron

(II.9) $$N^+ (r_0, E_0, \vec{\Omega}_0) = \frac{1}{S_0} \iiint \overline{N}_\infty \, d E \, d \vec{\Omega} \, d V$$

Cette fonction N^+ représente la teneur en neutrons du réacteur critique qui s'établit après la mise à feu avec un neutron dans les conditions $(\vec{r}_0 , E_0, \vec{\Omega}_0)$. Par definition, N^+ peut donc se concevoir comme fonction d'influênce (on d'importance), mise à part une constante de normalisation. Nous allons, dans un instant démontrer que cette fonction d'influence est bien l'adjointe de la densité neutronique N.

Jusqu'à présent nous n'avons analysé l'influence que pour le réacteur exactement critique, exempt de neutrons. Si le réacteur critique possède une puissance déterminée, différente de zéro, les raisonnements se font d'une manière complètement analogues aux précédents. L'introduction de neutrons dans le réacteur, perturbe la distribution neutronique régnante jusqu'à ce qu' après une période suffisamment longue, un état d'équilibre finisse de nouveau par s'établir. Il va de soi que le niveau de puissance, en fonction des conditions, sous lesquelles les neutrons furent introduits, va se trouver augmenté. Dans ce cas donc, l'argumentation de puissance s'établissant asymptotiquement qui sert à definir la fonction d'influence.

2. L'équation d'influence fonction du temps

Une fois définie la fonction d'influence et mise en évidence sa signification physique, reste à rechercher l'équation à lacuelle elle satisfait Pour cela, on peut s'appuyer sur un théorème de conservation pour l'influence.

L'influence de toute génération individuelle doit être égale à l'influence des neutrons de démarrage.

Ce théorème de démarrage résulte des raisonnements précédents dans lesquels, nous avons poursuivi les générations de neutrons depuis la mise à feu jusqu'à l'état d'équilibre. Ou peut en effet interrompre par la pensée le processus à une génération quelconque. Si donc on fixe la distribution de cette génération et qu'on mette de nouveau à feu le réacteur vide exactement avec cette distribution, le même résultat final asymptotique doit s'établir qu'en laissant se dérouler le processus initial.

Insérons dans le réacteur vide, au point $\vec{r}$, au total S_o neutrons appartenant au pinceau $(E.\ \vec{\Omega})$. L'influence de chaque neutron est d'après (II.9) proportionnelle à

$$N^+(\vec{r}, E, \vec{\Omega}) .$$

Parmi S_o neutrons, un nombre S_o' égal à

(II.-10) $$S_o' = S_o(1 - \Sigma|d\vec{r}|$$

atteignent le point $\vec{r} + d\vec{r}$, puisque, par (I.2) de p.

$$P(r) = n\sigma.\ e^{-n\sigma r} = \Sigma e^{-\Sigma r}$$

et par suite

$$P(\vec{r} + dv) = \Sigma e^{-\Sigma(\vec{r} + d\vec{r})} = P(\vec{r}) \quad (1 - \Sigma(dr))$$

Ces S_o' neutrons portent avec eux une influence égale à (fig. 3)

(II.11) $$S_o' N^+(\vec{r} + d\vec{r}, E, \vec{\Omega}) = S_o(1 - \Sigma\, dr)\, N^+(\vec{r} + d\vec{r}, E, \vec{\Omega})$$

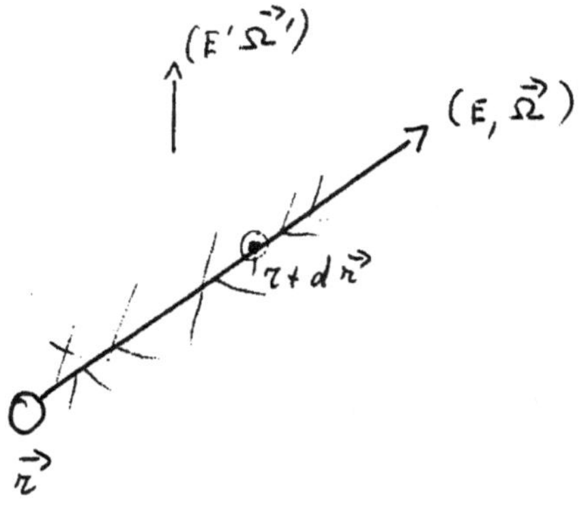

Fig. 3

Les neutrons d'énergie E qui partent du point $\vec{r}_0$ dans le pinceau $(E, \vec{\Omega})$, sont diffusés hors de ce pinceau sur le chemin qui mène au point voisin $\vec{r} + d\vec{r}$. Une partie d'entre eux se retrouve dans le pinceau diffusé $(E', \vec{\Omega}')$

Sur le parcours $d\vec{r}$, $S_0\, d\Sigma\, dr$ neutrons ont subi une collision. Chacun d'entre eux transporte avec lui une influence

$$N^+(\vec{r} + \varepsilon d\vec{r}, E', \vec{\Omega}')$$

où

$$0 \leq \varepsilon \leq 1,$$

puisque la collision a eu lieu sur le parcours $\vec{r}$ et $r + d\vec{r}$ (fig 2). Il nous faut encore la probabilité $\wp(E, \vec{\Omega} \to E', \vec{\Omega}')$ pour que l'un de ces neutrons aboutisse dans le pinceau $(E', \vec{\Omega}')$. D'après (I.7). de p. 20. , cette probabilité

(I. 8) $$\wp(E, \vec{\Omega} \to E'\, \Omega') = \frac{1}{v\Sigma} \Gamma(E, \vec{\Omega} \to E', \Omega)\, dE'\, d\Omega'$$

Au total, le pinceau initial perdra par collision; sur le parcours en question, l'influence

(II.12) $$\iint S_0\, \Sigma\, dr \quad N^+(\vec{r} + \varepsilon d\vec{r}, E', \vec{\Omega}') \times \wp(E, \vec{\Omega} \to E'\vec{\Omega}')$$

$$= \frac{S_0}{v}\, dr \int_0^{\infty}\!\!\int_{\Omega} \Gamma\, (E, \vec{\Omega} \to E', \bar{\Omega}')\, N^+(\vec{r} + \varepsilon d\vec{r}, E', \Omega')\, dt' d\vec{\Omega}'$$

D'après le théorème de conservation, il faut que l'influence emportée par des chocs, argumentée de l'influence parvenue au point $\vec{r} + d\,\vec{r}$ soit égale à l'influence totale des neutrons de source $S_0\, N^+(\vec{r}, E, \vec{\Omega})$:

(II. 13) $$S_0\, N^+(\vec{r}, E, \vec{\Omega}) = S_0(1 - \varepsilon\, dr)\, N^+(\vec{r} + d\vec{r}, E, \vec{\Omega})$$

$$+ \frac{S_0}{v}'\, dr \int_0^{\infty}\!\!\int_{\Omega} \Gamma\, (E, \vec{\Omega} \to E', \vec{\Omega}')\, N^+(\vec{r} + \varepsilon\, d\vec{r}, E', \vec{\Omega}')\, dE' d\vec{\Omega}'$$

Développons alors la fonction d'influence N^+ au point $(\vec{r} + d\vec{r})$ en série de Taylor, suivant les puissances de dr, divisons ensuite par S_o d r et passons à la limite dr -> 0. On obtient, de cette façon, l'équation integro différentielle de la fonction d'influence [(cf. (I-27) de pag. 26

(II. 14) $$0 = \vec{v}. \nabla N^+ - \gamma N^+ + \int_0^\infty N^+ \Gamma' dE' d\vec{\Omega}' ,$$

qui , à l'aide des opérateurs

(II. 15) $$\hat{H}^+ = - \vec{v} . \nabla + \gamma - \int_0^\infty \int_\Omega dE' d\vec{\Omega}' (\Gamma'_d + \Gamma'_i)$$

et

(II. 16) $$\hat{K}^+ = \frac{1}{4\pi} \nu(E) . v \Sigma_f(E) \int_0^\infty \int_\Omega dE' d\vec{\Omega}' f(E')$$

et compte tenu de (I-30) (de pag. 27) et de (I-15) (pag. 22), peut se mettre sous la forme

(II. 17) $$(\hat{K}^+ - \hat{H}) N^+ = 0$$

Comme l'indique la notation, ces opérateurs sont adjoints aux opérateurs (I-43) (I-44) (de pag. 51), resp (I-48) de p. 31) : c'est-à-dire pour toute founctions ψ et ϕ qui sont définis à l'intérieur du réacteur, l'on a la relation définition des opérateurs adjoints

(II-18) $$\int_{\text{Réacteur}} \int_0^\infty \int_\Omega \psi(\hat{K} - \hat{H}) \phi . d V d E d \Omega =$$
$$= \int_{\text{Réacteur}} \int_0^\infty \int_\Omega \phi(\hat{K}^+ - \hat{H}^+) \psi . d V d E d \Omega$$

La fonction d'influence N^+ est donc adjointe à la densité N dans le réacteur critique. Elle est univoquement déterminée par (II-18), compte tenu d'une condition aux limites. Cette condition aux limites résulte de la considération que des neutrons sur les frontières du réacteur $(\vec{r} = \vec{r}_{front.})$ n'ont pas d'influence sur la réaction ramifiée si leur direction de vol pointe vers l'éxterieur (direction $\vec{\Omega}_{ext}$). Cette partie du flux d'influence doit donc s'annuler

(II-19) $$N^{+}(\Omega_{front.}, E, \vec{\Omega}_{ext}) = 0$$

En outre, N^{+} doit être dans tout le volume du réacteur, continu , fini et positif.

L'équ. (II.19) représente la condition aux limites pour la fonction d'influence. Son sens est justement l'opposé de la condition aux limites (I-46) (p. 32) pour la densité neutronique N : tandis que des neutrons s'évadent hors du réacteur, l'influence entre dans le réacteur. Cette inversion trouve son expression dans signe contraire du terme flux de H^{+} ,

, La fonction d'influence dépendant du temps

Reste à procéder à une généralisation. D'aprés la définition donnée jusqu'a présent , la fonction d'influence est déterminée pas (II-17) où les opérateur $\hat{H}^{+}$ et $\hat{K}^{+}$ correspondent à l'état critique du réacteur. Modifions maintenant l'état du réacteur par exemple , en retirant lentement une barre de réglage (qui absorbe des neutrons) . Dans ce cas $\hat{H}^{+}$ et $\hat{K}^{+}$ vont renfermer le temps en tant que paramètre et il sera nécessaire d'introduire la fonction d'influence dépendant du temps.

Si l'on fait abstraction des neutrons retardés et des sources extérieures, l'équation des réacteurs dependant du temps (I-39) (p.30) a pour expression

(II-20) $$\frac{\partial N}{\partial t} = (\hat{K} - \hat{H})\ N ,$$

Formellement parlant, l'équation adjointe a pour forme

(II. 21) $$-\frac{\partial N^{+}}{\partial t} = (\hat{K} - \hat{H})\ N^{+}$$

Or la founction d'influence a une signification physique claire ; elle doit caractériser l'etat du réacteur, eu égard aux divèrses réactions ramifiées dont il peut être le siège lors du choix de divèrses conditions initiales. Or l'état du réacteur ne dépénd pas, de conditions initiales de quelque genre qu'elles soient, pour la dénsité neutronique, mais peut être modifié par des interventions extérièures ou à nouveau rétabli. On peut ainsi, par exemple, modifier l'état du réacteur considéré comme vide en retirant une barre de réglage et ramener le réacteur dans son état antérièur en renforçant la barre. Une fonction qui caractérise le réacteur ne doit donc pas dépendre explicitement du temps comme le N^+ défini par (II.21)

Il n'a va pas de même pour la densite neutronique N. Si, par exemple, un réacteur vide surcritique qui est caractérisé par une fonction d'influence constante dans le temps, est démarré d'une façon déterminée, la densité neutronique qui change très rapidement dans le temps sera décrite par (II-20). Les conditions de mise à feu servent de condition initiale.

Il résulte des considérations précédentes que (II-21) ne se prête pas à une définition physiquement admissible d'une fonction d'influence variable dans le temps. Il est plus raisonnable de continuer définir la fonction d'influence par l'équation (II.17) même lorsque l'etat du réacteur varie avec le temps ou n'est pas critique.

Chapter III. Les équations cinétiques des réacteurs

Les équations des réacteurs (I (39) (de p. 30) rapt. (I. 45) (de p. 31) décrivent le comportement de la densité neutronique $N(\vec{r}, t, E, \vec{\Omega})$ et caracterisent par conséquent l'état différentiel du réacteur qui dépend de la position de l'énergie et de la direction à l'instant t. Nous allons établir maintenant des relations qui décrivent l'allure intégrale du réacteur comme fonction du temps dans lesquelles les divèrses positions, énergies et directions dans le réacteur ne figurent plus. Pour les distinguer des équations des réacteurs ces relations à établir ont reçu le nom d'équations cinétiques des réacteurs.

Il faut tout d'abord faire le choix d'une grandeur qui prenne la place de N et qui puisse exprimer l'état intégral du réacteur. On pourrai choisir à cet effet la puissance du réacteur qui est proportionnelle au nombre de fissions par seconde ayant bien dans tout le réacteur , Pour obtenir la puissance du réacteur l'on a à multiplier N par $v \Sigma_f$ et à intégrer sur le volume du réacteur, sur toutes les énergies et toutes les directions

$$\int_R \int_0^\infty \int_\Omega N\, v\, \Sigma_f \, dV\, dE\, d\vec{\Omega}$$

En tant que grandeur intégrale qui ne dépend désormais que du temps, la puissance se prête fort bien à caractériser les processus dynamiques dont le réacteur est le siège.

Il est toutefois plus utile de choisir une autre grandeur, apparentée à la puissance, à savoir la teneur en neutrons, pondérée ou symbolique n(t) . Pour parvenir à cette grandeur il faut multiplier N par la fonction d'influence N^+ et intégrer ensuite sur le volume, les énergies et les directions :

$$\text{(III. 1)} \qquad n(t) = \int_R \int_0^\infty \int_\Omega N^+ (\vec{r}, E, \vec{\Omega})\, N(\vec{r}, t, E, \vec{\Omega}) \times dV\, dE\, d\vec{\Omega}$$

$$\equiv (N^+, N)\,, \text{ avec } (\phi, \psi) \equiv \int_V \int_0^\infty \int_\Omega \phi\psi\, dV\, dE\, d\vec{\Omega}$$

Quel est le sens physique de n (t) ? Comme la fonction N^+ caractérise l'influence d'un neutron en $(\vec{r}, t, E, \Omega)$. Le produit $N \cdot N^+ \, dV\, dE\, d\Omega$ mesure donc l'influence de tous les neutrons $(E; \vec{\Omega})$ dans l'élément de volume d V . L'intégrale (III. 1) mesure enfin l'influence totale de tous les neutrons se trouvant à l'instant t dans le réacteur, sur l'évolution ultérieure de la réaction en chaîne .

Le choix de (III-1) est fondé sur toute une sèrie de raisons 1) Tout d'abord il met en évidence la signification des diverses zones de réacteur, ainsi que les énergies et les directions neutroniques . 2) En deuxième lieu, cette fonction n joue un rôle decisif en théorie des perturbations. 3) Enfin, sous certaines conditions, n(t) admet la même variation avec le temps que la puissance du réacteur.

Il en est ainsi lorsque nous n'avons a faire appel qu'à la fonction d'influence indépendante du temps et que N peut se séparer en une partie fonction spatiale et une autre fonction du temps

Forme simplifiée des équations cinétiques

Pour parvenir à une équation différentielle valable pour n(t) , multiplions l'équation (I- 45) (p. 31) par $N^+(\vec{r}, t, E, \vec{\Omega})$ et intégrons sur V, E et Ω ; il vient

$$(N^+, \frac{\partial N}{\partial t}) = (1 - \beta)\ (N^+, \hat{K}_0 N) + \tag{III. 2}$$

$$+ \frac{1}{4\pi} \sum_i \lambda_i (N^+, f_i C_i) + (N^+, S) - (N^+, \hat{H} N)$$

et

$$(N^+, \frac{\partial f_i C_i}{\partial t}) = 4\pi\beta_i (N^+, \hat{K}_i N) - \lambda_i (N^+ f_i C_i) \tag{III-3}$$

Les opérateurs $\hat{H}$ et $\hat{K}$ contiennent le temps à titre de paramètre si l'état du réacteur change au cours du temps. Comme alors dans (II. 2) et (II-3) non seulement la densité mais aussi la fonction d'influence dépend , en vertu de (II. 17) p. 46) , du temps, ces rélations sont fort compliguées de peu d'intérêt pratique. On peut parvenir cependant à des équations essentiellement plus simples si l'ont tient compte du fait que dans

les cas pratiquement intéressants , la foncton d'influence dépéndant du temps ne diffère que très peu de sa valeur stationnaire N_o^+
Si l'on pose donc

(III 5) $N^+(\vec{r}, t, E, \vec{\Omega}) = N_o^+(\vec{r}, E, \vec{\Omega}) + \Delta N^+(\vec{r}, t, E, \Omega)$

l'on aura $\Delta N^+ \ll N_o^+$

et on pourra substituer N_o^+ à N^+ dans (II. 2) et (III-3) . Comme on doit analyser l'état critique d'un réacteur avant de passer aux problèmes cinétiques, il faut supposer connues la distribution de densité critique $\mathcal{N}(\vec{r}, E, \vec{\Omega})$ et par suite aussi, la fonction d'influence stationnaire.

Dès lors, (III. 2) prend la forme, en utilisant le spectre global f(E) d'après (I. 49) (de p. 32) et en rassemblant les neutrons produit au total dans l'expression $\hat{K} N$:

(III. 6)
$$\frac{\partial (N_o^+, N)}{\partial t} = -(N_o^+, \hat{H} N) + (N_o^+, \hat{K} N) - \sum_i \beta_i (N_o^+, \hat{K}_i N) + \frac{1}{4\pi} \sum_i \lambda_i (N_o^+, t, C_i) + (N_o^+, S) .$$

Ici, l'influence des neutrons extérieurs est donnée par

(III. 7)
$$s(t) = (N_o^+, S) ;$$

la contribution des neutrons retardés est décrite par

(III-8)
$$C_i(t) = \frac{1}{4\pi} (N_o^+, f_i C_i)$$

En posant encore

(III-9)
$$\Sigma_i = \frac{(N_o^+, \hat{K}_i N)}{(N^+, \hat{K} N)} \quad \text{avec} \quad \beta = \sum_i \Sigma_i \beta_i$$

il vient à l'aide de la production totale $\hat{K}N$

(III. 10)
$$\frac{\partial n}{\partial t} \left[\frac{(N_o^+, \hat{K} N)}{(N_o^+, \hat{H} N)} (1 - \Sigma\beta) - 1 \right] \frac{(N_o^+, \hat{H} N)}{(N_o^+, N)} n + \sum_i \lambda_i c_i + s$$

$\dfrac{(N^+_o, \hat{K} N)}{(N_o^+, \hat{H} N)}$ représente ici la production d'influence par seconde divise par la perte d'influence ayant lieu dans le même temps. D'après (I-48) (p. 32) à (K - H) N = 0 (K N = HN , ce rapport est égal à 1 pour le réacteur critique et > 1 pour le réacteur hypercritique. Cette grandeur peut être comme comme un facteur de multiplication effectif généralisé

(III. 11) $$k(t) = \frac{(N_o^+, \hat{K}N)}{(N_o^+, \hat{H}N)}$$

La quantité $\dfrac{(N_o^+ N)}{(N^+, \hat{H}N)}$ represénte la teneur en neutrons, divisée par la perte-influence pour séconder. L'inverse de cette grandeur ayant la dimension d'un temps, indique donc combien de temps il prendait jusqu'à ce que la teneur en neutrons soit épuisée par les pertes à partir de l'arrîe de la production ; elle revêt donc la signification d'une durée de vie généralisée $l(t)$ des neutrons dans le réacteur fini . Nous poserons donc

(III. 12) $$l(t) = \frac{(N_o^+, N)}{(N_o^+, \hat{H}N)}$$

Avec ces notions nouvelles, (III, 10) et (III-3) prennent leur forme définitive

(III. 13) $$\frac{dn}{dt} = \left[k(1-\Sigma\beta) - 1\right]\frac{n}{l} + \sum_i \lambda_i c_i + s$$

$$\frac{dc_i}{dt} = -\lambda_i c_i + \Sigma_i \beta_i \, k \frac{n}{l}$$

où N est supposé "factorise" en un produit d'une fonction de t et d'une fonction de r.

Les équations (II-13) portent le nom d'equations cinétiques du réacteur et expriment son allure dans le temps en fonction de $k(t)$ et $l(t)$. La discussion de ces équations constitue le problème centralé de la cinétique des réacteurs. Les conditions de validité de ces équations sont

1) Une seule matière fissile;

2) les β_i indépendants de l'énergie

3) les opérateurs $\hat{H}$ et $\hat{K}$ sont voisins de leur valeurs stationnaires.

Chapter IV. Cinétique des réacteurs à l'approximation de la théorie de la diffusion

Nous avons dévelonné dans ce qui précède (ch. I à III) la théorie du transport rigoureus. Or, déjà la solution de problèmes stationnaires est, dans le cadre de la théorie du transport, fort délicate, liée à des calculs fastidieux et ne conduit à des résultats sous forme close que dans des cas exceptionels. Pour les problèmes cinétiques, ces difficultés ne font que croître. Force est donc de faire appel, pour l'analyse de ces problèmes à une méthode d'approximation utilisable qui est connue sous le nom de théorie de la diffusion. Cette théorie de la diffusion, comme approximation à la théorie du transport, analyse la densité neutronique sans égard à la distribution angulaire. Lès résultats s'accorderont donc d'autant mieux avec ceux de la théorie exacte que le mouvement neutronique dépendra moins d'une direction privilégiée. L'approximation de la théorie de la diffusion fournira donc des résultats utiles pour des milieux homogènes à condition de se restrindre à des domaines qui sont éloignés d'au moins quelques longueurs de diffusion des frontières, des bords ou des sources

Théorie de la diffusion

1) Rallentissement de neutrons de fission jusqu'aux energies thermiques

2) Ralentissement des neutrons de fission en tenant compte de la possibilité de leur capture dans la région de résonance

3) Diffusion des neutrons thermiques

Passage à la théorie de la diffusion

Le passage de la théorie du transport à la théorie de la diffusion peut s'effectuer très simplement en supposant que la distribution de densité N possède la symmétrie sphérique , dans ce cas N n'est fonction du rayon Ω. Si l'on envisage la densité $N(r, t, E, \vec{\Omega})$ des neutrons mobiles dans la direction $\vec{\Omega}$ (fig. IV-1) on voit que N ne dépénd pas de φ , mais de ϑ ou de $\mu = \cos\vartheta$.

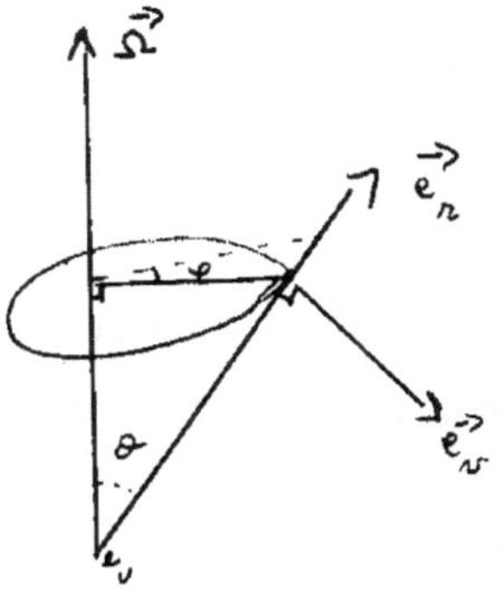

fig IV.I

Comme l'élément d'angle solide s'écrit

$$d\vec{\Omega} = |\sin\vartheta \, d\varphi \, d\varphi| = d\mu \, d\varphi ,$$

l'on a

$$\text{(IV-1)} \qquad \int_{=0}^{2\pi} N(r, t, E, \vec{\Omega}) \, d\mu \, d\varphi = 2\pi d\mu \times N(r, t, E, \vec{\Omega}) = \bar{N}(r, t, E, \mu) \, d\mu$$

Partons de la première des equations (I-45) (p. 31) en ne considérant qu'une seule matière fissile. Le terme pertes $\hat{H}N$ contient le terme $\vec{v}.\nabla N$, $(\vec{v} = \vec{\Omega} v)$ or d'après la fig. (IV.I)

$$\text{(IV-2)} \qquad \vec{v} \nabla N = \vec{v}\left(\vec{l}_2 \frac{\partial N}{\partial r} + \frac{\vec{e}_\vartheta}{r} \frac{\partial N}{\partial \vartheta}\right) = v\left(\mu \frac{\partial N}{\partial r} + \frac{1-\mu^2}{r} \frac{\partial N}{\partial \mu}\right)$$

L'intégration de (I-45) par rapport à φ , eu égard à (IV-2) et à (I-14) (p. 21) conduit à

$$\text{(IV-3)} \qquad \frac{\partial \bar{N}}{\partial t} = - v \left(\mu \frac{\partial N}{\partial r} + \frac{1-\mu^2}{r} \frac{\partial \bar{N}}{\partial \mu}\right), -\gamma \bar{N} + D + F + \bar{S}$$

où $\mathcal{D}$ est la contribution de diffusion (sans diffusion inélastique)

(IV-4) $$\mathcal{D} = \frac{1}{4\pi} \int_0^\infty dE' \int_{-1}^{+1} d\mu' \int_0^{2\pi} d\varphi' \int_0^{2\pi} d\varphi \, v'\Sigma_d \, N' \times \delta\left[E - (E' - \Delta E')\right] ;$$

et F est la source de fission

(IV-5) $$F = \frac{f_0}{2} \int_0^\infty \int_{-1}^{+1} (1-\beta')\nu'\Sigma_f \, v'\bar{N}(r, t, E', \mu') \, dE' \, d\mu' + \frac{1}{2} \sum_i \lambda_i f_i C_i$$

Afin de séparer la variation angulaire développons le flux neutronique $v \times \bar{N}$ en une série de fonctions de Legendre $P_n(\mu)$

(IV-6) $$v\bar{N}(r, t, E, \mu) = \sum_{n=0}^{\infty} \frac{2n+1}{2} P_n(\mu) \, \Phi_n(r, t, E)$$

où

(IV-7) $$\Phi_n(r, t, E) = \int_{-1}^{1} v\bar{N}(r, t, E, \mu) \, P_n(\mu) \, d\mu$$

Comme $P_0(\mu) = 1$, le premièr terme du développement est indépendant de la direction. Si nous poussons cette série jusqu'au deuxième terme $[P_1(\mu) = \mu]$, nous n'obtenons qu'une approximation grossière quant à la dépendance de $\vec{\Omega}$. Cette approximation ne vaut, d'après ce que nous avons dit en guise de préambule, que si les hypothèses consernant l'application de la théorie de la diffusion sont données. On posera donc à l'approximation de la théorie de la diffusion

(IV-8) $$v\bar{N} = \sum_{n=0}^{n=1} \frac{2n+1}{2} P_n \Phi_n = \frac{1}{2} P_0\Phi_0 + \frac{3}{2} P_1 \Phi_1 = \frac{1}{2}\Phi_0 + \frac{3}{2}\mu\Phi_1$$

Portons donc (IV-8) dans (IV-3) :

(IV-9) $$\frac{1}{v}\frac{\partial}{\partial t}\left(\frac{1}{2}\phi_o+\frac{3}{2}\mu\phi_1\right)=\frac{\partial}{\partial r}\left(\frac{1}{2}\mu\phi_o+\frac{3}{2}\mu^2\phi_1\right)-$$

$$-\frac{3}{2}\frac{1-\mu^2}{r}\phi_1-\frac{1}{v}\gamma\left(\frac{1}{2}\phi_o+\frac{3}{2}\mu\phi_1\right)+\mathcal{D}(\phi'_o,\phi'_1)+F(\phi'_o,\phi'_1)+$$

$$+\left(\frac{1}{2}S_o+\frac{3}{2}\mu\bar{S}_1\right)$$

Le flux de somme a été développé dans (IV-9) de la même façon que $v\bar{N}$

On peut multiplier (IV-9) successivement par les $P_n(\mu)$ et obtenir, grâce aux relations d'orthogonalités des $P_n(\mu)$ par intégration sur toutes les directions, un système d'équations différentielles pour ϕ_i. Multiplions d'abord (IV-9) par $P_o(\mu)=1$ et intégrons sur $d\mu$ de -1 à +1, il vient

(IV-10) $$\frac{1}{v}\frac{\partial\phi_o}{\partial t}=-\left(\frac{\partial}{\partial r}+\frac{2}{r}\right)\phi_1-\frac{1}{v}\gamma\phi_o+\int_{-1}^{+1}(\mathcal{D}+F)\,d\mu+\bar{S}_o$$

Multiplions ensuite par $P_2(\mu)=\mu$ et intégrons :

(IV-11) $$\frac{1}{v}\frac{\partial\phi_1}{\partial t}=-\frac{1}{3}\frac{\partial\phi_o}{\partial r}-\frac{1}{v}\gamma\phi_1+\int_{-1}^{+1}\mu(\mathcal{D}+F)\,d\mu+\bar{S}_1$$

Par (IV-4) et (IV-8) l'on a

(IV-12) $$\int_{-1}^{+1}\mathcal{D}\,d\mu=f_o\int_0^{\infty}(1-\beta')\nu'\Sigma'_f\,\phi'_o\,dE'+\sum_i\lambda_i f_i C_i\qquad\int_{-1}^{+1}\mu F\,d\mu=0$$

Le calcul des intégrales de diffusion (IV-12) nécessite l'introduction d'hypothèses concernant les processus de choc. Nous supposerons ici qu'il s'agit d'une diffusion purement classique des neutrons sur les noyaux de la matière du réacteur et que cette diffusion est isotrope dans le système du barycentre neutron-noyau. Si A est la masse du noyau, la perte d'énergie lors du choc neutron- A est

(IV-13) $$\Delta E' = E' \frac{1-\alpha}{2}(1-\bar{\mu}) \quad , \text{ avec } \alpha = \left(\frac{A-1}{A+1}\right)^2$$

où E' l'énérgie du neutron avant et E cette énergie après la collision.

Je donne le résultat final

(IV-14) $$\int_{-1}^{+1} \mathcal{D} \, d\mu = \frac{1}{2} \int_{E}^{E/\alpha} \Sigma'_d \; d'_o \phi'_o \, dE' \; ; \int_{-1}^{+1} \mathcal{D}\mu \, d\mu = \frac{1}{2} \int_{E}^{E/\alpha} \Sigma'_d \, d'_1 \phi'_1 \, dE'$$

où $d'_o = d_o(E'_o)$ et $d'_1 = d_1(E')$ sont des fonctions algébriques de E'.
En portant enfin (IV-12) et (IV-14) dans (IV-10) et (IV-11), on obtient l'approximation P_1 pour le flux $\phi_o(r, t, E)$:

(IV-15) $$\frac{1}{v}\frac{\partial \phi_o}{\partial t} = -\left(\frac{\partial}{\partial r} + \frac{\varepsilon}{r}\right)\phi_1 - \Sigma\phi_o + \int_{E}^{E/\alpha} \frac{\Sigma'_d \phi'_o}{1-\alpha}\frac{dE'}{E'} +$$

$$+ f_o \int_{0}^{\infty} (1-\beta')\nu' \Sigma'_f \phi_o \, dE' + \sum_i \lambda_i f_i C_i + \bar{S}_o \; ,$$

et son premier moment $\phi_1(r, t, E)$

(IV-16) $$\frac{1}{v}\frac{\partial \phi_1}{\partial t} = -\frac{1}{3}\frac{\partial \phi_o}{\partial r} - \Sigma\phi_1 + \int_{E}^{E/\alpha} \hat{\mu}'_o \frac{\Sigma'_d \phi_1}{1-\alpha}\frac{dE'}{E'} + \bar{S}_1$$

(Les quantités primée (') dependant de E') . La seconde équation (I-45) (de p. 31) peut s'intégrer elle aussi d'après (IV-1) . On peut en outre effectuer l'intégration de $\hat{K}_i \bar{N}$ par rapport à μ , après introduction de (IV-8), et obtenir ainsi comme condition complémentaire

(IV-17) $$\frac{\partial C_i}{\partial t} = -\lambda_i C_i + \int_{0}^{\infty} \beta'_i \nu' \Sigma_f \phi_o \, dE' \; .$$

Dans (IV-15) et (IV-16) figurent des intégrales de freinage $\int_{E}^{E/\alpha} \ldots$ dont la signification physique est la suivante : elles indiquent le nombre par cm^3 et par seconde des neutrons qui par freinage ont été portés à partir des bandes d'énérgie supérieures d E' , au cours

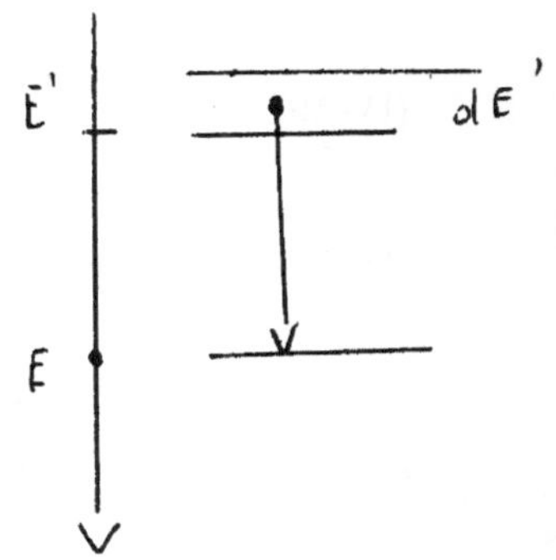

d'un seul choc à l'énérgie E. Ces intégrales de freinage ne permettent pas en général la solution simple du système (IV-15, 16) d'équations integro-différentielles couplées. En effet, comme les deux limites d'intégration dépendent de E, il est impossible de les transformer en équations différentielles par différentiation et découplage . Font exeption le freinage en hydrogène pur ($\alpha = 0$) ou dans un diffuseur de grand poids atomique (A>>1) où on peut poser $\alpha \simeq 1$.

Bien entendu, dans le cas géneral , le systeme précédent ne peut être résolu qu'approximativement, selon la situation existante . Je ne puis aborder ici que les cas les plus importants au point de vue cinétique, en procédant du plus simple vers le plus compliqué.

2. Cinétique des réacteurs monoénergétique.

a) Zone proche.

Pour les calculs d'orientation, on peut, en théorie cinétique, traiter le réacteur thermique en approximation grossière , comme monoénergétique en supposant que tous les neutrons sont animés de la même vitesse. Dans ce modèle fort primitif la production, la diffusion et l'absorption des neutrons se font à la méme énergie . Avec hypothèses $\Sigma'_d \phi'_o E'$ et $\Sigma'_d \phi_1 E'$. Dans (IV-15 à 17) sont indépendant, de l'énérgie et on peut effectuer les intégrations. Le flux ne dépend plus alors, dans l'équation ainsi obtenue que de r et t que nous désignerons par $\phi_{th}(r,t) = \phi(r,t)$. Il vient

(IV-18) $$\frac{1}{v}\frac{\partial\phi_o}{\partial t} = -\left(\frac{\partial}{\partial r}+\frac{2}{r}\right)\phi - \Sigma_a\phi_o + (1-\beta)\nu\Sigma_f\phi_o\nu + \sum_i \lambda_i C_i + S_o ,$$

(IV-19) $$\frac{1}{v}\frac{\partial\phi_1}{\partial t} = -\frac{1}{3}\frac{\partial\phi_o}{\partial r} - \Sigma_D\phi + S_1 ,$$

(IV-20) $$\frac{\partial C_i}{\partial t} = \beta_i\nu\Sigma_f\phi_o - \lambda_i C_i$$

Dans ces équations figurent la section d'absorption $a = \Sigma - \Sigma_d$ et le parcours libre moyen de diffusion λ_D défini par

(IV-21) $$\frac{1}{\lambda_D} \equiv \Sigma_D = \Sigma - \bar{\mu}_o\Sigma_d ,$$

qui , pour une absorption négligeable, se transforme en le parcours libre moyen de transport

(IV-22) $$\frac{1}{\lambda_{tr}} = \Sigma_{tr} = \Sigma_d(1-\bar{\mu}_o)$$

où $\bar{\mu}_o = \bar{\mu}\cos\bar{\theta}$, θ étant l'angle de diffusion dans le barycentre.

Pour simplifier les calculs, nous passons de la symétrie sphérique à la symétrie par rapport à un plan où ϕ ne dépendra que de x et de t. en nous restreignant à des réacteurs homogènes . Multiplions alors (IV-19) par (-v) et dérivons ensuite par rapport à x , dérivons de même , (IV-18) par rapport à t. On obtient alors une équation pour ϕ_o de second ordre en t. De même on obtient une équation pour ϕ, par élimination de ϕ_o :

(IV-20) $$\frac{\partial^2\phi_o}{\partial t^2} - \frac{v^2}{3}\frac{\partial^2\phi_o}{\partial x^2} + a_1\frac{\partial\phi_o}{\partial t} + a_2\phi_o + \mathcal{a}_o = 0$$
$$\frac{\partial^2\phi_1}{\partial t^2} - \frac{v^2}{3}\frac{\partial^2\phi_1}{\partial x^2} + a_1\frac{\partial\phi_1}{\partial t} + a_2\phi_1 + \mathcal{a}_1 = 0$$

Ici, les contributions des neutrons retardés et des sources exterieures sont désignées de façon abrégée par

$$(IV\text{-}21)\qquad Q_o \equiv - v\sum_i \lambda_i \left(\frac{\partial C_i}{\partial t} + v \Sigma_D C_i\right) + v^2\left(\frac{\partial S_1}{\partial x} - \Sigma_D S_o\right) - v\frac{\partial S_o}{\partial t}$$

$$Q_1 \equiv \frac{v^2}{3}\sum_i \lambda_i \frac{\partial C_i}{\partial x} + \frac{v^2}{3}\frac{\partial S_o}{\partial x} - \left(\frac{a^2}{\Sigma_D} S_1 + v\frac{\partial S_1}{\partial t}\right)$$

En outre

$$(IV\text{-}22)\qquad \begin{aligned} a_1 &\equiv v\left[\Sigma_a + \Sigma_D - (1-\beta)\nu\Sigma_f\right] \\ a_2 &\equiv v^2\Sigma_D\left[\Sigma_a - (1-\beta)\nu\Sigma_f\right] \end{aligned}$$

Le système (IV-20) n'est pas encore complètement découplé puisque Q_o et Q_1 dépendent encore de ϕ_o par (IV-17)

Pour étudier la distribution neutronique dans le voisinage immédiat, dit "zone proche" d'une source variable dans le temps; il faut faire appel au système (IV-20) .

Zone lointaine - Pour l'analyse des phénomènes dont les réacteurs sont le siège , à une distance r grande (comparée à la longueur du transport : $r \gg \lambda_{tr}$) (dite zone lointaine) on peut faire appel à l'équation de diffusion ordinaire

Les équations différentielles (IV-20) se distinguent des équations de la diffusion normalement utilisée par l'apparition de secondes dérivées temporelles ($\partial^2/\partial t^2$)La suppression de ces dérivées secondes s'obtient en faisant croître la vitesse v indéfiniment (v→0) tout en maintenant constant v/Σ_{tr} et $1/\Sigma_a v$.

Posons donc $\frac{\partial^2\phi}{\partial t^2} = 0$ dans la première équations (IV-20) et étudions la façon dont la seconde équation (IV-20) pour le premier moment ϕ_1 du flux se trouve modifie. Pour cela, retournons au système (IV-18) et (IV-19) (p. 59). On voit qu'on doit annuler le terme $\frac{\partial \phi_1}{\partial t}$ dans (IV-18) pour que, après élimination de ϕ_1 dans (IV-20), la dérivée seconde ne figure pas. De l'équation (IV-19) ainsi abrégée, on tire pour le premier moment du flux

$$\text{(IV-23)} \qquad \phi_1 = -\frac{1}{3\Sigma_D} \frac{\partial \phi_0}{\partial r} + 3D\, S_1$$

avec $\frac{\partial \phi_0}{\partial r} \to \nabla \phi$ pour la symétrie sphérique)

Il est alors tout indiqué d'introduire le vecteur de la densité de courant néutronique $\vec{J}(\vec{r}, t)$ par

$$\text{(IV-24)} \qquad \vec{J}(\vec{r}, t) = -D \nabla \phi_0(\vec{r}, t) \quad -D \;\text{grad}\; \phi_0(\vec{r}, t)$$

Si l'on supprime dans (IV-11) (p. 56) le terme $\partial \phi_1 / \partial t$ et qu'on élimine avec ce qui reste, ϕ_1 de l'équation (IV-10), il apparait l'opérateur

$$\left(\frac{\partial}{\partial r} + \frac{2}{r}\right) \frac{\partial}{\partial r},$$

qui se confond, dans le cas de symètrie sphérique, avec le laplacien $\nabla^2 \equiv \Delta$. Si l'on s'affranchit de la symétrie spherique que nous avons utilisée pour établir de manière simple l'approximation P_1, on obtient avec (IV-10) et $S_1 = 0$, l'équation de la diffusion monoénergetique indépendante du temps pour $\phi = \phi_0$

$$\text{(IV-25)} \qquad \frac{1}{v} \frac{\partial \phi}{\partial t} = \nabla \cdot (D \nabla \phi) - \Sigma_D \phi + (1-\beta) \nu \Sigma_f \phi + \sum_i \lambda_i \; C_i + S_0$$

Dans le terme source S_0 nous avons supprimé la barre qui résultait de l'intégration de (IV-1) p. 54). A cette équation il faut adjoindre l'équation des noyaux meres (IV-20) (de p.59)

(IV-26) $$\frac{\partial C_i}{\partial t} = \beta_i \nu \Sigma_f \phi - \lambda_i C_i$$

En partant de (IV-25) et (IV-26), il est aisé de formuler les équations cinétiques pour le réacteur monoénergétique, en posant.

(IV-27) $$\hat{K} = \nu \Sigma_f \quad \text{et} \quad \hat{H} = -\nabla.(D\nabla) + \Sigma_a \; ;$$

il vient

(IV-28) $$\begin{cases} \dfrac{1}{v}\dfrac{\partial \phi}{\partial t} = (1-\beta)\hat{K}\phi + \sum_i \lambda_i C_i + S_o - \hat{H}\phi \\ \dfrac{\partial C_i}{\partial t} = \beta_i \hat{K}\phi - \lambda_i C_i \end{cases}$$

Comme l'équation de la diffusion est autoadjointe, l'on a $\phi^+ = \phi$.

L'équation de la diffusion (IV-25) est une équation fort simplifiée non seulement par rapport à l'équation du transport mais aussi par rapport à l'approximation P_1 sous la forme (IV-18) et (IV-19) (de p.59)

La distribution de l'énergie dans le réacteur; j'ai montré, dans ce qui precède que sur l'exemple du réacteur monoénergétique, qu'on peut travailler dans la plupart des cas, avec une équation de diffusions qui est de premièr ordre dans le temps. L'Hypothese essentielle pour la validité de ce procedé est de se limiter à la zone lointaine, respectivement à des instants dans lesquels la majorité écrasante des neutrons de source ont dejà subi des chocs .

Le passage de la forme générale de l'approximation P_1 à l'équation de diffusion ordinaire a été effectué par la suppression de la dérivée temporelle de ϕ_1.

Le traitement monoénergétique du réacteur ne représente qu'une approximation grossière et n'est guère satisfaisante pour la plupart des buts en vue. C'est pourquoi on fait usage des équations dépendantes de l'énergie (IV-15) et (IV-16) (de p. 57) où l'on aura posé $\partial \phi_1 / \partial t = 0$.

Mais même sous cette forme simplifiée, ces équations (IV-15) (IV-16) ne peuvent être découplées et intégrées sous forme close que dans des cas tout à fait particuliers. Ici aussi des méthodes approchées sont de rigueur, telles que par exemple la méthode de diffusion des groupes. Cette méthode des groupes rassemble les neutrons ou divers groupes en fonction de leur énergie, à l'intérieur desquels ils sont considérés comme étant monoénergétiques. On obtient ainsi un système d'équations du type (IV-25) (IV-26) (de p. 61).

Méthode de diffusion à deux groupes

Appliquons cette méthode des groupes à deux groupes. Nous divisons pour cela toute la bande d'énergie des neutrons en le "groupe rapide" qui s'étend de l'énérgie de fission maximale E_0 (~ 2 M e v) à l'énergie, mettons, E_a et en le "groupe thermique" qui va de E_a à $E = 0$. Affectons l'indice 1 au groupe rapide, l'indice 2, an groupe thermique. On obtient ainsi pour le groupe rapide l'équation de diffusion.

(IV-29)
$$\frac{1}{v_1}\frac{\partial \phi}{\partial t} = \nabla\cdot(D_1 \nabla \phi_1) - \Sigma_{a_1}\phi_1 - q(\vec{r}, t, E_a) + (1-\beta_1)\nu_1 \Sigma_{f_1}\phi_1 + (1-\beta_2)\nu_2\Sigma_{f_2}\phi_2 + \sum_i \lambda_i C_i + S_1$$

avec

(IV-30)
$$q(\vec{r}, t, E) = \int_E^{E_0} \Sigma_d \phi' \; G(E' \to E)\, d E' ,$$

(avec $q(\vec{r}, t, E_0) = 0$) $G(E' \to E)$ étant la probabilité que la collision conduise à uné énérgie inférieure à E. En outre, on a admis que tous les neutrons de fission appartiennent au groupe rapide. Le facteur β est supposé constant dans le groupe rapide, respt. dans le groupe lent

Le flux "rapide" est donné par

(IV-31)
$$\phi_1(\vec{r}, t) = \int_E^{E_0} \phi(\vec{r}, t, E)\, d E$$

(où ϕ_1 est bien entendu dans cette section le flux rapide et non pas le premier moment du flux ! Les constantes du groupe sont définies par

$$1/v_1 = \int_{E_a}^{E_o} (\phi'/v')\, dE' \Big/ \int_{E_a}^{E_o} \phi'\, dE' \; ; \; \Sigma_{a_1} = \int_{E_a}^{E_o} \Sigma'_a \phi' dE' \Big/ \int_{E_a}^{E_o} \phi' dE \tag{IV-32}$$

$$D_1 = \int_{E_a}^{E_o} D' \nabla \phi'\, dE' \Big/ \int_{E_a}^{E_o} \nabla \phi\, dE' \quad \text{etc.}$$

Pour le calcul, on se sert de l'hypothèse d'un spectre d'énergie indépendant (au moins par domaine) de la position :

$$\phi(\vec{r}, t, E) = \phi(\vec{r}, t)\,.\, Z(E) \tag{IV-33}$$

De même, l'équation de diffusion pour le groupe thermique est

$$\frac{1}{v_2}\frac{\partial \phi_2}{\partial t} = \nabla.(D_2 \nabla \phi_2) - \Sigma_{a_2}\phi_2 + q(\vec{r}, t, E_a) + S_2 \tag{IV-34}$$

Ici $q(\vec{r}, t, 0) = 0$ et il n'apparait pas pratiquement, de neutrons de fission dotés d'énérgie thermique. Les constantes de groupe. doivent être formés de manière analogue à (IV-32)

La condition complémentaire prend ici la forme suivante

$$\frac{\partial C_i}{\partial t} = -\lambda_i C_i + \beta_{i1}\nu_1 \Sigma_{f_1}\phi_1 + \beta_{i2}\nu_2 \Sigma_{f_2}\phi_2 \tag{IV-35}$$

Formulation matricielle

Eu égard à (IV-30), on peut introduire dans (IV-29) et (IV-34) la définition

$$q(\vec{r}, t, E_a) \overset{\text{déf}}{\equiv} \Sigma_1 \phi_1 \tag{IV-36}$$

avec

$$\Sigma_1 \overset{\text{déf}}{\equiv} \frac{1}{\int_{E_a}^{E_o} \phi' d E'} \int_{E_a}^{E_{lim}} \Sigma_s \phi' G(E'- E)\, dE' \tag{IV-37}$$

qui définit une section efficace de freinage . Pour limite supérieure E_{lim}, il faut mettre la plus petite valeur de E_o et E_o / α

A l'aide de (IV-36) , il est possible de mettre les équations (IV-29) et (IV-39) sous une forme matricielle en posant :

(IV-38) $$\check{\phi}(\vec{r}, t) = \begin{pmatrix} \phi_1 \\ \phi_2 \end{pmatrix} ; \check{C}_i(\vec{r}, t) = \begin{pmatrix} C_i \\ 0 \end{pmatrix} ; \left(\check{\frac{1}{v}}\right) = \begin{pmatrix} \frac{1}{v_1} & 0 \\ 0 & \frac{1}{v_2} \end{pmatrix} ;$$

(IV-39) Matrice de perte et de diffusion : $$\hat{H} = \begin{pmatrix} \hat{J}_1 & 0 \\ -\Sigma_1 & \hat{J}_2 \end{pmatrix}$$

avec $$\hat{J}_1 = -\nabla\cdot(D\nabla) + \Sigma_{a_1} + \Sigma_1 ,$$
$$\hat{J}_2 = -\nabla\cdot(D_2\nabla) + \Sigma_{a_2}$$

(IV-40) Matrices de production $$\hat{K}_o = \begin{pmatrix} (1-\beta_1)\nu_1\Sigma_{f1} & (1-\beta_2)\nu_2\Sigma_{f2} \\ 0 & 0 \end{pmatrix}$$
$$\hat{K}_i = \begin{pmatrix} \beta_{i1}\nu_1\Sigma_{f1} & \beta_{i2}\nu_2\Sigma_{f2} \\ 0 & 0 \end{pmatrix}$$

Ceci permet de rassembler, avec une matrice de source $\check{S}$ constante de manière analogue , les équ. (IV-29), (IV-34) et (IV-35) en les équations des réacteurs d'après la méthode de diffusion à deux groupes

(IV-41) $$\left(\check{\frac{1}{v}}\right)\frac{\partial\check{\phi}}{\partial t} = \hat{K}_o\check{\phi} - \hat{H}\check{\phi} + \sum_i \lambda_i \check{C}_i + \check{S} ,$$
$$\frac{\partial \check{C}_i}{\partial t} = \hat{K}_i\check{\phi} - \lambda_i \check{C}_i$$

Pour passer aux équations cinétiques selon (III-13) (p. 52) , il faut connaître la fonction d'influence. Pour cela, formons les opérateurs adjoints

(IV-42) $$\hat{H}^{+} = \begin{pmatrix} \hat{J}_1 & -\Sigma_1 \\ 0 & J_2 \end{pmatrix} \quad \text{et} \quad \hat{K}^{+} = \begin{pmatrix} \nu_1\Sigma_{f_1} & 0 \\ \nu_2\Sigma_{f_2} & 0 \end{pmatrix}$$

[en supposant pour simplifier $\beta_{1i} = \beta_{2i} = \beta$ ce qui entraine

(IV-43) $\hat{K}_o = (1-\beta)\hat{K}$ et $\hat{K}_i = \beta_i \hat{K}$, avec

$$K = \begin{pmatrix} \nu_1 \Sigma_{f_1} & \nu_2 \Sigma_{f_2} \\ o & o \end{pmatrix}$$

lesquels appliqués à la fonction d'influence ϕ^+ liurent les équations différentielles pour la fonction d'influence stationnaire

(IV-44) $$(\hat{K} - \hat{H})\phi^+ = 0$$

Une fois résolu cette équation, on peut former les grandeurs n, s, c_i, k et l de manière analogue aux grandeurs du ch III et établir de ce chef les <u>équations cinétiques</u> d'après la méthode des deux groupes. Nous avons à définir en particulier

(IV-45)
$$\begin{cases} n(t) = \int \frac{1}{v}\phi^+\phi \, dV_i \quad c_i(t) = \int \phi^+ C_i \, dV; \\ s(t) = \int \phi^+ S \, dV; \quad k(t) = \dfrac{\int \phi^+ \hat{K}\phi \, dV}{\int \phi^+ \hat{H}\phi \, dV}; \\ l(t) = \dfrac{\int \frac{1}{v}\phi^+\phi \, dV}{\int \phi^+ \hat{H}\phi \, dV}. \end{cases}$$

Au lieu des $\Sigma_i \beta_i$ dans (III-13) (p. 52),l'on a l'expression :

(IV-46) $$\frac{(\phi^+, \hat{K}_i\phi)}{(\phi^+, \hat{K}\phi)} = \frac{\int \phi_1^+ (\beta_{i1}\nu_1\Sigma_{f_1}\phi_1 + \beta_{i2}\nu_2\Sigma_{f_2}\phi_2)\, d\bar{V}}{\int \phi_1^+ (\nu_1\Sigma_{f_1}\phi_1 + \nu_2\Sigma_{f_2}\phi_2)\, dV}$$

qui se confond avec β lorsque $\Sigma_{f_1} \approx 0$ (réacteur thermique) ou $\Sigma_{f_2} \approx 0$ (réacteur rapide) .

Faisons encore une remarque sur la relation entre densité neutronique et flux. Le flux rapide et le flux lent sont reliés aux densités correspondantes par les relations :

(IV-47) $$\phi_1 = v_1 N_1, \quad \phi_2 = v_2 N_2,$$

(2) $\hat{L}\,\phi = 0$ (équation homogène)

rendent extrémales les fonctionnelles suivantes ($L\,\delta = \delta\,L$

(3) $J = \int \psi_2\, \hat{L}\psi_1\, d\tau, \quad \delta J = 0$.

En effet en posant

(4) $\psi_1 = \phi_1 + \delta\phi_1 , \quad \psi_2 = \phi_2 + \delta\phi_2$,

où ϕ_1 et ϕ_2 sont deux solutions distinctes de l'équ. (2) et en utilisant la propriété (1), l'on a en effet

$$\delta J = J[\phi_i + \delta\phi] - J[\phi] = \int (\phi_2 + \delta\phi_2)\hat{L}(\phi_1 + \delta\phi)\, d\tau$$

$$-\int \phi_2\, \hat{L}\phi_1''\, d\tau = \int \phi_2\, \hat{L}\phi_{01}''\, d\tau + \int \delta\phi_2\, \hat{L}\phi_1\, d\tau +$$

$$+\int \phi_2\, \hat{L}\,\delta\phi_1\, d\tau + \int \delta\phi_2\, \hat{L}\,\delta\phi_1\, d\tau = \mathcal{O}[\delta\phi_1 \delta\phi_2] = \mathcal{O}[(\delta\phi)^2]\ \text{c.q.f.d.}$$

"$\int \delta\phi_1 \hat{L}\phi_2\, d\tau$"

En d'autres termes, lorsqu'on commet une erreur $\delta\phi$ sur les solutions de (2), la fonctionnelle J ne subit qu'une variation du second orde $\mathcal{O}[(\delta\phi)^2]$.

De même, envisageons l'équation avec second nombre ou équation inhomogène)

(5) $\hat{L}\phi_i = f_i$, (equation inhomogène)

$\hat{L}$ étant toujours un opérateur linèaire symétrique, possédant un inverse et où les f_i sont des fonctions ou sources données, . On peut alors poser

(6) $\phi_i' = \hat{L}^{-1} f_i - \phi_i$

et les ϕ_i' vérifient l'équation

Il n'en va pas de même pour la fonction d'influence ; l'on a au contraire

$$\text{(IV-48)} \qquad \phi_1^+ = N_1^+ \quad , \quad \phi_2^+ = N_2^+$$

ce qui résulte de ce qui suit. Par (IV-39) et (IV-40), l'on peut en effet mettre le flux et la densité stationnaires sous la forme

$$\text{(IV-49)} \quad \begin{cases} (\nu_1 \Sigma_{f_1} - \hat{J}_1)\phi_1 + \nu_2 \Sigma_{f_2} \phi_2 = 0 , \\ \Sigma_1 \phi_1 - \hat{J}_2 \phi_2 = 0 , \\ \text{soit} \\ (\nu_1 \Sigma_1 - \hat{J}_1) v_1 N_1 + \nu_2 \Sigma_{f_2} v_2 N_2 = 0 \\ \Sigma_1 v_1 N_1 - \hat{J}_2 v_2 N_2 = 0 . \end{cases}$$

Les équations adjointes correspondantes ont pour expression

$$\text{(IV-50)} \quad \begin{cases} (\nu_1 \Sigma_{f_1} - \hat{J}_1)\phi_1^+ + \Sigma_1 \phi_2^+ = 0 \\ \nu_2 \Sigma_{f_2} \phi_1^+ - \hat{J}_2 \phi_2^+ = 0 , \\ \text{soit } (\nu_1 \Sigma_{f_1} - \hat{J}_1) v_1 N_1^+ + \Sigma_1 v_1 N_2^+ = 0 \\ \nu_2 \Sigma_{f_2} v_2 N_1^+ - \hat{J}_2 v_2 N_2^+ = 0 \end{cases}$$

Dans les deux dernières équations (IV-50), on peut diviser la première par v_1 la seconde par v_2 et on trouve effectivement que les équations différentielles pour N_1^+ et N_2^+ se confondent avec les deux premières équations pour ϕ_1^+ et ϕ_2^+.

Dans le cadre des hypothèses restrictives que nous avons introduites, il est possible d'appliquer la cinétique des reacteurs, en principe à des substances modératrices quelconques . La difficulté principale qui se présente alors est la détermination des constantès de groupes, en particulier pour les modérateurs contenant de l'hydrogène. Cette difficulté se trouve réduite si l'on fait appel à la théorie de l'âge de Fermi pour la description du processus de freinage. Cette théorie dont va nous entretenir le Prof. Pignedoli conduit à une seule équation de diffusion, fort simplifiée il est vrai , qui représente assez bien le comportement de gros réacteurs thermiques pour certains buts.

T. Kahan

II Partie

Méthodes variationnelles en cinétique des réacteurs

1. Introduction . - L'impossibilité où l'on se trouve habituellement de résoudre de manière rigoureuse les équations différentielles attachées à des problèmes de diffusion, atomiques nucléaires ou autres, ont amené divers auteurs, (Schurniger, Hulthen , Kohn , etc) à proposer des méthodes de résolution approchées à l'aide de principes variationnels, d'aspect et de maniement très différents. Ces principes s'appliquant tous à des problèmes de même nature, nous avons tenté de voir, dès 1951 T.K. et G. Rideau .C.R. Acad. Sc. 233 1951, (1849), J. de Phys. 13 (1952) 326, T.K. G.R. et P. Roussoupon pos des méthodes d'approdimations variationnelles dans la théorie des collisions atomiques et dans la physique des piles nucléaires, memorial des Sciences Mathématiques, 1956)s'il n'existerait pas un principe plus général englobant tous les autres à titre de cas particuliers. C'est ainsi que nous avons été amenés à proposer le principe qui fait l'objet des présents exposés et qui a reçu de nombreuses applications principalement en théorie de la diffusion neutronique et en théorie des réacteurs nucléaires

2. Formulation du principe général.

Considérons un opérateur linéaire $\hat{L}$ symétrique, c'est-à-dire vérifiant la relation suivante (*)

$$(1) \qquad \int_\tau \varphi \hat{L}\, \theta\, d\tau = \int \theta\, \hat{L} \varphi\, d\tau$$

quelles que soient les fonctions φ et θ .

Les solutions satisfaisantes à

(*) C'est-à-dire $\langle \varphi | \hat{L} | \theta \rangle = \langle \theta | \hat{L} | \varphi \rangle \qquad \hat{L} = \hat{L}^T$ (T = transposé)

(7) $$\hat{L}\,\Phi_i = \hat{L}\,(\hat{L}^{-1} f_i - \Phi_i) = 0$$

Elle rendrons extrémales une fonctionnelle du type (3)

(8') $$J = \int \psi_2'\ \hat{L}\,\psi_1'\,d\tau = \int (\hat{L}^{-1} f_2 - \psi_2)\hat{L}(\hat{L}^{-1} f_1 - \psi_1)\,d\tau$$

$$= \int (\hat{L}^{-1} f_2 - \psi_2)(f_1 - \hat{L}\,\psi_1)\,d\tau =$$

$$= \int \left\{ \hat{L}^{-1} f_2 \cdot f_1 - \psi_2 f_1 - \hat{L}^{-1} f_2\, L\psi_1 + \psi_2\, \hat{L}\psi_1 \right\} d\tau$$

cont car f_i et f_2 sont des fonctions dérivées $f_2\,\psi_1$ par (1)

$$= \int \left\{ \psi_2\,\hat{L}\,\psi_1 - \psi_2 f_1 - \psi_1 f_2 \right\} d\tau$$

$$= \int \left\{ \psi_2\ (\hat{L}\,\psi_1 - f_1) - \psi_1\ f_2 \right\} d\tau$$

en omettant le terme $\int f_1 \hat{L}^{-1} f_2\,d\tau$ qui est constant puisque f_1 et 2 sont des fonctions derivées conues .

Les Φ_i rendent donc extrémale la fonctionnelle suivante

(8) $$J = \int \left\{ \psi_2\ (\hat{L}\psi_1 - f_1) - \psi_1\ f_2 \right\}, \quad \delta J = 0$$

Quand on substitue à ψ_1 et ψ_2 des solutions exactes de (5) , la fonctionnelle (8) se réduit à

$$-\int \psi_1\, f_2\, d\tau \qquad \text{(ou bien avec } \psi_2 \rightarrow \psi_1 \text{ à } -\int \psi_2\, f_1\, d\tau \text{)}$$

et l'on peut ainsi utiliser (8) pour le calcul approché de cette fonctionnelle.

Généralement , dans la conduite des calculs, il sera commode de transformer (5) en une équation intégrale, ce qui va nous amener à représenter $\hat{L}$ par un opérateur intégral dont le noyau sera symétrique quand $\hat{L}$ est symètrique .

3. Application aux problèmes de diffusion.

L'équation qui régit un problème de diffusion (scattering) en Mécanique quantique peut se réduire à la forme suivante ($d\tau \equiv d\, r \equiv d^3 r$)

$$\phi_1(\vec{r}) = \phi_1^0(\vec{r}) + \int K(\vec{r}, \vec{r}\,') \, V(\vec{r}\,') \phi_1(\vec{r}\,') \, d\tau \tag{9}$$

où le noyau $K(\vec{r}, \vec{r}\,')$ est généralement un noyau symétrique $[K(\vec{r}, \vec{r}\,') = K(\vec{r}\,', \vec{r})]$ et où $\phi_1^0(\vec{r})$ représente l'onde incidente $V(\vec{r})$ étant le potentiel du centre diffusion

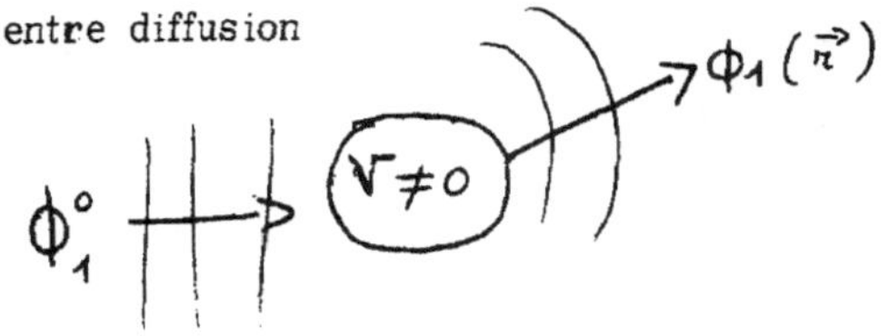

De même de nombreux problèmes en théorie du transport des neutrons , se posent sous la forme d'une équation intégrale du type de Fredholm

$$\phi(\vec{r}) = S(\vec{r}) + \lambda \int K(\vec{r}, \vec{r}\,') \phi(\vec{r}\,') \, d\vec{r}\,' \tag{9 bis}$$

où les limites sont données et le noyau $K(\vec{r}, \vec{r}\,')$ est réel, symétrique ou symétrisable et $S(\vec{r})$ le terme source, est une fonction donnée.

Faisons, dans (9), le changement de fonction inconnue ϕ_1 :

$$\phi_1'(\vec{r}) = \phi_1(\vec{r}) \sqrt{V(\vec{r})} \quad ; \quad \phi_1'^{0}(\vec{r}) = \phi_1^0(\vec{r}) \sqrt{V(\vec{r})} \tag{10}$$

et (9) prend alors la forme [1]

[1] Pour cet opérateur, j'utilise la notation $\hat{L} = \int F(\vec{r}, \vec{r}\,') \{ \; \} d\vec{r}\,'$ qui signifie $\hat{L}\phi = \int F(\vec{r}, \vec{r}\,') \{\phi(\vec{r}\,')\} d\vec{r}\,'$

$$\hat{L}\ \phi_1'(\vec{r}) = \phi_n^{o'}\ (\vec{r})$$

(11) où

$$\hat{L} = \int \left\{ \delta(\vec{r}-\vec{r}') - \sqrt{V(\vec{r})}\ K(\vec{r},\vec{r}')\sqrt{V(\vec{r}')} \right\} \left\{ \dots \right\} d\vec{r}'$$

En appliquant à ϕ_1' notre principe variationnnel relatif à (8) et en revenant au ϕ_1 initial, nous obtenons l'expression suivante, stationaire vis-à-vis des solutions de (9).

(12)
$$\begin{cases} J[\psi] = -\int \psi_2\ \phi_1^o\ V\ d\vec{r} - \int \psi_1\ \phi_2^o\ V\ d\vec{r} + \int \psi_1\ V \psi_2\ d\vec{r} \\ \qquad - \int \overline{\psi}_2(\vec{r})\ V\ K(\vec{r},\vec{r}')\ V(\vec{r}')\ \overline{\psi}_1\ (\vec{r}')\ d\vec{r}\ d\vec{r}' . \end{cases}$$

Quand $\overline{\psi}_1$ et $\overline{\psi}_2$ sont remplacées pàr des solutions ϕ_1 et ϕ_2 de (9), l'expression (12) a pour valeur

(13)
$$J[\phi] = -\int \phi_2^o\ \phi_1\ V\ d\vec{r},$$

Seule quantité qui ait un intérêt physique en théorie des collisions où elle représente l'amplitude de diffusion (cf. équ. (17))

L'opérateur $\hat{L}$, qui figure dans (11), doit avoir un inverse $\hat{L}^{-1}$ pour que les conclusions précédentes soient valables. Démontrons qu'il est bien ainsi. Posons pour simplifier l'écriture

(11-1)
$$\mathcal{H}(\vec{r},\vec{r}') = \sqrt{V(\vec{r})}\ K(\vec{r},\vec{r}')\sqrt{V(\vec{r}')}$$

On cherche un noyau $G(\vec{r},\vec{r}')$ tel que soit vérifiée l'équation

(II-2)
$$\int \mathcal{G}(\vec{r},\vec{r}'') \int \left\{ \delta(\vec{r}''-r') - \mathcal{H}(\vec{r}'',\vec{r}') \right\} d\vec{r}'' = \\ = \delta(\vec{r}-\vec{r}'),$$

car $\delta(\vec{r}-r')$ est le noyau de l'opérateur unité $\hat{I}$ $(\hat{I}\phi(\vec{r}) = \int \delta(\vec{r}-\vec{r}')\phi(r')\ d\vec{r}' = \phi(\vec{r})$. (11-2) fournit l'équation intégrale suivante pour $\mathcal{G}(\vec{r},\vec{r}'')$.

T. Kahan

(II-3) $$\mathcal{G}(\vec{r},\vec{r}') = \delta(\vec{r}-\vec{r}') + \int \mathcal{G}(\vec{r},\vec{r}'')\,\mathcal{H}(\vec{r}'',\vec{r}')\,d\vec{r}''\,.$$

La résolution en est immédiate car en appelant $\Gamma(\vec{r},\vec{r}')$ le noyau rèsolvant relatif an noyau $\mathcal{H}(\vec{r},\vec{r}')$, il vient

(II-4) $$\mathcal{G}(\vec{r},\vec{r}') = \delta(\vec{r}-\vec{r}') + \Gamma(\vec{r},\vec{r}')\,,$$

avec la forme bien connue du noyau rèsolvant,

(II.5) $$\Gamma(\vec{r},\vec{r}') = \mathcal{H}(\vec{r},\vec{r}') + \int \mathcal{H}(\vec{r},\vec{r}'')\,\mathcal{H}(\vec{r}'',\vec{r}') + \dots$$

Ceci montre l'existence d'un inverse à gauche. Mais $G(\vec{r},\vec{r}')$ est aussi un invêrse à droite puisque l'on vérifiê

(II.6) $$\int\left\{\delta(\vec{r}+\vec{r}'') - \mathcal{H}(\vec{r},\vec{r}')\right\}\mathcal{G}(\vec{r}'',\vec{r}')\,d\vec{r}''$$
$$= \delta(r-r') - \mathcal{H}(\vec{r},\vec{r}') + \Gamma(\vec{r},\vec{r}')$$
$$- \int \mathcal{H}(\vec{r},\vec{r}'')\,\Gamma(\vec{r}'',r')\,d\vec{r}'' =$$
$$= \delta(\vec{r}-\vec{r}')$$

D'après la définition même du noyau résolvant $\Gamma(\vec{r},\vec{r}')$.

4. Diffusion en Mécanique quantique.

Nous allors appliquer, d'abord à titre d'illustration, la théorie générale qui précède aux problèmes de diffusion de la mécanique ondulatoire. Dans ce cas nous avons à considérer l'équation

(14) $$\left[\nabla^2 + k^2 - V(\vec{r})\right]\Phi(\vec{r}) = 0\,,$$

et la solution de celle-ci qui se réduit à l'onde plane $\Phi_1^o = e^{i\vec{K}_1\vec{r}}$, $|\vec{K}_1| = k$, en l'absence du potentiel diffuseur $V(\vec{r})$. On sait que cette solution vérifie l'équation intégrale (cf. L. de Broglie, de la Mécanique ondulatoire à la théorie du noyau, Hermann, t. III p. 35(55) et

Mott and Massey, Theory of Atomic Collisions 2 ed. p. 116 (3) - T. Kahan, Revue de Physique Théorique Moderne, t. II etc.

$$(15) \qquad \phi_1(\vec{r}) = e^{i\vec{k}_1 . \vec{r}} - \frac{1}{4\pi} \int \frac{e^{ik|\vec{r}-\vec{r}'|}}{|\vec{r}-\vec{r}'|} V(\vec{r}') \phi_1(\vec{r}') d\vec{r}'$$

et admet la forme asymptotique

$$(16) \qquad \phi_1(\vec{r}) \underset{r \to \infty}{\longrightarrow} e^{i\vec{k}_1 \vec{r}} + f(\vec{k}_1, \vec{k}) \frac{e^{ikr}}{r},$$

où $f(\vec{k}_1, \vec{k})$ est l'amplitude de diffusion dans la direction $\vec{k}$ de l'onde incidente suivant $\vec{k}_1$, dont la valeur dans la direction $-\vec{k}_2$ est donnée par (cf. Mott et Massey cit p. 114 (30)]

$$(17) \qquad f(\vec{k}_1, -\vec{k}_2) = -\frac{1}{4\pi} \int e^{i\vec{k}_2 . \vec{r}} V(\vec{r}) \phi_1(\vec{r}) d\vec{r} .$$

Cette expression $f(\vec{k}_1, -\vec{k}_2)$ est, au facteur -4π près, la valeur stationnaire (V-S). de la quantité

$$(18) \qquad -4\pi J[\psi] = -4\pi f(\vec{k}_1 - \vec{k}_2) \; V.S. \left\{ \int e^{i\vec{k}_2 . \vec{r}} V(\vec{r}) \Psi_1(\vec{r}) d\vec{r} + \int e^{i\vec{k}_1 . \vec{r}} V(\vec{r}) \overline{\Psi}_2(\vec{r}) d\vec{r} - \int \overline{\Psi}_1 V \Psi_2 d\vec{r} . - \frac{1}{4\pi} \int \Psi_2(\vec{r}) . V(\vec{r}) \frac{e^{k|\vec{r}-\vec{r}'|}}{|\vec{r}-\vec{r}'|} V(\vec{r}') \overline{\Psi}_1(\vec{r}') d\vec{r} d\vec{r}' \right.$$

comme il résulte de nos formules (9) à (13). La méthode variationnelle fondée sur la formule (18) est une des formes que peut prendre le principe variationnel défini par (8).

Je vais indiquer maintenant la généralisation de notre méthode aux bosons et aux fermions.

Généralement, l'équation de diffusion (14) est écrite dans le barycentre de deux particules semblables. On sait qu'alors l'indiscernabilité des particules est responsable de phenomènes d'échange régis par la statistique particulière suive par les deux particules, phénomènes qui ont pour effet de modifier la valeur des sections efficaces obtenues quand on ne tient pas compte de cette indiscernabilité . Si le vecteur $\vec{k}_1$, qui donne la direction des ondes incidentes, est pris pour axe des z et si θ désigne l'angle entre $\vec{k}_1$ et $-\vec{k}_2$, la quantité qu'il est intéressant de connaitre est $f(\theta) \pm f(\pi-\theta)$, avec le signe + on le signe - suivant que l'on considère des bosons ou des fermions, où l'on a posé

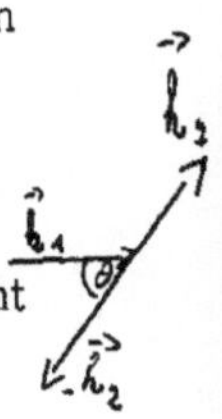

$$f(\vec{k}_1, +\vec{k}_2) = f(\pi-\theta)$$

A partir de maintenant, nous noterons $\phi_1^{(+)}(\vec{r})$ la solution de l'équation intégrale suivante relative à une onde incidente suivant $\vec{k}_1$:

$$(18\text{-}1) \qquad \phi_i^{(+)}(\vec{r}) = e^{i\vec{k}_i\vec{r}} - \frac{1}{4\pi}\int \frac{e^{k|\vec{r}-\vec{r}'|}}{|\vec{r}-\vec{r}'|} V(\vec{r}') \times \phi_1^{(+)}(\vec{r}')\, d\vec{r}' ,$$

tandis que la solution relative à une onde incidente suivant $-\vec{k}_1$ sera notée $\phi_1^{(-)}(\vec{r})$ et satisfera à l'équation intégrale

$$(18\text{-}2) \qquad \phi_1^{(-)}(\vec{r}) \; e^{-i\vec{k}_1\vec{r}} - \frac{1}{4\pi}\int \frac{e^{ik|\vec{r}-\vec{r}'|}}{|\vec{r}-\vec{r}'|} V(\vec{r}')\phi_1^{(-)}(\vec{r}')\, d\vec{r}'$$

Quand le potentiel est uniquement founction de la distance $|\vec{r}| = r$, on vérifie aisément la relation

$$(18\text{-}3) \qquad \phi_1^{(+)}(-\vec{r}) = \phi_1^{(-)}(\vec{r})$$

Si l'on tient compte de la statistique (effet d'échange), la solution exacte du problème de diffusion s'écrit :

$$(I8\text{-}4) \quad \begin{cases} \dfrac{1}{\sqrt{2}} \, (\phi_1^{(+)}(\vec{r}) + \phi_1^{(-)}(\vec{r}) \\ \qquad \text{(statistique de Bose-Einstein)}, \\[2mm] \dfrac{1}{\sqrt{2}} \, (\phi_1^{+}(\vec{r}) - \phi_1^{(-)}(\vec{r}) \\ \qquad \text{(statistique de Fermi- Dirac)}, \end{cases}$$

et il s'agit de tenir compte, dans le principe (18), de cet effet d'échange. Reprenons donc (18) terme à terme. A la fonction d'éssai $\overline{\psi}_2(\vec{r})$, que nous notons à partir de mainténant $\psi_1^{(+)}(\vec{r})$, associons $\psi_1^{(-)}(\vec{r})$ reliée à $\psi_2^{(+)}$ par

$$(18\text{-}5) \qquad \overline{\psi}_1^{(-)}(\vec{r}) = \psi_1^{(+)}(\vec{r}) \; .$$

En d'autres termes, les fonctions $\psi_1^{(+)}(\vec{r})$ et $\psi_1^{(-)}(\vec{r})$ sont des valeurs approchées des solutions exactes $\phi_1^{(+)}(\vec{r})$ et $\phi_1^{(-)}(\vec{r})$. On fait la même convention pour la fonction d'essai $\overline{\psi}_2(\vec{r})$. Il est donc nécessaire d'introduire les combinaisons

$$\frac{1}{\sqrt{2}} \, \overline{\psi}_1^{(+)} \pm \psi_1^{(-)}) \, , \, \frac{1}{\sqrt{2}} \, (\overline{\psi}_2^{(+)} \pm \overline{\psi}_2^{(-)})$$

(suivant la statistique adoptée) dans une formulation variationnelle devant tenir compte de l'échange.

Puisque j'ai à calculer $f(\theta) \mp f(\pi - \theta)$, j'aurai respectivement à ajouter on retrancher les valeurs de (18) relatives à $+\vec{k}_2$ et à $-\vec{k}_2$.

Soit

$$\begin{cases} \displaystyle \int e^{i\vec{k}_2 \cdot \vec{r}} \, V(r) \, \overline{\psi}_1^{(+)}(\vec{r}) \, d\vec{r} \; + \\[3mm] + \displaystyle \int e^{i\vec{k}_1 \vec{r}} \quad V(r) \, \overline{\psi}_2^{(+)}(\vec{r}) \, d\vec{r} - \int \overline{\psi}_1^{(+)} \, V \overline{\psi}_2^{(+)} \, d\vec{r} - \\[3mm] - \dfrac{1}{4\pi} \displaystyle \int \overline{\psi}_2^{(+)}(\vec{r}) \, V(r) \, \frac{e^{ik|\vec{r}-\vec{r}'|}}{|\vec{r}-\vec{r}'|} \, V(r') \overline{\psi}_1^{(+)}(\vec{r}') \, dr' \, d\vec{r} \; , \end{cases}$$

(18-6)
$$\int e^{-i\vec{k}_2\cdot\vec{r}}\, V(r)\, \overline{\Psi}_1^{(+)}(\vec{r})\, d\vec{r} + \int e^{i\vec{k}_1\cdot\vec{r}}\, V(r)\, \overline{\Psi}_2^{(-)}(\vec{r})\, d\vec{r} - \int \overline{\Psi}_1^{(+)}\, V\, \overline{\Psi}_2^{(-)}\, d\vec{r} - \frac{1}{4\pi}\int \overline{\Psi}_2^{(-)}(\vec{r})\, V(r)\, \frac{e^{ik|\vec{r}-\vec{r}'|}}{|\vec{r}-\vec{r}'|}\, V(r')\, \overline{\Psi}_1^{(+)}(\vec{r}')\, d\vec{r}\, d\vec{r}'$$

Etant donné la définition des fonctions,

$\overline{\Psi}^{(+)}(\vec{r})$ et $\overline{\Psi}^{(-)}(\vec{r})$, on vérifie aisément la saute d'égalités

(18-7)
$$\begin{cases} \int \left(e^{i\vec{k}_2\cdot\vec{r}} + e^{-i\vec{k}_2\cdot\vec{r}}\right) V(r)\, \overline{\Psi}_1^{(+)}(\vec{r})\, d\vec{r} = \\ = \pm \int \left(e^{i\vec{k}_2\cdot\vec{r}} + e^{-i\vec{k}_2\cdot\vec{r}}\right) V(r)\, \overline{\Psi}_1^{(-)}(\vec{r})\, d\vec{r} = \\ = \frac{1}{2}\int \left(e^{i\vec{k}_2\cdot\vec{r}} \pm e^{-i\vec{k}\cdot\vec{r}}\right) V(r)\left(\overline{\Psi}_1^{(+)}(\vec{r}) \pm \overline{\Psi}_1^{(-)}(\vec{r})\right) d\vec{r}, \end{cases}$$

(18-8)
$$\begin{cases} \int e^{i\vec{k}_1\cdot\vec{r}}\, V(r)\left(\overline{\Psi}_2^{(+)}(\vec{r}) \pm \overline{\Psi}_2^{(-)}(\vec{r})\right) d\vec{r} = \\ = \pm \int e^{-i\vec{k}_1\cdot\vec{r}}\, V(r)\left(\overline{\Psi}_2^{(+)}(\vec{r}) \pm \overline{\Psi}_2^{(-)}(\vec{r})\right) d\vec{r} \\ = \frac{1}{2}\int \left(e^{i\vec{k}_1\cdot\vec{r}} \pm e^{-i\vec{k}_1\cdot\vec{r}}\right) V(r)\left(\overline{\Psi}_2^{(+)}(\vec{r}) \pm \overline{\Psi}_2^{(-)}(\vec{r})\right) d\vec{r}, \end{cases}$$

(18-9)
$$\begin{cases} \int \overline{\Psi}_1^{(+)}(\vec{r})\, V(r)\left(\overline{\Psi}_2^{(+)}(r) \pm \overline{\Psi}_2^{(-)}(\vec{r})\right) d\vec{r} = \\ = \pm \int \overline{\Psi}_1^{(-)}(\vec{r})\, V(r)\left(\overline{\Psi}_2^{(+)}(\vec{r}) \pm \overline{\Psi}_2^{(-)}(r)\right) d\vec{r} = \\ = \frac{1}{2}\int \left(\overline{\Psi}_1^{(+)}(\vec{r}) \pm \overline{\Psi}_1^{(-)}(\vec{r})\right) V(r)\left(\overline{\Psi}_2^{(+)}(\vec{r}) \pm \overline{\Psi}_2^{(+)}(\vec{r})\right) d\vec{r}, \end{cases}$$

$$\int (\overline{\Psi}_2^{(+)}(\vec{r}) \pm \overline{\Psi}_2^{(-)}(\vec{r})) \frac{e^{ik|\vec{r}-\vec{r}'|}}{|\vec{r}-\vec{r}'|} V(r)\, V(r')\, \Psi_1^{(+)}(\vec{r}')\, d\vec{r}\, d\vec{r}' =$$

$$(18\text{-}10) \qquad = \frac{1}{2} \int (\overline{\Psi}_2^{(+)}(\vec{r}) \pm \Psi_2^{(-)}(\vec{r}))\, V(r) \frac{e^{ik|\vec{r}-\vec{r}'|}}{|\vec{r}-\vec{r}'|} V(\vec{r}')\, (\overline{\Psi}_1^{(+)}(\vec{r}') \pm$$

$$\pm \Psi^{(-)}(\vec{r}'))\, d\vec{r}\, d\vec{r}' .$$

La quantité $\frac{1}{\sqrt{2}} (e^{i\vec{k}_i \cdot \vec{r}} \pm e^{-\vec{k}_i \vec{r}})$ représente l'onde incidente suivant la direction $\vec{k}_2$ quand on tient compte de la statistique. De même pour la quantité $\frac{1}{\sqrt{2}} (e^{i\vec{k}_i \vec{r}} + e^{-i\vec{k}_i \vec{r}})$. Donc, finalement, on obtient les deux principes variationnels ci-dessous, directement applicables , soit aux bosons , soit aux fermions :

$$f(\theta) + f(\pi - \theta) =$$

$$(18\text{-}11) \quad \begin{cases} = \text{V.S.} \Big\{ \int \frac{e^{i\vec{k}_2 - r} + c^{-i\vec{k}_2 \cdot \vec{r}}}{\sqrt{2}} V(r)\, \Xi_2^{(+)}(\vec{r})\, d\vec{r} \; + \\ + \frac{e^{i\vec{k}_1 \cdot \vec{r}} + e^{-i\vec{k}_1 \vec{r}}}{\sqrt{2}} V(r)\, \Xi_2^{(+)}(\vec{r})\, d\vec{r} \; - \\ - \int \Xi_1^{(+)}(\vec{r})\, V(r)\, \Xi_2^{(+)}(\vec{r})\, d\vec{r} \; - \\ - \frac{1}{4\pi} \int \Xi_2^{(+)}(\vec{r})\, V(r) \frac{e^{ik|\vec{r}-\vec{r}'|}}{|\vec{r}-\vec{r}'|} V(r)\, \Xi_1^{(+)}(\vec{r}')\, d\vec{r}\, d\vec{r}' \Big\} \end{cases}$$

$$f(\theta) - f(\pi - \theta) =$$

$$(18\text{-}12) \quad = \text{V.S.} \Big\{ \int \frac{e^{i\vec{k}_2 \cdot \vec{r}} - e^{-i\vec{k}_2 \cdot \vec{r}}}{\sqrt{2}} V(r)\, \Xi_1^{(-)}(\vec{r})\, d\vec{r}$$

$$+ \int \frac{e^{ik_2 \cdot r} - e^{i\vec{k}_2 \cdot \vec{r}}}{\sqrt{2}} V(r)\, \Xi_2^{(-)}(\vec{r})\, d\vec{r} \; -$$

$$- \int \Xi_2^{(-)}(r)\ V(r)\ \Xi_2^{(-)}(\vec{r})\,d\vec{r}\ -$$

$$- \frac{1}{4\pi}\int \Xi_2^{(-)}(\vec{r})\ V(r)\ \frac{e^{ik|\vec{r}-\vec{r}'|}}{|\vec{r}-\vec{r}'|}\ V(r')\,\Xi_1^{(-)}(\vec{r})\,d\vec{r}\,d\vec{r}'$$

Dans les formules précédentes, $\Xi^{(+)}$ et $\Xi^{(-)}$ remplacent les quantités

$$\frac{1}{\sqrt{2}}\,(\Psi^{(+)}(\vec{r}) + \Psi^{(-)}(\vec{r}) \quad \text{et} \quad \frac{1}{\sqrt{2}}\,(\Psi^{(+)}(\vec{r}) - \Psi^{(-)}(\vec{r})\,)$$

et représentent les fonctions essai qui doivent être utilisées . On notera qu'elles sont assujetties à la condition d'être paires dans le cas de la statistique de Bose- Einstein et à la condition d'être impaires dans le cas de la statistique de Fermi - Dirac

Nous allons montrer maintenant comment (18) permet d'arriver aux méthodes proposés par Kohn (Phys Rev. 74 (1948) 1763 - Hulthén , X^{o} Congrès des Mathématiciens scandinaves – Copenhague 1946) .

5 Méthode de Kohn - A l'aide de la fonction δ de Dirac, on peut mettre l'opérateur $\hat{L} = \nabla^2 + k^2$ sous forme d'opérateur intégral dont le noyau $G(\vec{r}, \vec{r}')$ est évidemment symétrique . Ceci étant, l'équation (14) s'écrit

$$(19) \quad \begin{cases} V(\vec{r})\ \phi(\vec{r}) = \int G(\vec{r}, \vec{r}')\ \phi(\vec{r}')\ d\vec{r}' \\ \text{avec}\ \ \hat{L} = (\nabla^2 + k^2) = \int G(\vec{r}, \vec{r}')\{\cdots\}\,d\vec{r}\ , \end{cases}$$

En posant

$$(20) \qquad \phi_1(\vec{r}) = \phi_1^{o}(\vec{r}) + \frac{1}{\sqrt{V(\vec{r})}}\ \phi_1'(\vec{r})$$

on obtient, pour les $\phi_1'(r)$, l'équation intégrale suivante

(21) $$\phi'_1(\vec{r}) = -\varphi'^{o}_1(\vec{r}) + \int \frac{1}{\sqrt{V(\vec{r})}}\, G(\vec{r}\,\vec{r}')\, \frac{1}{\sqrt{V(r')}}\, \phi'_1(\vec{r}')\, dr'$$

où j'ai posé

(22) $$\phi_1^{o}(\vec{r}) = \sqrt{V(r)}\,\phi_1^{o}(\vec{r}) .$$

En écrivant notre principe variationnel relatif aux ϕ'_1, on obtient une méthode de calcul approché pour la quantité

(22) $$-\int e^{i(\vec{k}_1+\vec{k}_2)\vec{r}}\, V(\vec{r})\, d\vec{r} - 4\pi\, f(k_1, -k_2) ,$$

si bièn qu'en revenant aux ϕ initiaux, l'amplitude de diffusion $f(\vec{k}_1, -\vec{k}_2)$ est, au facteur 4π près, valeur stationnaire de la quantité

(23) $$4\pi\, f(\vec{k}_1, -\vec{k}_2) = \text{V.S.}\left\{\int \overline{\Psi}_2(\vec{r})\left[\nabla^2 + k^2 - V(\vec{r})\right] \times \overline{\Psi}_1(\vec{r})\, d\vec{r} - \int e^{i\vec{k}_2\cdot\vec{r}}\left[\nabla^2 + k^2\right]\overline{\Psi}_1(\vec{r})\, d\vec{r}\right.$$

où j'ai utilisé

(24) $$\left[\nabla^2 + k^2\right] = \int G(\vec{r}\,\vec{r}')\,\{\ldots\}\, d\vec{r}' .$$

Pour le calcul de la dernière intégrale de (23), on utilise la formule de Green :

$$\int \left(f\frac{\partial g}{\partial n} - g\frac{\partial f}{\partial n}\right) dS = \int \left(f\nabla^2 g - g\nabla^2 f\right) d\tau ,$$

$\overline{\Psi}_1$ étant une fonction d'essai que l'on supposera avoir la forme asymptotique

(25) $$\overline{\Psi}_1 = e^{i\vec{k}_1\cdot\vec{r}} + f(\vec{k}_1, \vec{k})\,\frac{e^{ikr}}{r}$$

On conduira le calcul de la façon indiquée dans le memoire de Kohn (cor. cit. p. 1766) . pour retrouver exactement le principe variationnel qu'il à proposé.

Il a lieu de remarquer que le principe (18) que nous proposons, bien que découlant de la même méthode, fournit néanmoins une précision supérieure a celle obtenue par la méthode de Kohn . En effet, en prenant pour fonction d'èssai simplement l'approximation d'orde zéro (onde plane) , le principe de Kohn fournit la première approximation de Born, tandis que notre principe (18) donne directement jusqu'à la seconde approximation.

6. Méthode relative à l'équation radicale de la diffusion.

Nous allons maintenant déduire les méthodes relatives à l'équation radicale de l'équation de la diffusion directement de notre principe variationnel général. A cet effet, prenons pour les fonctions $\overline{\Psi}_1$ et $\overline{\Psi}_2$ qui figurent dans l'expression (18) un devèloppement de la forme

$$\overline{\Psi}_1(\vec{r}) = \sum_{n=0}^{\infty} (2n+1)\, i^n P_n(\cos\theta)\, \Lambda_n(|\vec{r}|) ,$$

(26)

$$\overline{\Psi}_2(\vec{r}\,') = \sum_{0}^{\infty} (2n+1)\, i^n P_n(\cos \widehat{\vec{k}_2 \vec{r}})\, \Lambda_n(|\vec{r}|) ,$$

où $\Lambda_n(\vec{r})$ a la forme asymptotique

(27) $$\Lambda_n(|\vec{r}|) \underset{r\to\infty}{\longrightarrow} \frac{e^{in'_n}}{kr} \sin\left(kr - \frac{n\pi}{2} + n'_n\right), \quad (|\vec{r}| = r)$$

la direction $\vec{k}_1$ étant prise comme axe des z et la direction $\vec{k}_2$ ayant pour cordoonnés polaires (Θ , 0)

Nous aurons également à utiliser le dévelop-pement suivant (*)

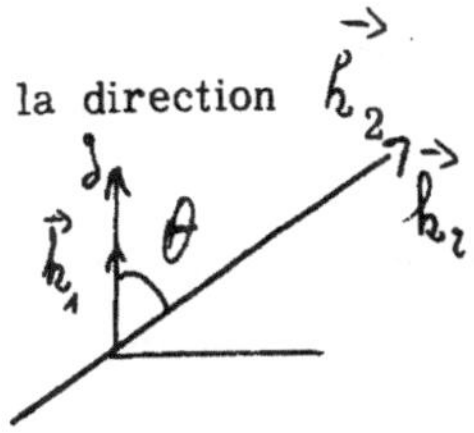

(*) Mott-Sneddon, Wave mechanics and its application Oxford, 1948 p. 386

$$
(28)\quad
\begin{cases}
-\dfrac{1}{4\pi}\,\dfrac{e^{ik|\vec r-\vec r'|}}{|\vec r-\vec r'|} = -\dfrac{1}{4\pi}\displaystyle\sum_0^\infty \dfrac{2n+1}{k}\,\dfrac{\mathcal{S}_n(kr_>)}{r_>}\,\dfrac{\zeta_n(kr_<)}{r_<}\,P_n\cos(\vec r,\vec r') \\[2ex]
\zeta_n(kr) = \sqrt{\dfrac{\pi k r}{2}}\; J_{n+1/2}(kr) \\[2ex]
\mathcal{S}_n(kr) = \sqrt{\dfrac{\pi k r}{2}}\Big\{ i\; J_{n+1/2}(k\,r) + (-1)^n\; J_{-n-\frac12}(k\,r)\Big\},
\end{cases}
$$

ainsi que l'expression

$$
P_n(\cos \widehat{\vec r,\vec r'}) = P_n(\cos\theta)\,P_n(\cos\theta') +
$$

$$
+\;2\sum_{m=1}^{n}\frac{(n-m)!}{(n+m)!}\,P_n^m(\cos\theta)\,P_n^m(\cos\theta')\cos m(\varphi-\varphi'),
$$

(θ ,) étant les coordonnées polaires de , (θ', ') celles de r' . Bien entendu le potentiel V(r) , qui va intervenir, ne dépend, par hypothèse, que de la distance r à l'origine (symètriie spherique) .

Ceci étant, (18) se met sous la forme

$$
(30)\quad
\begin{aligned}
&4\pi\sum_0^\infty (2n+1)(-1)^n\,P_n(\cos\Theta) \\
&\times\Big[\frac{2}{k}\int_0^\infty \sqrt{\frac{\pi k r}{2}}\;J_{n+1/2}(k'r)\,\Gamma_n(r)\,V(r)\,dr\;- \\
&-\int_0^\infty (\Gamma_n(r))^2\,V(r)\,dr\;- \\
&-\frac{1}{k}\int\!\!\int \Gamma_n(r)\,V(r)\,\mathcal{S}_n(k\,r_>)\,\zeta_n(k\,r_<)\,V(r')\,\Gamma_n(r')\,dr\,dr'\Big],
\end{aligned}
$$

où j'ai posé

$$
\Lambda_n(r) \equiv \frac{\Gamma_n(r)}{r}
$$

Quand Ψ_1 et Ψ_2 sont des solutions exactes Φ_1 et Φ_2 de l'équation d'onde , il est clair que Λ_n devient $L_n(r)$, solution de l'équation différentielle radicale, relative à la valeur n du moment cinétique . On obtiendra des Ψ_1 et Ψ_2 différant peu les Φ_1, et Φ_2 en prenant dans (26) des Λ_n différant peu des $L_n(r)$. La variation des $\overline{\Psi}_2$ et $\overline{\Psi}_2$ ainsi obtenue n'est pas la variation la plus générale, mais c'est la seule qui importe dans le cas de la symétrie sphérique que nous envisageons ici. La prèmière variation de (30) qui résulte de cette variation radicale, doit être nulle et, par suite de l'orthogonalité des polynômes de Legendre, la première variation de chacune des quantités entre crochets de 30 est nulle. Quand Γ_n (r) se confond avec $G_n(r) = L_n(r)$, cette quantité entre corchets devient égale à

$$-\frac{e^{2i\eta_n}-1}{2ik^2},$$

où η_n est le déphasage habituel (*) Donc, l'expression entre crochets provenant de (30) :

$$-\frac{e^{2i\eta_n}-1}{2ik^2} = \text{V.S.}\left\{\frac{2}{k}\int_0^\infty \sqrt{\frac{\pi k r}{2}}\, J_{n+1/2}(kr)\Gamma_n(r)\, V(r)\, dr - \right.$$
$$(30) \qquad - \int_0^\infty (\Gamma_n(r))^2\, V(r)\, dr -$$
$$- \frac{1}{k}\int\int \Gamma_n(r)\, V(r)\, \mathcal{f}_n(k\, r_>)\, \mathcal{g}_n(k\, r_<)$$
$$\left. \times\ V(r')\Gamma_n(r')\, dr\, dr' \right\}.$$

fournit une méthode variationnelle de calcul approchée des déphasages. En même temps, (31) donne l'équation intégrale que doit vérifier $G_n(r)$, à

(*) Remarquons que la fonction $G_n(r)$ introduite ici diffère de celle que l'on trouve dans Mott et Massez , cit, ch. II et ch VII, en ce qu'elle a la forme asymptothique :

$$\frac{e^{i\eta_n}}{k}\sin(kr - n\,\pi/2 + \eta_n).$$

à savoir

$$(32) \qquad G_n(r) = \frac{1}{k}\sqrt{\frac{\pi kr}{2}}\; J_{n+1/2}(kr) - \frac{1}{k}\int_0^\infty \mathcal{J}_n(kr_>)\, \xi(kr_<)\, V(r')\, G_n(r')\, dr' .$$

Nous allons maintenant établir en détail les calculs conduisant à la formule (30) . Procédons terme à terme .

a) Calcul du terme :

$$-\frac{1}{4\pi}\int \overline{\Psi}_2(\vec{r})\, V(r)\, \frac{e^{ik|\vec{r}-\vec{r}'|}}{|\vec{r}-\vec{r}'|}\, V(r')\, \overline{\Psi}_1(\vec{r}')\, \overline{\Psi}_1(\vec{r}')\, d\vec{r}\, dr'$$

Nous effectuons d'abord l'intégration sur $d\vec{r}'$ en tenant compte du fait que $\phi_1(\vec{r})$ ne dèpend pas de φ'. Ce qui donne

$$-\frac{1}{4\pi}\int \frac{e^{ik|\vec{r}-\vec{r}'|}}{|\vec{r}-\vec{r}'|}\, V(r')\, \Psi_1(\vec{r}')\, d\vec{r}' =$$

(30-1) $$= \sum_{n'=0}^{\infty}\sum_{n=0}^{\infty} -\frac{1}{2k}\,(2n'+1)\,(2n+1)\, i^n P_{n'}(\cos\theta)$$

$$\times \int_0^\infty \frac{\mathcal{J}_{n'}(kr_>)\,\xi_{n'}(kr_<)}{r_>\, r_<}\, V(r')\, \Lambda_n(r')\, r'^2\, d\vec{r}'$$

$$\times \int_0^\pi P_{n'}(\cos\theta')\, P_n(\cos\theta)\, \sin\theta'\, d\theta' .$$

En utilisant les relations d'orthogonalité des polynômes de Legende (*) il vient

---

(*) Valiron : Théorie des founctions (Masson) p. 207

$$-\frac{1}{4\pi}\int \frac{e^{ik|\vec{r}-\vec{r}'|}}{|\vec{r}-\vec{r}'|} V(r') \overline{\Psi}_1(\vec{r}') \, d\vec{r} =$$

$$(30\text{-}2) \qquad = \sum_{n=o}^{\infty} - \frac{2n+1}{k} i^n P_n(\cos\theta)$$

$$\times \int_0^{\infty} \frac{\varphi_n(kr_<)\,\zeta_n(kr_>)}{r_< r_>} V(r') \Lambda_n(r'^2) \, dr' .$$

Nous devons maintenant effectuer l'intégration sur $d\vec{r}$. La variable n'apparaitra que dans $\overline{\Psi}_2(\vec{r})$ et ainsi en effectuant d'abord l'intégration sur φ nous ferons disparaître les termes de $\overline{\Psi}_2(\vec{r})$ qui contiennent $\cos m\varphi$ et il reste

$$\frac{1}{4\pi}\int \overline{\Psi}_2(\vec{r}) V(r) \frac{e^{ik|\vec{r}-\vec{r}'|}}{|\vec{r}-\vec{r}'|} V(r') \overline{\Psi}_1(\vec{r}') \, d\vec{r}\, d\vec{r}' =$$

$$(30\text{-}3) \qquad = -\frac{2\pi}{k} \sum_{n=o}^{\infty} \sum_{n'=o}^{\infty} (2n+1)(2n'+1)\, i^{n+n'} P_n(\cos\Theta) \times \int_0^{\infty} \Lambda_n(r) V(r) \frac{\varphi_{n'}(kr_>)\,\zeta_{n'}(kr_<)}{r_> r_<}$$

$$\times V(r') \Lambda_{n'}(r') r^2 r'^2 \, dr\, dr'$$

$$\times \int_0^{\pi} P_n(\cos\theta) P_{n'}(\cos\theta) \sin\theta \, d\theta =$$

$$= -\frac{4\pi}{k} \sum_0^{\infty} (2n+1)(-1)^n P_n(\cos\Theta)$$

$$\times \int_0^{\infty} \Lambda_n(r) V(r) \frac{\varphi_n(kr_>)\,\zeta_n(kr_<)}{r_> r_<}$$

$$\times V(r) \Lambda_n(r') r^2 r'^2 \, dr\, dr' .$$

6 <u>Calcul du terme</u>

$$-\int \overline{\Psi}_1(\vec{r}) V(r) \overline{\Psi}_2(\vec{r}) \, d\vec{r} .$$

$\overline{\Psi}_1(\vec{r}) V(r)$ ne contenant pas φ, en effectuant d'abord l'intégration sur φ on élimine les termes de $\overline{\Psi}_2(\vec{r})$ qui contiennent cos $m\varphi$, et il reste

$$(30\text{-}4) \qquad -\int \Psi_1(\vec{r})\, V(r)\, \overline{\Psi}_2(\vec{r})\, d\vec{r} =$$

$$= -2\pi \sum_{n=o}^{\infty} \sum_{n'=o}^{\infty} (2n+1)(2n'+1)\, i^{n+n'}\, P_{n'}(\cos \Theta)$$

$$\times \int_o^{\infty} \Lambda_n(r)\, V(r)\, \Lambda_{n'}\, r^2\, dr$$

$$\times \int_o^{\pi} P_n(\cos\theta)\, P_{n'}(\cos\theta)\, \sin\theta\, d\theta =$$

$$= -4\pi \sum_{n=o}^{\infty} (2n+1)\,(-1)^n\, P_n(\cos \Theta)$$

$$\times \int_o^{\infty} (\Lambda_n(r))^2\, V(r)\, r^2\, dr\,,$$

Toujours en utilisant les relations d'orthogonalité des polynômes de Legendre.

Pour le calcul des deux termes restants, on doit utiliser le développement bien connu (*)

$$(30\text{-}5) \qquad e^{i\vec{k}.\vec{r}} = \sum_{n=o}^{\infty} (2n+1) i^n\, P_n(\cos \widehat{\vec{k}.\vec{r}})$$

$$\times \sqrt{\frac{\pi}{2kr}}\; J_{n+1/2}(kr)\,,$$

et en employant le développement (29) des polynômes de Legendre , la quantité

$$c) \qquad \int e^{i\vec{k}.\vec{r}}\, V(r)\, \Psi_1(\vec{r})\, d\vec{r}$$

(*) L. de Broglie , loc. cit p. 20
T. Kahan , Precis. . de Physique Théorique t.II (P. UF.)

devient, après élimination des termes contenant φ par intégration sur φ:

(30-6)
$$\int e^{i\vec{k}_2\cdot\vec{r}}\, V(r)\, \overline{\Psi}_1(\vec{r})\, d\vec{r} =$$

$$= 2\pi \sum_{n=o}^{\infty} \sum_{n'=o}^{\infty} (2n+1)(2n'+1)\, i^{n+n'} P_n(\cos \Theta)$$

$$\times \int_0^{\infty} \sqrt{\frac{\pi}{2kr}}\, J_{n+1/2}(kr)\, V(r)\, \Lambda_n(r)\, r^2\, dr$$

$$\times \int_0^{\pi} P_n(\cos\theta)\, P_{n'}(\cos\theta)\, \sin\theta\, d\theta =$$

$$= 4\pi \sum_{o}^{\infty} (2n+1)(-1)^n\, P_n(\cos \Theta)$$

$$\times \int_0^{\infty} \sqrt{\frac{\pi}{2kr}}\, J_{n+1/2}(kr)\, V(r)\, \Lambda_n(r)\, r^2\, dr .$$

Le calcul du terme

d)
$$\int e^{i\vec{k}_i\cdot\vec{r}}\, V(r)\, \overline{\Psi}_2(\vec{r})\, d\vec{r}$$

s'effectue de la même façon . Là encore, les termes contenant φ et provenant du développement de $\overline{\Psi}_2(\vec{r})$ s'eliminent par intégration sur φ et l'utilisation des relations d'orthogonalité des polynômes de Legendre donne

(30-7)
$$\int e^{i\vec{k}_i\cdot\vec{r}}\, V(r)\, \overline{\Psi}_2(\vec{r})\, d\vec{r} =$$

$$= 4\pi \sum_{n=o}^{\infty} (2n+1)(-1)^n\, P_n(\cos \Theta)$$

$$\times \int_0^{\infty} \sqrt{\frac{\pi}{2kr}}\, J_{n+1/2}(kr)\, V(r)\, \Lambda_n(r)\, r^2\, dr ,$$

Si bien que l'expression (18) devient finalement

$$(30\text{-}8)\qquad 4\pi \sum_0^\infty (2n+1)(-1)^n P_n(\cos \Theta)$$

$$\times \left[\frac{2}{k} \int_0^\infty \sqrt{\frac{\pi kr}{2}}\, J_{n+1/2}(kr)\,\Gamma_n(r)\, V(r)\, dr \; - \right.$$

$$- \int_0^\infty (\Gamma_n(r))^2\, V(r)\, dr \; -$$

$$\left. - \frac{1}{k} \int_0^\infty \Gamma_n(r)\, V(r)\, \mathcal{J}_n(kr_>)\, \xi(kr_<)\, V(r')\, \Gamma_n(r')\, dr\, dr' \right]$$

7. Méthode de Schwinger déduite de notre principe général (3)

Nous allons montrer maintenant que la méthode de Schwinger(*) se déduit de notre principe général (3) en donnant une forme convenable à l'équation intégrale (32) . A cet effet, introduisons une nouvelle fonction $\mathcal{G}_n(r)$ reliée à G(r) par

$$\mathcal{G}_n(r) \equiv k\, G_n(r)\, \frac{e^{-i\eta_n}}{\cos \eta_n}$$

Cette nouvelle fonction vérifie alors l'équation intégrale

$$(34)\qquad \mathcal{G}_n(r) = \sqrt{\frac{\pi kr}{2}}\, J_{n+1/2}(kr) \; -$$

$$- \frac{(-1)^n}{k} \int_0^\infty \sqrt{\frac{\pi kr_>}{2}}\, J_{n+1/2}(kr_>)$$

$$\times \sqrt{\frac{\pi kr_<}{2}}\, J_{-n-1/2}(kr_<)\, V(r')\, \mathcal{G}_n(r')\, dr' ,$$

d'où l'on tire :

$$(35)\qquad \operatorname{tg} \eta_n = - \int \sqrt{\frac{\pi kr}{2}}\, J_{n+1/2}(kr)\, V(r) \cdot \mathcal{G}_n(r)\, dr .$$

(*) J. Schwinger Conférences inédites, 1947. Blattand Jackson, Phys. Rev, 76 (1949) 21 .

On peut alors écrire (34) sous la forme

(36) $$\mathcal{G}_n(r) = \int \mathcal{R}(r,r')\, V(r')\, \mathcal{G}_n(r')\, dr' ,$$

après avoir posé

(37) $$\mathcal{R}(r,r') \equiv \frac{1}{\operatorname{tg}\eta_n} \sqrt{\frac{\pi kr}{2}}\, J_{n+1/2}(kr) \sqrt{\frac{\pi kr'}{2}}\, J_{n+1/2}(kr') + \frac{(-1)^n}{k} \sqrt{\frac{\pi kr_<}{2}}\, J_{n+1/2}(k r_<) \sqrt{\frac{\pi kr_>}{2}}\, J_{-n-1/2}(k r_>) .$$

On symètrise finalement le noyau précédent en posant

(38) $$\mathcal{K}_n(r) = \sqrt{V(r)}\; \mathcal{G}_n(r) ,$$

et l'équation prend la forme

$$\hat{\mathcal{L}}\, \mathcal{K}_n = 0$$

avec

(39) $$\hat{\mathcal{L}} = \int \left\{ \delta(r-r') + \sqrt{V(r)}\, \mathcal{R}(r,r') \sqrt{V(r')} \left\{ \dots \right\} dr' \right. .$$

En appliquant alors le principe (3), on conclut que l'expression

(40) $$\operatorname{cotg} \eta_n = \text{V.S.} \left\{ \int \mathcal{G}_n^2(r)\, V(r)\, dr - \frac{(-1)^n}{k} \iint (r)\, V(r) \sqrt{\frac{\pi kr_<}{2}}\, J_{n+1/2}\, kr \times \sqrt{\frac{\pi kr}{2}}\, J_{-n-1/2}(kr_>)\, V(r')\, \mathcal{G}(r')\, dr\, dr' + \frac{1}{k} \dots \left[\sqrt{\frac{\pi kr}{2}}\, J_{n+1/2}(kr)\, V(r)\, \mathcal{G}_n(r)\, dr \right]^2 \right\}$$

a une valeur stationnaire pour les solutions exactes du problème de diffusion. (40) fournit la méthode de calcul approché de cotg η_n proposée par Schwinger (*)

(*) cf. par exemple, Blatt-Jackson, loc. cit. p. 21, J. Schwinger Loc. cit.

8 Généralisation[1] à des opérateurs non symétriques du principe variationnel général

Considérons un opérateur linéaire $\hat{L}$ et l'équation homogène (sans second membre)

(41) $$\hat{L}\bar{\Psi} = o$$

avec sa transposée hermétique[2]

(42) $$\hat{L}^{+}\phi = o$$

Nous allons montrer en appliquant notre technique générale que la quantité

(43) $$J = \int \phi'^{+} \; \hat{L} \; \bar{\psi}' d\tau \, ,$$

nulle quand ϕ', $\bar{\psi}'$ viennent coincider avec le soutions ϕ et $\bar{\Psi}$ de (41 et (42) respectivement, ne diffère de zéro que par des termes infiniment petits du second ordre quand ϕ' et $\bar{\psi}'$ diffèrent de ϕ et $\bar{\psi}$ par des infiniment petits du premier ordre . Posons en effet ,

$$\phi' = \phi + \delta\phi, \quad \bar{\psi}' = \bar{\psi} + \delta\bar{\psi}$$

Il vient alors

$$\delta J = \int \phi'^{+} \hat{L} \; \bar{\psi}' d\tau =$$

$$= \int (\phi + \delta\phi)^{+} \; \hat{L} \; (\psi + \delta\psi) \, d\tau =$$

$$= \int (\delta\phi)^{+} \; \hat{L} \psi \; d\tau + \int \phi^{+} \hat{L} \, \delta\psi \; d\tau$$

La première intégrale est nulle par (41) . Quant à la seconde, elle peut s'écrire

(1) P. Roussopoulos. C.R. Accad. Sc. Paris t. 236 (1953)1859) Kahan - Rideau-Roussopoulos, loc. cit. 22

(2) l'opérateur adjoint $\hat{L}^{+}$ adjoint à $\hat{L}$ est défini par

(43-1) $$\int \psi^{+} \; L \, \phi \, d\tau = \int H^{+} \psi^{+} \phi \, d\tau \langle \psi | \hat{L} | \phi \rangle = \overline{\langle \phi | \hat{L}^{+} \psi \rangle}$$

$$\int \phi^+ \hat{L}\, \delta\psi \, d\tau = \left[\int \delta\psi^+ \hat{L}^+ \phi \, d\tau \right]^+ = 0$$

en appliquant la définition du transposé hermitique (adjoint) d'un opérateur et l'équation (42).

Le cas plus compliqué de l'équation inhomogène ou avec second membre où $\overline{\psi}$ et ϕ sont solutions de

(44) $$\hat{L}\,\overline{\Psi} = f,$$

(45) $$\hat{L}^+ \varphi = g$$

se ramène au précédent en posant

$$\overline{\Psi} = \overline{\Psi} - (\hat{L}^{-1})\, f, \qquad \phi = \varphi - (\hat{L}^{-1})^+ \, g,$$

où $\overline{\psi}$ et ϕ vérifient maintenant des équations du type (41) et (42).

Il vient alors

(46) $$J = \int \phi' \hat{L} \overline{\Psi}' \, d\tau = \int (\varphi' - (\hat{L}^{-1})^+ g)\, \hat{L}(\psi' - (L^{-1})^+ f)\, d\tau$$

$$= \int \Big[\phi' \hat{L} \psi' - \phi' \hat{L}\, (\hat{L}^{-1})^+ f -$$

$$- (\hat{L}^{-1})^+ g(\hat{L}\, \psi') - (\hat{L}^{-1})^+ g\hat{L}(\hat{L}^{-1})^+ f \Big] \, d\tau^{(1)} =$$

$$= \int \Big[\phi' \hat{L} \psi' - \phi' f - \phi \hat{L} \psi' + \phi f \Big] \, d\tau$$

$$= \int \Big[\phi \hat{L} \psi' - \phi' f - g\psi' + \phi f \Big]$$

Finalement, nous pouvons écrire

(47) $$\left.\begin{matrix} \int g^+ \psi \, d\tau \\ \int \varphi^* f d\tau \end{matrix}\right\} = \text{V.S.} \left\{ \int \varphi'^* f d\tau + \int \psi' g^+ d\tau - \int \varphi'^+ \hat{L} \psi' d\tau \right\} =$$

$$= \text{V.S.} \left\{ (\varphi'^+, f) + (\psi', g^+) - (\varphi^+, \hat{L}\psi') \right\}$$

(1) $(\hat{L}^{-1})^+ = (\hat{L}^+)^{-1}$.

En effet de $1 = 1^+ = (\hat{L}\, L^{-1})^+ = (\hat{L}^{-1})^+ \hat{L}^+$, on tire $(\hat{L}^{+1})^{-1} = (L^{-1})^+$. Si $\hat{L}$ est hermitique $(\hat{L}^+)^{-1} = \hat{L}^{-1}$.

La notation V.S. signifiant comme précédement, que la quantité au premier nombre de (47) est la valeur stationnaire (ou extrémale) de la fonctionnelle du second membre.

Les formules précédentes se simplifient quand $\hat{L}$ est un opérateur hermitique ($\hat{L}^+ = \hat{L}$) car nous aurons à considérer alors comme équations

$$\hat{L}\,\psi_1 = f_1, \; L^+\psi_2 = L\psi_2 = f_2,$$

(47) s'écrivant dans ce cas

$$(48) \qquad \left.\begin{matrix}\int f_2^* \psi_1 \, d\tau \\ \int \psi_2^* f_1 \, d\tau\end{matrix}\right\rangle = \text{V.S.}\left\{\int \psi_1' f_2^+ \, d\tau + \int \psi_2^{1+} f_1 \, d\tau - \int \left[\psi_2'\right]^+ \hat{L}\psi_1' \, d\tau\right\}$$

Remarquons que le principe variationnel (47) contient à titre de cas particulier notre principe variationnel pour opérateur symètrique (8). Ce principe s'écrirait

$$(49) \qquad \left\{\begin{matrix}\int \psi_1 f_2 \, d\tau \\ \int \psi_2 f_1 \, d\tau\end{matrix}\right\rangle = \text{V.S.}\left\{\int \psi_2' f_1 d\tau + \int \psi_1' f_2 \, d\tau - \int \psi_2' \; \hat{L}\psi_1' \, d\tau\right\},$$
$$= \text{V.S.}\left\{(\psi', f_1) + (\psi_1' f_2) - (\psi'_2, \hat{L}\psi_1)\right\},$$

ψ'_1 et ψ'_2 étant des formes approchées de ψ_1, ψ_2 solutions des équations différentielles

$$(50) \qquad \hat{L}\psi_1 = f_1 \; , \qquad \hat{L}\,\psi_2 = f_2 \; ,$$

$\hat{L}$ étant un opérateur symétrique tel que $\int \varphi_1 \hat{L}\,\varphi_2 \, d\tau = \int \varphi_2 \hat{L}\varphi_1 \, d\tau$. Nous montrerons aisément que (49) est identique à (47) en envisageant $\hat{L}$ sous forme d'un opérateur intégral

$$(51) \qquad \hat{L}\,\Phi = \int L \; (\tau, \tau')\,\Phi(\tau')\, d\tau'$$

Si $\hat{L}$ est la combinaison de dérivées de divers ordres il suffit alors de les remplacer par des opérateurs intégraux dont les noyaux sont des dérivées de divers ordres de la fonction δ, de Dirac (cf. p. 7,11). Ceci étant, on déduira aisément que le noyau propre au transposé hermitique de $\hat{L}$ sera

$$\int d\tau' \left\{ \quad \right\} L^{+} (\tau', \tau),$$

et que le noyau d'un opérateur métrique sera lui-même une fonction symétrique de τ et τ'. Des lors, la seconde équation (50) qui s'écrit

$$\int \hat{L}(\tau, \tau') \psi_2 (\tau') d\tau' = f_2(\tau) = \int d\tau' \hat{L} (\tau', \tau) \psi_2(\tau')$$

pourra se mettre sous la forme

$$\int d\tau' \psi_2(\tau')^{+} \hat{L}^{+}(\tau', \tau) = f_2(\tau)^{+},$$

ce qui permet de conclure que, dans le cas des opérateurs symétriques, les deux équations

$$\hat{L} \psi_2 = f_2 , \qquad L^{+} \psi_2^{+} = f_2^{+}$$

sont équivalentes.

Il suffit maintenant d'écrire le principe variationnel (47) en partant du compte d'équation

$$L \psi_1 = f_1 , \quad L^{+} \psi_2^{+} = f_1^{+}$$

pour retrouver le principe variationnel (49).

On peut généraliser notre principie variationnel pour des systèmes d'équation. Envisageons par exemple le système homogéne

(52) $$\hat{L}_j (\psi_j) = 0 ,$$

les L_j étant des opérateurs différentielles ou linéaires. E notation vectorielle, (52) peut s'écrire

(53) $$\hat{\mathcal{L}}(\vec{\psi}) = 0$$

où

$$\hat{\mathcal{L}} = \begin{pmatrix} \hat{L}_1 & & & & \\ & L_2 & & & \\ & & L_j & & \\ & & & \cdot & \\ & & & & \cdot L_n \end{pmatrix}$$

qui est du type (41) . On peut alors montrer que l'expression

(53) $$J = \int (\vec{\phi}, \hat{\mathcal{L}}) \vec{\psi}' \, d\tau$$

est extrémale où verifiě l'équation adjointe à (53) .

Considérons, à titre d'application, l'opérateur symétrique

(55) $$\hat{L} \; \frac{d}{dx_i} \left(a^{ij} \frac{d}{dx_j}\right) + c \;, \qquad (i, j = 1, 2, 3, \ldots, n)$$

où les a^{ij} et c sont des fonctions des coordonnées x_i et $a^{ij} = a^{ji}$ une fo nction telle que (56) $\hat{L}\varphi = 0$ avec les conditions aux limites

(57) $$\left(a^{ij} \frac{d\varphi}{dx_j} \; i\right)_S = 0$$

C'est le cas simple où $\overline{\psi} = 0$ dans (41) et (43) prend la forme

(58) $$J = -\int \varphi \hat{L} \varphi \, d\tau = -\int \varphi \left[\frac{d}{dx_i}\left(a^{ij} \frac{d\varphi}{dx_j}\right) + c\varphi\right] d\tau =$$

$$= -\int \frac{d}{dx_i}\left(a^{ij} \frac{d\varphi}{dx_j}\right) d\tau + \int a^{ij} \frac{d\varphi}{dx_j} \frac{d\varphi}{\partial i} \, d\tau - \int c\varphi^2 \, d\tau =$$

$$= -\int a^{ij} \frac{d\varphi}{dx_j} n_i \, dS + a^{ij} \frac{d\varphi}{dx_j} \frac{d\varphi}{dx_i} \, d\tau - \int c\varphi^2 \, d\tau$$

$$= a^{ij} \frac{d\varphi}{dx_j} \frac{d}{\partial x_j} - c\varphi^2 \Big] \, d\Sigma \;,$$

conpte tenu de la condition aux limites (57).

Soit maintenant le système

(59) $$\frac{d}{dx_j}\left(a^{ij}_{\alpha\beta}\frac{d\varphi^{\alpha}}{dx_i}\right) c_{\alpha\beta}\,\mu^{\alpha} = 0$$

où

$$i, j = 1, 2, 3, \dots, n \quad \text{et} \quad \alpha, \beta = 1, 2, 3, \dots, m,$$

et

(60) $$a^{ij}_{\alpha\beta} = a^{j\,o}_{\alpha\beta} = a^{ij'}_{\beta\alpha}, \quad c_{\alpha\beta} = c_{\beta\alpha}$$

avec les conditions aux limites

(61) $$\left(a^{ij}\frac{d\varphi^{\alpha}}{dx_i}n_j\right)_S = 0$$

Dans ce cas, le principe variationnel à appliquer est du type (54). Contentons-nous d'énoncer le principe sans le démontrer :

(62) $$J = \int\left[a^{ij'}_{\alpha\beta}\frac{d\varphi^{\alpha}}{dx_i}\frac{d\varphi^{\beta}}{d_j} - c_{\alpha\beta}\varphi^{\alpha}\varphi^{\beta}\right]d\tau$$

(58) et (62) jouent un rôle important dans la méthode de l'hypercycle lorsque la métrique n'est pas définie positive (*)

9 Principe variationel général, comme extension du principe de Ritz-Rayleigh

L'inconvénient des principes variationels de la forme (47) ou (48) est de fournir une approximation dépendant de la normalisation particulière adoptée pour les fonctions d'essai. De plus, appliquée, à la mécanique quantique, ils fournissent une méthode de calcul qui ne diffère pas essentiellement du calcul des perturbations dont on sait, par ailleurs, les dif-

(*) cf. J.L. Synge, The Hypercycle in Mathematical Physics, p. 292 et (Cambridge Unniversity Press, 1957)

L. Cairo et T. Kahan, Variational Techniques in Electromagnetique, p. 58 / Bmacjoe Mp, dp, ; 1965

ficultés . Aussi donnerons-nous maintenant un principe variationnel général; extension du principe Ritz-Rayleigh qui dans son application à la mécanique quantique , échappe aux objections précédentes. Un cas particulier a déjà été. étudié de façon d'étaillée par Schwinger, Blatt et Jackson (loc .cit.) et utilisé pour la détermination des déphasages dans un champ de forces centrales V(Ω) .

Rappelons d'abord brièvement ce qu'est le principe de Ritz-Rayleigh[1] Nous allors • le déduire au demeurant par la suite (65) (83) de notre principe variationnel géneral Le problème posé est celui de la détermination de la plus petite valeur propre de

$$\hat{A}\psi = \lambda \hat{B}\psi,$$

où $\hat{A}$ et $\hat{B}$ sont des opérateurs hermitiques définis positif . En utilisant cette dernière hypothèse, on peut poser

$$\Phi = \hat{B}^{1/2}\psi ,$$

de sorte que nous sommes ramenés à

$$\hat{B}^{-1/2}\,\hat{A}\,\hat{B}^{-\frac{1}{2}}\Phi - \lambda\,\Phi \quad ,$$

le problème de valeur propre la plus simple possible. On vérifie aisément que

$$\hat{A}' = \hat{B}^{-\frac{1}{2}}\,A\,\hat{B}^{-\frac{1}{2}} \quad ,$$

est comme un opérateur hermitique défini positif. On peut alors démontrer que

$$(63) \qquad \lambda = \text{V.S.}\ \frac{(\Phi,\hat{A}'\Phi)}{(\Phi,\Phi)}$$

---

[1]Pour les détails cf. par exemple comant -Hilbert, Méthoden der mathematischen Physik ,ou Schwinger Lect. Nots Haward U . 1947) .

En effet

$$\delta \frac{(\phi, \hat{A}'\phi)}{(\phi, \phi)} = \frac{1}{(\phi, \phi)^2} \Big\{ (\phi, \phi) \Big[(\delta\phi, \hat{A}'\phi) + $$

$$+ (\phi, \hat{A}'\delta\phi) - (\delta\phi, \phi) + (\phi, \delta\phi) \Big] (\phi, \hat{A}'\phi \Big\},$$

est en tenant compte de ce que $\hat{A}'\phi = \lambda\phi$ donne

$$\lambda = \frac{(\phi, \hat{A}'\phi)}{(\phi, \phi)}$$

pour les solutions exactes ϕ , on trouve, après l'utilisation de l'hermiticié, une première variation nulle. Notons que dans cette démonstration, aucun usage n'à été fait du caractère d'opérateur défini positif de A' . Ceci n'entre en jeu que

dans la suite des démontrations permettant de conduire à la convergence du procédé vers la plus basse valeur propre d'opérateur.

Ceci rappelé, envisageons de nouveau des équations de la forme

(64) $$\hat{L}\,\overline{\Psi} = f, \quad \hat{L}^{+}\phi = f_2 ,$$

la quantité que nous voulons calculer étant

(65) $$Q = \int f_2^{+} \overline{\Psi}\, d\tau = \int f_1 \phi^{+}\, d\tau$$

on peut alors r'écrire les équations (64) sous la forme

(66) $$\hat{L}\,\overline{\Psi} = f_1 = \frac{\int f_1(\tau)\, f_2^{+}(\tau')\, \overline{\Psi}(\tau_1')\, d\tau'}{\int f_2^{+}(\tau')\, \overline{\Psi}(\tau')\, d\tau'} =$$

$$= \frac{1}{Q} \int f_1(\tau)\, f^{+}(\tau')\, \overline{\Psi}'(\tau')\, d\tau'$$

$$\hat{L}^{+} = \frac{1}{Q^{+}} \int f_2(\tau)\, f_1^{+}(\tau')\, \overline{\Phi}(\tau')\, d\tau'$$

les opérateurs intégraux du second membre étant bien conjugués hermitiques d'après ce qui a éte dit plus haut.

Sous cette forme (66), la détermination de $1/Q$ est ramenée à la résolution d'un problème de valeurs propres généralisées,

(68) $$\hat{L}\,\overline{\psi} = \lambda \hat{M}\,\overline{\psi}$$
$$L^{+} = \lambda\, \hat{M}^{+}\phi$$

avec l'opérateur (69) $\lambda\hat{M} \equiv \frac{1}{Q}\int f_1(\tau)\; f^{+}(\tau')\{\ldots\}\, d\tau'$

problème au quel on peut évidemment appliquer le principe de Ritz-Rayleigh, tout au moins d'un point de vue formel, car le fait que les opérateurs intégraux dans (68) ne sont plus obligatoirement définis positifs nous obligent à renoncer aux développements classiques sur la convergence de la méthode.

Nous obtiendrons donc

(69) $$\frac{1}{Q} = \text{V.S.}\left\{ \frac{\int \phi'^{+}\hat{L}\,\overline{\psi}'\, d\tau}{\left(\int \phi'^{+}\varphi_1\, d\tau\right)\left(\int \overline{\psi}' f_2^{+}\, d\tau\right)} \right\}$$

qui pour le cas particulier où L est hermitique s'écrit

(70) $$\frac{1}{Q} = \text{V.S.}\left\{ \frac{\int \overline{\psi}'^{+}\hat{L}\,\overline{\psi}'\, d\tau}{\left|\int \psi'^{+} f_1\, d\tau\right|^{2}} \right\},$$

car dans ce cas la seconde équation de (64) était $\hat{L}\phi = f_1$ on doit prendre $\phi' \equiv \overline{\psi}'$. C'est un principe de la forme (70) qu'ont utilise Schwringer Blatt et Jackson dans le calcul des déphasages, l'opérateur $\hat{L}$ étant alors un opérateur intégral symétrique réel, donc hermitique.

10. Unicité ou multiplicité des principes variationnels.

Nous avons pu associer au même type d'équations deux principes variationnels (47) et (70), sensiblement différents, par le calcul de la même quantité. Aussi peut-on se demander si les principes (47) et (70) sont les deux seuls que l'on puisse associer à des équations du type (44) (45), ou si, au contraire, il serait possible d'en trouver un plus grand nombre. C'est cette seconde partie de l'alternative qui se trouve réalisée.

En effet, posons

(71) $$\alpha = \int \phi^{+} f \, d\tau, \quad \beta = \int \phi'^{+} \hat{L}^{\xi} \bar{\psi}' \, d\tau, \quad \gamma = \int \bar{\psi}' g^{+} \, d\tau,$$

et considérons une fonction analytique

$$F(\alpha, \beta, \gamma)$$

à laquelle nous imposons d'être stationnaire vis - à vis des variations ϕ' et $\bar{\psi}'$ autour des solutions exactes de (44) et (45) et de se réduire à la valeur correcte de α quand ϕ' et $\bar{\psi}'$ coïncident avec ces solutions exactes.

Il suffit pour cela d'imposer la condition

$$F(\alpha, \alpha, \alpha) = \alpha$$

quel que soit α, mais ceci entraîne

$$\frac{\partial F(\alpha, \alpha, \alpha)}{\partial \alpha} + \frac{\partial F(\alpha, \alpha, \alpha)}{\partial \beta} \delta\beta/\delta\alpha + \frac{\partial F(\alpha, \alpha, \alpha)}{\partial \gamma} \delta\gamma/\delta\alpha = 1,$$

et en tenant compte de $\delta\beta = \delta\alpha + \delta\gamma$, la condition de stationnarité donne les variations $\delta\alpha$, $\delta\gamma$ étant arbitraires,

$$\frac{\partial F(\alpha, \alpha, \alpha)}{\partial \alpha} + \frac{\partial F(\alpha, \alpha, \alpha)}{\partial \beta} = \frac{\partial F(\alpha, \alpha, \alpha)}{\partial \beta} + \frac{\partial F(\alpha, \alpha, \alpha)}{\partial \gamma} = 0$$

d'où résulte

$$\frac{\partial F(\alpha, \alpha, \alpha)}{\partial \alpha} = - \frac{\partial F(\alpha, \alpha, \alpha)}{\partial \beta} = \frac{\partial F(\alpha, \alpha, \alpha)}{\partial \gamma} = 1$$

Mais F étant supposé analysique, il existe au moins une valeur α_0 de

$\alpha_1 \beta_1 \gamma$ autour de laquelle elle est développable en série de Taylor .

En portant les développements obtenues dans les dernières équations écrites, on obtient des relations entre les divers coefficients de la série telle qu'un arbitraire illimité subsiste dans le choixe de ces coefficients.

Le résultat est tout autre si nous cherchons un principe variationnel $F(\alpha, \beta, \gamma)$ indépendant des normes particulières des foncctions d'essai Ψ' et Φ'. Si en effet la norme de Φ' est multipliée par μ, celle de Ψ' par μ', α est changé en $\mu\alpha$, γ en $\mu'\gamma$ et β en $\mu'\beta$.

L' indépendance vis-à-vis de la norme, entraîne que soit vérifiée quels que soient μ et μ'

$$F(\mu\alpha, \mu\mu'\beta, \mu'\gamma) = F(\alpha, \beta, \gamma).$$

choisissons en particulier $\mu' = \frac{1}{\gamma}$, $\mu = \frac{1}{\alpha}$, il vient

$$F(\alpha, \beta, \gamma) = F(1, \frac{\beta}{\alpha\gamma}, 1)$$

et $F(\alpha, \beta, \gamma)$ doit être une fonction de la seule quantité $\alpha\gamma/\beta$ et la condition que $F(\alpha, \beta, \gamma) = \alpha$ quand $\alpha = \beta = \gamma$ conduit à prendre

$$F(\alpha, \beta, \gamma) = \frac{\alpha\gamma}{\beta} ,$$

c'est à dire le principe (70), p.99. On voit donc que ce principe se distingue de l'ensemble de tous les autres possibles ce qui sans doute, a fait son intérêt et son succès.

11. Principes variationnels généraux de la théorie des collisions en mécanique quantique.

En écrivant sous différentes formes, convenablement adaptées les equations fondamendales de la théorie de la diffusion [1], nous avons pur deduire tout un ensemble de principes variationnels, appblicables spécifiquement à la théorie des collisions (cf. aussi plus haut § 4, p. 74 Ces principes ont déjà été énoncés par Schwinger [2], Schwinger et Lippmann[3], Goldberger[4], mais aussi sans autre justification qu'une simple vérification :(ça'colle")on trouvera dans notre Mémorial des Sciences Mathématiques la théorie détaillée et la justification rigoureuse de ces diverses méthodes.

12. Opérateur linéaires particulièrs en physique des réacteurs nucléaires

Les équations linéaires qui décrivent des systèmes intéressants en physique des réacteurs nucléaires peuvent toujours être mises sous la forme

$$\hat{L}\,\phi = f \quad , (\hat{L} = K - \hat{H}) \tag{72}$$

où ϕ représente la distribution d'état inconnue, $\hat{L}$ est un opérateur linéaire et f' est vecteur donné, (décrivant d'ordinaire une source extérieure S si $\hat{L}$ est relié à la conservation des neutrons).

Le cas homogène $f = 0$ est inclus, pour vu que l'opérateur $\hat{H}$ implique un paramètre indéterminée λ pour assurer l'existence d'une solution ; ce cas peut être connu comme limite (singulière) de l'équation in-homogène lorsque f tend vers zéro et le paramètre s'approche d'une valeur propre.

---

(1) T. Kahan G. Rideau, P. Roussoupoulos loc. cit. p. 14 à 22, p. 28 a 49, etc.

(2) loc. cit.

(3) Phys. Rev. 79 (1950) 469)

(4) Phys. Rev. 84 (1951), 929.

Bien que la détermination de ϕ fournira toute l'information sur le système d'écrit par (72), l'on s'intéresse d'ordinaire à la moyenne pondérée

(73) $$(g^{+}, \phi) \equiv \int g^{+} \phi \, d\tau$$

où g^{+} est un certain vecteur donné, d'habitude une quantité du genre section efficace, plutôt qu'un vecteur complet ϕ. Comme il est en général possible de deviner grossièrement la forme de la solution en s'appuyant sur des raisons physiques, le problème de calcul revient à convertir cette information de médiocre qualité en une seule pièce déinformation précise en ce qui concerne la valeur de (g^{+}, ϕ).

Notre principe variationnelle permet de le faire. Définissons le vecteur adjoint ϕ^{+} correspondant à ϕ comme solution de l'équation adjointe

(74) $$\hat{L}^{+} \phi^{+} = g^{+}$$

et considérons, de nouveau, notre fonctionnelle

(75) $$J[\phi'^{+}, \phi'] = (g^{+}, \phi') + (\phi'^{+}, f) - (\phi'^{+}, \hat{L}, \phi')$$

où les arguments ϕ', ϕ'^{+} sont maintenant à regarder comme des réacteurs arbitraires indépendants. Si, soit ϕ'^{+}, soit ϕ', soit l'un et l'autre vérifient (73) et (74) la fonctionnelle $J[\phi' = \phi, \phi'^{+} = \phi^{+}]$ se réduit à

(76) $$(\phi^{+}, f) = (g^{+}, \phi)$$

Pour des variations petites des arguments ϕ', ϕ'^{+} autour des solutions exactes ϕ, ϕ^{+}, l'on a de nouveau :

(77) $$\begin{aligned} \delta J = (g^{+}, \delta\phi) + (\delta\phi^{+}, f) \\ - (\delta\phi^{+}, \hat{L}\phi) - (\hat{L}^{+}\phi^{+}, \delta\phi) = 0 \end{aligned}$$

Il en résulte, une fois de plus, que des erreurs du premier ordre dans les fonctions d'essai Φ'^{+} et ϕ' (mettons de l'ordre de 10^{-1}) n'éntrainent que les erreurs du second ordre dans(g^{+}, ϕ') (donc de l'ordre de $(10^{-1})^2 = 10^{-2}$) ; de plus, le meilleur choix parmi toute une classe de fonctions d'essai peut se fonder sur la propriété de stationnarité de J . Remarquons que si la classe (2) -

est en effet l'équation differentielle inhomogène (*)

(a) $$\hat{L}\,\bar{\Psi}(x) = f(x) \; ,$$

liée ci à certaines conditions aux limites. Par inversion

(b $$\Psi(x) = \hat{L}^{-1} f(x) \; .$$

ou

(c) $$f(x) = \int f(x')\,\delta(x - x)\,dx'$$

d'où , puis que $\hat{L}$ n'agit que sur les coordonnées non primées (x)

(d) $$\bar{\Psi}(x) = \hat{L}^{-1} f(x) = \hat{L}^{-1}\int f(x')\,\delta(x'-x)\,dx' =$$

$$= \int f(x')\,\hat{L}^{-1}\,\delta(x'-x)\,dx' \; ,$$

(e) $$\overset{\text{dif}}{\equiv} \int f(x')\,G(x', x)\,dx'$$

G(x, x') étant la fonction de Green de (a) , avec

(f) $$G(x', x) \overset{\text{dif}}{\equiv} \hat{L}^{-1}\,\delta(x' - x)$$

d'où

(g) $$\hat{L}^{-1}\,G(x', x) = \delta(x' - x)$$

Si par exemple $\hat{L} = \frac{\partial}{\partial x}$, on peut mettre $\delta(x'-x)$ sous la forme d' une intégrale de Fourier

(*) Carir-Kahan loc. cit. D (Paris et Blanche (London)

$$\delta(x'-x) = \frac{1}{2\pi} \int e^{ik(x-x')}\, dk .$$

alors

$$\hat{L}^{-1} e^{ikx} = \frac{e^{ikx}}{ik}$$

(en effet $\hat{L}\,\hat{L}^{-1} e^{ikx} = e^{ikx} = \hat{L}\,\frac{e^{ikx}}{ik} = e^{ikx}$)

Il vient alors pour la fonction de Green

$$G(x,x') = \hat{L}^{-1}\delta(x'-x) = \frac{1}{2\pi}\int \hat{L}^{-1} e^{ik(x-x')}\, dk = \frac{1}{2\pi}\int \frac{e^{ik(x-x')}}{ik}\, dk$$

permise de fonction, d'essai n'inclut pas la solution exacte, l'approximation "optimale" choisie par le principe variationnel sera relative à la fonction de poids w(x) dans le produit scalaire, puisque w(x) définit l'importance relative aux différents points de l'espace des phases $(\vec{r}, \vec{v})$ des écarts à la solution exact (1). L'utilité de la fonctionnelle (75) réside dans le fait qu'elle fournit une estimation variationnelle d'une moyenne pondérée quelconque de la solution.

En particulier, si on choisit g^{+} successivement comme une série de vecteur de base, ou peut évaluer les composantes correspondantes de ϕ et, par conséquant, le vecteur solution complet.

En plus, notre principe comprend la plupart des principes variation-

(1) Tout opérateur linéaire nous l'avons vu, est représentable (moyennant des fonctions δ) sous forme d'un opérateur intégral (note sur p 61 - bis)

(a) $\hat{H}\phi = \int w(r')\, H(r,r')\, \phi(r')\, dr'$

où w(r') est la fonction de poids, avec (b) $H^{+}(r,r') = H(r',r)$
La fonction de poinds peut être éliminée, en posant

(b) $\bar{\phi}(r) = \sqrt{w(r)}\;\phi(r)$

(d) $\bar{H}(r,r') = \sqrt{w(r)}\;\; H(r,r')\sqrt{w(r')}$

sans changer les symetrie du noyau H. Au demeurant, la liberté dans le choix de le fonction d'essai peut être utilisée le cas échéant pour symétriser les équations du système.

nels utilisés pour les problèmes linèaires, titre de cas particuliers. Si l'opérateur est auto -adjoint (hermitique) $\hat{L}^+ = \hat{L}$ et $g^+ = f$, alors $\phi^+ = \phi$ et (75) se réduit à :

(78) $$J_1[\phi'] = 2\,(f\,,\phi') - (\phi',\ \hat{L}\phi')$$

fonctionnelle utilisée par Kourganoff [(1)] et indépendament par Huang[(2)] pour la solution du problème de Milne

Pour le cas homogéne, $f = 0$, si l'on remplace L par $\lambda\hat{I} - \hat{L}$ où $\hat{I} \equiv \hat{1}$ est l'opérateur unité, alors

(79) $$J_2[\phi'] = (\phi',\hat{L}\,\phi') - \lambda(\phi',\phi')$$

La condition que J_2 est extrémale est justement la condition que $(\phi', \hat{H}\phi)$ soit extrémale assujettie à la condition supplémentaire que (ϕ, ϕ) soit maintenu constant, (la valeur propre λ apparaissant comme multiplicateur de Lagrange) et elle est équivalente au principe de Ritz-Rayleigh que

(80) $$J_3[\phi'] = \frac{(\phi',\ \hat{L}\,\phi')}{(\phi\ ,\ \phi)}$$ (dit quotient de Rayleigh)

soit stationnaire.

On peut aussi admettre des fonctions d'essai de la forme $c\,\phi'$ et $c^+\phi'^+$, où c et c^+ sont des amplitudes scalaires indépendantes à choisir de façon à rendre J stationnaire [(3)]

Enportant dans (75), il vient

---

(1) Kourganoff, V. Basic Methods in Transfer Problem, Oxford U. P. (1952) p. 141.

(2) Huang, S. Phys. Rev. 88(1952) 50

(3) Davison B, Neutron Transport Theory Oxford U.P. (1952) p. 141.

En portant dans (75), il vient

$$(81) \qquad J\left[\phi'^{+}, \phi'\right] = c\,(g^{+}, \phi') + c^{+}(\phi'^{+}, f) - c\,c^{+}\,(\phi'^{+}, \hat{H}\,\phi') .$$

En égalant à zéro les dérivées de J par rapport à c et c^{+} conduit à ce qui suit comme choix optimal pour les amplitudes c et c^{+}(1)

$$(82) \qquad c = \frac{(\phi'^{+}, f)}{\phi'^{+}, \hat{H}\phi'} , \quad c^{+} \frac{(g^{+}\,\phi')}{(\phi'^{+}, \hat{H}\phi')}$$

En portant (82) dans (81) conduit à une fonctionnelle indépendante de la normalisation

$$(83) \qquad J_{L}\left[\phi'^{+}, \phi'\right] = \frac{(\phi'^{+}, f)(g^{+}, \phi')}{(\phi'^{+}, \hat{H}\,\phi)} \quad \text{(quotient de Rayleigh généralisé)}$$

(83) à été utilisé par Schwinger [2] dans le calcul de diffusion, et par Francis , et al. [3] dans des problèmes neutroniques; Goertzel[4] a donné une autre démonstration de (83) . Il définit le scalaire λ par

$$(84) \qquad \lambda = 1/(g^{+}, \phi)$$

et il multiplie (72) c.à.d. $\hat{L}\phi = f$, pour $1 = \lambda(g^{+}, \phi)$ pour obtenir

$$(85) \qquad \hat{L}\,\phi = \lambda\, f(g^{+}, \phi) .$$

dont l'équation adjointe s'écrit

---

(1) l'on a : $\dfrac{\partial J}{\partial c} = (g^{+}, \phi') - c^{+}\,(\phi'^{+}, \hat{H}\,\phi') = 0,$

$\dfrac{\partial J}{\partial c^{+}} = (\phi'^{+}, f) - c(\phi'^{+}, \hat{H}\,\phi') = 0$

(2) Phys. Rev; 78 (1950) 135, H. Levine et J. Schwinger, Phys Rev. 75 (1949)1423

(3) Francis , N, Stewart J. Bohl, L. et Krieger T. , Seconde conferénce de Genève , Paper A conf. 15/ P/ 627 (1947).

(4) Goertzel, G. Discussion at the Gathriburg Conference on Neutron Thermalisation , may 1958.

T. Kahan

(86) $$\hat{L}^{+} \phi^{+} = \lambda g^{+} (\phi^{+}, f)$$

Comme ces équations sont mises sous la forme avec λ comme valeur propre, le quotient de Rayleigh généralisé (83) peut être utilisé immédiatement pour fournir une estimation variationnelle de $1/\lambda = (g^{+}, \phi)$. Nous voyons donc que la démonstration indiquée par Goertzel en 1958, n'est autre chose, formule par formule, que notre démonstration 68) à (70) (p. 99) parue dès 1956 [1]

Si $\hat{L} = \hat{I} - \hat{K}$, de sorte que les équations du système deviennent

$$\hat{H} \phi = f = (\hat{J} - \hat{K})\phi = \phi - \hat{K}\phi, \quad \phi = \hat{K}\phi + f,$$

$g^{+} = f$ et $K^{+} = K$, alors (83) peut s'écrire

(87) $$J_5[\phi'] = \frac{(\phi', \phi') - (\phi', \hat{K}\phi)}{(\phi', f)^2}$$

(puisqu'une fonctionnelle différente de zéro sera stationnaire si son inverse l'est) ; c'est là le principe utilisé par Marshak [2] pour le problème de Milne

Pour le cas d'une équation aux valeurs avec $\hat{H}$ remplacé par $\lambda\hat{I} - \hat{L}$, $f = g^{+} = 0$, et $\hat{H}^{+} = \hat{H}$, et prenant $c\phi'$ comme fonction d'essai, l'on obtient

(88) $$\begin{cases} J[\phi'] = c^2 \; (\phi, \; L, \phi') \, c^2 \, \lambda(\phi, \phi) \\ \dfrac{\partial J}{\partial c} = 2c \left[(\phi, \hat{H}\phi) - \lambda(\phi, \phi) \right] = 0 \end{cases}$$

Comme l'amplitude doit être différente de zéro pour une solution non triviale, il en résulte

---

(1) Kahan - Rideau - Roussopoulos, loc. cit (1956) p. 24. 25 .

(2) Cairo - Kahan Blackie, loc. cit. p. 53
Phys. Rev. 71 (1947) 688 .

$$\lambda = \frac{(\phi_1, \hat{L}\phi)}{(\phi, \phi)},$$

ce qui est le quotient de Rayleigh obtenu précédemmentcf. (80)) .

13. Lagrangien et équations d'Euler.

L'analyse précédente (δ 12) présuppose que les équations linéaires ($\hat{L}\phi = f$)p. 102 et (74) ($L^+\phi^+ = g^+$)(p. 103) sont données et on en conclut l'allure stationnaire de la fonctionnelle (75) (p. 103) $J[\phi', \phi] = (g^+, \phi') + (\phi'^+, f) - (\phi'^+ \hat{L}^+ \phi')$ La reciproque est aussi vraie et conduit à l'interprétation de J comme le lagrangien du système à partir duquel on peut obtenir des équations de définition (équation d'Euler) ui appendice (-A I à .A.II.) Supposons en effet que la fonctionnelle (75) soit stationnaire pour de faibles variations indépendantes $\delta\phi^+$ et $\delta\phi$ autour des solutions ϕ^+, et ϕ . Par l'équation (77)(p. 103).

$$(g^+ - \hat{L}^+ \phi^+, \delta\phi) + (\delta\phi^+, f - \hat{L}\phi) = 0 \tag{90}$$

Comme $\delta\phi^+$ et $\delta\phi$ sont arbitraires, les composantes des facteurs restants s'annulent dans toutes les directions; le caractère stationnaire de J entraine dans les équations

$$\hat{L}\phi = f, \quad \hat{L}^+\phi^+ = g^+ . \tag{91}$$

Par conséquent, étant donné les équations linèaires décrivant le système physique, on peut écrire immédiatement une forme lagrangienne J de la quelle les équations peuvent être déduites et qui fournirà une estimation pour toute fonctionnelle linèaire desirée de la solution.

9. Signification et importance de la fonnction adjointe

De nombreux travaux ont été consacrés ces dernières années

pour interpréter et utiliser la notion de fonction adjointe , en theorie de la diffusion neutronique , en particulier en physique des réacteurs nucleaires.

Nous allons, dans ce qui suit mettre en relief la significa- tion de la fonction adjointe pour un système arbitraire. On obtient de la sorte un point de vue plus général.

Pour le système réactif linéaire, nous avons vu que, dans l'état stationnaire $(\partial \phi / \partial t = 0)$, l'équation prend la forme

(92) $$o = \hat{H} \phi - \lambda \hat{K} \phi + S$$

et s'interprète comme une équation de continuité où ϕ représente le flux neutronique, S une source extérieure, $\hat{K}$ un opérateur de production et $\hat{H}$ un opérateur de pertes (absorption et fuite). Le paramètre est une constante dèterminant la multiplication (neu- tronique) du système et sera pour l'instant supposé inférieure à la valeur nécessaire pour la criticalité λ_{cr} $\lambda < \lambda_{cr}$; la source extérieure ne sera donc pas nulle partout.

L'équation inhomogène (92) a un terme de somme $(S \equiv f)$. Comme les conditions aux limites non nulles peuvent se traiter comme des sources singulières, notre équation (92) se mettra sous la forme générale

(93) $$\hat{L} \phi = f$$

avec $\hat{L} = \lambda \hat{K} - \hat{H}$ et $f = S$, avec des conditions aux limites homogènes . La fonctionnelle duale

(94) $$J = \int_a^b dx \left[\phi^+ (\hat{L} \phi - f) - g^+ \phi \right]$$

donne , par variation de ϕ^+ et ϕ, l'équation (93) et l'équation adjointe

(95) $$\hat{L}^+ \phi^+ = g^+$$

avec $\hat{L}^+$ défini par $<\phi^+, \hat{L}\,\psi> = < L^+ \phi^+, \psi>$

L'équation (93) est dite parfois l'équation du système, et ϕ , la fonction du système .

Remarquons que (94) peut se mettre sous la forme

(96) $$J = \int_a^b dx \left[\phi\,[\hat{L}^+ \phi^+ - g^+) - f\,\phi^+\right]$$

Lorsque (93) et (95) sont satisfaites la valeur de la fonctionnelle. J est donnée par l'une des deux expressions

(97) $$-\int_a^b dx\; g^+(x)\phi(x) = -(g^+,\phi)$$

(98) $$-\int_a^b dx\; f(x)\;\phi^+(x) = -(f,\phi^+)$$

$$\Big\} = J$$

Comme en général le choix du terme inhomogène g^+ dans l'équation adjointe est dans une mesure très large, arbitraire (en théorie des réacteurs, on prend pour g^+ la section efficace, correspondant à un certain processus physique donné) Nous disposons de beaucoup de liberté pour choisir une grandeur intéressante J, reliée au système , qu'on désire obtenir avec une précision du second ordre.

Pour mettre en évidence la signification de la fonction adjointe dans le cas général, considérons l'effet d'une source unité

(99) $$f(x) = \delta(x - x_0)$$

En portant dans (98) il vient

(100) $$J = - \phi^+(x_0)$$

En rapprochant l'équation (100) de l'équation (98), on voit que la fonction adjointe est la contribution d'une source unité à la quantité intéressante J. Dans ce sens, la fonction adjointe est une fonction de Green pour une certaine quantité reliée au système (en l'occurence J), la quantité étant dèterminée par le choix de l'inhomogénéité adjointe g^+.

Lorsque l'équation (93) $\hat{L}\phi = f$ est un problème de valeur initiale, alors l'équation adjointe (95) ; $\hat{L}^+ \phi^+ = g^+$ doit être un problème de valeur finale.

Soit par exemple l'équation

(100-1) $$\hat{L}(x, t)\phi = f(x, t)$$

et la fonctionnelle

(100-2) $$J = \int_0^T dt \int_a^b dx \left[\phi^+(\hat{L}\phi - f) - g^+(x, t)\phi(x, t)\right]$$

La valeur de la fonctionnelle est chacune des deux suivantes

(100-3) $$J = -\int_0^T dt \int_a^b dx \; g^+(x, t)\phi(x, t) = -\int_0^t dt < g^+, \phi >$$

(100-4) $$J = -\int_0^T dt \int_a^b dx \; f(x, t)\phi^+(x, t) = -\int_0^T dt < f, \phi^+ > .$$

Soit f la source unité

(100-5) $$f(x,\ t) = \delta(t-t_o)\,\delta(x-x_o)$$

L'équation (100-4) devient alors

(100-6) $$J = -\phi^{+}(x_o,\ t_o)\ .$$

Dans un système physique, une source telle que (100-5) peut affecter l'avenir , mais non pas le passé. Avec cette source (100-5) , $\phi(x,\ t)$ doit être nul pour des temps precédents t_o (pour $t < t_o$, $\phi(x,\ t) = 0$), et l'équation (100-3) devient

(100-7) $$J = -\int_{t_o}^{T} dt \int_a^b dx\ g^{+}(x,\ t)\,\phi(x,\ t)$$

La fonction adjointe est donc une contribution intégrée à la quantité intéressante J pour l'intervalle de temps $t_o < t \leq T$. Si l'inhomogénéité adjointe est une condition de valeur finale , alors

(100-8) $$g^{+}(x,\ t) = \phi^{+}_{T}\ \delta(t - T)$$

et l'équation (100-7) devient

(100-9) $$J = -\int_a^b dx\ \phi^{+}_{T}(x)\,\phi(x,\ T)$$

et la fonction adjointe est la contribution à une certaine quantité telle que J, à l'instant final T .

Dans un sens, la fonction adjointe ϕ^{+} peut être concue comme offrant une description plus fondamentale du système que la fonction du système ϕ . La fonction nous dit comment le système (réacteur) va se comporter pour une source extérieure particulière donnée. La fonction adjointe, d'autre part , nous dit comment le réacteur répond à des sources en général .

On peut présenter ce résultat sous une forme différente.

Tant que le système est sous critique $(\lambda < \lambda_{cr})$, la solution formelle de (92) peut s'écrire

(101) $$\phi = \hat{L}^{-1} f = (\hat{H} - \lambda \hat{K})^{-1} f$$

Dans la représentation coordonnée , on utilise les fonctions propres de l'opérateur coordonné x, dont chacune peut être étiquetée par la valeur correspondante de x (c.à.d ϕ_x) plutôt que pour un indice entier n (c'est à dire ϕ_n)[1] :

(102) $$\phi_x(x') = \delta(x-x')$$

On a alors pour vecteur arbitraire ψ :

(103) $$\psi_x = \int dx' \delta(x-x') \psi(x') = \psi(x)$$

(104) $$\hat{L}_{xx'} = dx''' \delta(x-x''') \; dx'' \; L^{+}(x''', x'') \delta(x''-x') = \hat{L}(x, x')$$

L(x, x') étant ici le noyau de l'opérateur $\hat{L}$. C'est dire que la représentation d'opérateurs abstraits tels que $\hat{L}$, et de vecteur abstraits tels que ψ sous la forme d'opérateurs intégraux et de fonctions intégrales peuvent être considérée comme une représentation de composantes dans laquelle des fonctions delta sont utilisées comme vecteur de base (cf. 102) . Si la sommation dans ce cas intégration formelle sur des indices répétés est sous entendue , les équations (93) et (95) peuvent s'écrire

(105) $$\hat{L}_{xx'} \phi_{x'} = f_x , \quad \phi^{+}_{x'} \hat{L}_{x', x} = g^{+}_x$$

puisque $L^{+}_{xx'} = L_{x'x}$

[1] Alors pour ψ quelconque $\psi = \Sigma a_n \phi_n = \Sigma (\phi^{+}, \phi_n) \phi_n$

Dès lors (101), s'écrit dans cette représentation coordonnée comme

(106) $$\phi_x = (\hat{H} - \lambda \hat{K})^{-1}_{xx'} f_x ,$$

Cette représentation en termes de source (f) du flux ϕ_x au point x, dû à une source externe au point x' (c. à. d $\delta(x-x')$) fois la source extérieure présente au point x' sommé sur x' ; $(\hat{H} - \lambda K)^{-1}_{xx'}$ est justement la fonction de Green G(x, x') du système : (c. à. d $\phi_x = \int f(x') \, G(x, x') \, dx'$) . La solution adjointe de (95) s'écrit de la même façon sous la forme

(107) $$\phi^+_x = (H^+ - \lambda \hat{K})^{-1}_{xx'} g^+_{x'}$$

$$= \left[(\hat{H} - \lambda \hat{K})^{-1}_{xx'}\right]^+ g^+_{x'} = (\hat{H} - \lambda K^{-1})^{-1}_{x'x} \, g^+_{x'}$$

puisque $(\hat{L}^+)^{-1} = (\hat{L}^{-1})^+$

En effet, de $\hat{I} = (\hat{L} \hat{L}^{-1})^+ = (\hat{L}^{-1})^+ \hat{L}^+$ on tire $(\hat{L}^+)^{-1} = (\hat{L}^{-1})^+$.

En faisant appel à la même interprétation que tout à l'heure, pour la fonction de Green avec les arguments permutés, le flux adjoint ϕ^+_x est évidemment donné par le flux en x' dû à une source unité en x, multiplié par la section efficace g^+ en x' et sommé sur x' (c. à. d) $\phi^+_x = \int (x', x) \, g^+_{x'} \, , dx'$) .

C'est dire que le flux adjoint en x est précisément la vitesse avec laquelle le processus physique, représenté par la section efficace g^+ se produit dans l'entier espace de phase du système lorsque l'on introduit une source unité en x. Il convient de faire observer que toute quantité du genre section efficace peut être utilisée pour la source adjointe et que chacune donnera lieu à une solution adjointe distincte.

La déduction précédente demeure valable seulement pour un système sous critique . Pour discuter le cas homogène avec f = o , nous allons le considérer comme un cas limite lorsque la source extérieure s'annule tandis que le paramètre λ tend vers la plus basse valeur propre. Pour effectuer ce passage à la limite, introduisons les vecteurs propres du système homogène

(108) $$\hat{H}\,\phi_n = \lambda_n\,\hat{K}\phi_n, \quad H^+\phi_n^+ = \lambda_n\,\hat{K}^+\phi_n^+$$

L'opérateur $\hat{H} - \lambda\hat{K}$ a pour résolution spectrale

(109) $$\hat{H} - \lambda\hat{K} = \sum_n (\lambda_n - \lambda)\,\hat{K}\phi_n\,(\phi_n^+\,,\;)$$

et son inverse s'écrit (1)

(110) $$(\hat{H} - \lambda K)^{-1} = \sum \frac{\phi_n}{\lambda_n - \lambda}\,(\phi_n^+\,,\;)\hat{K}^{-1}$$

comme le développement de Schmidt usuel pour l'opérateur résolvant. La solution de (92) (de p.110) c.à. d (92) $\hat{H}\phi = \lambda\hat{K}\phi + f$, peut s'écrire

(93) $$\phi = \sum_n \frac{(\phi_n^+\,,\,\hat{K}^{-1}\,f)}{\lambda_n - \lambda}\,\phi_n$$

où $(\phi_n^+\,,\,\hat{K}^{-1}\,f)$ est l'intégrale du flux qui engendrerait des neutrons à la même vitesse que la source externe $\hat{K}^{-1}\,f$, fois son importance par

---

(1) $$\hat{H}\,\psi = \hat{H}\sum a_n\phi_n\ , \text{ avec } \psi = \sum_n a_n\phi_n\ ;$$

Si $\hat{H}\phi_n = \lambda_n\phi_n$, $(\phi_n^+\,,\phi_m) = \delta_{nm}$

$$\psi = \sum_n \phi_n\,(\phi_n^+\,,\psi)$$

$$\hat{H}\psi = \hat{H}\sum_n \phi_n\,(\phi_n^+\,,\psi) = \sum_n \hat{H}\phi_n\,(\phi_n^+\,,\psi) =$$

$$= \sum_n \lambda_n\,\phi_n(\phi_n^+\,,\psi)\ , \text{ d'ou}$$

$$\hat{H} = \sum_n \phi_n\,(\phi_n^+,\;)$$

rapport à la source adjointe.

On peut maintenant envigaser le cas où λ tend vers la valeur propre la plus basse λ_0 correspondant à la solution de l'état permanent pour le système critique . Si nous faisons maintenant décroître la source extérieure à la même vitesse que $\lambda_0 - \lambda_1$, alors tous les termes impliquant des modes supérieurs au premier mode s'annulleront et la solution pourra s'écrire comme (1)

(94) $$\phi \cong \frac{(\phi_0^+ , \hat{K}^{-1} f)}{\lambda_0 - \lambda} \phi_0$$

Physiquement, le réacteur se comporte de plus en plu comme un système sans absorption pour les neutrons dans le mode le plus bas ϕ_0 si bien qu'un neutron donné et ses descendants (ou sa progéniture) dans ce mode échappent aux pertes pendant une longue période, alors que des neutrons dans les modes supérieurs sont pendus sensiblement à la même vitesse indépendanmment de la valeur de λ. A la criticabilité, il n'y a pas de perte nette de neutrons dans le mode le plus bas, et son amplitude croitrait linéairement avec le temps, si la source extérieure ne s'était annulée elle aussi.

L'équation (94) devient exacte dans la mesure où λ_0 s'approche de λ. Il en résulte que la forme du flux près de la criticabilite sera celle du mode le plus bas, indépendant de la distribution de source qui ne déterminera que l'amplitude relative . Par conséquent, la fonction d'influence (importance neutronique relative à n importe quel processus physique qui peut se produire dans le réacteur sera la même fonction $\phi_0^+(x)$ à un facteur d'échelle près , en d'autres termes, les différentes fonctions d'influence correspondant à chuque processus physique dans le système sous-critique tendent tous à se confondre (aux

(1) En effet $\lambda_0 - \lambda \to 0$, $f \to 0$ avec $\lambda_0 - \lambda_i \neq 0$, si $i \neq 0$.

amplitudes relatives près) lorsque le réacteur devient critique. Ceci est raisonnable pour des raisons physiques car un réacteur critique ne peut entretenir le mode le plus bas dans l'état personnel ; à mesure que le système s'approche de la criticabilite, es neutrons avec leur progéniture persistent assez long temps pour oublier la distribution de source. L' "importance" d'un neutron relativement aux différentes positions dans l'espace des phases d'un systeme critique peut donc se mesurer par n'importe quel processus physique et pas seulement en termes de densité de fission ou; de niveau de puissance .

I A. Application des méthods variationnelles à l'équation de diffusion à deux dimensions

Le formalisme qui vient d'être présenté peut être utilisé , par exemple, pour ramener de manière systématique l'équation de Boltzmann aux diverses formes simplifiées généralement mises en ouvre en physique des réacteurs, en partant avec la dépendance complète vis - à vis de la position de l'énergie et de l'angle de diffusion .

Au lieu de procéder ainsi, nous allons, avec Salengul, considérer une application choisie de façon à illustrer les genres d'approximations qu'on peut utiliser pour réduire un lagrangien donné à une forme plus simple, bien que les résultats finaux présentent un certain intérêt intrinsèque.

Le motif et l'intérêt principal qu'il y a à réduire le lagrangien multivariant est d'établir un certain type de compromis entre les buts opposés, d'une part d'une simple description qui diminuera la quantité d'information mise en jeu et par conséquent le temps consacré aux calculs, et d'autre part, d'une précision suffisante pour représenter les traits physiques essentiels du problème .

A titre de cette application, qui conduit au type d'approximation la plus simple, considérons un problème à deux dimensions dans lequel la composition est une fonction arbitraire de deux coordonnées. Cette situation peut se présenter dans le problème d'une répartition plus uniforme de la puissance dans un réacteur en redistribuant le combustible ou en changeant les propriétés du coeur (partie centrale) comme fonction de la position. L'effet sur la réactivité (1) et la distribution de puissance (en particulier si le réflecteur entoure complètement le coeur) se trouve compliqué du fait de la non séparabilité de l'équation de diffusion.

L'équation de diffusion à un groupe peut s'écrire

(95) $$-\nabla . D\nabla \phi + \Sigma_a \phi = \frac{1}{K} \nu\Sigma_f \phi ,$$

où k est la constante de multiplication du système (2) qui est introduit comme un paramètre commode à valeur propre. Si le système est thermique et sensiblement homogène pour les neutrons rapides, ϕ peut être considéré comme le flux thermique, à condition de faire le remplacement

(96) $$\nu \Sigma_f \rightarrow \Sigma p \quad e^{-B^2\tau} \nu\Sigma_f$$

où B^2 (3) est le facteur de forme géométrique et Σ_1 p et τ sont respectivement le facteur local dit de fission rapide, la probabilité de fuite de résonance et l'âge neutronique. Comme (95) est auto-adjointe, son

(1) Pour un réacteur infiniment grand R = C (C = criticalité)

(2) La réactivité est la quantité $\delta C = C-1$ on écrit aussi $\delta C = \delta k / k$, $h = \nu\Sigma_f / \Sigma_a$

$(k = (N_o^+ R N)(N_o^+, \hat{H} N)$. $\hat{K}$ =op de production $\hat{H}$ = op de diffusion et de pertes

(3) Les Anglo-Saxons appellent B^2 le "Bakling" de la distribution ϕ_n

De $\Delta\psi_n(\vec{r}) + B_n^2 \psi_n(\vec{r})$, il résulte $B_2^2 = -\Delta\psi_n/\psi_n$ a le second membre(flamvage) est une mesure de la convexité de la surface ψ_n (x, y, z)

lagrangien peut s'écrire

(97) $$J\left[\phi\right] = \int d r \left[D\ (\nabla\phi)^2 + \Sigma_a\ \phi^2 - \frac{1}{k}\nu\Sigma_f\ \phi^2\right]$$

On a omis les termes provenant de l'intégration par parties qui fournit le terme de fuite ; l'espace fonctionnel ne comporte que les fonctions qui vérifient des conditions aux limites, de sorte que des variations de flux aux limites ne fournissent aucune contribution. Pour fixer les les idées, supposons que le réacteur soit uniforme suivant la direction des z et que sa section droite soit dans le plan (x, y) soit rectangulaire

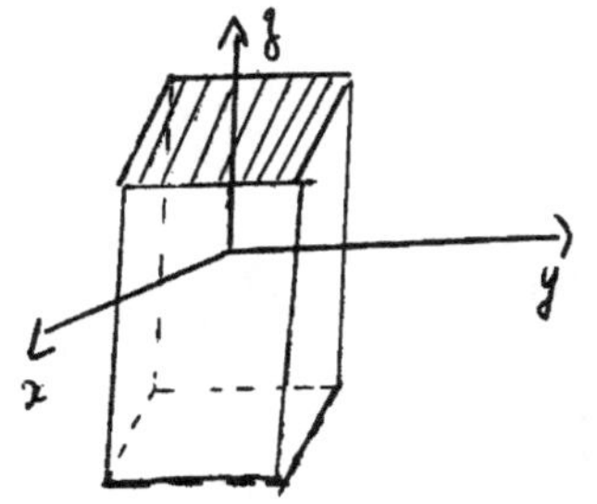

Pour approcher en détail la distribution de flux, choisissons un produit de founctions dépendant separément de x et de y

(98) $$\phi(x,y) = \phi_1(x) \cdot \phi_2(y) .$$

Comme l'équation de diffusion est homogène, il sera commode de normer les flux moyens dans les directions x et y de la façon suivante

(99) $$\int dx\ \phi_1^2\ (x) = 1\ , \quad \int dy \phi_2^2 (y) = 1,$$

ou l'intégration s'étend sur le réacteur entier

Le lagrangien réduit est alors

(100) $$J\left[\phi_1\ ,\ \phi_2\right] = \int dx \int dy \left[\phi_1^2\ \ D\ (\frac{d\phi_2}{dy})^2 + \right.$$

$$+\phi_2^2 D\left(\frac{d\phi_1}{dx}\right)^2 + \phi_1^2 \Sigma_a \phi_2^2 - \frac{1}{k} \phi_1^2 \nu\Sigma_f \phi_2^2\Big] ,$$

et les équations d'Euler correspondantes sont

$$(101) \qquad - D_1 \frac{d^2\phi_1}{dx^2} + (\Sigma_{a1} + D_1 B_1^2)\phi_1 = \frac{1}{k_1} \nu \Sigma_{f1} \phi_1 ,$$

$$(102) \qquad - D_2 \frac{d^2\phi_2}{dy^2} + (\Sigma_{a2} + D_2 B_2^2)\phi_2 = \frac{1}{k_2} \nu \Sigma_{f2} \phi_2$$

où les coefficients sont donnès par

$$(103) \qquad D_1(x) = dy\, \phi_2^2 (y) D(x, y) , \quad D_2(y) = \int dx\, \phi_1^2 (x) D(x, y) ,$$

$$(104) \qquad \Sigma_{a1}(x) = \int dy\, \phi_2^2 (y) \Sigma_a(x, y) , \quad \Sigma_{a2}(y) = \int dx\, \phi_1^2(x) \Sigma_a (x, y)$$

$$(105) \qquad \nu\Sigma_{f1}(x) = \int dy\, \phi_2^2 (y) \nu\Sigma_f (x, y) , \quad \nu\Sigma_{f2}(y) = \int dx\, \phi_1^2 (x) \nu\Sigma_f (x, y)$$

$$(106) \qquad B_1^2(x) = \frac{\int dy \left(\frac{d\phi_2}{dy}\right)^2 D(x, y)}{\int dy\, \phi_2^2 (y)\, D(x, y)} , \quad B_2^2(y) = \frac{\int dx \left(\frac{d\phi_1}{dx}\right)^2 D(x, y)}{dx\, \phi_1^2 (x)\, D(x, y)}$$

Comme les deux équations sont couplées par les moyennes des sections efficaces, on peut admettre une valeur de $\phi_2(y)$, calculer $B_1^2(x)$, puis admettre une valeur de ϕ_1^2 et calculer B_2^2, recalculer ϕ_2 et ainsi de suite. Pourvou que ce procédé itératif converge, il définira un ensemble de flux self-consistants $\phi_1(x)$ et $\phi_2(y)$ qui fournira la meilleure approximation produit, $\phi_1 \phi_2$ pour le flux réel. Les facteurs de forme B_1^2 et B_l^2 donnent les fuites appropriées dans les directions transverses correspondantes. Le procédé est l'analogue exact de la méthode de Hartree-Fock pour le calcul du champ self-constant en physique atomique.

T. Kahan

Ayant obtenu des solutions self-consistantes des équations (101) et (102) , on détermine une valeur de la constante de multiplication du système k par chacune d'elles. Pour établir l'identité des deux valeurs propres calculées de cette façon , multiplions les deux équations respectivement par ϕ_1 et ϕ_2 et intégrons-les ; il vient

$$(107) \qquad k_1 = \frac{\int dx\, \phi_1^2 \nu \Sigma_{f1}}{\int dx\; D_1 \left(\frac{d\phi_1}{dx}\right)^2 + \int dx\, \phi_1^2 \; (\Sigma_{a1} + D_1 B_1^2)}$$

$$(108) \qquad k_2 = \frac{\int dy\, \phi_2^2 \nu \Sigma_{f2}}{\int dy\; D_2 \left(\frac{d\phi_2}{dy}\right)^2 + \int dx\, \phi_2^2 \; (\Sigma_{a2} + D_2 B_2^2)}$$

En faisant usage des définitions (103) à (106) , il vient $k_1 = k_2$ et que les deux sont égaux à la valeur de k obtenue en posant $J[\phi_1 , \phi_2]$ égal à zéro : $J = 0$ il s'ensuit que les valeurs calculées pour tout ces équations sont cohérentes comme il se doit. La dérivation variationnelle ass que la distribution de flux résultante est la meilleure solution (comme estimée p l'effet des erreurs sur la valeur propre)à l'intérieur des limitations imposées par la condition de séparabilité.

II. Applications du principe variationnel au problème de Milne .

Nous avons vu que maint problème de la théorie du transport des neutrons se pose sous la forme d'une équation intégrale du type de Fredholm

$$(1) \qquad \phi(x) = \lambda \int L\; (x, x')\, \phi(x')\; dx' \; + S(x) \; ,$$

où les limites sont données , et où le noyau $L(x,x')$est réel symètrique ou symétrisable,

Si la solution exacte de (1) est la plupart du temps inconnue, il est relativement aisé de trouver des solutions approchées $\phi_a(x)$.

Si l'on pose alors :

(2) $$\phi(x) = \phi_a(x) + \psi(x) \quad ,$$

le problème revient à trouver des valeurs raisonnablement correctes pour $\psi(x)$. Portant (2) dans(1), il vient

(3) $$\psi(x) = \lambda \int L(x, x')\psi(x')\, dx' + f(x)$$

où

(4) $$f(x) = \lambda \int L(x,x')\, \phi_a(x')\, dx' - \phi_a(x) + S(x)$$

Si d'autre part, $G(x_o \to x)$ est la fonction de Green de (1) c.à.d. $\left\{\phi(x) = \int G(x, x_o)\phi(x_o)\, dx_o\right\}$ l'on a

(4) $$G(x_o, x) = \lambda \int L(x,x')\, G(x_o, x')\, dx' + \lambda\, L(x, x_o)$$

A l' de notre opérateur intégrale

$$L \quad (x-x') \quad - \quad L(x, x') \; \ldots \; dx \; ,$$

les équations (3) et (4) prennent la forme

(3') $$L \quad (x_o) = f(x_o) \; ,$$

(4') $$L \;\; G(x_o, \; x) = \quad L(x, x_o)$$

Notre théorème montre que les deux quantités

$$(x_o)L(x, x_o)dx_o \quad \text{et} \quad f(x_o)G(x_o \; , \; x)\; dx_o$$

sont égales à la même valeur stationnaire (S. V)

$$\frac{\lambda\int\psi(x_0)\,L(x,x_0)\,dx_0}{\int f(x_0)G(x,x_0)dx_0}\ \text{V.S.}\ \frac{\int G'(x\to x_0)f(x_0)dx_0\int\psi'(x_0)L(x,x_0)dx_0}{\int G'(x\to x_0)\hat{L}\psi'(x_0)\,dx_0} \tag{5}$$

où $G'(x,-x_0)$ et $\psi'(x_0)$ sont des valeurs approchés pour G et ψ. Nous savons par ailleurs que la fonctionnelle (5) garde sa propriété stationnaire indépendamment de la signification particulière de $\psi(x_0)$, $f(x_0)$, $G(x\to x_0)$ et $L(x, x_0)$.

Un cas particulier important est l'estimation de

$$\lim_{x\to\infty}\int L(x,x_0)\psi(x_0)\,dx_0 .$$

Si

$$\lim_{x\to\infty} L(x,x_0)=0 \quad \text{et} \quad \lim_{x\to\infty} S(x)=0,$$

l'on aura

$$\lim_{x\to\infty} G(x_0,x)=\mu(x_0) \tag{5 bis}$$

où μ est un facteur constant à déterminer dans chaque cas particulier par des considérations physiques. Dans ces conditions

$$\begin{aligned}\lim_{x\to\infty}\int\psi(x_0)\,L(x,x_0)\,dx_0 &= \mu\int f(x_0)\phi(x_0)\,dx_0 \\ &= \mu\int f(x_0)\,\phi_a(x_0)dx_0+\mu\, f(x_0)\psi(x_0)\,dx_0\end{aligned} \tag{6}$$

Le dernier terme de (6) peut être évalué par exemple par notre principe variationnel mis sous la forme particulière de Schwinger

$$\int f(x_0)\,\psi(x_0)\,dx_0 = \text{V.S.}\left\{\frac{\left[\int\psi(x_0)f(x_0)dx_0\right]^2}{\psi'(x_0)\hat{L}\,\psi'(x_0)dx_0}\right\} \tag{7}$$

Il ne nous a pas été possible de déterminer la direction de l'erreur dans (5) (par excès ou par défaut). Toutefois dans le cas de l'équation (7), on a montré [1, 2, 3, 4] que, sous des conditions particulières imposées à $\hat{L}$ et $f(x_0)$, la fonctionnelle (7) est un maximum.

Nous notons aussi que notre fonctionnelle (5) nous fournit une estimation de $\psi(x)$ pour chaque point dans l'intervalle d'intégration, en contraste aux méthodes exposées dans (2), (3) et (4).

Le problème de Milne.

Nous allons appliquer avec J. Devooght, la fonctionnelle (5) à la détermination du flux de neutrons (ou de photons) dans un milieu semi-infini diffusant de manière isotrope et sans capture, qui est limite par la vide et qui entretient un courant constant venant de l'infini : c'est le problème de Milne.

La situation physique de ce problème de Milne est assez bien reproduite par la surface extérieure d'un réacteur nucléaire sans écran

Il se pose aussi en astrophysique sous la forme du transfert du rayonnement de l'intérieur du soleil vers sa surface.

Ce phénomène est essentiellement régi par les mêmes équations qui gouvernent le transport des neutrons dans les réacteurs.

En réalité, c'est à propos des problèmes soulevés par le transfert radiatif au sein des étoiles que la plupart des recherches ont été entre prises sur ce problème majeur de Milne.

---

(1) J. Devooght, Phys. Rev. 111 (1958) 665

(2) J. Le Caine, Phys Rev. 72 (1947) 564

(3) R. E. Marschak, Phys. Rev. (71) (1947) 694

(4) B. Davison, Phys. Rev. (71) (1947) 694

T. Kahan

Ici nous avons $\lambda = 1$; $S(x) = 0$, $L(x,x') = \frac{1}{2} E_1(|x-x'|)$ et l'approximation de diffusion nous donne $\phi_a(x) = x$, c'est à dire $f(x) = \frac{1}{2} E_3(|x|)$, où

$$E_n(|x|) = \int_1^\infty \frac{c^{-t|x|}}{t^n} dt$$

Comme $\lim\limits_{x \to \infty} L(x, x') = 0$, on pourrait appliquer aussi bien (7) avec (cf. (3))

$$\psi(\infty) = \lim_{x \to \infty} \int L(x,x') \psi(x') \, dx' \tag{8}$$

puisque $\lim\limits_{x \to \infty} f(x) = 0$

et obtenir

$$\psi(\infty) = \frac{\mu}{2} \int_0^\infty E_3(x) \, x \, dx + \tag{9}$$

$$+ \mu \, V.S \left\{ \frac{\left[\frac{1}{2} \int_0^\infty E_3(x) \, \psi'(x) \, dx \right]^2}{\int_0^\infty \psi'(x) \, dx \left[\psi'(x) - \frac{1}{2} \int_0^\infty E_1(|x-x'|) \psi'(x') dx' \right.} \right.$$

Des considérations physiques (cf (3) et B. Davison, Neutron Transport Theory (Oxford U. Pr, New York, 1957, p. 210) montrent que $\mu = 3$. Notons que la solution (9) n'est autre chose qu'une estimation variationnelle bien connue de la longueur dite d'extrapolation en théorie du transport.

Si d'autre part, nous voulons déterminer $\psi(\infty)$ par (5), nous prendrons $\psi'(x) = z_0$ et en remplaçant $G(x, x_0)$ par [cf. equ. (2)]

(2') $\phi(x_0) = \phi_a(x_0) + \psi(x_0) = x_0 + z_0$, nous obtenons

$$(10)\quad \psi(\infty) = \text{V.S.} \lim_{x \to \infty} \left\{ \frac{\int_0^{\infty} (x_0+z_0)\frac{1}{2} E_3(x_0)dx_0 \int_0^{\infty} \frac{1}{2} z_0 E_1(|x-x_0|)\, dx_0}{\int_0^{\infty} (x_0+z_0)\left[z_0 - \frac{1}{2} \int_0^{\infty} E_1(|x-x'_0|)\, z_0\, dx'_0 \right] dx'_0} \right.,$$

d'où il résulte par intégration

$$\psi(\infty) = \frac{3+z_0}{4+6z_0}$$

Le résultat dépend de la normalisation choisie pour $\psi(x)$ [à - travers $\phi(x)$] mais si l'on prend une valeur compatible avec un schéma itératif, il faut prendre $z_0 = \psi(\infty)$ ce qui donne immédiatement

$$\psi(\infty) = \frac{1}{\sqrt{2}} = 0,7071$$

qui differe de moins de 0,5 % de la valeur exacte 0,7104 .

Il reste à déterminer $\psi(x)$ pour tout point du milieu à l'aide de la fonctionnelle (5) . Comme la fonction de Green pour l'approximation de diffusion dans un demi-espace sans capture est

$$2\pi\,|x+x_0| - 2\pi\,|x-x_0|,$$

nous choisirons la fonction de Green approchée suivante

$$(11)\qquad G'(x, x_0) = 2\pi\,|x+x_0| - 2\pi\,|x-x_0| + 4\pi\psi(\infty) .$$

qui jouit de la propriété suivante :

$$\lim_{x \to \infty} G'(x, x_0) = 4\pi\left[x_0 + \psi(\infty)\right],$$

c'est-à-dire $\mu = 4\pi$. cf (5 bis) p. 124) . L'association de l'équation

(5) avec (3) conduit à

$$(12)\ \psi(x) \simeq \frac{1}{2} E_3(x) + \frac{\int_0^\infty \left[|x+x_0| - |x-x_0| + 2\psi(\infty)\right] \frac{1}{2} E_3(x_0) dx_0 \int_0^\infty \frac{1}{2} E_1(|x-x_0|) dx_0}{\int_0^\infty \left[|x+x_0| - |x-x_0| + 2\psi(\infty)\right] \frac{1}{2} E_2(x_0) dx_0}$$

Des intégrations directes mais longues conduisent au résultat final :

$$(13)\qquad \psi(x) = \frac{1}{2} E_3(x) + \psi(\infty)\left[1 - \frac{1}{2} E_2(x)\right] \left[1 - \frac{12 E_5(x)}{4\psi(\infty)+3}\right] \left[1 - \frac{6E_4(x)}{3\psi(\infty)+2}\right]^{-1},$$

qui admet la forme asymptotique correcte

$$\psi(x) = \psi(\infty)\left[1 - \frac{1}{2} E_2(x)\right] + \frac{1}{2} E_3(x) .$$

La valeur sur la frontière (interface libre)

$$\psi(0) = {}^{7}/_{12} = 0,584$$

diffère de 1 % de la valeur exacte 0,577 .

L'erreur n'excède jamais 1,5 % en d'autres points . On peut présumer que l'erreur plus grande près de l'interface libre est due à la forme assez incorrecte de la fonction de Green à la frontière x=0

Conclusions

Pour terminer , je voudraise souligner que le nouveau principe variationnel (5) dont la précision parait suffisante pour la plu part des objectifs à atteindre , jouit de deux propriétés intéressantes majeures .

1. Nous n'avons fait aucune hypothèse sur le développement de la

fonction d'essai $\psi'(x)$ en série de fonctions, et néanmoins, nous obtenons avec Devooght une précision satisfaisante avec la fonction d'essai grossièrement approchée $\psi'(x) = z_0$.

On pourrait obtenir une précision bien plus grande si l'on avait fait appel à un développement de $\psi'(x)$ en une série de fonctions $E_n(x)$ avec des coefficients inconnus .

2 Nous n'avons pas à calculer un extrémum et nous sommes dispensés de calculer les coefficients de $E_n(x)$, ce qui est long et fastidieux .

En général, la méthode qui vient d'être exposée pourrait aussi bien s'appliquer à des problèmes en théorie du transport dont la solution exacte est inconnue, tel que le problème de la basse ou de la sphère, contrairement aux méthodes classiques (cf. (2) à (4) et Weinberg et Wigner , The Physical Theory of Neutron Chain Reactor) .

The University of Chicago 1958) valables pour des densités asymptotiques. Nous avons établi une détermination directe de $\psi(x)$ par (5) et non pas par $\psi(\infty)$ qui, au demeurant , serait impossible dans le cas où tous les points du milieu réactionnel sont à des distances finies. Les fonctions de Green approchées $G'(x; x_0)$ pourraint toujours être prises sous forme de fonction de Green appartenant à l'approximation de diffusion qui sont tout à fait faciles à construire pour un grand nombre de problèmes.

Signalons, pour terminer, un travail intéressant de Koshin et Brooks, Journal for Mathematical Physics, sur la généralisation de nos méthodes et ses applications à la théorie du transport des neutrons

T. Kahan

Signalons encore : M. D. Kostin et M. Brooks " generalization of the variational method of Kahan, Rideau and Roussopoulos . II . A variactional Principle for linear operators and its application to Neutron - Transport theory" Journal of Mathematical Physics . Vol. 8 . n. 1. January 1967.

T. Kahan

APPENDICE 1

1 Rappel des notions fondamentales en calcul des variations.

L'objet du calcul des variations est de trouver des fonctions y(x) qui rendent stationnaires ou extrémales des fonctionnelles du tipe $J[y(x)]$ J (y) . Une fonctionnelle $J[y]$ est une variable qui revêt une valeur numérique particulière pour chaque fonction y(x) qui y est portée Example simple

$$J[y] = \int_a^b y(x)\,dx . \tag{A.1}$$

Chaque fonction y(x) fournit une seule valeur numérique de la fonctionnelle $J[y]$. Pour cette raison , une fonctionnelle est parfois appelée "fonction d'une fonction" .

Soit donnée la fonctionnelle

$$J = J[y] , \tag{A.2}$$

le problème fondamental du calcul des variations est de trouver donc une fonction y(x) telle que des accroissements du premier ordre $\delta y(x)$ dans cette fonction $(y \to y + \delta y)$ induisent seulement des accroissements du second ordre dans la fonctionnelle J $(J \to J + \delta^2 J)$; en d'autres termes lorsque

$$y(x) \to y(x) + \delta y(x) , \tag{A.3}$$

$$J[y] \to J + 0 \quad [\max \delta y(x)]^2 \tag{A.4}$$

En portant (A.3) dans (A.2) et en imposant (A.4), on obtient une équation déterminant y(x) . Cette équation porte le nom d'équation d'Euler du problème,. Dans la section suivante, nous établirons les équations d'Euler pour plusieurs problèmes.

T. Kahan

2. Problème simple . Le cas où la fonctionnelle est de la forme

(A.5) $$J = \int_a^b F(x,\ y,\ y')\, dx \qquad (y' = dy/dx)$$

est un des problèmes les plus simples du calcul des variations. Faisant varier y(x) comme dans A.3, on obtient, au premier ordre en δy :

(A.6) $$\delta J = \int_a^b \left(\frac{\partial F}{\partial y}\, \delta_y + \frac{\partial F}{\partial y'}\, \delta y'\right) dx\ ,$$

(car $\delta x = 0$

où

(A.7) $$\delta y' = \frac{d}{dx}\,(\delta y)\ ,$$

En intégrant par parties, il vient (1)

(A.8) $$\delta J = \int_0^b \left(\frac{\partial F}{\partial y} - \frac{d}{dx}\,\frac{\partial F}{\partial y'}\right) dx\ \delta y + \frac{\partial F}{\partial y'}\,\delta y\ \Big|_a^b$$

Si la fonction y(x) est assujettie à vérifier les conditions (aux limites

(A.9) $$y(a) = y_a\ ,\quad y(b) = y_b\ ,$$

alors $\delta y\,(a)$ et $\delta y(b)$ doivent être nuls et les termes aux limites de

(1) $$\int_a^b \frac{\partial F}{\partial y'}\,\delta y'\, dx = \int_a^b \frac{\partial F}{\partial y'}\,\frac{d(\delta y)}{dx}\, dx = \frac{\partial F}{\partial y'}\,\delta y\Big|_a^b - \int \frac{d}{dx}\left(\frac{\partial F}{\partial y'}\right)\delta y\ dx$$

(A. 8) sont nuls. Si non , y(x) doit satisfaire aux conditions aux limites naturelles

(A. 10) $$\frac{\partial F}{\partial y'}(a) = \frac{\partial F}{\partial y'}(b) = 0 \qquad \left(\frac{\partial F}{\partial y'}(a) = \frac{\partial F(x, \ldots)}{\partial y'}\Big|_{x=a}\right)$$

Comme $\partial y(x)$ est arbitraire (sauf à satisfaire, le cas échéant à certaines conditions aux limites), y(x) doit vérifier par (A. 8) l'équation

(A. 11) $$\frac{\partial F}{\partial y} - \frac{d}{dx}\frac{\partial F}{\partial y'} = 0$$

pour que la variation du premier ordre en J soit nulle. L'équation (A. 11) , équation différentielle ordinaire, est dite "équation d'Euler du problème .

Pour déterminer si cette valeur stationnaire ou extrémale est un minimum, un maximum ou un "col", on peut étudier la variation second $\delta^2 J = \delta(\delta J)$ Si cette variation seconde est définie positive, définie négative, ou de signe indétérminé, alors la valeur extrémale est un minimum, maximum ou un col respectivement.

3. L'effet des termes aux limites. Il est souvent intéressant d'obtenir des équations avec des conditions aux limites autres que les équations (A. 9) ou (A. 10) . La fonctionnelle peut être modifiée par l'addition de termes de limite, conduisant ainsi à une équation d'Euler avec des conditions aux limites différentes. Supposons par exemple que la fonctionnelle (A. 5) . Soit modifiée ainsi :

(A. 12) $$J = \int_a^b F(x, y, y')\,dx - g_1[y(a)] + g_2[y(b)]$$

Par le procédé précédent, on aboutit de nouveau à l'équation d'Euler (A. 11) , mais les conditions aux limites naturelles sont modifiées ainsi :

(A.13) $$\frac{\partial F}{\partial y'}(a) + \frac{\partial g_1}{\partial y}(a) = 0 ,$$

(A.14) $$\frac{\partial F}{\partial y}(b) + \frac{\partial g_2}{\partial y}(b) = 0 .$$

4. Problèmes avec plusieurs variables dépendantes .

Si la fonctionnelle dépend de plusieurs fonctions $y(x)$, $y_2, \dots y_n$ sous la forme

(A.15) $$j = \int_a^b dx\, F(x, y_1, y_1', y_2, y_2', \dots, y_n, y'_n) ,$$

on obtient un système d'équations d'Euler simultané

(A.16) $$\frac{\partial F}{\partial y_i} - \frac{d}{dx}\frac{\partial F}{\partial y'_2} = 0, \qquad (i = 1, 2, 3, \dots, n)$$

avec les termes de limite .

(A.17) $$\frac{\partial F}{\partial y'_i}\delta y_i \Big|_a^b = 0 \qquad (i = 1, 2, 3, \dots, n)$$

5. Problèmes avec plusieurs variables indépendantes.

Si la fonctionnelle J est définie sur plusieurs dimensions sous la forme

(A.18) $$J: \int_{a_1}^{b_1} dx_1 \int_{a_2}^{b_2} dx_2 \dots \int_{a_n}^{b_n} dx_n \quad F(x_1, x_2, \dots x_n, y, y_{x_1}, y_{x_2}, \dots, y_{x_n})$$

où

(A.19) $$y_{xi} \equiv \frac{\partial y}{\partial x_i} , \quad (x = 1, 2, \dots, n) ,$$

alors l'équation d'Euler est- l'équation aux dérivées partielles

(A.20) $$\frac{\partial F}{\partial y} - \frac{\partial}{\partial x_1}\frac{\partial F}{\partial y_{x1}} - \frac{\partial}{\partial x_2}\frac{\partial F}{\partial y_{x2}} - \dots - \frac{\partial}{\partial x_n}\frac{\partial F}{\partial y_{xn}} = 0 .$$

6. Fonctionnelle, avec des dérivés d'ordre supérieur.

Soit

(A.21) $$J = \int_a^b dx \quad F(x, y, y^{(1)}, y^{(2)}, \dots, y^{(n)})$$

où

(A.22) $$y^{(i)} \equiv \frac{\partial^i y}{\partial x^i} ,$$

alors n intégrations par parties seront nécessaires pour obtenir l'équation d'Euler qui sera

(A.23) $$\frac{\partial F}{\partial y} - \frac{d}{dx}\frac{\partial F}{\partial y^{(1)}} + \dots + (-1)^n \frac{d^n}{dx^n}\frac{\partial F}{\partial y^{(n)}} = 0$$

c'est là, en général, une équation d'ordre 2n.

7. Problèmes avec contraintes; multiplicateurs de Lagrange.

Supposons que l'équation (A.5) ; $J = \int_a^b dx \; F(x,y,y')$ soit à rendre stationnaire assujettie à la contrainte

(A.24) $$\int_a^b dx \quad E\;(x, y, y') = 0$$

Cette contrainte nous interdit de prendre des variations complètement arbitraires en y(x), de sorte que (A.11) n'est pas l'équation d'Euler pour ce problème.

Le procédé du multiplicateur de Lagrange revient à multiplier l'équation (A. 24) par une constante arbitraire λ et à ajouter le terme résultant à la fonctionnelle J ; l'on obtient alors

(A. 25) $$J = \int_a^b dx \left[F(x, y, y') + \lambda E(x, y, y')\right] .$$

En faisant varier y , l'on a

(A. 26) $$\delta J = \int_a^b dx\, \delta y \left[\frac{\partial}{\partial y}(F+\lambda E) - \frac{d}{dx}\frac{\partial(F+\lambda E)}{\partial y'}\right] .$$

Puisque λ est un paramètre arbitraire , il peut être choisi de façon à satisfaire à l'équation

(A. 27) $$\frac{\partial}{\partial y}(F + \lambda G) - \frac{d}{dx}\frac{\partial}{\partial y'}(F+\lambda G) = 0 .$$

On peut résoudre l'équation (A. 27) pour $y(x, \lambda)$ et faire appel à (A. 24) pour résoudre pour λ . Le paramètre λ choisi de cette façon est appelé "multiplicateur de Lagrange" . Un autre type de contrainte souvent appliquée à la fonctionnelle (A. 5) est

(A. 28) $$V(x, y, y') = 0$$

Le procédé est similaire, sauf que le multiplicateur est une fonction de x. L'équation d'Euler devient

(A. 29) $$\frac{\partial}{\partial y}\left[F + \lambda(x)\, V\right] - \frac{d}{dx}\left\{\frac{\partial}{\partial y'}\left[F + \lambda(x)\, V\right]\right\} = 0$$

Les équations (A. 28) et (A. 29) sont résolues simultanément pour $y(x)$ et $\lambda(x)$.

CENTRO INTERNAZIONALE MATEMATICO ESTIVO
(C.I.M.E.)

C. CATTANEO

" SULLA CONDUZIONE DEL CALORE"

In questa conferenza sono state svolte considerazioni contenute nei seguenti lavori :

" Sulla conduzione del calore" , Atti Semin. Mat. Fis . della Università di Modena , 3 , 3 (1948)

" Sur la conduction de la chaleur" , C.R. de l'Acad. des SC. de Paris, 247 , p. 431 (1958) .

Corso tenuto a Varenna (Como) dal 19 al 27 settembre 1966

CENTRO INTERNAZIONALE MATEMATICO ESTIVO

(C.I.M.E.)

Cataldo Agostinelli

"FORMULE DI GREEN PER LA DIFFUSIONE DEL CAMPO MAGNETICO IN UN FLUIDO ELETTRICAMENTE CONDUTTORE"

Corso tenuto a Varenna dal 17-al 19 settembre 1966

FORMULE DI GREEN PER LA DIFFUSIONE DEL CAMPO MAGNETICO IN UN FLUIDO ELETTRICAMENTE CONDUTTORE.

di

Cataldo Agostinelli

(Università - Torino)

1. E' noto come in un mezzo elettricamente conduttore, di conducibilità finita, e soggetto a un campo magnetico, questo, col tempo, si diffonde nel mezzo e il coefficiente di diffusività magnetica η, analogo al coefficiente di condicibilità termica e alla viscosita cinematica nei fluidi viscosi in movimento, è inversamente proporzionale alla conducibilità elettrica.

Ora in questa lezione mi propongo di stabilite per la diffusione del campo magnetico in un fluido elettricamente conduttore, dotato di un assegnato movimento, delle formule integrali che generalizzano quelle ben note di Green.

Queste formule, che sono così feconde nei diversi rami della Fisica matematica, sono già state come si sa impiegate da tempo nella idrodinamica pura. In effetti H. A. Lorents aveva ottenuto dei risultati interessanti nel caso dei moti permanenti e poi G. W. Oseen aveva dedicato un'ampia memoria nello sviluppo di una teoria generale del movimento dei fluidi viscosi mediante l'applicazione delle formule generalizzate di Green [1].

Recentemente, in una memoria in corso di stampa nella Rivista della Associazione Italiana di Meccanica Teorica e Applicata (A.I.M.E.T.A) queste formule sono state da me estese al caso del movimento di un fluido viscoso incomprensibile, elettricamente conduttore, in cui si ge-

---

[1] G. W. Oseen, Sur les formules de Green generalisées qui se preséntent dans l'hydrodynamique et sur quelques-unes de leurs applications. "Acta Mathematica", B. 34, 1911, pp. 205-284, B. 35, 1912, pp. 97-192.

nera un campo magnetico [2], mentre la Prof.ssa Maria Teresa Vacca ne ha fatto l'applicazione al caso piano [3].

Qui, come ho già detto, mi limiterò a considerare soltanto le formule generalizzate di Green per la diffusione del solo campo magnetico.

2. Nel caso in cui siano trascurabili la corrente di spostamento e la corrente di convezione, le equazioni di Maxwell e l'equazione che esprime la legge generalizzata di Ohm, risultano

$$\mathrm{rot}\,\vec{H} = \vec{I}\,, \quad \mathrm{rot}\;E = -\mu\frac{\partial \vec{H}}{\partial t}\,, \quad (\mathrm{div}\;\vec{H} = 0).$$

$$\vec{I} = \sigma(E + \mu\,\vec{v}\wedge\vec{H})\;.$$

Prendendo il rotore di ambo i membri della prima di queste equazioni, eliminando quindi la densità di corrente $\vec{I}$ e il campo elettrico $\vec{E}$, servendosi delle altre due, e osservando che

$$\mathrm{rot}\;\mathrm{rot}\;\vec{H} = \mathrm{grad}\,\mathrm{div}\;\vec{H} - \Delta_2\vec{H} = -\,\Delta_2\,\vec{H}$$

si ha

$$-\Delta_2\,\vec{H} = \mu\sigma\,\mathrm{rot}\,(\vec{v}\wedge\vec{H}) - \mu\sigma\frac{\partial \vec{H}}{\partial t}\;.$$

Dividendo per $\mu\sigma$ e ponendo $\eta = \frac{1}{\mu\sigma}$, con η coefficiente di diffusività magnetica, si ottiene

---

[2] C. Agostinelli, Sulle formule integrali di Green in Magnetoidrodinamica.

[3] M. T. Varra, Sui moti magnetoidrodinamici piani di un fluido viscoso incomprensibile elettricamente conduttore. "Rendiconti del Seminario Matematico dell'Università e del Politecnico di Torino", Vol. 25°, 1965-66.

(1) $$\eta \Delta_2 \vec{H} - \frac{\partial \vec{H}}{\partial t} = \mathrm{rot}\,(\vec{H} \wedge \vec{v}) \,,$$

che è l'equazione della diffusione del campo magnetico, con

(2) $$\mathrm{div}\, \vec{H} = 0 \,.$$

Alle equazioni (1) e (2), dove il secondo membro della (1) lo considereremo come noto, associeremo le equazioni aggiunte [4]

(3) $$\eta \Delta_2 \vec{H}' + \frac{\partial \vec{H}'}{\partial t} - \mathrm{grad}\, \mathcal{U}' = 0$$

(4) $$\mathrm{div}\, \vec{H}' = 0,$$

con la condizione

(5) $$\Delta_2 \mathcal{U}' = 0 \,,$$

essendo il vettore $\vec{H}'$ e lo scalare U' delle incognite da determinare

Le equazioni (3), (4), (5) ammettono le solutioni:

(6)
$$\vec{H}'_1 = \mathrm{rot\,rot}\,(\psi \vec{i}) = \mathrm{grad}\, \frac{\partial \psi}{\partial x} - \Delta_2 \psi . \vec{i} \,, \quad U'_1 = \frac{\partial}{\partial x}\left(\frac{\partial \psi}{\partial t} + \eta \Delta_2 \psi\right)$$
$$\vec{H}'_2 = \mathrm{rot\,rot}\,(\psi \vec{j}) = \mathrm{grad}\, \frac{\partial \psi}{\partial y} - \Delta_2 \psi . \vec{j} \,, \quad U'_2 = \frac{\partial}{\partial y}\left(\frac{\partial \psi}{\partial t} + \eta \Delta_2 \psi\right)$$
$$\vec{H}'_3 = \mathrm{rot\,rot}\,(\psi \vec{k}) = \mathrm{grad}\, \frac{\partial \psi}{\partial z} - \Delta_2 \psi . \vec{k}, \quad U'_3 = \frac{\partial}{\partial z}\left(\frac{\partial \psi}{\partial t} + \eta \Delta_2 \psi\right),$$

dove $\vec{i}, \vec{j}, \vec{k}$ sono i versori di una terna di assi cartesiani ortogonali O(x y z) di riferimento, con la condizione che la funzione ψ

---

[4] Si osservi che se si assumano come equazioni aggiunte, associate alle (1) e (2) le

$$\eta \Delta_2 \vec{H}^* + \frac{\partial \vec{H}^*}{\partial t} = 0 \,, \quad \mathrm{div}\, \vec{H}^* = 0 \,,$$

basta porre $$\vec{H}^* = \vec{H}' - \mathrm{grad} \int U' \, dt \,,$$

per avere le equazioni (3) e (4), con la condizione (5).

verifichi l'equazione

$$\Delta_2\left(\eta\Delta_2\psi+\frac{\partial\psi}{\partial t}\right)=0. \tag{7}$$

Considereremo di questa equazione una soluzione che dipende soltanto dalla distanza $r=\sqrt{(x-x_0)^2+(y-y_0)^2+(z-z_0)^2}$ di un punto $P(x\,y\,z)$ da un punto $P_0(x_0\,y_0\,z_0)$, e dal tempo t.

Se poniamo $\Delta_2\psi=\Phi$, la (7) diventa

$$\eta\Delta_2\Phi+\frac{\partial\Phi}{\partial t}=0, \tag{7'}$$

Essendo Φ funzione soltanto di r e di t, si può scirvere

$$\frac{\partial^2}{\partial r^2}(r\Phi)+\frac{1}{\eta}\frac{\partial}{\partial t}(r\Phi)=0. \tag{8}$$

Ora se $F(r,t)$ é soluzione di questa equazione si riconosce facilmente che è anche soluzione la

$$\Phi=\frac{1}{r}\frac{\partial}{\partial r}(r\,F).$$

Dalla teoria del calore si ha che è soluzione della (8) la funzione

$$F=\frac{1}{r\sqrt{t_0-t}}\,e^{\frac{r^2}{4\eta(t_0-t)}}$$

e pertanto avremo

$$\Delta_2\psi\equiv\frac{1}{r}\frac{\partial^2}{\partial r^2}(r\psi)=\Phi=\frac{1}{r}\frac{\partial}{\partial r}(r\,F)$$

da cui si deduce

$$\frac{\partial}{\partial r}(r\psi)=r\,F$$

e quindi

$$(9) \qquad \psi(r,t) = \frac{1}{r\sqrt{t_o - t}} \int_{r'}^{r} e^{-\frac{\xi^2}{4\gamma(t_o-t)}} \, d\xi ,$$

che è la soluzione errata della (7), dove r' è una costante arbitraria che assumeremo come raggio di una sferetta S_o con centro nel punto $P_o(x_o\, y_o\, z_o)$. Essa è definita per $o \leq t < t_o$, ed è per definizione identicamente nulla per $t > t_o$.

In base al valore (9) della funzione ψ, le componenti H'_{1x}, H'_{1y}, H'_{1z} del vettore H'_1, definito dall prima delle (6), risultano:

$$H'_{1x} = -\left(\frac{\partial^2\psi}{\partial y^2} + \frac{\partial^2\psi}{\partial z^2}\right) = -\frac{1}{r^2}\left[1 - 3\frac{(x-x_o)^2}{r^2}\right](\psi - E) + \left[1 - \frac{(x-x_o)^2}{r^2}\right]\frac{E}{2\gamma(t_o-t)}$$

$$(10) \qquad H'_{1y} = \frac{\partial^2\psi}{\partial x \partial y} = \frac{3(x-x_o)(y-y_o)}{r^4}(\psi - E) - \frac{(x-x_o)(y-y_o)}{r^2}\,\frac{E}{2\gamma(t_o-t)}$$

$$H'_{1z} = \frac{\partial^2\psi}{\partial x \partial z} = \frac{3(x-x_o)(z-z_o)}{r^4}(\psi - E) - \frac{(x-x_o)(z-z_o)}{r^2}\,\frac{E}{2\gamma(t_o-t)}$$

dove per semplicità si e posto

$$(11) \qquad E(r,t) = \frac{e^{-\frac{r^2}{4\gamma(t_o-t)}}}{\sqrt{t_o - t}} ,$$

e per la funzione scalare U'_1 si ha il valore

$$(12) \qquad U'_1 = \frac{r'(x-x_o)}{2\gamma^3(t_o-t)}\, E(r', t) .$$

Le componenti degli altri due vettori H'_2, H'_3, e i valori delle funzioni scalari corrispondenti U'_2, U'_3, si ottengono dalle (10) e (12) con opportuno scambio delle coordinate.

Sulla sfera S_o, cioè per $r = r'$, si ha $\psi = 0$, e le quantità H'_{1x}, H'_{1y}, H'_{1z}, U'_1 assumono i valori

$$H'_{1x} = \frac{1}{r'^2}\left[1 - 3\frac{(x-x_o)^2}{r'^2}\right]E(r', t) + \left[1 - \frac{(x-x_o)^2}{r'^2}\right]\frac{E(r', t)}{2\gamma(t_o-t)}$$

$$(13) \qquad H'_{1y} = -\frac{3(x-x_o)(y-y_o)}{r'^4}E(r', t) - \frac{(x-x_o)(y-y_o)}{r'^2}\frac{E(r', t)}{2\gamma(t_o-t)}$$

$$H'_{1z} = -\frac{3(x-x_o)(z-z_o)}{r'^4}E(r', t) - \frac{(x-x_o)(z-z_o)}{r'^2}\frac{E(r', t)}{2\gamma(t_o-t)},$$

$$(14) \qquad U'_1 = \frac{x - x_o}{2\,r'^2}\,\frac{E(r', t)}{t_o - t} .$$

3. Ciò premesso consideriamo, nel campo in cui si muove il fluido, un dominio D (t) limitato da un superficie S(t), in generale variabile col tempo. Sia P_o (x_o, y_o, z_o) un punto interno a questo dominio, e sia S_o una sferetta con centro in P_o e raggio r' sufficientemente piccolo in modo che questa sfera nell'intervallo di tempo $0 \leq t \leq t_o$ sia sempre tutta contenuta nel dominio D.

Formiamo ora mediante le equazioni (1) e (3), la seguente combinazione

$$(15) \qquad \gamma(\Delta_2\vec{H} \times \vec{H}' - \Delta_2\vec{H}' \times \vec{H}) - \frac{\partial}{\partial t}(\vec{H} \times \vec{H}') + \operatorname{grad} U' \times \vec{H} = \operatorname{rot}(\vec{H} \wedge \vec{v}) \times \vec{H}',$$

ottenuta moltiplicando la (1) scalarmente per $\vec{H}'$, la (3) scalarmente per $\vec{H}$, e sottraendo quindi membro a membro.

Integrando ambo i membri della (15) rispetto al dominio D'(t) limitato dalla superficie S(t) e dalla sfera S_o, abbiamo

$$(16)\quad \int_{D'(t)} \gamma(\Delta_2 \vec{H} \times \vec{H}' - \Delta_2 \vec{H}' \times \vec{H})\, d\tau - \int_{D'(t)} \frac{\partial}{\partial t}(\vec{H} \times \vec{H}')\, d\tau + \int_{D'(t)} \operatorname{grad} U' \times H\, d\tau =$$

$$= \int_{D'(t)} \operatorname{rot}\,(\vec{H} \wedge \vec{v}) \times \vec{H}'\, d\tau\ .$$

Applichiamo ora le note formule di trasformazione degli integrali di volume in integrali di superficie , osservando che

$$\operatorname{grad}\ U' \times H = \operatorname{div}\ (U'\vec{H})\ ,$$

e ricordando inoltre che se $F(x,y,z,t)$ è una funzione derivabile, definita in una regione dello spazio contenente il dominio $D(t)$ limitato da una superficie $S(t)$, variabile col tempo, sussiste la relazione [5]

$$\int_{D(t)} \frac{\partial F}{\partial t}\, d\tau = \frac{d}{dt}\int_{D(t)} F\, d\tau + \int_{S(t)} F\, V_n\, d\,S,$$

dove V_n è la componente, secondo la normale interna , della velocità con cui si spostano i punti di $S(t)$.

Dalla (16) si ha allora

$$-\int_{S(t)+S_o} \gamma\left(\frac{d\vec{H}}{dn} \times \vec{H}' - \frac{d\vec{H}'}{dn} \times \vec{H}\right) d\,S - \frac{d}{dt}\int_{D'(t)} \vec{H} \times \vec{H}'\, d\tau -$$

$$-\int_{S(t)+S_o} (\vec{H} \times \vec{H}' . V_n + U'\vec{H} \times \vec{n})\, d\,S = \int_{D'(t)} \operatorname{rot}(\vec{H} \wedge \vec{v}) \times \vec{H}'\, d\tau.$$

Integrando ancora rispetto al tempo da o a t_o, e osservando che per $t \to t_o$ è $\vec{H}' = 0$, si ottiene

---

(5) Cfr. E. Goursat, Course d'Analyse , t. I , p. 666 (Paris,Gauthier-Villars, 1933) .

$$(17)\quad -\int_0^{t_0} dt \int_{S(t)} \left[\eta\left(\frac{d\vec{H}}{dn}\times\vec{H}' - \frac{d\vec{H}'}{dn}\times\vec{H}\right) + V_n \vec{H}\times\vec{H}' + U'\vec{H}\times\vec{n}\right] dS -$$

$$-\int_0^{t_0} dt \int_{S_0} \left[\eta\left(\frac{d\vec{H}}{dn}\times\vec{H}' - \frac{d\vec{H}'}{dn}\times\vec{H}\right) + V_n \vec{H}\times\vec{H}' + U'\vec{H}\times\vec{n}\right] dS +$$

$$+\int_{D'(o)} (\vec{H}\times\vec{H}')_{t=0}\, d\tau = \int_0^{t_0} dt \int_{D'(t)} \mathrm{rot}\,(\vec{H}\wedge\vec{v})\times\vec{H}'$$

In questa equazione occorrerà sostituire al vettore $\vec{H}'$ e alla funzione scalare U', succesivamente i vettori $\vec{H}'_1, \vec{H}'_2, \vec{H}'_3$, e le funzioni scalari U'_1, U'_2, U'_3, definiti dalle (6) e passare al limite per $r' \to o$.

Riferiamoci al caso di $\vec{H}' = \vec{H}'_1$ e $U' = U'_1$, dove i valori delle componenti di $\vec{H}'_1$ e della funzione U'_1 sono dati dalle relazioni (10) e (12).

Si riconosce intanto che

$$(18)\qquad \int_{S_0} \eta\, \frac{d\vec{H}}{dn}\times\vec{H}'_1\, dS = 0 .$$

Si osservi per questo che

$$\frac{d\vec{H}}{dn}\times\vec{H}'_1 = \frac{dH_x}{dn}H'_{1x} + \frac{dH_y}{dn}H'_{1y} + \frac{dH_z}{dn}H'_{1z},$$

e che sulla sfera S_0 risulta

$$\frac{dH_x}{dn}H'_{1x} = \left(\frac{\partial H_x}{\partial x}\frac{x-x_0}{r'} + \frac{\partial H_x}{\partial y}\frac{y-y_0}{r'} + \frac{\partial H_x}{\partial z}\frac{x-x_0}{r'}\right)H'_{1x} \quad, \text{ ecc.}$$

dove H'_{1x}, H'_{1y}, H'_{1z} hanno i valori espressi dalle (13).

Sostituendo nell'integrale (18), e applicando il teorema della media sugli integrali, si vede subito che quell'integrale è nullo.

Analogamente, poichè sulla sfera S_0 è

$$V_n = V_x\frac{x-x_0}{r'} + V_y\frac{y-y_0}{r'} + V_z\frac{z-z_0}{r'},$$

si trova che

$$\int_{S_0} V_n \cdot \vec{H} \times \vec{H}'_1 \cdot dS = 0 . \tag{19}$$

Per quanto riguarda il limite dell'integrale

$$I = \int_0^{t_0} dt \int_{S_0} \left(\eta \frac{d\vec{H}'_1}{dn} - U'_1 \vec{n} \right) \times \vec{H} \cdot dS, \tag{20}$$

osserviamo intanto che si ha

$$\frac{d\vec{H}'_1}{dn} \times \vec{H} = H_x \frac{dH'_{1x}}{dn} + H_y \frac{dH'_{1y}}{dn} + H_z \frac{dH'_{1z}}{dn} ,$$

$$U'_1 \cdot \vec{H} \times \vec{n} = U'_1 (H_x \cos nx + H_y \cos ny + H_z \cos nz)$$

e sulla sfera S_0 risulta

$$\frac{dH'_{1x}}{dn} = \frac{\partial H'_{1x}}{\partial x} \frac{x-x_0}{r'} + \frac{\partial H'_{1x}}{\partial y} \frac{y-y_0}{r'} + \frac{\partial H'_{1x}}{\partial z} \frac{z-z_0}{r'} =$$

$$= \frac{3}{r'^3}\left[3 \frac{(x-x_0)^2}{r'^2} - 1\right] E(r',t) + \frac{1}{2\eta r'}\left[3 \frac{(x-x_0)^2}{r'^2} - 1\right] \frac{E(r',t)}{t_0-t} - \frac{r'}{4\eta^2}\left[1 - \frac{(x-x_0)^2}{r'^2}\right] \frac{E(r',t)}{(t_0-t)^2} ,$$

$$\frac{dH'_{1y}}{dn} = \frac{9(x-x_0)(y-y_0)}{r'^5} E(r',t) + \frac{3(x-x_0)(y-y_0)}{2\eta r'^3} \frac{E(r',t)}{t_0-t} + \frac{(x-x_0)(y-y_0)}{4\eta^2 r'} \frac{E(r',t)}{(t_0-t)^2}$$

$$\frac{dH'_{1z}}{dx} = \frac{9(x-x_0)(z-z_0)}{r'^5} E(r',t) + \frac{3(x-x_0)(z-z_0)}{2\eta r'^3} \frac{E(r',t)}{t_0-t} + \frac{(x-x_0)(z-z_0)}{4\eta^2 r'} \frac{E(r',t)}{(t_0-t)^2} ,$$

$$U'_1 \vec{n} \times \vec{H} = U'_1 \left(H_x \frac{x-x_0}{r'} + H_y \frac{y-y_0}{r'} + H_z \frac{z-z_0}{r'}\right) .$$

Si trova che

$$\lim_{r'\to 0}\int_0^{t_0} dt\int_{S_0}\left\{\eta\left(H_y\frac{dH'_{1y}}{dx}+H_z\frac{dH'_{1z}}{dn}\right)-U'_1\left(H_y\frac{y-y_0}{r'}+H_z\frac{z-z_0}{r'}\right)\right\}dS=0$$

e pertanto il limite dell'integrale (20) si riduce a

$$\lim_{r'\to o} I=\lim_{r'\to o}\int_o^{t_o} dt\int_{S_o}\left(\eta\frac{dH'_{1x}}{dn}-U'_1\frac{x-x_o}{r'}\right)H_x\,dS=$$

$$=\lim_{r'\to o}\int_o^{t_o} H_x(x_o,y_o,z_o,t)\,dt\int_{S_o}\left\{\frac{3\eta}{r'^3}\left[3\frac{(x-x_o)^2}{r'^2}-1\right]E(r',t)+\frac{1}{2r'}\left[3\frac{(x-x_o)^2}{r'^2}-\right.\right.$$

$$\left.\left.-1\right]\frac{E(r',t)}{t_o-t}-\frac{r'}{4\eta}\left[1-\frac{(x-x_o)^2}{r'^2}\right]\frac{E(r',t)}{(t_o-t)^2}-\frac{(x-x_o)^2}{2r'^3(t_o-t)}E(r',t)\right\}dS=$$

$$=-\lim_{r'\to o}\int_o^{t_o} H_x(x_o,y_o,z_o,t)\left\{\frac{2\pi}{3\eta}\frac{r'^3E(r',t)}{(t_o-t)^2}+\frac{2\pi}{3}r'\frac{E(r',t)}{t_o-t}\right\}dt=$$

$$=-\lim_{r'\to o}\int_o^{t_o}\left[H_x(x_o,y_o,z_o,t)-H_x(x_o,y_o,z_o,t_o)\right]\left\{\frac{2\pi}{3\eta}\frac{r'^3E(r',t)}{(t_o-t)^2}+\right.$$

$$\left.+\frac{2\pi}{3}r'\frac{E(r',t)}{t_o-t}\right\}dt-H_x(x_o,y_o,z_o,t_o)\lim_{r'\to o}\int_o^{t_o}\left\{\frac{2\pi}{3\eta}\frac{r'^3E(r',t)}{(t_o-t)^2}+\right.$$

$$\left.+\frac{2\pi}{3}r'\frac{E(r',t)}{t_o-t}\right\}dt=$$

$$=-H_x(x_o,y_o,z_o,t_o)\lim_{r'\to o}\int_o^{t_o}\left\{\frac{2\pi}{3\eta}\frac{r'^3E(r',t)}{(t_o-t)^2}+\frac{2\pi}{3}r'\frac{E(r',t)}{t_o-t}\right\}dt\,.$$

Ma risulta

$$\lim_{r'\to o}\int_o^{t_o} r'^3\frac{E(r',t)}{(t_o-t)^2}dt=4\eta\sqrt{\pi\eta},\qquad \lim_{r'\to o}\int_o^{t_o} r'\frac{E(r',t)}{t_o-t}dt=2\sqrt{\pi\eta},$$

si ha perciò infine

$$(21)\qquad \lim_{r'\to o}\int_o^{t_o} dt\int_{S_o}\left(\eta\frac{d\vec{H}'_1}{dn} - U'_1\cdot\vec{n}\right)\times\vec{H}\cdot dS = -4\pi\sqrt{\pi\eta}\,H_x(x_o,y_o,z_o,t_o).$$

Consideriamo ora il

$$(22)\qquad \lim_{r'\to o}\int_o^{t_o} dt\int_{S(t)} U'_1\cdot\vec{H}\times\vec{n}\,dS = \lim_{r'\to o}\int_o^{t_o} dt\int_{S(t)}\vec{H}\times\vec{n}\cdot\frac{r'(x-x_o)}{2r^3}\,\frac{E(r',t)}{t_o-t}dS.$$

Posto

$$f(t) = \int_{S(t)}\vec{H}\times\vec{n}\;\frac{x-x_o}{r^3}\,dS$$

possiamo scrivere

$$\int_o^{t_o} dt\int_{S(t)}\vec{H}\times\vec{n}\cdot U'_1\,dS = \frac{1}{2}\left\{\int_o^{t_o} r' f(t_o)\,\frac{E(r',t)}{t_o-t}dt + \right.$$

$$\left. +\int_o^{t_o}\left[f(t) - f(t_o)\right]\frac{r'E(r',t)}{t_o-t}dt\right\},$$

e passando al limite per $r'\to o$, si ottiene

$$\lim_{r'\to o}\int_o^{t_o} dt\int_{S(t)}\vec{H}\times\vec{n}\cdot U'_1\,dS = \frac{1}{2}f(t_o)\lim_{r'\to o}\int_o^{t_o}\frac{r'E(r',t)}{t_o-t}dt = \sqrt{\pi\eta}\,f(t_o),$$

cioè

$$(23)\qquad \lim_{r'\to o}\int_o^{t_o} dt\int_{S(t)}\vec{H}\times\vec{n}\cdot U'_1\,dS = \sqrt{\pi\eta}\int_{S(t_o)}\vec{H}\times\vec{n}\;\frac{x-x_o}{r^3}\,dS.$$

I limiti degli integrali che figurano nella (17) estesi al dominio $D'(t)$, tendono agli stessi integrali estesi all'intero dominio $D(t)$; così pure i limiti dei rimanenti integrali estesi all superficie $S(t)$, tendono agli stessi integrali, dove però le componenti $H'_{1x}, H'_{1y}, H'_{1z}$ del vettore

$\vec{H}'_1$ vanno calcolate per $r' = 0$, cioè mediante le (10), ove si ponga

$$(24) \qquad \psi = \frac{1}{r\sqrt{t_o - t}} \int_o^r e^{-\frac{\xi^2}{4\eta(t_o - t)}} d\xi .$$

L'equazione (17), per $\vec{H}' = \vec{H}'_1$ e $U' = U'_1$, al limite per $r' \to o$ porge quindi :

$$(25) \qquad 4\pi\sqrt{\pi\eta}\, H_x(x_o, y_o, z_o, t_o) = \int_{D(o)} (\vec{H} \times \vec{H}'_1)_{t=o}\, d\tau - \sqrt{\pi\eta} \int_{S(t_o)} \vec{H} \times \vec{n}\, \frac{x - x_o}{r^3}\, dS -$$

$$- \int_o^{t_o} dt \int_{S(t)} \left[\eta \left(\frac{d\vec{H}}{dn} \times \vec{H}'_1 - \frac{d\vec{H}'_1}{dn} \times \vec{H} \right) + V_n \vec{H} \times \vec{H}'_1 \right] dS -$$

$$- \int_o^{t_o} dt \int_{D(t)} \mathrm{rot}\, (\vec{H} \wedge \vec{v}) \times \vec{H}'_1\, d\tau .$$

Due formule analoghe alla (25) si hanno per le componenti H_y, H_z del campo magnetico.

Osserviamo che risulta

$$\vec{H} \times \vec{H}'_1 = \vec{H} \times \mathrm{rot}\ \mathrm{rot}\, (\psi \vec{i}) = \mathrm{div} \left[\mathrm{rot}\, (\psi \vec{i}) \wedge \vec{H} \right] + \mathrm{rot} (\psi \vec{i}) \times \mathrm{rot}\, \vec{H} =$$

$$= \mathrm{div} \left[\vec{H} \times \mathrm{grad}\, \psi \,.\, \vec{i} - H_x\, \mathrm{grad}\, \psi \right] + \mathrm{rot}\, \vec{H} \wedge \mathrm{grad}\, \psi \times \vec{i} ,$$

e pertanto il primo integrale del secondo membro della (15) si trasforma nel modo seguente

$$(26) \qquad \int_{D(o)} (\vec{H} \times \vec{H}'_1)_{t=o}\, dt = - \int_{S(o)} \left(H \times \mathrm{grad}\, \psi \,.\, \cos nx - H_x \frac{d\psi}{dn} \right)_{t=o} dS +$$

$$+ \int_{D(o)} \left(\mathrm{rot}\, \vec{H} \wedge \mathrm{grad}\, \psi \cdot \times \vec{i} \right)_{t=o} d\tau ,$$

dove figura la corrente di conduzione $\vec{I} = \text{rot}\ \vec{H}$ all'istante iniziale, ed è

$$\text{grad}\,\Psi = \frac{\partial\psi}{\partial r}\ \text{grad}\ r = \left[\frac{E(r,t)}{r} - \frac{1}{r^2}\int_0^r E(\xi, t)\,d\xi\right]\ \text{grad}\ r.$$

Così pure, essendo

$$\text{rot}\,(\vec{H}\wedge\vec{v})\times\vec{H}'_1 = \text{div}\left[(\vec{H}\wedge\vec{v})\wedge\vec{H}'_1\right] + \text{rot}\ \vec{H}'_1\times\vec{H}\wedge\vec{v}$$

si ha

$$(27') \quad \int_{D(t)} \text{rot}\,(\vec{H}\wedge\vec{v})\times\vec{H}'_1\,d\tau = -\int_{S(t)} (\vec{H}\wedge\vec{v})\wedge\vec{H}'_1\times\vec{n}\ dS + \int_{D(t)} \text{rot}\ \vec{H}'_1\times\vec{H}\wedge\vec{v}\,d\tau,$$

dove è

$$\text{rot}\ \vec{H}'_1 = -\,\text{rot}(\Delta_2\Psi\cdot\vec{i}) = \vec{i}\wedge\text{grad}\ \Delta_2\Psi = \vec{i}\wedge\frac{E(r,t)}{4\eta^2(t_0-t)^2}\ r\ \text{grad}\ r$$

4. Se il campo si estende a tutto lo spazio, sotto determinate condizioni gli integrali di superficie tendono a zero e gli integrali di volume risultano convergenti. In tal caso la formula (25), tenendo conto delle (26) e (27), porge

$$(28) \quad 4\pi\sqrt{\pi\eta}\ H_x(x_0, y_0, z_0, t_0) = \iiint\left(\vec{I}\wedge\text{grad}\,\Psi\times\vec{i}\right)_{t=0} d\tau\ -$$

$$-\int_0^{t_0} dt\iiint\left[\vec{i}\wedge\frac{E(r,t)}{4\eta^2(t_0-t)^2}\ r\ \text{grad}\ r\right]\times\vec{H}\wedge\vec{v}\ d\tau$$

e analogamente

$$4\pi\sqrt{\pi\eta}\quad H_y(x_0, y_0, z_0, t_0) = \iiint(\vec{I}\wedge\text{grad}\,\Psi\times\vec{j})_{t=0}\ d\tau\ -$$

$$-\int_0^{t_0} dt\iiint\left[\vec{j}\wedge\frac{E(r,t)}{4\eta^2(t_0-t)^2}\ r\ \text{grad}\ r\right]\times\vec{H}\wedge\vec{v}\ d\tau$$

(28')

$$4\pi\sqrt{\pi\eta}\, H_z(x_o,y_o,z_o,t_o) = \iiint (\vec{I}\wedge \operatorname{grad}\psi \times \vec{k})_{t=o}\, d\tau -$$

$$-\int_o^{t_o} dt \iiint \left[\vec{k}\wedge \frac{E(r,t)}{4\eta^2(t_o-t)^2}\, r \operatorname{grad} r \right] \times \vec{H}\wedge\vec{v}\, d\tau,$$

dove gli integrali tripli sono estesi a tutto lo spazio. Se inizialmente la corrente di conduzione $\vec{I}$ è assegnata in un dominio limitato D_o, i primi integrali dei secondi membri delle formule precedenti saranno estesi al dominio D_o.

Le formule (28) e (28') si possono ovviamente conglobare nell'unica formula vettoriale

$$4\pi\sqrt{\pi\eta}\, H(x_o,y_o,z_o,t_o)= \iiint (\vec{I}\wedge \operatorname{grad}\psi)_{t=o}\, d\tau - \int_o^{t_o} dt \iiint \frac{E(r,t)}{4\eta^2(t_o-t)^2}\, r \operatorname{grad} r \wedge(\vec{H}\wedge\vec{v})\, d\tau,$$

che è un'equazione integrale che definisce il campo magnetico $\vec{H}$ in ogni punto $P_o(x_o,y_o,z_o,t_o)$, e in ogni istante t_o, per mezzo dei valori della corrente di conduzione $\vec{I}$ all'istante iniziale in un dato dominio e dei valori della velocità del mezzo fluido che occupa tutto lo spazio.

CENTRO INTERNAZIONALE MATEMATICO ESTIVO

(C. I. M. E.)

A. PIGNEDOLI

" TRANSFORMATIONAL METHODS APPLIED TO SOME ONE-DIMENSIONAL PROBLEMS CONCERNING THE EQUATIONS OF THE NEUTRON TRANS-PORT THEORY".

Corso tenuto a Varenna dal 17 al 19 settembre 1966

TRANSFORMATIONAL METHODS APPLIED TO SOME ONE-DIMENSIONAL PROBLEMS CONCERNING THE EQUATIONS OF THE NEUTRON TRANSPORT THEORY

by

Antonio PIGNEDOLI (Università - Bologna)

1. Introductory remarks.

There is a very great bibliography concerning the mathematical theory of slowing down and diffusion of the neutrons within the moderating medium of an atomic fission pile. The methods used for such study must allow for the differences in the reactivity that take place in the reactors chiefly during the starting and because of the sinking and the prising of the control bars, so doing a quantitative scheme of the behaviour of the fast neutrons produced in the nuclear fission, to reach, by slowing in the moderating medium, the thermal energy.
The neutrons, endowed with such energy, are "trapped" in turn , by the uranium nuclei producing new nuclear fissions, and so on, by the well known chain reaction.
The methods used for the study of the neutron diffusion in a moderator recall or upon the research, in opportune conditions , of the solution of the Maxwell-Boltzmann integro-differential equation of the transport theory , or on the research of opportune supplementary conditions of solutions of the partial differential equations, of diffusion, derived from the application of these so called "phenomenological theory".
This lecture is concerning the transport theory point of view and, in particular, the application of transformational methods to some one-dimensional problems.
As well known, a neutron is a heavy uncharged elementary particle.
Of the forces which act upon it the nuclear forces are by far the most important and they are only one that need be taken into a account under the conditions in which one is interested in the diffusion of neutrons.

Since these forces have an extremely short range , it follows that:

1) the motion of a neutron can be described in terms of it collisions with atomic nuclei and with other freely moving neutrons;

2) these collisions are well-defined events;

3) between such collisions a neutron moves with a constant velocity, that is in a straight line with a constant speed;

4) the mutual collisions of freely moving neutrons may safely be neglected and only the collision of neutron with atomic nuclei with a surrounding medium need be taken into account;

5) for a neutron travelling at a given speed through a given medium the probability of neutron collision per unit path lenght is a constant;

6) the neutron or neutrons emerging from a collision do so at the point of space where the collision took place.

We will indicate with N the neutron density in a medium (the neutron density N in a medium is a function of the position, denoted by the vector $\vec{r}$, the direction of the neutron $\vec{\Omega}$, its velocity v and the time t) .

Let $\Sigma(v)$ be the total macroscopic cross section for all processes; $\Sigma(v)$ is the inverse mean free path. Let c(v) be the mean number of secondary neutrons produced per collision. The quantity $c(v)\ \Sigma(v)$ is the mean number of secondary of unit path. Let $c(v')\ f\ (v', \vec{\Omega} \to v, \vec{\Omega})\ dv\, d\Omega$ be the mean number of neutrons produced in the velocity range dv and cone $d\Omega$ when a neutron of velocity v' and direction $\vec{\Omega}$ undergoes a collision with a stationary nucleus. Let $S(\vec{r}, v, \vec{\Omega}, t)$ be the source strenght of neutrons in the particular volume element. The rate of change

of $N(\vec{r}, v, \vec{\Omega}, t)$ with time is equal to the number of neutrons scattered to the velocity v and direction $\vec{\Omega}$ from other directions and velocity less the loss due to leakage and scattering out of v and $\vec{\Omega}$.
The transport equation is therefore :

$$1) \quad \frac{\partial N(\vec{r}, v, \vec{\Omega}, t)}{\partial t} = - v \vec{\Omega} \cdot \operatorname{grad} N - v \Sigma(v) N + \iint v' c(v') \Sigma(v') f(v', \vec{\Omega}' \rightarrow v, \vec{\Omega}) N(\vec{r}, v', \vec{\Omega}', t) dv' d\Omega' + S(\vec{r}, v, \vec{\Omega}, t) .$$

The physical meaning of this equation is the following . The rate of change $\frac{\partial N}{\partial t}$ is equal to the neutrons scattering to v and $\vec{\Omega}$ plus sources minus the leakage and the scattering out of v and $\vec{\Omega}$.
In the case of the isotropic scattering, the density of the neutrons at $\vec{r}$ is

$$N_0(\vec{r}, v') = \int N(\vec{r}, \vec{\Omega}', v') d\Omega' ,$$

and the following integral equation is obtained

$$(2) \quad v N_0(\vec{r}, v) = \int \frac{\exp(-\rho\Sigma)}{4\pi\rho^2} \left[S(\vec{r}', v, t - \frac{\rho}{v}) + \int v' \Sigma_s(v') f(v' \rightarrow v) . N_0(\vec{r}', v', t - \frac{\rho}{v}) dv' \right] dV$$

(dV = volume element) .

If the neutron distribution does not vary with time, then $N(\vec{r}, v, \vec{\Omega})$ satisfies the equation :

$$(3) \quad v\vec{\Omega} . \operatorname{grad} N + v \Sigma(v) N = \iint v'(c(v') \Sigma(v') f(v', \vec{\Omega}' \rightarrow v, \vec{\Omega}) N(\vec{r}, v', \vec{\Omega}') dv' d\Omega' + S(\vec{r}, \vec{\Omega}, v) .$$

This equation does not have, generally, a solution and the following modified equation is considered :

$$(4) \qquad v\vec{\Omega}\cdot \text{grad}\ N + (\alpha + v\Sigma)\, N = $$
$$= \iint v'c(v')\ \Sigma(v')\ f(v',\vec{\Omega}' \longrightarrow v,\vec{\Omega})\ N(\vec{r}\ ,v',\ \vec{\Omega}')\, dv'\, d\Omega',$$

(α is a constant) .

This equation has solutions for certain eigenvalues α_i . The eigenfunctions of the equation satisfy to the following conditions :

a) Continuity at the boundary between two media ;

b) $N_i = 0$ at the free surface for all incoming directions.

A general solution of (1) is :

$$(5) \qquad N(\vec{r}\ , v,\ \vec{\Omega}, t) = \sum_i a_i\, N_i(\vec{r}\ ,\ v,\ \vec{\Omega})\ \exp\ (\alpha_i t) \quad .$$

(We assume that the N_i are a complete set) .
For large t , it is :

$$(6) \qquad N(\vec{r},\ \vec{\Omega},\ v, t) = N_o(\vec{r},\ v,\ \vec{\Omega})\ \exp\ (\alpha_o t)\ ,$$

where α_o is the maximum of the α_i' s .

($\alpha_o < 0$) subcritical system, $\alpha_o = 0$ critical system,

$\alpha_o > 0$ supercritical system) .

The coefficients a_i of the series are determined from the boundary conditions which may be specified at some particular time.

2. The one group theory.

Consider the time-dependent equation (1). If all neutrons have the same velocity v_0, then it is :

$$(7)\quad \begin{cases} N(\vec{r}, \vec{\Omega}, v, t) = \delta(v-v_0)\, N(\vec{r}, \vec{\Omega}, t), \\ f(v', \vec{\Omega}' \rightarrow v, \Omega) = \delta(v'-v_0)\, f(\vec{\Omega}' \rightarrow \vec{\Omega}), \\ S(\vec{r}, \vec{\Omega}, v, t) = \delta(v-v_0)\, S(\vec{r}, \vec{\Omega}, t). \end{cases}$$

Let :

$$(8)\quad \begin{cases} N(\vec{r}, \vec{\Omega}, t) = \int N(\vec{r}, \vec{\Omega}, v, t)\, dv, \\ f(\vec{\Omega}' \rightarrow \vec{\Omega}) = \int f(v', \vec{\Omega}' \rightarrow v, \vec{\Omega})\, dv, \\ S(\vec{r}, \vec{\Omega}, t) = \int S(\vec{r}, \vec{\Omega}, v, t)\, dv. \end{cases}$$

Integrating the transport equation over v, results in the following one velocity group equation :

$$(9)\quad \frac{\partial N(\vec{r}, \vec{\Omega}, t)}{\partial t} = - v_0 \vec{\Omega} \cdot \text{grad } N - v_0 \Sigma N +$$

$$+ v_0 c(v_0)\, \Sigma(v_0) \int N(\vec{r}', \vec{\Omega}', t)\, f(\vec{\Omega}' \rightarrow \vec{\Omega})\, d\Omega' + S(\vec{r}, \vec{\Omega}, t)$$

[$v_0 N = \psi(\vec{r}, \vec{\Omega}, t)$ = angular distribution of the neutron flux.

If we assume :

A) time independence of N ;

B) No variation of Σ and c with neutron velocity ;

C) f $(\vec{\Omega}' \rightarrow \vec{\Omega})$ independent of v; and integrate (1) without the term $\frac{\partial N}{\partial t}$, we obtain :

$$(10)\qquad \vec{\Omega}\cdot \operatorname{grad}\ \psi(\vec{r},\vec{\Omega}) + \Sigma\,\psi(\vec{r},\vec{\Omega}) = $$
$$= c\,\Sigma \int \psi(\vec{r},\vec{\Omega}')\ f(\vec{\Omega}'\longrightarrow\vec{\Omega})\,d\Omega' + S(\vec{r},\vec{\Omega}),$$

where :

$$(11)\qquad \begin{cases} \psi(\vec{r},\vec{\Omega}) = \int vN(\vec{r},v,\vec{\Omega})\,dv, \\ S(\vec{r},\vec{\Omega}) = \int S(\vec{r},v,\vec{\Omega})\,dv. \end{cases}$$

3. Solution of the one group transport equation in an infinite uniform medium in the case of the time-independence and with a plane source at x = 0.

We consider now the one group time indipendent transport equation for an infinite uniform medium with a plane source at $x = 0$.

That is the equation :

$$(1)\qquad \mu\frac{\partial\psi(x,\mu)}{\partial x} + \Sigma\,\psi(x,\mu) = \frac{1}{2}\,c\,\Sigma\int_{-1}^{+1}\psi(x,\mu)\,d\mu + \frac{\delta(x)}{4\pi},$$

where $\delta(x)$ is the Dirac "Delta distribution ".

We apply to this equation the method of the Fourier transform. We indicate with $\pi(\tau,\mu)$ the Fourier transformation of $\psi(x,\mu)$, that is we put :

$$\pi(\tau,\mu) = \int_{-\infty}^{+\infty}\psi(x,\mu)\ \exp(-i\,\tau\,x)\,dx.$$

We have

$$(2)\ i\,\tau\,\mu\,\Pi(\tau,\mu) + \Sigma\,\pi(\tau,\mu) = \frac{c\,\Sigma}{2}\int_{-1}^{+1}\pi(\tau,\mu)\,d\mu + \frac{1}{4\pi}.$$

Now we indicate with $\pi_0(\tau)$ the μ indipendent quantity at the second member of (2) ; and we have ;

$$\pi(\tau,\mu)=\pi_o(\tau)\quad/(\Sigma+i\tau\mu). \tag{3}$$

Substituting back in the equation (2) we have :

$$(\Sigma+i\tau\mu)\ \frac{\pi_o(\tau)}{\Sigma+i\tau\mu}=\frac{1}{2}c\Sigma\int_{-1}^{+1}\frac{\pi_o(\tau)}{\Sigma+i\tau\mu}d\mu+\frac{1}{4\pi}$$

$$\pi_o(\tau)=\frac{1}{4\pi\left[1-\frac{c\Sigma}{\tau}\ \mathrm{arctg}\ \frac{\tau}{\Sigma}\right.}\ . \tag{4}$$

From the Fourier inversion fourmula we obtain :

$$\psi(x,\mu)=\frac{1}{8\pi^2}\int_{-\infty}^{+\infty}e^{i\tau x}(\Sigma+i\tau\mu)^{-1}\left[1-\frac{c\Sigma}{2i\tau}\lg\left(\frac{\Sigma+i\tau}{\Sigma-i\tau}\right)\right]^{-1}d\tau\ . \tag{5}$$

The total flux is obtained by integration over μ ;

$$\Phi(x)=2\pi\int_{-1}^{+1}\psi(x,\mu)\,d\mu\ ,$$

that is :

$$\Phi(x)=\frac{1}{4\pi}\int_{-\infty}^{+\infty}e^{i\tau x}\left[1-\frac{c\Sigma}{2i\tau}\log\left(\frac{\Sigma-i\tau}{\Sigma+i\tau}\right)\right]^{-1}\cdot\ (i\tau)^{-1}\log\left(\frac{\Sigma+i\tau}{\Sigma-i\tau}\right)d\tau\ . \tag{6}$$

The integral can be evaluated by the method of residues .

The integrand has a simple pole where the term in the denominator vanishes , i.e. where :

$$c\ \Sigma\ \log\frac{\Sigma+i\tau}{\Sigma-i\tau}=2i\tau\ . \tag{7}$$

If $c<1$ the poles are at $\tau=\pm\, i\,K$, where K is given by

$$K/\Sigma=\mathrm{tgh}\left(\frac{K}{c\Sigma}\right). \tag{8}$$

(The principal rooth of this equation is real). An examination of the integrand reveals a singularity at $\tau = +i\Sigma$. This is of the form $\log z$ for $z \to 0$. To avoid this singularity the plane is cut and the deformed contour is taken along the imaginary axis to $i\Sigma$ and back again . We have :

I) an asymptotic part of the solution : that is arising from the residue at $\tau = iK$

II) a transient part which is only important near to the source; that arises from the contributions of the integral around the cut (fig. 1) .

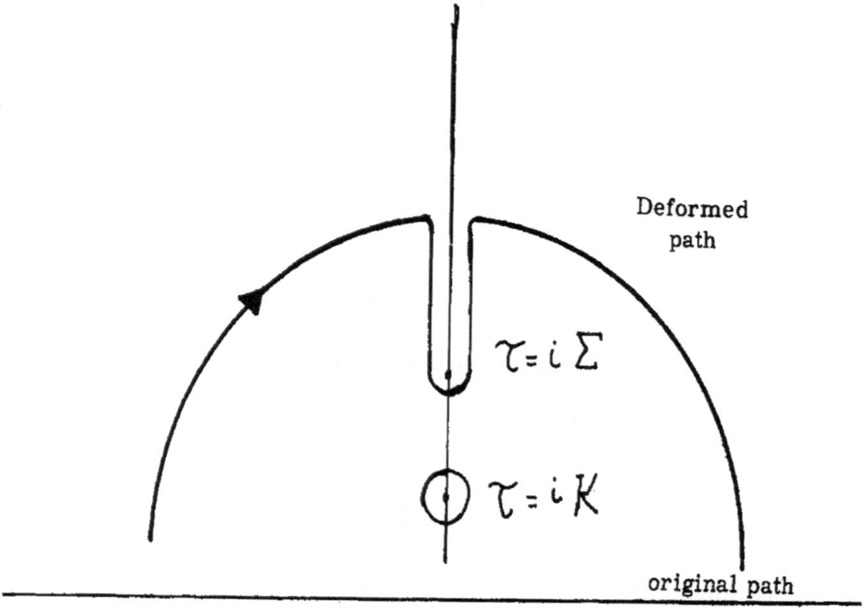

Fig. 1 .

For $x < 0$, the integral can be evaluated by taking a path of integration in the lower half of the complex plane. A contribution then arises from the pole $\tau = -iK$.

We have therefore a first part of the solution, that is :

(9) $$\Phi_1 = \frac{2(1-c)}{c} \cdot \frac{\Sigma^2 - K^2}{K^2 - \Sigma^2(1-c)} \cdot \frac{K}{2(1-c)\Sigma} \exp(-K|x|).$$

This part dominates in Φ at large values of x.

A second part of the solution corresponds to $c > 1$, that is :

(10) $$\Phi_2 = \int_0^\infty \frac{2\Sigma^2(\eta+1)\exp\left[-(\eta+1)\Sigma|x|\right] d\eta}{\left[2\Sigma(\eta+1) - c\Sigma \log(2\eta^{-1}+1)\right]^2 + \pi^2 c^2 \Sigma^2}.$$

When x is small, the major contribution to the integral comes from large values of η, i.e. the integral is given approximately by :

(11) $$\int_0^\infty \frac{\exp(-(\eta+1)|x|\Sigma)\, d\eta}{2(\eta+1)} = \frac{1}{2} F_1(\Sigma|x|).$$

This is important near the source $x = 0$ and decreases rapidly as $\exp(-\Sigma x)$ when x tends to ∞

4. Solution of the time-dependent transport equation without sources in a semi-infinite medium.

Now we consider the time-dependent Boltzmann integro-differential equation for the case of the neutron transport in a semi-infinite medium without sources and for the one-group theory with non-isotropic collisions. That is we consider the integro-differential equation :

(1) $$\frac{1}{v} \frac{\partial N(x,\mu,t)}{\partial t} + \mu \frac{\partial N(x,\mu,t)}{\partial x} = \eta\sigma_s \int_{-1}^{+1} N(x,\mu',t) \frac{1+3P\mu\mu'}{2} d\mu' - (\sigma_s + \sigma_c)\eta N(x,\mu,t),$$

where η = number per unit of volume of collision centers for the neutron ;

σ_c = cross section of capture of the neutron ;

σ_1 = cross section of scattering.

P = constant

and

$$f(\vec{\Omega}' \longrightarrow \vec{\Omega}) = (1+3P\mu\mu')/2 .$$

Putting

$$\eta\sigma_c = \Lambda^{-1}, \quad v\eta\sigma_c = \tau^{-1},$$

we obtain the equation :

$$(2) \qquad \frac{\partial N(x,\mu,t)}{\partial t} + v\mu\frac{\partial N(x,\mu,t)}{\partial x} = \left(-\frac{v}{\Lambda}+\frac{1}{\tau}\right) N(x,\mu,t) + \frac{v}{\Lambda}\int_{-1}^{+1} N(x,\mu',t)\,\frac{1+3P\mu\mu'}{2}\,d\mu', \quad (\Lambda, v, \tau, P \text{ constants})$$

We have the following boundary condition:

$$(I) \qquad N(x,\mu,t) = 0 \qquad \text{for } x = 0, \quad \mu < 0 .$$

We put now :

$$(3) \qquad \omega = \frac{v}{\Lambda} + \frac{1}{\tau}, \quad \omega_o = \frac{v}{\Lambda} .$$

Writing

$$(4) \qquad N(x,\mu,t) = \exp(-\omega t)\, F(x,\mu,t) ,$$

we obtain :

$$(5) \qquad \frac{\partial F(x,\mu,t)}{\partial t} + v\mu\frac{\partial F(x,\mu,t)}{\partial x} = \omega_o \int_{-1}^{+1} F(x,\mu',t)\,\frac{1+3P\mu\mu'}{2}\,d\mu' ,$$

with

$$F(x, \mu, t) = 0 \qquad \text{for } x = 0, \ \mu < 0 .$$

We apply a $\mathcal{L}_{t,p} \mathcal{L}_{x,q}$ (double $\mathcal{L}$-transformation) and we put :

$$\varphi(p, \mu) = \int_{-\infty}^{+\infty} \exp(-pt) \ F(0, \mu, t) \, dt ,$$

$$\psi(q, \mu) = \int_{-\infty}^{0} \exp(qx) \, N(x, \mu, 0) \, dx ,$$

$$\Phi(p, q, \mu) = \mathcal{L}_{t,p} \mathcal{L}_{x,q} \left[N(x, \mu, t) \right] = \int_{-\infty}^{0} e^{qx} \, dx \int_{0}^{+\infty} e^{-pt} N(x, \mu, t) dt ,$$

We obtain the integral equation :

$$(6) \quad -\psi(q, \mu) + p\Phi(p, q, \mu) + v\mu \left[\varphi(p, q, \mu) - q\Phi(p, q, \mu) \right] = = \omega_o \int_{-1}^{+1} \Phi(p, q, \mu') \frac{1+3P\mu\mu'}{2} \, d\mu ,$$

with

$$(7) \qquad \varphi(p, \mu) = 0 \ \text{ for } \mu < 0.$$

One considers $F(0, \mu, t)$ as a given function and $\varphi(p, \mu)$ is a known function. The function $\psi(q, \mu)$ is dependent from the initial values of $F(x, \mu, t)$ and will be determined when the function φ is given . In order to solve the integral equation (6) we write :

$$(8) \qquad \varphi = \sum_{m=1}^{\infty} \frac{\varphi_m(\mu)}{p^m} , \qquad (9) \qquad \psi = \sum_{n=1}^{\infty} \frac{\psi_n(\mu)}{q^n}$$

$$(10) \qquad \Phi = \sum_{1}^{\infty} {}_{mn} \frac{\Phi_{mn}(\mu)}{p^n \, q^n}$$

and we observe that we shall have :

$$(11) \qquad \varphi_m(\mu) = 0 \qquad \text{for } \mu < 0 , \quad (m=1,2,3,\ldots\ldots) .$$

Substituting in the integral equation (6), we obtain :

$$(12) \quad \sum_{m,n=1}^{\infty} \frac{\Phi_{mn}(\mu)}{p^{m-1} q^{n}} - \sum_{n-1}^{\infty} \frac{\Psi_n(\mu)}{q^n} + v\mu \left[\sum_{m=1}^{\infty} \frac{\varphi_m(\mu)}{p^m} - \sum_{m,n=1}^{\infty} \frac{\Phi_{mn}(\mu)}{p^m q^{n-1}} \right]$$

$$= \omega_o \sum_{m,n=1}^{\infty} \frac{1}{p^m q^n} \int_1^{+1} \Phi_{m,n}(\mu') \frac{1+3P\mu\mu'}{2} d\mu' .$$

Therefore we have necessarily :

$$(13) \quad \Phi_{m1} = \varphi_m(\mu) , \quad (m = 1, 2, 3, \ldots\ldots)$$

$$(14) \quad \Phi_{1n} = \Psi_n(\mu) , \quad (n = 1, 2, 3, \ldots\ldots)$$

$$(15) \quad \Phi_{m+1,n}(\mu) - v\mu \, \Phi_{m,n+1}(\mu) =$$

$$= \omega_o \int_{-1}^{+1} \Phi_{mn}(\mu) \frac{1+3P\mu\mu}{2} d\mu' , \quad (m,n = 1, 2, 3, \ldots\ldots) .$$

The equations (13) give the coefficients Φ_{m1} of the double series Φ in terms of the φ_m. The equation (15) is recurrent and gives $\Phi_{m,2}$, $\Phi_{m,3}, \ldots, \Phi_{m,n}, \ldots$ in terms of the φ_m.

When we have $\Phi_{m,n}$, we have also Φ_m, and the equation (14) gives the functions Ψ_n in terms of the functions φ_m.

Applying the theorems on the $\mathcal{L}$-transformation of the series, we obtain , after the calculations :

$$(16) \quad \Phi(p, q, \mu) = \mathcal{L}_{t,p} \mathcal{L}_{x,q} \left[\sum_{1\ mn}^{\infty} \Phi_{mn}(\mu) \frac{t^{m-1}}{(m-1)!} \cdot \frac{(-x)^{n-1}}{(n-1)!} \right] ,$$

that is (unicity of the L-transformation):

$$(17) \qquad F(x, \mu, t) = \sum_{1}^{\infty}{}_{mn} \quad \Phi_{mn}(\mu) \frac{t^{m-1}}{(m-1)!} \cdot \frac{(-x)^{n-1}}{(n-1)!} \; .$$

One demonstrates immediately that it is

$$F(x, \mu, t) = 0 \text{ for } \quad x = 0, \qquad \mu < 0 \; .$$

It is easily possible to demostrate also that for $F(x, \mu, t)$ exists an exponential majorant series

$$H\left(1 + 3P\frac{\omega_o}{\omega}\right) e^{\omega\left(t - \frac{x}{v}\right)}$$

Finally we obtain the density function :

For $\mu > 0$:

$$(18) \quad N(x, \mu, t) = e^{-\omega t} \sum_{m=1}^{\infty} \frac{t^{m-1}}{(m-1)!} \left\{ \sum_{m=0}^{\infty} \frac{\varphi_{m+n}(\mu)}{(v\mu)^n} \frac{(-x)^n}{n!} + \right.$$

$$\left. + \; 3P\omega_o \sum_{m=1}^{\infty} \frac{\alpha_{m+2n-2,\, 2n-2}}{v^{2n-1}} \frac{x^{2n-1}}{(2n-1)!} \right\}$$

For $\mu < 0$:

$$(19) \quad N = 3P\omega_o e^{-\omega t} \sum_{mn=1}^{\infty} \frac{\alpha_{m+2n-2,\, 2n-2}}{v^{2n-1}} \frac{t^{m-1}}{(m-1)!} \frac{x^{2n-1}}{(2n-1)!} \; .$$

And it is

$$N(0, \mu, t) = 0 \quad \text{for} \quad \mu < 0$$

So our problem is solved .

A. Pignedoli

BIBLIOGRAPHY

Hopf, E Mathematical problems of radiative equilibrium, Cambridge Tracts, No 31.

Davison, B, Neutron transport theory, Oxford University Press, 1957.

Marshak, R, Rev. of modern Physics, 19, No 3, 1947.

Lecaine, J, Phys. Review , 72, 564, 72, 1947.

Lecaine, J, Canad. J. Res, A, 28, 242, 1950.

Case, K., Annals of Physics, 91, No 1, 1960.

Pignedoli, A. Atti del Seminario mat. e fis. della Università di Modena , 1947.

Tait, J. H., Neutron transport theory, Longmans, London, 1964.

CENTRO INTERNAZIONALE MATEMATICO ESTIVO

(C.I.M.E.)

A. PIGNEDOLI

ON THE RIGOROUS ANALYSIS OF THE PROBLEM OF THE NEUTRON TRANSPORT IN A SLAB GEOMETRY AND ON SOME OTHER RESULTS.

Corso tenuto a Varenna dal 17 al 19 settembre 1966

ON THE RIGOROUS ANALYSIS OF THE PROBLEM OF THE NEUTRON TRANSPORT IN A SLAB GEOMETRY AND ON SOME OTHER RESULTS.

by

Antonio Pignedoli

1. Introduction.

This lecture is concerning the problem of the neutrons distribution in a slab geometry [1]. The statement of the problem is the following. The neutron density is satisfying the integro-differential Boltzmann equation :

$$(1)\quad \frac{1}{v}\frac{\partial N(x,\mu,t)}{\partial t}+\mu\frac{\partial N(x,\mu,t)}{\partial x}+\Sigma N(x,\mu,t)=\frac{c\Sigma}{2}\int_{-1}^{+1}N(x,\mu',t)d\mu'$$

where the symbols are usual (see my first lecture in this Course) . The physical conditions are the following :

a) the cross section Σ is constant ;

b) the production of particles is isotropic ;

c) the slab is surrounded by a vacuum, so that no particle may enter the slab from outside ;

d) at the time $t = 0$ a particle distribution $f(x,\mu)$ exists inside the slab.

The supplementary conditions of the problem are :

$$(2)\quad \begin{cases} N(a,\mu,t)=0\ ,\quad \mu<0\ ,\quad t>0\ ; \\ N(-a,\mu,t)=0\ ,\quad \mu>0\ ,\quad t>0\ ; \\ N(x,\mu,0)=f(x,\mu)\ ;\ -a\le x\le a,\ -1\le\mu\le 1. \end{cases}$$

We ask for the resulting time-dependent particle distribution. The equation (1) may be simplified somewhat by writing ·

---

(1) Important works in this direction have been done by :

J. Lehner and C. M. Wing, Communications on pure and applied mathematics, vol. 8, 1955.
idem Duke Mathem. Journal, 1956.

R. L. Bowden and C. D. Williams, Journal of mathematical Physics, vol. 11, 1964.

$$N(x, \mu, t) = \exp(-c \Sigma t) \chi(x, \mu, t)$$

and choosing, for convenience, $v = 1$, $\Sigma = 1$.

We have :

$$(3) \qquad \frac{\partial \chi(x, \mu, t)}{\partial t} = -\mu \frac{\partial \chi(x, \mu, t)}{\partial x} + \frac{c}{2} \int_{-1}^{+1} \chi(x, \mu', t) d\mu'$$

with the supplementary conditions :

$$(4) \qquad \begin{cases} \chi(a, \mu, t) = 0 , & \mu < 0 , \; t > 0 ; \\ \chi(-a, \mu, t) = 0 , & \mu > 0, \; t > 0 ; \\ \chi(x, \mu, 0) = f(x, \mu), & -a \leq x \leq a, \; -1 \leq \mu \leq 1 . \end{cases}$$

In order to obtain a formal solution, we can write the equation (3) in the form :

$$(5) \qquad \frac{\partial \chi(x, \mu, t)}{\partial t} = A \chi(x, \mu, t) ,$$

where A is the time-independent operator :

$$(6) \qquad A. = -\mu \frac{\partial}{\partial x} . + \frac{c}{2} \int_{-1}^{+1} . d\mu'$$

A formal application of the Laplace transformation with respect to t gives :

$$(7) \qquad \lambda \bar{\chi} - f(x, \mu) = A \bar{\chi} ,$$

where

$$(8) \qquad \bar{\chi} = \bar{\chi}(x, \mu, \lambda) = \int_0^{+\infty} e^{-\lambda t} \chi(x, \mu, t) \, dt =$$

$$= \mathcal{L}_{t,\lambda} \chi(x, \mu, t) .$$

Obviously the equation (7) can be written :

$$(9) \qquad (\lambda - A) \bar{\chi} = f .$$

If $\lambda - A$ were now just a complex-valued function, the solution to (9) would be simply :

(10) $$\bar{\chi} = (\lambda - A)^{-1} f.$$

But A is actually an operator, and it remains the problem of finding $(\lambda - A)^{-1}$ when it exists and determining its properties.

Let us for a moment press the formalism still further. Applying the inverse transform operator to (10) yields:

(11) $$\chi(x, \mu, t) = (2\pi i)^{-1} \int_{b-i\infty}^{b+i\infty} \left[(\lambda - A)^{-1} f\right] e^{\lambda t} \, d\lambda.$$

The evaluation or estimation of (11) depends on a knowledge of the behavior of $(\lambda - A)^{-1} f$. The singularities of this function are of particular importance. Thus we must study the operator $(\lambda - A)^{-1}$; this implies we must investigate the spectrum of the operator A. Let us suppose that $(\lambda - A)^{-1} f$ is a "well-behaved" function of λ except for certain values λ_j where the operator A fails to exist. This latter event will happen when there are functions $\psi_j(x, \mu)$ such that:

(12) $$(\lambda_j - A)\psi_j = \lambda_j \psi_j - A\psi_j = 0,$$

that is when ψ_j is an eigenfunction of A belonging to an eigenvalue λ_j.

If there were infinitely many λ_j, then we should have a formal expansion as the following:

(13) $$\chi(x, \mu, t) = \sum_{j=1}^{\infty} e^{\lambda_j t} g_j(x, \mu).$$

But there is not an infinite number of eigenvalues λ_j, and one demonstrates that the solution (13) must be replaced by:

(14) $$\chi(x, \mu, t) = \sum_{j=1}^{n} e^{\lambda_j t} g_j(x, \mu) + \xi(x, \mu, t),$$

where $\xi(x, \mu, t)$ is, in a sense, small compared to the other terms of (14). There is, of course, always at least an eigenvalue λ_1 belonging to the operator A. Therefore the sum $\sum_{j=1}^{n} e^{\lambda_j t} g_j(x, \mu)$ is never empty and the sum contributes the dominant term for large t. Thus one can determinate the asymptotic time behavior of χ by assuming $\chi = e^{\lambda t} g(x, \mu)$ and then finding the value of λ of largest real part for which this expression satisfies the equation

$$(15) \qquad \frac{\partial \chi}{\partial t} = -\mu \frac{\partial \chi}{\partial x} + \frac{c}{2} \int_{-1}^{+1} \chi(x, \mu', t)\, d\mu' .$$

But we shall expose a rigorous analysis of the problem, on which there is an important literature.

2. The rigorous analysis of the problem.

The action of the operator A is defined by the equation

$$(1) \qquad \frac{\partial \chi}{\partial t} = A \chi .$$

Now we choose the Hilbert space H of complex valued functions $g(x, \mu)$, defined and square integrable in the Lebesgue sense over the rectangle $|x| \leq a, |\mu| \leq 1$:

$$(2) \qquad \int_{-1}^{+1} d\mu \int_{-a}^{+a} |g(x, \mu)|^2 \, dx < +\infty .$$

We define the usual inner product of two functions g and h in H:

$$(3) \qquad (g, h) = \int_{-1}^{+1} d\mu \int_{-a}^{+a} g(x, \mu)\, \overline{h(x, \mu)} \, dx ,$$

and the usual norm :

$$\|g\| = \sqrt{(g,g)} = \sqrt{\int_{-1}^{+1} d\mu \int_{-a}^{+a} g(x,\mu)\, \overline{g(x,\mu)}\, dx}. \tag{4}$$

Write :

$$A = -D + cJ, \quad D. = \mu \frac{\partial}{\partial x} \cdot, \quad J. = \frac{1}{2} \int_{-1}^{+1} .\, d\mu' . \tag{5}$$

Let d_D the domain of the operator D be the set of functions g, $g \in H$ absolutely continuous in x for each μ in $|\mu| \leq 1$ and such that $Dg \in H$. Call d_J the domain of the operator J, the set of all $g \in H$ such that Jg exists for each x in $|x| \leq a$ and $Jg \in H$.

Finally define d_A as the linear manifold of all functions g belonging to both d_D and d_J and such that

$$\begin{cases} g(a,\mu) = 0, & -1 \leq \mu < 0, \\ g(-a,\mu) = 0, & 0 < \mu \leq 1. \end{cases} \tag{6}$$

Hence A is an operator from the Hilbert space H into itself with domain d_A.

Concerning the spectrum of the operator A, one can first demonstrate that there are no eigenvalues in the left half of the lambda plane. We have the following :

<u>Theorem 1.</u>

<u>There is no function</u> $\psi \in d_A$ <u>satisfying the equation</u> $A\psi = \lambda\psi$ for some λ with $\mathrm{Re}(\lambda) < 0$.

In order to establish if there exist eigenvalues of A with $\mathrm{Re}(\lambda) \geq 0$, $\mathrm{Im}(\lambda) \neq 0$, we consider the function :

$$\Phi(x) = \int_{-1}^{+1} \psi(x,\mu')\, d\mu' \quad \text{in} \quad L_2(-a,a), \tag{7}$$

and we demonstrate the following theorems :

Theorem 2.

Suppose $\psi(x,\mu) \in d_A$ satisfies $A\psi = \lambda\psi$ with $Re(\lambda) \geq 0$, $\lambda \neq 0$. Then the function $\Phi(x)$ satisfies :

(8) $$\Phi(x) = \frac{c}{2} \int_{-a}^{+a} E(\lambda |x-x'|)\, \Phi(x')\, dx' ,$$

where

$$E(u) = \int_1^{\infty} \exp(-ut) \frac{1}{t}\, dt , \quad Re(u) \geq 0 , \quad u \neq 0 .$$

Theorem 3.

There is no function $\psi \in d_A$ satisfying $A\psi = \lambda\psi$ with $Re(\lambda) \geqslant 0$, $Im(\lambda) \neq 0$

It remains to examine the possibility of λ real non-negative. We shall write $\lambda = \beta$ when λ is real . The case $\beta = 0$ is easy. The equation

$$A\psi = \lambda\psi$$

becomes

$$A\psi = -\mu \frac{\partial \psi}{\partial x} + \frac{c}{2}\psi = 0 ,$$

which yields :

(9) $$\psi(x,\mu) - \psi(-a,\mu) = \frac{c}{2\mu} \int_{-a}^{x} \Phi(x')\, dx' , \quad |x| \leq a .$$

Choose x such that the right side of (9) is not zero, $x = x_1$, $-a < x_1 < a$. Because of the $\frac{1}{\mu}$ factor, the right side is then not integrable over μ, $|\mu| \leq 1$. But the left side of (9) is integrable over μ since $\psi \in d_A$. Hence $\beta = 0$ is not an eigenvalue. Suppose now $\beta > 0$. By theorem 2, every solution of $A\psi = \lambda\psi$ yields a function Φ satisfying (8) .

Conversely if $\Phi \in L_2$ satisfies (8), then the corresponding ψ is a solution of $A\psi = \lambda\psi$.

We have the following :

Theorem 4

A necessary and sufficient condition that $\psi(x,\mu)$ be a solution of $A\psi = \lambda\psi$ corresponding to $\beta = 0$ is that there be an L_2 - solution Φ to

$$\Phi(x) = \frac{c}{2} \int_{-a}^{+a} E(\lambda|x-x'|)\Phi(x')\,dx' ,$$

for that β value.

We have also the following :

Theorem V.

The point spectrum $P\sigma$ A of A consists of a finite but nonempty set of points $\beta_1 > \beta_2 > \ldots > \beta_m$ all lying on the positive λ -axis.

Theorem VI.

The eigenvalues of the operator A are of finite multiplicity . Now we consider the adjoint A^* of A [(2)].

It is a matter of relatively easy computation to prove that the operator A^* adjoint to A is given by :

$$A^{*}. = \mu \frac{\partial}{\partial x} . + \frac{c}{2} \int_{-1}^{+1} . \, d\mu' .$$

[(2)] cfr. F. Riesz-B. Sz. Nagy, Functional Analysis, F. Nugar, New York, 1954

M. H. Stone, Linear transformation in Hilbert Space and their applications to Analysis, Amer. Mathem. Society Colloquium publication n. 15, New York, 1932

A. E. Taylor, An introduction to functional analysis, J. Wiley, New York, 1958

The domain $d_{A^*} \subset H$ consists of those functions $\psi^*(x,\mu)$ belonging to both d_D and d_J and such that

$$\psi\,(a,\mu) = 0, \mu > 0\ ; \psi(-a,\mu) = 0\ ,\ \mu < 0\ .$$

Clearly the operator A is not self-adjoint (this fact complicates considerably the problem).

If $\psi(x,\mu)$ is an eigenfunction of A, with eigenvalue λ, then $\psi^*(x,\mu) = \overline{\psi(-x,\mu)}$ is an eigenfunction of A^* with eigenvalue $\bar{\lambda}$. The converse is also true.

Theorem 1

The residual spectrum $R\sigma A$ of A is empty.

Theorem 2.

The continuous spectrum of A contains the half-plane $\mathrm{Re}\,(\lambda) \leqslant 0$.

Let Γ be the set of points in the right half of the λ-plane exclusive of the points of $P\sigma A$. The points of Γ are either in $C\sigma A$ (const. spectrum) or $\mathfrak{M}A$ (restr. Spectrum) and hence $(\lambda - A)^{-1}$ exists and its domain d_A is dense in H. One demonstrates the following:

Theorem.

The resolvent set of A contains Γ. The resolvent set is : Re $(\lambda) > 0$ deleted by the point spectrum.

Summary theorem.

The linear operator A of

$$A\psi = \lambda\psi$$

a) is non self-adjoint ;

b) decomposes the spectral plane as follows :

Point spectrum : a finite nonempty point set lying on $\lambda > 0$;

Residual spectrum : empty ;

Continuous spectrum : $\mathrm{Re}(\lambda) \leqslant 0$;

Resolvent set $\mathrm{Re}(\lambda) > 0$ deleted by the point spectrum

3. An integral representation of the solution of the transport equation.

Considering the integro-differential equation :

$$(1) \qquad \frac{\partial \chi(x,\mu,t)}{\partial t} = -\mu \frac{\partial \chi(x,\mu,t)}{\partial x} + \frac{c}{2} \int_{-1}^{+1} \chi(x,\mu',t)\, d\mu'$$

whit the supplementary conditions :

$$(2) \qquad \begin{cases} \chi(a,\mu,t) = 0 \,, \mu < 0 \,, \quad t > 0 \,; \\ \chi(-a,\mu,t) = 0 \,, \mu > 0 \,, \quad t > 0 \,; \\ \chi(x,\mu,0) = f(x,\mu) \,, \; -a \leq x \leq a; \; -1 \leq \mu \leq 1 \,, \end{cases}$$

and having determined how the operator A decomposes the spectral plane , we are in condition to return to our original problem, that is the solution of the aforesaid problem. We write :

$$(3) \qquad \chi(x,\mu,t) = T(t)\, f(x,\mu) \,.$$

Symbolically we have :

$$T(t) = \exp\, (At) \,;$$

more properly $T(t)$, $t \geqslant 0$ is a semigroup of operators generated by A. Such a semigroup exists provided : A is closed; d_A is dense in H ; $\|R_\lambda\| < (\lambda - K)^{-1}$, $\lambda > K$ for some $K > 0$.

Effectively the operator A generates a semigroup $T(t)$ of bounded operators for $t \geqslant 0$.

It follows from known theorems[3] that the solution χ is unique in H,

---

[3] E. Hille, Functional Analysis and Semi-groups, Amer Math.Soc. coll. publ. Vl. 31, New York , 1948

R. F. Phillips, Perturbation Theory for semi-groups of linear operators, trans. of the Amer. Mathem. Soc., Vol. 74, 1953 .

that

$$(4) \qquad \lim_{t \to 0} \| \chi(x,\mu,t) - f(x,\mu) \| = 0 ,$$

and that

$$(5) \qquad \chi(x,\mu,t) = \lim_{\omega \to \infty} \frac{1}{2\pi i} \int_{b-i\omega}^{b+i\omega} e^{\lambda t} R_\lambda f \, d\lambda \ , \ t>0, b>\beta_1, f \in d_A$$

The integral is to be considered the strong limit of Riemann sums. It converges uniformly for $0 < t_0 \leqslant t \leqslant t_1 \leqslant \infty$ so that χ is continuous in t for fixed (x,μ). If we fix (x,μ) then $R_\lambda f$ is a complex valued continuous function of λ on the integration path so that the integral may be considered as an ordinary Riemann integral. We now wish to move the line of integration in (5) to the left so as to pick up the contributions from the singularities at β_j $(j = 1, 2, \dots, m)$. To accomplish this we must study the behavior of $R_\lambda f$.

We restrict f to d_A. Then, for fixed x and μ, $R_\lambda f$ is an analytic function of λ for $Re(\lambda) > 0$ except at the points β_j $(j = 1, 2, \dots, m)$.

Let β_j be an eigenvalue of multiplicity s, with linearly independent eigenfunctions $\Psi_{j,1}, \Psi_{j,2}, \dots, \Psi_{j,s}$.

Let the adjoint functions be $\Psi^*_{j,k}(x,\mu) = \overline{\Psi_{j,k}(-x,\mu)}$ $j = 1, 2, \dots, m$; $k = 1, 2, \dots, s$.

One demonstrates that it is :

$$(6) \qquad \chi(x,\mu,t) = \lim_{\omega \to \infty} \frac{1}{2\pi i} \int_{-i\omega}^{+i\omega} e^{\lambda t} R_\lambda f \, d\lambda \ +$$

$$+ \sum_{j=1}^{m} e^{\beta_j t} \sum_{k=1}^{s_j} (f, \Psi^*_{j,k}) \Psi_{j,k}(x,\mu) \ , \ 0 < \delta < \beta_m .$$

(one shifts the line of integration to the left of β_m but not as far as the imaginary axis, picking up the residues at β_j, $j = 1, 2, 3, \ldots, m$. And estimating the integral, one obtains:

$$(7) \quad \chi(x,\mu,t) = \sum_{j=1}^{m} e^{\beta_j t} \sum_{k=1}^{s_j} (f, \psi^*_{j,k})\, \psi_{j,k}(x,\mu) + \xi(x,\mu,t)$$

where for almost all (x,μ), $|\xi|$ is dominated as follows (setting $\delta = \frac{1}{t}$):

$$|\xi(x,\mu,t)| < D(x,\mu)\, t^2 \log^2 t\,,$$

where D depends on f and is finite for almost all (x,μ):

4. Some other results on the transport theory.

From the somewhat practical viewpoint, it is desirable to obtain more information about the eigenvalues of the operator A.
Results in this direction have been obtained by S. Schlesinger [(4)]. But we will not consider this direction of research in this lecture. We consider some other results in the transport theory.
G. Pimbley [(5)] has studied the multienergy state time-dependent slab problem. This is especially important from a physicist's viewpoint because practical calculations of particle behavior (e.g. of neutron behavior in reactors) generally rely on replacing the continuous energy dependence by

(4) S. Schlesinger, Approximating eigenvalues and eigenfunctions of symmetric kernels, Journal of the Society of industrial and applied mathematics, vol. 5, 1957.
idem, Some eigenvalue problems in the theory of neutrons, Los Alamos scientific laboratory reports, LA 1908, 1955.

(5) G. Pimbley, Solution of an initial value problem for the multi-velocity neutron transport equation with a slab geometry, Journal of Mathem. and Mech., vol. 8, 1959.

a discrete dependence .

Pimbley considered the equations :

$$(8)\quad \frac{1}{v_i}\frac{\partial N_i(x,\mu,t)}{\partial t}+\mu\frac{\partial N_i}{\partial x}+\sum_i N_i=\frac{1}{2}\sum_{j=1}^{n} c_{ij}\int_{-1}^{+1} N_j(x',\mu',t)\,d\mu'$$

$i=1,2,3,\dots,n$

with

$$(9)\quad \begin{cases} N_i(a,\mu,t)=0\ , & \mu<0\ ;\\ N_i(-a,\mu,t)=0\ , & \mu>0\ ;\\ N_i(x,\mu,0)=f_i(x,\mu)\ . \end{cases}$$

It was discovered that, while some results carried over quite completely, it was required in other cases to considerably restrict the matrix c_{ij} in order to preserve the analogy .

For the time-dependent transport problem for bounded geometries, some very general results have been obtained by K. Jörgens [(6)].

Let $\mathfrak{D}$ and $\mathfrak{V}$ be bounded measurable sets in position space (x_1, x_2, x_3) and velocity space (v_1, v_2, v_3) respectively. Let $\mathfrak{V}$ be bounded away from zero (the particle velocity cannot become arbitrarily small, and "trapping" is avoided) . The transport equation can be written in such a general case in the form :

$$(10)\quad \frac{\partial N}{\partial t}=-\sum_{j=1}^{3} v_j\frac{\partial N}{\partial x_j}-\Sigma(x,v)\sqrt{v_1^2+v_2^2+v_3^2}\,N+KN=AN\quad (x=x_1,x_2,x_3)\ .$$

Let $N\in H$, the space of square integrable functions on $\mathfrak{D}\times\mathfrak{V}$ and let $\Sigma(x,v)$ be non-negative, bounded and measurable on $\mathfrak{D}\times\mathfrak{V}$. In Jörgens analysis K is a quite general bounded linear operator on H. In must cases K in an integral operator:

6) K.Jörgens , An asymptotic expansion in the theory of neutron transport, Comm. in pure and applied mathem., vol. 11, 1958.

$$(10') \quad K\,N = \int_{\mathcal{V}'} k(x, v, v')\, N(x, v', t)\, dv'$$

which describes the scattering of particles of velocity (v'_1, v'_2, v'_3) into velocity (v_1, v_2, v_3) . It is further assumed that no particles can enter $\mathfrak{D}$ from outside .

Under the conditions stated, the problem

$$\frac{\partial N}{\partial t} = AN , \qquad N(x, v, 0) = f(x, \mathcal{V})$$

has a unique solution

$$N(x, v, t) = e^{At} f(x, v) = T(t)\, f(x, v) .$$

$(\Sigma(x, v) = \Sigma(x_1, x_2, x_3, v_1, v_2, v_3)$ etc.) .

With certain additional restrictions, the operator $T(t)$ is completely continuous for $t > 3\, t_o$ where t_o is the maximum length of time that a particle takes in crossing $\mathfrak{D}$. This is true, for example, in the n velocity group theory.

It is also true when $\mathcal{V}$ is a three-dimensional set K is the form ($*$) and k is bounded and integrable over $\mathfrak{D} \times V \times V'$.

When $T(t)$ is completely continuous it has a discrete spectrum $e^{\lambda_j t}$ (plus possibly the point zero) where the λ_j are the eigenvalues of A.

The spectrum of A may be empty .

The spectrum is certain nonempty for the sphere problem(7)

The formal series expansion of the solution function N is complicated.

---

(7) R. Van Norton, New York Univ. Report, 1960

J. Lehner, Comm. on pure and appl. mathem , 1962 .

ɔroblem of the convergence of this series, in the case that the spec- is infinite, remains insolved. Asymptotic results holding for large 'e been obtained by Jörgens [8]

5. Researchs using the method of distributions in sense of L. Schwartz.

Consider the simple time-indep. transport operator :

$$\mu \frac{\partial N}{\partial x} + N = \frac{c}{2} \int_{-1}^{+1} N(x,\mu')\, d\mu' , \quad c = \text{const} \quad (\Sigma = 1 \text{ for convenience}).$$

ssume :

$$N(x,\mu) = g(x)\, h(\mu) .$$

ituting in (1) we have readily :

$$N(x,\mu) = \exp\left(-\frac{x}{\nu}\right) \Phi_\nu(\mu) ,$$

$$\left(1 - \frac{\mu}{\nu}\right) \Phi_\nu(\mu) = \frac{c}{2} \int_{-1}^{+1} \Phi_\nu(\mu')\, d\mu' .$$

ν is a yet arbitrary . Let us require that :

$$\int_{-1}^{+1} \Phi(\mu')\, d\mu' = 1 .$$

Jörgens , An asymptotic expansion in the theory of neutron transport , Communic. in pure and appl. Mathem. vol. 11, 1958 .

Grosswald, Neutron transport in spherically simmetric systems, Journal of Mathem. and Mechanics, 1961 , vol. 10 .

Then manipulating we have :

$$\Phi_\nu(\mu) = \frac{c\,\nu}{2\nu - \mu} \;,$$

$$(5) \quad 1 = \int_{-1}^{+1} \Phi_\nu(\mu')\, d\mu' = \frac{\gamma\nu}{2}\int_{-1}^{+1} \frac{d\mu}{\nu-\mu} = \nu c \;\mathrm{tgh}^{-1}\left(\frac{1}{\nu}\right).$$

The equation (5) has two roots $\pm\,\nu_0$. For $c > 1$, ν_0 is pure imaginary ; for $0 < c < 1$, $\nu_0 > 1$. K. M. Case[9] has observed that other solutions may be obtained assuming that $\Phi_\nu(\mu)$ is a distribution in the sense of L. Schwartz [10]

Using this idea we can write :

$$(6) \qquad \Phi_\nu(\mu) = \frac{c}{2}\; P\, \frac{\nu}{\nu-\mu} + F(\nu)\,\delta(\mu-\nu)$$

where P ind. that the Cauchy princ. value is to be taken whenever an integral involving $\Phi_\nu(\mu)$ occurs.

For ν not in $(-1,+1)$ the procedure leads again to two roots $\pm\,\nu_0$.

However for $-1 < \nu < 1$ ( ) gives

$$1 = \frac{c}{2}\,\nu\; P\int_{-1}^{+1} \frac{d\mu}{\nu-\mu} + F(\nu)$$

from which $F(\nu)$ may be found .

Thus for any ν in $(-1,+1)$ there is a solution of the type (6) .

Case has also examined some time-dipendent cases in this way .

---

(9) K. M. Case, Elementary solutions of the transport equation and their applications Annals of Physics, vol. 9. 1960 .

(10) L. Schwartz , Theorie des distributions, vol. I, Hermann, Paris , 1950

P.P. Abbati Marescotti[11] has studied the problem of the slab in the multigroup theory and in the case of non isotropic scattering, that is the non-stationary problem

$$(8)\quad \frac{\partial f_i}{\partial t} = A f_i \;,\ \text{in the rectangle } R:\ |x| \leq a, |\mu| \leqslant 1,\quad (i = 1, 2, \ldots, n),$$

$$0 \leq t < +\infty ;$$

$$f_i(x, \mu, 0) = \psi_i(x, \mu) ; \quad f_i(a, \mu, t) = 0,\ \mu < 0 ;$$

$$f_i(-a, \mu, t) = 0,\ \mu > 0 .$$

It is solved the problem of determining $f_i \in L_2(R)$

One gives here a summary of the results :

a) Using the theory of the semi-groups, one demonstrates a theorem of e-xistence and uniqueness ;

b) The determination of the resolvent $R(\lambda, A)$ is reduced to a solution of a system of integral equations of Fredholm of the second kind ;

c) One finds the solution in the form of a Neumann series :

$$(9)\qquad f_i(x, \mu, t) = \exp(-\Sigma_i v_i t)\, f_i(x - v_i \mu t, \mu, 0) + \sum_0^{\infty}{}_r\ F_i^{(r)}\ (x, \mu, t)\ ,$$

where the first term at the second member represents the neutrons of the ith group without interactions in the time interval $0 \longmapsto t$ and the second term the number of neutrons with r+1 interactions in the same time interval.

Abbati Marescotti[12] has also solved the problem of the transport in a homogeneous sphere in stationary conditions, with spherical simmetry, isotropic scattering and with unknown source term S(r) : that is the problem:

---

(11) see P.P. Abbati-Marescotti, Atti Acc. Scienze di Bologna, 1962.

(12) P.P. Abbati-Marescotti, Atti Scienze di Torino, 1962 - 63.

$$(10)\quad \mu \frac{\partial f}{\partial r} + \frac{1-\mu^2}{r}\frac{\partial f}{\partial \mu} + \sigma f = \frac{c}{2}\int_{-1}^{+1} f(r,\mu')\,d\mu' + S(r),$$

$$0 \le r \le r_0,\ (r_0 = \text{radius of the sphere}),$$

$$f(r_0,\mu) = 0,\ \mu \le 0;\quad f(r_0,\mu) = g(\mu),\ \mu \ge 0.$$

The function $g(\mu)$ is assumed absolutely continuous in the interval $0 \mapsto 1$. The problem of determining $f(r,\mu)$ and $S(r)$ is reduced with a unique unknown function

$$(11)\qquad \psi(r) = \frac{c}{2}\int_{-1}^{+1} f(r\ ,\mu')\,d\mu' + S(r).$$

This function is determined by solving an integral equation of Volterra of the first kind. One studies also the dependence between f and S. Finally we shall say that the author [13] has studied the approximate solution of the problem in a spherical medium with $c = c(r) \in Bv$, $\Sigma = \Sigma(r) \in Bv$ (the symbols are usual) without regular assumptions concerning g.

The problem is reduced to determining $f(r,\mu)$ and $S(r)$ satisfying the conditions

$$(12)\qquad \int_{-1}^{+1}\int_0^{r_0} r^2 \left|f(r,\mu)\right| dr\ d\mu < +\infty,\quad \int_0^{r_0} r^2 \left|S(r)\right| dr < +\infty;$$

(the quantity of matter or radiation is finite). One determines necessary and sufficient conditions for g; in the case of Σ = const in every part in which the interval is divided, one solves a chain of integral equations; in general one constructs an approximate solution, demonstrating the convergence of the method.

---

[13] P. P. Abbati-Marescotti, Atti della Acc. delle Scienze di Bologna, 1964.

A. Pignedoli

BIBLIOGRAPHY

B. Davison, Neutron Transport theory, Oxford at the Clarendon Press, 1957.

G. Milton Wing, An Introduction to transport theory, J. Miley and Sons, New York-London 1962

K. M. Case, F. de Hoffman and G. Placzak, Introduction to the theory of neutron diffusion, Government Printing Office, 1954.

CENTRO INTERNAZIONALE MATEMATICO ESTIVO

(C.I.M.E.)

Giorgio Sestini

"PRINCIPI DI MASSIMO PER LE SOLUZIONI DI EQUAZIONI PARABOLICHE"

Corso tenuto a Varenna (Como) dal 19 al 27 settembı

1966

PRINCIPI DI MASSIMO PER LE SOLUZIONI DI EQUAZIONI PARABOLICHE

di

Giorgio Sestini

(Università-Firenze)

Introduzione.

Questo seminario ha l'unico scopo di richiamare l'attenzione sull'importanza dei così detti "principi di massimo" nella risoluzione di problemi al contorno per l'equazioni paraboliche , ad es. della diffusione, qualunque sia la questione fisica che ha condotto a quel problema e il tipo di grandezza caratteristica del processo considerato.

E' ben noto che, dopo aver superato le notevoli difficoltà di tradurre in equazioni un problema fisico , troppo spesso anche a spese del suo stesso significato fisico, si presenta la questione della risoluzione del problema matematico cui si è giunti. Questa risoluzione, specialmente se si tratta di problema non lineare, si rivela spesso così difficile da preferire una valutazione approssimata della soluzione, in modo da potere controllare, attraverso al confronto con dati sperimentali, fino a che punto il problema matematico resti aderente al problema fisico considerato.

In ogni caso, almeno in un primo tempo, si cerca di assicurare l'esistenza e l'unicità della soluzione del problema analitico in esame, seguendo di regola una via che permette anche l'istituzione di un algoritmo costruttivo della ricercata soluzione, appartenente ad una fissata classe di funzioni. Due metodi assai usati sono quello delle approssimazioni successive, generalmente applicato dopo aver trasformato il problema da differenziale in integrale o integro-differenziale, oppure quello proprio dell'analisi funzionale che conduce a considerare una conveniente trasformazione di un opportuno spazio funzionale in sé, alla quale si possa applicare il teorema del punto unito.

In tali metodi, sia per dimostrare la convergenza del processo iterativo, sia per assicurare l'esistenza del punto unito per la trasformazione

funzionale individuata, è di fondamentale interesse il conoscere "a priori" valutazioni approssimate della ricercata soluzione.
I principi di massimo danno un fondamentale aiuto a questo ordine di problemi. Di rilevante interesse è poi il loro impiego nella dimostrazione di teoremi di unicità.

Sulla scorta degli studiatissimi problemi di tipo ellittico e, in particolare delle proprietà di massimo delle funzioni armoniche, molti studi sono stati fatti per giungere a stabilire anche per le soluzioni di equazioni di tipo parabolico analoghi teoremi, atti a caratterizzare le proprietà di speciali classi di soluzioni di problemi al contorno di questo tipo.

Conviene subito osservare che se anche tali classi di soluzioni possono sembrare assai ristrette dal punto di vista analitico, esse tuttavia sono del tutto soddisfacenti a descrivere vaste categorie di fenomeni fisici del tipo di quello classico della diffusione del calore.

I primi risultati ottenuti riguardavano, come è ben naturale, le soluzioni di particolari problemi lineari. Successivamente sono state rilevale analoghe proprietà anche per classi più generali di soluzioni di problemi parabolici anche quasi lineari. Per i problemi al contorno non lineari, a quanto ne so, moltissimo resta ancora da fare. In ciò sta il carattere di "seminario" di questa lezione.

Per una sistematica, ampia e moderna trattazione delle equazioni paraboliche, accompagnata da una ricca bibliografia, è molto utile ed interessante la lettura del recente libro di A. FRIEDMAN [1] (¹).

I " <u>principi di massimo</u>" .

Seguendo L. NIRENBERG [2], i "principi di massimo" si distinguono in "deboli" e in "forti".

(¹) I numeri in parentesi quadra si riferiscono alla bibliografia posta la termine della relazione.

G. Sestini

I principi deboli affermano che la soluzione di un problema parabolico, in un conveniente dominio e appartenente ad una certa classe di funzioni, ha i suoi massimi sul contorno del campo di definizione.

Quelli forti affermano che se una soluzione di un problema parabolico, in un certo dominio e appartenente ad una certa classe, ha un massimo interno al campo di definizione, essa è costante.

E' evidente che questo secondo tipo include i principi deboli come immediata conseguenza.

E' quasi superfluo accennare che principi analoghi valgono per i minimi.

In generale le dimostrazioni di tali principi vengono date nel caso di due variabili, ed io, nell'esporne alcuni, mi atterrò alla regola, essendo immediatamente estendibili ad equazioni paraboliche in un numero qualsivoglia di variabili. Del resto dal punto di vista fisico-matematico le equazioni in due variabili coprono una vasta classe di problemi non stazionari (quando una delle due variabili sia il tempo) nei quali la dipendenza dal posto della grandezza caratteristica ricercata è affidata ad una sola variabile, cosa che si verifica tutte le volte che si ha a che fare con campi dotati di simmetria (problemi piani, cilindrici o sferici), nei quali la coordinata spaziale può interpretarsi come distanza del generico punto, in cui si cerca il valore della grandezza ad un certo istante, da un piano o da una retta o da un punto.

Il "principio di massimo" di M. GEVREY (1913)

M. GEVREY [3], che con E.E. LEVI [4] può considerarsi il fondatore dello studio sistematico dei problemi di tipo parabolico, ha stabilito che, assegnata l'equazione parabolica lineare:

(1) $$Z_{xx} + a(x,y)\, Z_x + b(x,y)\, Z_y + c(x,y)\, Z = 0 \quad , \quad Z_a = \frac{\partial Z}{\partial a} ,$$

con $a(x,y)$, $b(x,y)$, $c(x,y)$ funzioni continue di x ed y in una regione finita R del piano x,y, ogni soluzione regolare di (1) (cioè continua con le derivate che compaiono in (1)), non può avere in R né massimi positivi ,nè minimi negativi, quando sia $c < 0$.

Infatti le necessarie condizioni per l'esistenza ad es. di un massimo positivo in un punto $P_o \equiv (x_o, y_o) \in R$:

$$Z_{x_o, x_o} < 0 \,, \quad (Z_x)_{x_o, y_o} = (Z_y)_{x_o, y_o} = 0 \,, \quad c\ Z < 0 \,,$$

rendono insoddisfatta la (1) .

Nel caso in cui sia $b < 0$ (che è il caso dell'equazione della diffusione) , qualunque sia il segno di c (se è $c \geq 0$ basta operare in (1) il cambiamento di funzione incognita $Z = U \exp ky$ e scegliere la costante in modo che risulti $kb + c < 0$) resta facilmente provato che per ogni campo R' interno ad un contorno regolare γ (nel senso di essere continuo, semplice e tale da formare con una caratteristica che lo incontri uno o più contorni privi di punti doppi) il valore assunto da Z in un punto $P \in R'$ risulta compreso tra il massimo positivo e il minimo negativo dei valori che la Z assume sulla parte di γ al di sotto della caratteristica per P.

Da questa proprietà discende subito, nelle condizioni specificate, il teorema di unicità per la soluzione di (1) con assegnate condizioni al contorno.

Il "principio di massimo" di M. PICONE (1929) .

Un altro principio debole si deve a M. PICONE [5] . Si consideri l'equazione di tipo parabolico :

$$E(Z) = \sum_{h,k} a_{hk}(P)\, Z_{x_h x_k} + \sum_h b_h(P)\, Z_{x_h} - z_t + c(P) = f(P) \,, \tag{2}$$

con $\sum_{hk} a_{hk} \lambda_h \lambda_k$ forma quadratica definita positiva, per λ_i reali, essendo $a_{hk} = a_{kh}$ funzioni continue di $P \equiv (x_1, x_2, \ldots, x_n, t) \in R$, dove R è un qualunque dominio limitato e connesso di uno spazio euclideo E_{n+1} ad n+1 dimensioni.

Si indichi con FR la frontiera di R e con $F_{-t}R$ la parte di FR, se esiste, per i cui punti esiste la normale interna ad R avente verso opposto a quello dell'asse della variabile t e tale che un intorno circolare, convenientemente piccolo, di un suo punto P appartiene a $F_{-t}R$ (un dominio sferico non ha evidentemente $F_{-t}R$; un dominio parallellepipedo, con gli spigoli laterali paralleli all'asse della variabile t, ha come $F_{-t}R$ la base superiore). Nell'ipotesi che anche b_h, c, f siano funzioni continue di $P \in R$ e Z (P) soluzione regolare (nel solito senso) di (2), si ha il seguente teorema; se $f(P) \leq C$ (≥ 0) ed è $Z(P) \geq 0$ (≤ 0) per $P \in FR - F_{-t}R$, allora è $Z(P) \geq 0$ (≤ 0) per $P \in R$.
Discendono immediatamente, come corollari del teorema, la unicità della soluzione di (2), nelle ipotesi dichiarate, che prende assegnati valori su $FR-F_{-t}R$ e, nel caso dell'equazione omogenea (f=0), se $c \leq 0$ il seguente principio di massimo debole : nelle ipotesi dichiarate, ogni soluzione regolare di E(Z) = 0 raggiunge il massimo dei valori del suo modulo su $FR-F_{-t}R$.

Il criterio di massimo viene poi esteso al caso di campi illimitati, facendo alcune ipotesi sul campo e sul comportamento della ricercata soluzione per $P \to \infty$.

L'elegante dimostrazione del teorema si basa essenzialmente su di un Lemma di algebra (teor. di Moutard) che ci limitiamo a ricordare: Siano $\sum_{hk} a_{hk} \lambda_h \lambda_k$, $\sum_{ij} \alpha_{ij} \xi_i \xi_j$ due forme quadratiche definite o semidefinite (positive o negative) . La somma

$$\sum_{hk} a_{hk} \alpha_{hk}$$

è non negativa o non positiva secondo che le due forme, se non son nulle, hanno lo stesso segno o segno opposto.

Il "principio forte di massimo" di L. NIRENBERG (1953) [2].

Sia $Z(x_1, x_2, \dots, x_n; t_1, t_2, \dots, t_m)$ una funzione di $n+m$ variabili in un dominio D ad $n+m$ dimensioni e consideriamo l'operatore differenziale :

$$(3) \qquad L(Z) = \sum_{hk} a_{hk} Z_{x_h x_h} + \sum_{ij} \alpha_{ij} Z_{t_i t_j} + \sum_h b_h Z_{x_h} + \sum_i \beta_i Z_{t_i} ,$$

ellittico nelle variabili x_h e parabolico nelle variabili t_i, cioè tale che la forma quadratica $\sum_{hk} a_{hk} \lambda_h \lambda_k$ è definita positiva, mentre la forma quadratica $\sum_{ij} \alpha_{ij} \xi_i \xi_j$ è soltanto semidefinita positiva per λ_h e ξ_i reali.

Supposti i coefficienti a_{hk}, α_{ij}, b_h, β_i funzioni continue di $P \in D$ e $Z(P) \in C^2$, si ha il seguente teorema : Se Z è tale che rislulti $L(Z) \geq 0$ e in un punto $P_0 \in T$ la Z raggiunge il suo valore massimo, allora è $Z(P) = Z(P_0)$ in tutta la parte di iperpiano t_i = costante, passante per P_0 e appartenente a T.

Accenneremo alle linee della dimostrazione nel caso di due variabili e cioè per l'operatore :

$$L'(Z) = A\, Z_{xx} + B\, Z_{tt} + aZ_x + bZ_t , \qquad A > 0 , \quad B \geq 0 .$$

Si dimostra da prima che, se in $P_0 (x_0, t_0) \in T$ la Z assume il suo valore massimo M , allora è ancora $Z = M$ in ogni punto della caratteristica $t = t_0$, appartenente a T.

Questo teorema è una quasi immediata conseguenza del seguente Lemma: se Z assume il valore massimo in punto P' della circonferenza di un cerchio appartenente a T, l'ascissa x' di P' coincide con quella del centro del cerchio.

La proprietà vale anche per un contorno ellittico, avente gli assi paralleli agli assi x e t.

Ci riferiremo ora al caso B=0 e b = -1, cioè al caso che interessa i problemi di diffusione. Quanto ora ricordato implica che se in T è:

$$(4) \qquad L''(Z) = A Z_{xx} + a Z_x - Z_t \geq 0$$

e in $P_0 \in T$ la $Z(P_0)$ raggiunge il suo valore massimo, allora si ha $Z(P) = Z(P_0)$, per ogni P appartenente alla caratteristica $t = t_0$ passante per P_0 e appartenente a T.

Premesso questo si indichi, per ogni $P \in T$, con $S(P)$ l'insieme dei punti $Q \in T$, che possono essere collegati con P mediante curve semplici, appartenenti a T, sulle quali, nel senso da Q a P, la t è non decrescente. Ebbene se per $P_0 \in T$ è $Z(P_0)$ massima, allora si ha $Z(P) = Z(P_0)$ per $P \in S(P_0)$; in questa affermazione, valida naturalmente anche per l'operatore $L(Z)$ definito in (3), sta il principio forte di massimo.

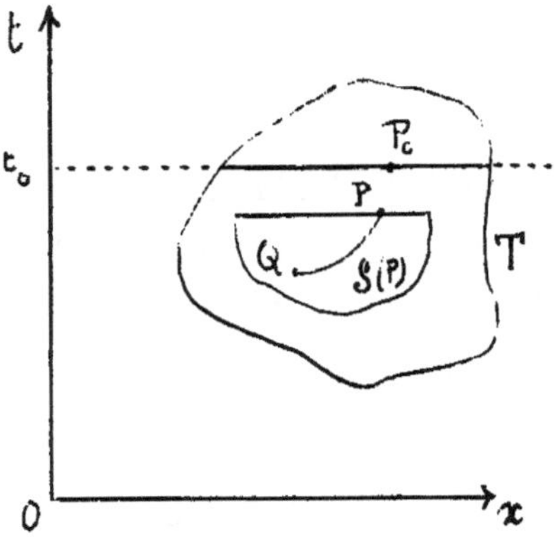

Ne segue in particolare che la soluzione regolare della equazione $L''(Z) = 0$ può raggiungere i suoi valori massimi (o minimi) sul contorno del campo di definizione e da ciò la possibilità di limitazioni "a priori" della cercata soluzione per un particolare problema al contorno associato all'equazione $L''(Z) = 0$ o ad altra più generale dei tipi considerati $(L(Z) = 0,\ L'(Z) = 0)$.

Generalizzazione di T. KUSANO (1953).

Di questo molto importante "principio forte di massimo" si hanno generalizzazioni di A. FRIEDMAN [6] e di T. KUSANO [7]. Accenneremo soltanto a quella di KUSANO , perchè riguarda una equazione quasilineare e cioè del tipo :

$$\sum_{hk} a_{hk} (P, t, Z, \text{grad } Z) Z_{x_h x_k} - Z_t = f(P, t, Z, \text{grad } Z), \tag{5}$$

con a_{hk} ed f funzioni definite in un dominio $D' \equiv [(P, t) \in D, \ |Z| < \infty, \ \|\text{grad } Z\| < \infty]$ e limitate in ciascun sottoinsieme compatto di D'.

Il principio forte di massimo per la (5) , la cui dimostrazione ricalca passo a passo quella del criterio di NIREMBERG , è il seguente :

se esiste una funzione semicontinua, positiva H(P, t, Z , grad Z) tale che :

$$\sum_{hk} a_{hk} \lambda_h \lambda_k \geq H \|\lambda\|^2 \ , \text{ per ogni } \lambda_h \text{ reale}$$

e Q (P, t, Z, grad Z) $\in$ D' ; se inoltre la f(P, t, Z, grad Z) è lipschitziana rispetto a Z e gradZ e si ha $f(P, t, Z, 0) \geq 0$ per $Z \geq 0$, allora, se in un punto $P_0 \in D$ la $Z(P_0)$ è massima , si ha $Z(P) = Z(P_0)$ per $P \in S(P_0)$, essendo S(P) l'insieme appartenente a D, già definito.

Conclusione.

Questa una rapida e molto sommaria scorsa sui cosidetti "principi di massimo" per le equazioni di tipo parabolico e un rapidissimo cenno sul loro utile impiego nella risoluzione di problemi al contorno originati da questioni fisico-matematiche, ad es. del tipo di quelle che hanno formato oggetto di studio in questo Corso. Il fatto che tra il principio di GEVREY e quello di KUSANO intercorrano cinquanta anni giusti, avvalora l'importanza e l'attualità della questione, e, a mio avviso, giustifica l'aver-

G. Sestini

ne fatto oggetto di un seminario.

BIBLIOGRAFIA .

1 A. FRIEDMAN. Partial differential equations of parabolic type, Prentice-Hall Inc. , Englewood Cliffs N. J. 1964 .

2. L. NIRENBERG , A strong maximum principle for parabolic equations , Comm. Pure Appl. Math. 6, 167-177 (1953).

3 M. GEVREY , Equations aux dérivées partielles du tipe parabolique, J. Math. Pures Appl. , (6) 9, 306 - 471 , 1913 .

4. E. E. LEVI , Sull'equazione del calore, Ann. Mat. Pura Appl. (3), 14, 187-264 , 1907 .

5 M. PICONE , Sul problema della propagazione del calore in un mezzo privo di frontiera , conduttore, isotropo omogeneo, Math. Ann. 101,701-712 , 1929.

6. A. FRIEDMAN. Remarks on the maximum principle for parabolic equations and its applications, Pacific J. Math. 8, 201-211, 1958.

7. T. KUSANO, On the maximum principle for quasi-linear parabolic equations, Proc . Japan Acad., 39, 211-216 , 1963.

Stampa: Editoriale Grafica - Roma - Tel. 5890154